W9-BXG-284

APPLIED ALGEBRA FOR THE COMPUTER SCIENCES

Prentice-Hall
Series in Automatic Computation

AHO, ed., *Currents in the Theory of Computing*
AHO and ULLMAN, *The Theory of Parsing, Translation, and Compiling*, Volume I: *Parsing*; Volume II: *Compiling*
ANDREE, *Computer Programming: Techniques, Analysis, and Mathematics*
ANSELONE, *Collectively Compact Operator Approximation Theory and Applications to Integral Equations*
BATES and DOUGLAS, *Programming Language/One*, 2nd ed.
BLUMENTHAL, *Management Information Systems*
BRENT, *Algorithms for Minimization without Derivatives*
BRINCH HANSEN, *Operating System Principles*
BRZOZOWSKI and YOELI, *Digital Networks*
COFFMAN and DENNING, *Operating Systems Theory*
CRESS, et al., *FORTRAN IV with WATFOR and WATFIV*
DAHLQUIST, BJÖRCK, and ANDERSON, *Numerical Methods*
DANIEL, *The Approximate Minimization of Functionals*
DEO, *Graph Theory with Applications to Engineering and Computer Science*
DESMONDE, *Computers and Their Uses*, 2nd ed.
DRUMMOND, *Evaluation and Measurement Techniques for Digital Computer Systems*
ECKHOUSE, *Minicomputer Systems: Organization and Programming (PDP-11)*
FIKE, *Computer Evaluation of Mathematical Functions*
FIKE, *PL/1 for Scientific Programmers*
FORSYTHE and MOLER, *Computer Solution of Linear Algebraic Systems*
GEAR, *Numerical Initial Value Problems in Ordinary Differential Equations*
GILL, *Applied Algebra for the Computer Sciences*
GORDON, *System Simulation*
GRISWOLD, *String and List Processing in SNOBOL4: Techniques and Applications*
HANSEN, *A Table of Series and Products*
HARTMANIS AND STEARNS, *Algebraic Structure Theory of Sequential Machines*
JACOBY, et al., *Iterative Methods for Nonlinear Optimization Problems*
JOHNSON, *System Structure in Data, Programs, and Computers*
KIVIAT, et al., *The SIMSCRIPT II Programming Language*
LAWSON and HANSON, *Solving Least Squares Problems*
LORIN, *Parallelism in Hardware and Software: Real and Apparent Concurrency*
LOUDEN AND LEDIN, *Programming the IBM 1130*, 2nd ed.
MARTIN, *Computer Data-Base Organization*
MARTIN, *Design of Man-Computer Dialogues*
MARTIN, *Design of Real-Time Computer Systems*
MARTIN, *Future Developments in Telecommunications*
MARTIN, *Programming Real-Time Computing Systems*
MARTIN, *Security, Accuracy, and Privacy in Computer Systems*
MARTIN, *Systems Analysis for Data Transmission*

MARTIN, *Telecommunications and the Computer*
MARTIN, *Teleprocessing Network Organization*
MARTIN and NORMAN, *The Computerized Society*
MCKEEMAN, et al., *A Compiler Generator*
MEYERS, *Time Sharing Computation in the Social Sciences*
MINSKY, *Computation: Finite and Infinite Machines*
NIEVERGELT, et al., *Computer Approaches to Mathematical Problems*
PLANE and MCMILLAN, *Discrete Optimization: Integer Programming and Network Analysis for Management Decisions*
POLIVKA and PAKIN, *APL: The Language and Its Usage*
PRITSKER and KIVIAT, *Simulation with GASP II: A FORTRAN-based Simulation Language*
PYLYSHYN, ed., *Perspectives on the Computer Revolution*
RICH, *Internal Sorting Methods Illustrated with PL/I Programs*
SACKMAN and CITRENBAUM, eds., *On-Line Planning: Towards Creative Problem-Solving*
SALTON, ed., *The SMART Retrieval System: Experiments in Automatic Document Processing*
SAMMET, *Programming Languages: History and Fundamentals*
SCHAEFER, *A Mathematical Theory of Global Program Optimization*
SCHULTZ, *Spline Analysis*
SCHWARZ, et al., *Numerical Analysis of Symmetric Matrices*
SHAH, *Engineering Simulation Using Small Scientific Computers*
SHAW, *The Logical Design of Operating Systems*
SHERMAN, *Techniques in Computer Programming*
SIMON and SIKLOSSY, eds., *Representation and Meaning: Experiments with Information Processing Systems*
STERBENZ, *Floating-Point Computation*
STOUTEMYER, *PL/1 Programming for Engineering and Science*
STRANG and FIX, *An Analysis of the Finite Element Method*
STROUD, *Approximate Calculation of Multiple Integrals*
TANENBAUM, *Structured Computer Organization*
TAVISS, ed., *The Computer Impact*
UHR, *Pattern Recognition, Learning, and Thought: Computer-Programmed Models of Higher Mental Processes*
VAN TASSEL, *Computer Security Management*
VARGA, *Matrix Iterative Analysis*
WAITE, *Implementing Software for Non-Numeric Application*
WILKINSON, *Rounding Errors in Algebraic Processes*
WIRTH, *Systematic Programming: An Introduction*
YEH, ed., *Applied Computation Theory: Analysis, Design, Modeling*

APPLIED ALGEBRA FOR THE COMPUTER SCIENCES

AUTHUR GILL

University of California
Berkeley, California

PRENTICE-HALL, INC., *Englewood Cliffs, New Jersey*

Library of Congress Cataloging in Publication Data

GILL, ARTHUR

Applied algebra for the computer sciences.
(Prentice-Hall series in automatic computation)
Includes bibliographies and index.

1. Algebra, Abstract. 2. Machine theory. I. Title.
QA162.G55 512'.02 75-2110
ISBN 0-13-039222-7

10 9 8 7 6

Printed in the United States of America

PRENTICE-HALL INTERNATIONAL, INC., *London*
PRENTICE-HALL OF AUSTRALIA PTY, LTD., *Sydney*
PRENTICE-HALL OF CANADA, LTD., *Toronto*
PRENTICE-HALL OF INDIA PRIVATE LIMITED, *New Delhi*
PRENTICE-HALL OF JAPAN, INC., *Tokyo*
PRENTICE-HALL OF SOUTHEAST ASIA (PTE.) LTD., *Singapore*

To Jonathan and Leori

CONTENTS

PREFACE xiii

A NOTE TO THE READER xv

1 SETS 1

1-1 Sets 1
1-2 Subsets and Power Sets 3
1-3 Complement, Union, and Intersection 6
1-4 Venn Diagrams 7
1-5 Membership Tables 10
1-6 Basic Laws of Set Operations 13
1-7 Partitions 15
1-8 The Minest Normal Form 16
1-9 The Maxset Normal Form 19
1-10 More on Minset and Maxset Normal Forms 21

2 RELATIONS 25

2-1 Cartesian Products 25
2-2 Relations 26
2-3 Composition of Relations 30
2-4 Relation Matrices of Composite Relations 31
2-5 Properties of Relations 36
2-6 Equivalence Relations 39
2-7 Partial Orderings 43

3 FUNCTIONS 49

3-1 Functions 50
3-2 Injections, Surjections, and Bijections 52
3-3 Composition of Functions 54
3-4 Identity and Inverse Functions 58
3-5 Permutations 61
3-6 Cardinality 64
3-7 The Characteristic Function 68
3-8 The Peano Postulates and Finite Induction 71
3-9 Examples of Proof by Induction 74
3-10 Recursive Definitions 77
3-11 Some Properties of Integers 83

4 ALGEBRAIC SYSTEMS 91

4-1 Operations 91
4-2 Algebraic Systems 94
4-3 Some Typical Postulates 96
4-4 Integral Domains 99
4-5 The Algebra of Sets 101
4-6 The Duality Principle 103
4-7 Homomorphism and Isomorphism 105
4-8 Congruence Relations 110
4-9 Direct Product of Algebraic Systems 118

5 PROPOSITIONS 121

5-1 Propositions 121
5-2 Negation, Disjunction, and Conjunction 122
5-3 Implication 125
5-4 Equivalence 128
5-5 The Algebra of Propositions 131
5-6 Theorem Proving 134
5-7 Quantifiers 139

6 LATTICES AND BOOLEAN ALGEBRAS 142

6-1 Posets 142
6-2 Lattices 145

6-3 Basic Laws of Lattices 148
6-4 The Lattice of Partitions 151
6-5 Distributive Lattices 156
6-6 Complemented Lattices 158
6-7 Boolean Algebras 161
6-8 Atomic Representation of Boolean Algebras 165
6-9 The Boolean Algebras $\mathcal{B}_2^r$ 170
6-10 Boolean Expressions 173

7 COMBINATIONAL AND SEQUENTIAL NETWORKS 179

7-1 Combinational Networks 179
7-2 Standard Combinational Networks 183
7-3 Switching Algebra 187
7-4 Normal Forms of Transmission Functions 191
7-5 Synthesis of Combinational Networks 195
7-6 Complementary and Dual Transmission Functions 201
7-7 Functional Completeness 205
7-8 Sequential Networks 211
7-9 Finite-State Machines 214
7-10 Sequential Networks and Finite-State Machines 221

8 LANGUAGES AND AUTOMATA 229

8-1 Sets of Strings 229
8-2 Languages 232
8-3 Phrase-Structure Languages 236
8-4 Regular Languages and Finite-State Automata 240
8-5 Regular Sets 249
8-6 Recognizers and Turing Machines 259

9 GROUPS 264

9-1 Binary Algebras 264
9-2 Semigroups and Monoids 266
9-3 Groups 270
9-4 Some Group Properties 275
9-5 Subgroups 279
9-6 Cosets 281

10 RINGS AND FIELDS 287

10-1 Rings 287
10-2 Ideals 292
10-3 Fields 296
10-4 Polynomials over Rings 302
10-5 Polynomials over Fields 305
10-6 Polynomial Ideals 308
10-7 Extension Rings and Fields 316
10-8 Galois Fields 322
10-9 Vector Spaces 327
10-10 Matrices over Fields 331

11 CODES 338

11-1 The Communication Channel Model 338
11-2 Linear Codes 342
11-3 Error Correction with Linear Codes 348
11-4 Hamming Codes 353

12 GRAPHS 360

12-1 Preliminary Definitions 360
12-2 Connectivity and Traversability 367
12-3 Trees 374
12-4 Bipartite Graphs 382
12-5 Planar Graphs 387
12-6 Directed Graphs 395
12-7 Puzzles and Games 403

INDEX 417

INDEX TO THEOREMS, EXAMPLES, AND ALGORITHMS 429

PREFACE

The purpose of this book is to cover, in a mathematically precise manner, a variety of concepts, results, techniques, and applications of modern algebra that are of particular use to beginning students in the computer sciences. Abstract topics such as sets, relations, functions, Boolean algebras, groups, rings, and fields are interspersed in this book with computer-oriented applications such as combinational and sequential networks, formal languages, automata, and codes. The object of this interspersal is to provide the beginner with motivation for delving into abstract mathematics, and to lay the groundwork for a number of applied areas which, as a computer-sciences major, the student will pursue in the future.

This book opens with three chapters on sets (with a traditional treatment of subsets, power sets, set operations, Venn diagrams, partitions, normal forms), relations (covering Cartesian products, composition of relations, graphs, partial orderings), and functions (including permutations and characteristic functions). The third chapter also includes a discussion of cardinality, the Peano postulates, and an extensive coverage of mathematical induction and recursion; it closes with some basic properties of integers. Chapter 4 introduces the general concept of an algebraic system, with the integral domain and the algebra of sets serving as examples. Also covered are homomorphism, congruence relations, and direct products of algebraic systems. The next two chapters undertake a detailed study of the algebra of propositions (with rudiments of theorem proving), of lattices, and of Boolean algebras. Chapter 7 is devoted entirely to applications—in particular, to combinational networks (switching algebra, transmission functions and their normal forms, complementary and dual networks, functional completeness) and sequential networks (analysis and synthesis of finite-state machines). Applications are pursued further in Chapter 8, which introduces formal languages—in particular, regular languages and finite-state automata—and computability. The following two chapters return to abstract algebra, covering semigroups, monoids, groups, subgroups, cosets, rings, ideals, fields, polynomials, extension and Galois fields, and a brief introduction to vector spaces. Chapter 11 illustrates the application of material covered in the preceding two chapters to the design of error-correcting and error-detecting codes. Resuming a topic introduced in Chapter 2, Chapter 12 delves into the details of graphs (connectivity and traversability, trees, bipartite graphs, planar graphs, and directed graphs) and closes with a discussion of applications to optimization problems, puzzles, and games.

Each chapter in this book starts with an introduction that previews the material in the chapter. In chapters that deal with purely mathematical topics, the introduction also explains in what way these topics are important to the computer scientist and hints at various applications to which they are relevant. Some general references are listed at the end of each chapter, for readers who wish to pursue the subject matter further. Each section in each chapter ends with a set of exercises whose purpose is to complement and illustrate the material in the text. Some of these problems involve program writing and require prior knowledge of some higher-level programming language, such as FORTRAN.

This book (which includes, essentially, all of the material listed under Course B3, Introduction to Discrete Structures, in ACM Curriculum 68) is intended for third-year undergraduate students in computer science. It can serve as a textbook for a one-term course where material on propositions, groups, rings, fields, and codes (Chapters 5, 9, 10, 11) can be deleted, or for a two-term course where all chapters would be covered to lesser or greater extent. These courses should not have any formal prerequisites (save high-school algebra), although the book occasionally assumes some prior familiarity (albeit very superficial) with combinatorics, matrix operations, and number systems. The chapter on codes also assumes rudimentary knowledge of the notion of probability. This book can serve as prerequisite for courses in logical design, automata theory, formal languages, and coding theory.

The author is indebted to Professor Lotfi A. Zadeh of the Department of Electrical Engineering and Computer Sciences in Berkeley, for constant encouragement and inspiration. Also, thanks are due Professors Elwyn Berlekamp, Manuel Blum, Gene Lawler, and Phil Spira (all from Berkeley) for many helpful discussions. Finally, I would like to thank my wife for her patience and understanding while I was writing this book.

ARTHUR GILL

A NOTE TO THE READER

In each chapter of this book, *sections* are numbered by chapter, then section, i.e., 1-1, 1-2, 1-3, 6-1, 6-2, 6-3, and so on.

Within each section, *figures, tables, theorems, algorithms*, and *examples* are double numbered by chapter and sequence. *Equation* numbers appear in parentheses at right.

To facilitate cross-referencing, chapter and section numbers appear at the top of the pages.

The end of every theorem proof, algorithm, and example is indicated by the symbol □.

At the end of the book the reader will find a subject index as well as special indexes for theorems, algorithms, and examples that appear in the text.

1 SETS

In this chapter we introduce the most fundamental mathematical concept —that of a *set*. Also introduced are the related concepts of *subset, power set,* and *partition*. Various operations (specifically—the *complement*, *union*, and *intersection* operations) are considered as means by which new sets can be generated from given ones, and basic laws are developed that characterize these operations. *Venn diagrams* and *membership tables* are introduced as useful tools in the manipulation and analysis of sets. Finally, we show how generated sets can be expressed in the so-called *minset normal form* and *maxset normal form*, whose importance will become apparent in later chapters.

To the computer scientist, the notion of a set is indispensable. In later chapters set, subset, power set, partition, etc. appear constantly in the study of finite-state machines and formal languages. Later we also study the relevance of membership tables and the normal forms to applied areas such as propositional calculus (of fundamental importance to theorem proving) and switching algebra (the basis of logical design).

1-1. SETS

The concept of a *set* cannot be defined precisely. Intuitively, it connotes an aggregate, or a collection, or a class of objects that are called the *elements* of a set. If an element a belongs to a set A, we write $a \in A$; otherwise, we write $a \notin A$. The notation $a_1, a_2, \ldots, a_n \in A$ is equivalent to $a_1 \in A, a_2 \in A, \ldots, a_n \in A$.

Examples of sets that appear frequently in this book include:

$\mathbb{N}$ = the set of *positive integers* or *natural numbers* (1, 2, 3, . . .)
$\mathbb{Z}$ = the set of *nonnegative integers* (0, 1, 2, . . .)

$\mathbb{I}$ = the set of *integers* (0, 1, −1, 2, −2, . . .)
$\mathbb{P}$ = the set of *prime numbers* (A prime number is a positive integer greater than 1 that is divisible only by itself and by 1.)
$\mathbb{Q}$ = the set of *rational numbers* (A rational number is a number expressible in the form i/j, where $i, j \in \mathbb{I}$ and $j \neq 0$.)
$\mathbb{R}$ = the set of *real numbers* (which includes all rational as well as irrational numbers)
$\mathbb{C}$ = the set of *complex numbers* (which includes all numbers of the form $a + ib$, where $a, b \in \mathbb{R}$ and $i = \sqrt{-1}$)
$\mathbb{N}_m(m \geq 1)$ = the set of positive integers between 1 and m, inclusively $(1, 2, \ldots, m)$
$\mathbb{Z}_m(m \geq 0)$ = the set of nonnegative integers between 0 and $m - 1$, inclusively $(0, 1, \ldots, m - 1)$

The number of elements in a set A is called the *cardinality* (or *order*) of A and is denoted by $\#A$. When $\#A$ is a finite number, A is said to be a *finite set*; otherwise, A is an *infinite set*. The sets $\mathbb{N}$, $\mathbb{Z}$, $\mathbb{I}$, $\mathbb{P}$, $\mathbb{Q}$, $\mathbb{R}$, and $\mathbb{C}$ are all infinite. Sets $\mathbb{N}_m$ and $\mathbb{Z}_m$ are finite (with cardinality m). Cardinality is discussed more fully in Sec. 3-6.

It is useful to be able to talk about a set that contains no elements whatsoever. For example, the set of all positive integers that satisfy the equation $x^2 = 8$ has no elements. Such a set is called an *empty set* (or *null set*, or *void set*), and is denoted by $\varnothing$. The cardinality of $\varnothing$ is defined as 0.

There are two common methods for specifying the elements of a set. The first method simply lists the elements in an arbitrary order (with no duplications). If $a_1, a_2, \ldots, a_n$ are the elements of A, then we write

$$A = \{a_1, a_2, \ldots, a_n\}$$

For example, the set of all integers whose absolute value does not exceed 3 can be written as

$$X = \{-3, -2, -1, 0, 1, 2, 3\} \tag{1-1}$$

This method, of course, is applicable only to finite sets.

When the number of elements in A is very large or infinite, a more practical method for specifying A is to spell out a *defining condition*, denoted by $P(a)$, such that $a \in A$ if and only if $P(a)$ holds. Essentially, $P(a)$ prescribes a rule, or formula, that enables us (at least in principle) to decide whether or not a is in A. The general form of this specification is

$$A = \{a \mid P(a)\}$$

and is read:

A is the set of all elements a such that $P(a)$

For example, set X specified in (1-1) can also be written as

$$X = \{x \mid x \in \mathbb{I}, -3 \leq x \leq 3\}$$

(Another way of writing it is $X = \{x \in \mathbb{I} \mid -3 \leq x \leq 3\}$ or, when $\mathbb{I}$ can be taken for granted, simply $X = \{x \mid -3 \leq x \leq 3\}$.)

Note that the formulation of a defining condition does *not* automatically guarantee the existence of a set. For example, consider the following situation: Barber Joe shaves all and only those men in town who do not shave themselves; then A is defined as the set of all those men in town who are shaved by Joe. A little reflection reveals that, whether or not Joe shaves himself, A must be a set such that Joe $\in A$ and at the same time, Joe $\notin A$. This is an absurdity, of course, and refutes the assumption that A exists. In this book we do not worry about such hypothetical cases, relying, instead, on our intuition to satisfy us that all sets under discussion indeed exist.

Two sets A and B are said to be *equal* (denoted $A = B$) if every element of A is also an element of B, and conversely. That is, $A = B$ if $a \in A$ implies $a \in B$, and $b \in B$ implies $b \in A$ (otherwise, we write $A \neq B$). For example, the sets $\{-3, -2, -1, 0, 1, 2, 3\}$ and $\{x \mid x \in \mathbb{I}, -3 \leq x \leq 3\}$ are equal.

PROBLEMS

1. List the elements in these sets:

(a) The set of letters in the word Mississippi

(b) The set of prime numbers less that 20

(c) $\{r \mid r \in \mathbb{R}, r^2 + r - 6 = 0\}$

(d) $\{q \mid q \in \mathbb{Q}, q^2 - 1 = 15 \text{ and } q^3 = 60\}$

(e) $\{i \mid i \in \mathbb{I}, i^2 - 10i - 24 < 0 \text{ and } 5 \leq i \leq 15\}$.

2. Express the finite set $\{a_1, a_2, a_3, a_4, a_5\}$ in a defining condition form.

3. Express this set in a defining condition form:

$$A = \{0, 2, 4, 6, 8, \ldots, 98, 100, 250\}$$

4. Show that this set does not exist:

$$A = \{S \mid S \text{ is a set such that } S \notin S\}$$

(This is known as *Russell's paradox.*)

1-2. SUBSETS AND POWER SETS

A set A is said to be a *subset* of a set B if every element of A is also an element of B (that is, if $a \in A$ implies $a \in B$). When this is the case, we write $A \subset B$ (or $B \supset A$) and say that A is *included* in B (otherwise, we write A

$\not\subset B$ or $B \not\supset A$). If $A \subset B$, B is called a *superset* of A. If $A \subset B$ but $A \neq B$, A is said to be a *proper subset* of A. For example, the set $\{1, 2, 3\}$ is a proper subset of the set $\{x \mid x \in \mathbb{I}, -3 \leq x \leq 3\}$.[1]

Clearly, $A = B$ if and only if $A \subset B$ and $B \subset A$. Also, for any set A, $\varnothing \subset A$ and $A \subset A$.

It is convenient to assume that an empty set is a subset of every set. This assumption leads to:

THEOREM 1-1.

The empty set is unique.

Proof Suppose there are two empty sets, $\varnothing_1$ and $\varnothing_2$. Since $\varnothing_1$ and $\varnothing_2$ are included in every set, $\varnothing_1 \subset \varnothing_2$ and $\varnothing_2 \subset \varnothing_1$, which implies $\varnothing_1 = \varnothing_2$. □

It is quite legitimate for a set to have other sets as elements. For example, the set of all sets of distinct positive integers whose sum is 6; that is, the set

$$\{\{6\}, \{1, 5\}, \{2, 4\}, \{1, 2, 3\}\}$$

has sets as elements.

In any discussion that concerns sets of sets, it is often convenient to use the notation $\{A_i\}_{i \in K}$ to denote the set of all sets A_i such that $i \in K$; that is,

$$\{A_i\}_{i \in K} = \{A_i \mid i \in K\}$$

In this context, K is referred to as an *index set*. For example, the set of sets

$$\{A_0, A_1, A_2, A_3, A_4\} \tag{1-2}$$

can be denoted by $\{A_i\}_{i \in K}$ where $K = \{0, 1, 2, 3, 4\}$ is the index set.

When $K = \{i \mid i \in \mathbb{I}, i_a \leq i \leq i_b\}$, the notation $\{A_i\}_{i \in K}$ can be replaced by the notation $\{A_i\}_{i=i_a}^{i_b}$. For example, the set (1) can be written as $\{A_i\}_{i=0}^{4}$.

Given a set A, the set of all subsets of A (including the empty set $\varnothing$) is called the *power set* of A and is denoted by $\mathbf{2}^A$; that is,

$$\mathbf{2}^A = \{S \mid S \subset A\}$$

For example, if $A = \{1, 2, 3\}$, then

$$\mathbf{2}^A = \{\varnothing, \{1\}, \{2\}, \{3\}, \{1, 2\}, \{1, 3\}, \{2, 3\}, \{1, 2, 3\}\}$$

[1] In some texts, $A \subseteq B$ means A is a subset of B, while $A \subset B$ means A is a proper subset of B. In this book we do not use this dual notation.

THEOREM 1-2.
If A is a finite set with cardinality $\#A$, then

$$\#(\mathbf{2}^A) = 2^{\#A}$$

Proof Let $\#A = n$. The number of ways that i distinct elements can be selected from n elements is given by

$$\binom{n}{i} = \frac{n!}{i!\,(n-i)!}$$

Hence, the number of distinct subsets of A (including $\varnothing$) is

$$\#(\mathbf{2}^A) = 1 + \binom{n}{1} + \binom{n}{2} + \cdots + \binom{n}{n}$$

From the binomial theorem we know that

$$(x + y)^n = x^n + \binom{n}{1}x^{n-1}y + \binom{n}{2}x^{n-2}y^2 + \cdots + \binom{n}{n}y^n$$

Setting $x = y = 1$, we have

$$2^n = (1 + 1)^n = 1 + \binom{n}{1} + \binom{n}{2} + \cdots + \binom{n}{n}$$

Hence, $\#(\mathbf{2}^A) = 2^n$. □

PROBLEMS

1. Which of these statements are incorrect?
   (a) $\{a\} \in \{\{a\}\}$ (b) $\{a\} \subset \{\{a\}\}$
   (c) $\{a\} \in \{\{a\}, a\}$ (d) $\{a\} \subset \{\{a\}, a\}$
2. Consider the sets $\mathbb{N}$, $\mathbb{Z}$, $\mathbb{I}$, $\mathbb{P}$, $\mathbb{Q}$, $\mathbb{R}$, $\mathbb{C}$, $\mathbb{N}_m$, and $\mathbb{Z}_m$ defined in Sec. 1-1. Which of these sets are proper subsets of others in the same list?
3. What is the power set of $\{\varnothing, a, \{a\}\}$?
4. Let $A = \{a\}$. Find the power sets of A and of $\mathbf{2}^A$.
5. Show that this set does not exist:

$$A = \{S \mid S \text{ is a set}\}$$

   (This is known as *Cantor's paradox*.) (*Hint:* Is A in A?)

1-3. COMPLEMENT, UNION, AND INTERSECTION

In many discussions, all subsets under consideration are included in a single fixed set that is called the *universal set* (or *universe*) and is denoted by U. The universal set may be identified once and for all at the beginning of a discussion, and henceforth may be assumed to constitute the superset of every set mentioned in the discussion. For example, if U is declared as the set $\mathbb{N}$ of positive integers, then, unless otherwise specified, every set under discussion is assumed to be a set of positive integers (for example, the set $\{n \mid n^2 - 8n + 15 = 0\}$ is understood to be the set $\{3, 5\}$).

The *complement* of a set A, denoted by A', is the set of all elements of U that are not in A; that is,

$$A' = \{u \mid u \notin A\}$$

For example, if $U = \mathbb{Z}$ and $A = \{2k \mid k \in \mathbb{Z}\}$, then A' is the set of all odd integers. Clearly, in all cases, $U' = \varnothing$ and $\varnothing' = U$.

Given two sets, A and B, the *complement of A relative to B*, denoted by $A - B$, is the set of all elements of A that are not in B; that is,

$$A - B = \{a \mid a \in A, a \notin B\}$$

For example, if A is the set $\mathbb{P}$ of prime numbers and B the set of odd numbers, then $A - B = \{2\}$. In all cases when $A = U$, then $A - B$ is simply B'.

Given two subsets, A and B, in a universal set U, new subsets in U can be formed from the so-called union and intersection of A and B. The *union* of sets A and B, denoted by $A \cup B$, is a set consisting of all elements of U that are either in A or in B;[1] that is,

$$A \cup B = \{u \mid u \in A \text{ or } u \in B\}$$

The *intersection* of sets A and B, denoted by $A \cap B$, is a set consisting of all elements of U that are in both A and B; that is,

$$A \cap B = \{u \mid u \in A \text{ and } u \in B\}$$

For example, if $U = \mathbb{N}$, $A = \mathbb{P}$, and B is the set of all odd numbers, then $A \cup B$ consists of all odd numbers and 2, and $A \cap B$ consists of all prime numbers except 2.

Sets A and B are said to be *disjoint* if they have no common elements; that is, if $A \cap B = \varnothing$. For example, the set of even numbers and the set of odd numbers are disjoint.

[1] A phrase such as "either (alternative 1) or (alternative 2)" will always mean "either (alternative 1), or (alternative 2), or *both alternatives*." More generally, "(alternative 1), or (alternative 2), or . . . , or (alternative r)" will mean "(alternative 1), or (alternative 2), or . . . , or (alternative r), or *any combination of these alternatives*."

Let $\{A_i\}_{i=1}^{r}$ be a set of subsets of some universal set U. Any set that can be created by applying the operations $'$, $\cup$, and $\cap$ to $\varnothing$, U, $A_1, A_2, \ldots,$ and A_r, is called a *set generated by* $A_1, A_2, \ldots, A_r$. For example, the sets $\varnothing$, $B \cap C'$, $((A \cup B') \cap C)' \cup A'$, and U are all sets generated by A, B, C.

PROBLEMS

1. Let $U = \{1, 2, 3, 4, 5\}$, $A = \{1, 4\}$, $B = \{1, 2, 5\}$, and $C = \{2, 4\}$. Determine the sets:
(a) $A \cap B'$
(b) $(A \cap B) \cup C'$
(c) $A \cup (B \cap C)$
(d) $(A \cup B) \cap (A \cup C)$
(e) $(A \cap B)'$
(f) $A' \cup B'$
(g) $(B \cup C)'$
(h) $B' \cap C'$
(i) $(A \cap B) \cap C$
(j) $\mathbf{2}^A \cap \mathbf{2}^C$
(k) $\mathbf{2}^A - \mathbf{2}^C$

2. Given these subsets of $\mathbb{I}$:
$A = \{1, 2, 7, 8\}$
$B = \{i \mid i^2 \leq 50\}$
$C = \{i \mid i \text{ divides } 30\}$
$D = \{i \mid i = 2^k, k \in \mathbb{I}, 0 \leq k \leq 6\}$
determine the sets:
(a) $A \cup (B \cup (C \cup D))$
(b) $A \cap (B \cap (C \cap D))$
(c) $B - (A \cup C)$
(d) $(A' \cap B) \cup D$

3. Consider these subsets of $\mathbb{N}$:
$A = \{n \mid n < 12\}$
$B = \{n \mid n \leq 8\}$
$C = \{n \mid n = 2k, k \in \mathbb{N}\}$
$D = \{n \mid n = 3k, k \in \mathbb{N}\}$
$E = \{n \mid n = 2k - 1, k \in \mathbb{N}\}$
Express the following sets as sets generated by A, B, C, D, E:
(a) $\{2, 4, 6, 8\}$
(b) $\{3, 6, 9\}$
(c) $\{10\}$
(d) $\{n \mid n \text{ even}, n > 10\}$
(e) $\{n \mid n = 3, \text{ or } n = 6, \text{ or } n \geq 9\}$
(f) $\{n \mid n \text{ even and } n \leq 10, \text{ or } n \text{ odd and } n > 9\}$
(g) $\{n \mid n \text{ is a multiple of } 6\}$

4. Let U be some universal set with subsets A and B. List all sets generated by A, B.

1-4. VENN DIAGRAMS

Many relationships involving complements, union, and intersection of sets can be readily visualized by means of diagrams where the universal set U is represented by a rectangular area, and subsets of U are represented by circular regions within that rectangle. Such diagrams are called *Venn diagrams* (after John Venn, 1834–1883, the English mathematician). Figures 1-1(a), (b), and (c) show Venn diagrams where the cross-hatched regions in U represent the sets A', $A \cup B$, and $A \cap B$, respectively.

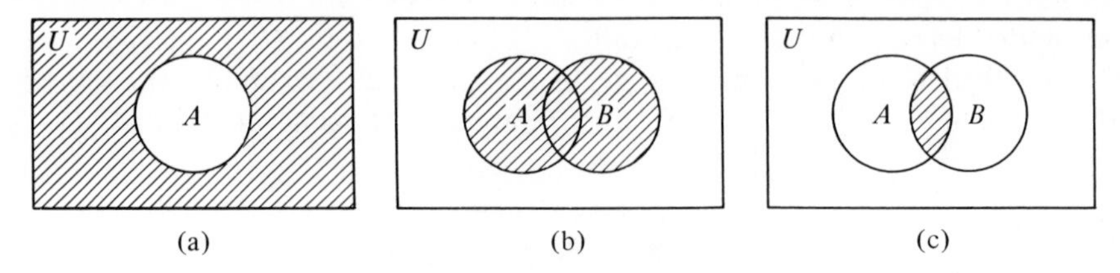

Fig. 1-1 Venn diagrams displaying: (a) A', (b) $A \cup B$, (c) $A \cap B$

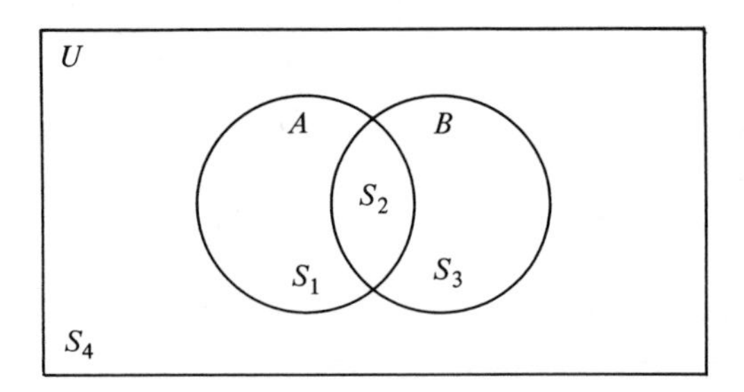

Fig. 1-2 Venn diagram displaying two subsets

Figure 1-2 shows a Venn diagram that is often found convenient in deriving relationships involving two subsets of U, say, A and B. These subsets may divide U into as many as four disjoint subsets that, in the figure, are represented by the regions S_1, S_2, S_3, and S_4. From this figure it is clear, for example, that

$$A - B = A \cap B' \tag{1-3}$$

[since $A - B = S_1$ and $A \cap B' = (S_1 \cup S_2) \cap (S_1 \cup S_4) = S_1$]. As another example, the identity

$$(A \cup B)' = A' \cap B'$$

can be verified by noting that $(A \cup B)' = (S_1 \cup S_2 \cup S_3)' = S_4$ and $A' \cap B' = (S_3 \cup S_4) \cap (S_1 \cup S_4) = S_4$.

Relationships involving cardinalities of sets can also be readily derived from Venn diagrams. For example, the identity

$$\#(A \cup B) = \#A + \#B - \#(A \cap B)$$

can be inferred from the fact (using Fig. 1-2) that

$$\begin{aligned}\#(A \cup B) &= \#S_1 + \#S_2 + \#S_3 \\ &= (\#S_1 + \#S_2) + (\#S_3 + \#S_2) - \#S_2 \\ &= \#A + \#B - \#(A \cap B)\end{aligned}$$

A more elaborate example is:

Example 1-1

In a class of 170 students, 120 students have taken Spanish, 80 students have taken French, and 60 students have taken Russian; 50 students have taken both Spanish and French, 25 students have taken both Spanish and Russian, and 30 students have taken both French and Russian; 10 students have taken all three languages. How many students have not taken any of the three languages? Denoting the sets of students who have taken Spanish, French, and Russian by S, F, and R, respectively, we have

$$\#S = 120,\ \#F = 80,\ \#R = 60$$
$$\#(S \cap F) = 50,\ \#(S \cap R) = 25,\ \#(F \cap R) = 30$$
$$\#((S \cap F) \cap R) = 10$$

From this information we can compute the number of elements in each region in the Venn diagram of Fig. 1-3, and hence derive the number (namely, 5) of students not taking any of the three languages. □

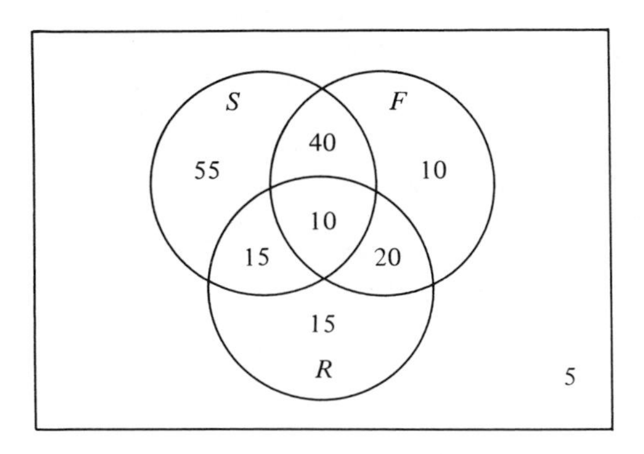

Fig. 1-3 Venn diagram for Example 1-1

PROBLEMS

1. A, B, and C are subsets of U. Prove these identities, using Venn diagrams:
   (a) $(A \cup B) \cap (A' \cup C) = (A \cap C) \cup (A' \cap B)$
   (b) $(A \cup B) \cap (A \cup C) = A \cup (B \cap C)$
   (c) $((A \cap B') \cup (B \cap C')) \cup ((B' \cap C) \cup (A' \cap B))$
   $= (A \cap B') \cup ((B \cap C') \cup (A' \cap C))$
   $= (A \cap C') \cup ((B' \cap C) \cup (A' \cap B))$
   (d) $A - (B \cup C) = (A - B) \cap (A - C)$
   (e) $A - (B \cap C) = (A - B) \cup (A - C)$

2. A, B, and C are subsets of U. Using Venn diagrams, show that these statements are equivalent:
(a) $A \subset B$ (b) $A \cap B = A$
(c) $A \cap B' = \varnothing$ (d) $A' \cup B = U$
(e) $B' \subset A'$

3. A, B, and C are subsets of U. Is the following true or false?

$$(A \cup B) \cap (B \cup C)' \subset A \cap B'$$

4. A, B, and C are subsets of U. Are the following statements true or false?
(a) $A \cap B = A \cap C$ implies $B = C$.
(b) $A \subset B$ if and only if $A \cup B = B$.
(c) $A \subset B$ if and only if $A \cap B = A$.
(d) $A \cap (B - C) = \varnothing$ if and only if $A \subset C$.
(e) $(A - B) \cup C = A$ if and only if $B \subset C$.

5. The following data have been compiled on a group of 80 job applicants: 30 are minors, 38 are males, 38 are U.S. citizens, 11 are male minors, 15 are male U.S. citizens, 13 are minor U.S. citizens, 3 are male minor U.S. citizens, and 12 are neither minor, nor male, nor U.S. citizens. Are these data correct?

6. The following data have been compiled on a class of 60 students: (i) There are 35 undergraduate students, 14 of whom are mathematics majors. (ii) There are 2 graduate students who are mathematics majors. (iii) All mathematics majors take programming. (iv) 9 students take logical design, 3 of whom are mathematics majors and 7 of whom are undergraduates. (v) 21 undergraduate and 4 graduate students take programming. (vi) No graduate but 5 undergraduate students take both logical design and programming.
(a) How many students take neither programming nor logical design?
(b) How many mathematics majors do not take logical design?
(c) How many undergraduate students take either programming or logical design but not both?

1-5. MEMBERSHIP TABLES

The complement of a set A can, alternatively, be defined as follows:

If $u \notin A$, then $u \in A'$;
if $u \in A$, then $u \notin A'$.

The union of sets A and B can be defined as follows:

If $u \notin A$ and $u \notin B$, then $u \notin A \cup B$;
if $u \notin A$ and $u \in B$, then $u \in A \cup B$;
if $u \in A$ and $u \notin B$, then $u \in A \cup B$;
if $u \in A$ and $u \in B$, then $u \in A \cup B$.

Similarly, the intersection of A and B can be defined by the following:

If $u \notin A$ and $u \notin B$, then $u \notin A \cap B$;
if $u \notin A$ and $u \in B$, then $u \notin A \cap B$;
if $u \in A$ and $u \notin B$, then $u \notin A \cap B$;
if $u \in A$ and $u \in B$, then $u \in A \cap B$.

These definitions can be summarized in *membership tables* such as shown in Table 1-1, where the digits 0 and 1 in a column labeled S stand for $u \notin S$ and $u \in S$, respectively.

Table 1-1(a) MEMBERSHIP TABLE FOR A'

A	A'
0	1
1	0

Table 1-1(b) MEMBERSHIP TABLE FOR $A \cup B$

A	B	$A \cup B$
0	0	0
0	1	1
1	0	1
1	1	1

Table 1-1(c) MEMBERSHIP TABLE FOR $A \cap B$

A	B	$A \cap B$
0	0	0
0	1	0
1	0	0
1	1	1

The idea of a membership table, for sets generated by A and B, can be extended to sets generated by any subsets $A_1, A_2, \ldots, A_r$ of U. In the general membership table constructed for a set S generated by $A_1, A_2, \ldots, A_r$, the first r columns are labeled $A_1, A_2, \ldots, A_r$, and the last column is labeled S. A digit 0 in a column labeled A_i stands for $u \notin A_i$, whereas a 1 stands for $u \in A_i$. If, under the conditions indicated in the first r columns of the kth row, it is found that $u \notin S$, a 0 is entered in the kth row of column S. If, under these conditions, it is found that $u \in S$, a 1 is entered. Thus, the membership table consists of 2^r rows, which correspond to the 2^r possible membership/nonmembership conditions of u in $A_1, A_2, \ldots, A_r$. To facilitate discussion, a row containing digits $\delta_1, \delta_2, \ldots, \delta_r$ (where each δ_i is either 0 or 1) in the columns labeled $A_1, A_2, \ldots, A_r$, respectively, will be referred to as row $\delta_1\delta_2 \ldots \delta_r$.

Example 1-2

Table 1-2 illustrates the construction of the set

$$S = ((A \cap B) \cup (A' \cap C)) \cup (B \cap C)$$

generated by A, B, C. The first three columns list all 8 possible membership/nonmembership conditions of u in A, B, and C, while the last column lists the corresponding membership/nonmembership condition of u in S. The last column is arrived at by successively constructing intermediate membership tables for the sets A', $A \cap B$, $A' \cap C$, $B \cap C$, and $(A \cap B) \cup (A' \cap C)$. Each column beyond the third is constructed from preceding columns by direct reference to the membership tables that define $'$, $\cup$, and $\cap$ (see Table 1-1). For example, when entries in columns A, B, and C are 1, 1, and 0, respectively (row 110), the entry in column A' is 0 [by Table 1-1(a)]; the entry in column $A \cap B$ is 1 [by Table 1-1(c)]; the entry in column $A' \cap C$ is 0 [by Table 1-1(c)]; the entry in column $B \cap C$ is 0 [by Table 1-1(c)]; the entry in column $(A \cap B) \cup (A' \cap C)$ is 1 [by Table 1-1(b)]; and the entry in column $((A \cap B) \cup (A' \cap C)) \cup (B \cap C)$ is 1 [by Table 1-1(b)]. □

Table 1-2 MEMBERSHIP TABLE FOR $((A \cap B) \cup (A' \cap C)) \cup (B \cap C)$

A	B	C	A'	$A \cap B$	$A' \cap C$	$B \cap C$	$(A \cap B) \cup (A' \cap C)$	$((A \cap B) \cup (A' \cap C)) \cup (B \cap C)$
0	0	0	1	0	0	0	0	0
0	0	1	1	0	1	0	1	1
0	1	0	1	0	0	0	0	0
0	1	1	1	0	1	1	1	1
1	0	0	0	0	0	0	0	0
1	0	1	0	0	0	0	0	0
1	1	0	0	1	0	0	1	1
1	1	1	0	1	0	1	1	1

Note that a column labeled $\varnothing$ in any membership table is 0 throughout, and a column labeled U is 1 throughout.

Suppose columns labeled S and T in a membership table are identical. Hence, $u \in S$ implies $u \in T$, and $u \in T$ implies $u \in S$. Thus, $S \subset T$ and $T \subset S$, and hence $S = T$. For example, we can infer that

$$(A \cap B) \cup (A' \cap C) = ((A \cap B) \cup (A' \cap C)) \cup (B \cap C)$$

because the columns $(A \cap B) \cup (A' \cap C)$ and $((A \cap B) \cup (A' \cap C)) \cup (B \cap C)$ in Table 1-2 have identical elements. Thus, membership tables can serve to prove the equality (or inequality) of sets generated by subsets of U. As the number of subsets becomes larger, verification (or refutation) of set equalities by Venn diagrams becomes increasingly cumbersome, and the systematic construction of membership tables becomes a highly practical substitute.

PROBLEMS

1. Prove the identities of Problem 1 Sec. 1-4 by using membership tables. [In parts (d) and (e) make use of identity (1-3)].

2. How can the fact that $S \subset T$ be recognized from the membership tables of S and T? Use membership tables to prove or disprove that $(A \cup B) \cap (B \cup C)' \subset A \cap B'$.

3. $A_1, A_2, \ldots, A_r$ are subsets of U. At most, how many distinct sets can be generated by $A_1, A_2, \ldots, A_r$?

1-6. BASIC LAWS OF SET OPERATIONS

Using either the Venn diagram or the membership table approach, a variety of identities and relationships involving the complements, union, and intersection of sets can be derived, as demonstrated in the preceding section. Table 1-3 lists several basic identities (or laws), that are referred to in a later chapter as the postulates of a mathematical structure called the algebra of sets. Table 1-4 lists some additional useful identities. In both tables, sets A, B, and C are arbitrary subsets of a universal set U.

The operations of union and intersection have been defined for two sets only. When two or more operations are performed in succession, parentheses

Table 1-3 SET LAWS

Commutative Laws	
1. $A \cup B = B \cup A$	1′. $A \cap B = B \cap A$
Associative Laws	
2. $A \cup (B \cup C) = (A \cup B) \cup C$	2′. $A \cap (B \cap C) = (A \cap B) \cap C$
Distributive Laws	
3. $A \cap (B \cup C) = (A \cap B) \cup (A \cap C)$	3′. $A \cup (B \cap C) = (A \cup B) \cap (A \cup C)$
Identity Laws	
4. $A \cup \varnothing = \varnothing \cup A = A$	4′ $A \cap U = U \cap A = A$
Complement Laws	
5. $A \cup A' = U$	5′. $A \cap A' = \varnothing$

Table 1-4 ADDITIONAL SET LAWS

Involution Law	
1, 1'. $(A')' = A$	
Idempotent Laws	
2. $A \cup A = A$	2'. $A \cap A = A$
Null Laws	
3. $A \cup U = U$	3'. $A \cap \varnothing = \varnothing$
Absorption Laws	
4. $A \cup (A \cap B) = A$	4'. $A \cap (A \cup B) = A$
De Morgan's Laws	
5. $(A \cup B)' = A' \cap B'$	5'. $(A \cap B)' = A' \cup B'$

must be used in order to avoid ambiguity. For example, the parentheses in $A = B \cup (C \cap D)$ specify that A is obtainable by this sequence of operations: (i) Obtain the intersection of C and D, say, E; (ii) obtain the union of B and E. The result, in general, is quite different from that obtainable from $(B \cup C) \cap D$; hence, the parentheses-free expression $B \cup C \cap D$ is ambiguous.

The associative laws in Table 1-3 state that, for all A, B, and C, $A \cup (B \cup C) = (A \cup B) \cup C$, and $A \cap (B \cap C) = (A \cap B) \cap C$. Hence, the absence of parentheses in expressions of the form $A \cup B \cup C$ and $A \cap B \cap C$ does *not* cause any ambiguity. More generally, the associative laws permit the use of the following parentheses-free notation for the union and intersection of any sets $A_1, A_2, \ldots, A_r$:

$$A_1 \cup A_2 \cup \ldots \cup A_r = \{u \mid u \in A_1, \text{ or } u \in A_2, \ldots, \text{ or } u \in A_r\} = \bigcup_{i=1}^{r} A_i$$

$$A_1 \cap A_2 \cap \ldots \cap A_r = \{u \mid u \in A_1, \text{ and } u \in A_2, \ldots, \text{ and } u \in A_r\} = \bigcap_{i=1}^{r} A_i$$

When the set of subscripts i is not necessarily $\{1, 2, \ldots, r\}$ but constitutes an arbitrary index set K, these forms can be written as $\bigcup_{i \in K} A_i$ and $\bigcap_{i \in K} A_i$, respectively.

PROBLEMS

1. Using membership tables, verify the laws of Tables 1-3 and 1-4.

2. A, B, and C are subsets of U. Prove these identities, using the laws of Tables 1-3 and 1-4:

(a) $(B \cup C) \cap A = (B \cap A) \cup (C \cap A)$

(b) $(B \cap C) \cup A = (B \cup A) \cap (C \cup A)$

(c) $(A \cap B) \cup (A \cap B') = A$
(d) $B \cup ((A' \cup B) \cap A)' = U$

3. Use De Morgan's laws to prove that the complement of $(A \cap B') \cup (A' \cap (B \cup C'))$ is $(A' \cap B) \cap (A \cup B') \cap (A \cup C)$.

4. (a) A_i are sets of real numbers, defined as:

$$A_0 = \{a \mid a < 1\}$$

$$A_i = \left\{a \mid a \leq 1 - \frac{1}{i}\right\} \qquad (i = 1, 2, \ldots)$$

Prove that

$$\bigcup_{i=1}^{\infty} A_i = A_0$$

(*Hint:* Prove that $\bigcup_{i=1}^{\infty} A_i \subset A_0$ and $A_0 \subset \bigcup_{i=1}^{\infty} A_i$.)

(b) B_i are sets of real numbers defined as:

$$B_0 = \{b \mid b \leq 1\}$$

$$B_i = \left\{b \mid b < 1 + \frac{1}{i}\right\} \qquad (i = 1, 2, \ldots)$$

Prove that

$$\bigcap_{i=1}^{\infty} B_i = B_0$$

1-7. PARTITIONS

Let $\pi = \{A_i\}_{i \in K}$ (where K is an arbitrary index set) be a set of *nonempty* subsets of A. The set π is called a *partition* of A if every element of A is in exactly one of the A_i; that is, if

(a) $A_i \cap A_j = \varnothing$ whenever $i \neq j$;
(b) $\bigcup_{i \in K} A_i = A$.

(The A_i are sometimes called the *blocks* of the partition.)

For example, if A_i is defined as the set of all integers which, upon division by 5, leave the remainder i, then set $\{A_0, A_1, A_2, A_3, A_4\}$ is a partition of the set of integers $\mathbb{I}$. In Fig. 1-2, set $\{S_1, S_2, S_3, S_4\}$ is a partition of U.

A partition $\bar{\pi} = \{\bar{A}_i\}_{i \in \bar{K}}$ is a *refinement* of partition $\pi = \{A_j\}_{j \in K}$ if every $\bar{A}_i$ is a subset of some A_j. If, in addition, at least one $\bar{A}_i$ is a *proper* subset of some A_j, then $\bar{\pi}$ is said to be a *proper refinement* of π. The process of partitioning U can be viewed as that of drawing boundary lines across the Venn diagram area that represents U. Partition $\bar{\pi}$ is a proper refinement of partition π if $\bar{\pi}$ is formed by adding at least one new boundary line to the lines already existing for π.

PROBLEMS

1. Let $\{A_1, A_2, \ldots, A_r\}$ be a partition of set A. Show that $(A_1 \cap B, A_2 \cap B, \ldots, A_r \cap B\}$ is a partition of set $A \cap B$.

2. In how many distinct ways can an n-element set be partitioned into two blocks?

1-8. THE MINSET NORMAL FORM

Let $\{A_i\}_{i=1}^r$ be a set of subsets in a universal set U. A set of the form $\bigcap_{i=1}^r \hat{A}_i$, where each $\hat{A}_i$ may be either A_i or A_i', is called a *minset generated by* $A_1, A_2, \ldots, A_r$. There are 2^r such minsets.

THEOREM 1-3

The set of all minsets generated by $A_1, A_2, \ldots, A_r$ constitutes a partition of U.

Proof To prove the theorem, it suffices to prove that every element of U is in exactly one minset. If $u \in U$, then $u \in A_1$ or $u \in A_1'$; $u \in A_2$ or $u \in A_2'$; ... ; $u \in A_r$ or $u \in A_r'$. Hence, u is in some minset $\bigcap_{i=1}^r \hat{A}_i$. Now consider the intersection T of two distinct minsets. Because these minsets are distinct, there is some i such that T is included in both A_i and A_i' and hence in $A_i \cap A_i' = \varnothing$. Hence, $T = \varnothing$; therefore no $u \in U$ can belong to two minsets simultaneously. □

Figure 1-4 illustrates the partition of U formed by the eight minsets generated by A, B, C.

It is convenient to denote a minset $\bigcap_{i=1}^r \hat{A}_i$ by $M_{\delta_1\delta_2\ldots\delta_r}$, where each δ_i is either 0 or 1, according to the following rule: $\delta_i = 0$ if $\hat{A}_i = A_i'$, and $\delta_i = 1$ if $\hat{A}_i = A_i$. In this manner, the subscript of $M_{\delta_1\delta_2\ldots\delta_r}$ uniquely describes

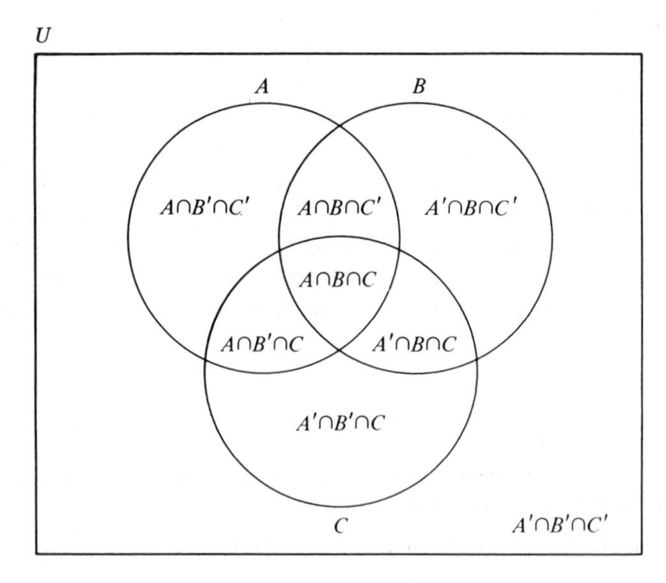

Fig. 1-4 Partition formed by minsets generated by A, B, C

the minset that it intends to represent. For example, minset $A \cap B \cap C'$ generated by A, B, C is denoted by M_{110}.

Envision an arbitrary minset $M_{\delta_1\delta_2\ldots\delta_r} = \bigcap_{i=1}^{r} \hat{A}_i$ and its membership table. By definition of intersection, $u \in M_{\delta_1\delta_2\ldots\delta_r}$ if and only if $u \in \hat{A}_1$, $u \in \hat{A}_2, \ldots, u \in \hat{A}_r$. Hence, the column labeled $M_{\delta_1\delta_2\ldots\delta_r}$ has exactly one 1. This 1 appears in the row that contains 1, 1, . . . , 1 in the columns labeled $\hat{A}_1, \hat{A}_2, \ldots, \hat{A}_r$, respectively: hence, in the row that contains $\delta_1, \delta_2, \ldots, \delta_r$ in the columns labeled $A_1, A_2, \ldots, A_r$, respectively. Thus, the column labeled $M_{\delta_1\delta_2\ldots\delta_r}$ has 1 in row $\delta_1\delta_2\ldots\delta_r$, and 0s in all other rows. For example, the column labeled M_{110} in the membership table for M_{110} has 1 in row 110, and 0s elsewhere.

Now, consider the union $M_{\delta_1\delta_2\ldots\delta_r} \cup M_{\delta_1'\delta_2'\ldots\delta_r'}$ of two distinct minsets generated by $A_1, A_2, \ldots, A_r$ and envision a membership table whose columns are labeled $A_1, A_2, \ldots, A_r, M_{\delta_1\delta_2\ldots\delta_r}, M_{\delta_1'\delta_2'\ldots\delta_r'}$, and $M_{\delta_1\delta_2\ldots\delta_r} \cup M_{\delta_1'\delta_2'\ldots\delta_r'}$. The last column can be constructed from the two preceding columns by direct invocation of Table 1-1(b), which defines the $\cup$ operation. From Table 1-1(b), it follows that column $M_{\delta_1\delta_2\ldots\delta_r} \cup M_{\delta_1'\delta_2'\ldots\delta_r'}$ would contain a 1 only in those rows where the column labeled $M_{\delta_1\delta_2\ldots\delta_r}$ or the column labeled $M_{\delta_1'\delta_2'\ldots\delta_r}$ contains a 1: hence, only in rows $\delta_1\delta_2\ldots\delta_r$ and $\delta_1'\delta_2'\ldots\delta_r'$. More generally, consider the l distinct minsets $M_{\delta_{11}\delta_{12}\ldots\delta_{1r}}, M_{\delta_{21}\delta_{22}\ldots\delta_{2r}}, \ldots, M_{\delta_{l1}\delta_{l2}\ldots\delta_{lr}}$ and the membership table for the set

$$S = M_{\delta_{11}\delta_{12}\ldots\delta_{1r}} \cup M_{\delta_{21}\delta_{22}\ldots\delta_{2r}} \cup \ldots \cup M_{\delta_{l1}\delta_{l2}\ldots\delta_{lr}}$$

The column labeled S would have 1 in rows $\delta_{11}\delta_{12}\ldots\delta_{1r}, \delta_{21}\delta_{22}\ldots\delta_{2r}, \ldots, \delta_{l1}\delta_{l2}\ldots\delta_{lr}$, and 0 in all other rows. For example, Table 1-5 shows how the membership table for the set

$$(A' \cap B' \cap C) \cup (A' \cap B \cap C) \cup (A \cap B \cap C') \cup (A \cap B \cap C)$$

Table 1-5 MEMBERSHIP TABLE FOR $(A' \cap B' \cap C) \cup (A' \cap B \cap C) \cup (A \cap B \cap C') \cup (A \cap B \cap C)$

A	B	C	$A' \cap B' \cap C = M_{001}$	$A' \cap B \cap C = M_{011}$	$A \cap B \cap C' = M_{110}$	$A \cap B \cap C = M_{111}$	$M_{001} \cup M_{011} \cup M_{110} \cup M_{111}$
0	0	0	0	0	0	0	0
0	0	1	1	0	0	0	1
0	1	0	0	0	0	0	0
0	1	1	0	1	0	0	1
1	0	0	0	0	0	0	0
1	0	1	0	0	0	0	0
1	1	0	0	0	1	0	1
1	1	1	0	0	0	1	1

can be constructed from the membership tables of minsets $A' \cap B' \cap C$, $A' \cap B \cap C$, $A \cap B \cap C'$, and $A \cap B \cap C$.

THEOREM 1-4

Every set S generated by $A_1, A_2, \ldots, A_r$ is equal to $\varnothing$ or to a union of distinct minsets generated by $A_1, A_2, \ldots, A_r$.

Proof If $S \neq \varnothing$, consider a set of minsets selected as follows: $M_{\delta_{i1}\delta_{i2}\ldots\delta_{ir}}$ is in this set if and only if column S in the membership table of S contains 1 in row $\delta_{i1}\delta_{i2}\ldots\delta_{ir}$. Denote the union of all these minsets by T. If a column labeled T is added to the membership table of S, then, by the preceding discussion, it would be identical to the column labeled S. Hence, $S = T$. □

When a set is expressed as $\varnothing$ or as a union of distinct minsets, it is said to be in a *minset normal form* or *minset canonical form*. The fact that every set can be expressed in this form is of great importance in set theory and related fields.

The proof of Theorem 1-4 suggests the following procedure for constructing a minset normal form:

ALGORITHM 1-1

To find a minset normal form of a given set S generated by $A_1, A_2, \ldots, A_r$:

(i) Construct the membership table for S. If the column labeled S contains no 1s, then $S = \varnothing$. Otherwise:

(ii) Suppose column S contains 1s in rows $\delta_{11}\delta_{12}\ldots\delta_{1r}, \delta_{21}\delta_{22}\ldots\delta_{2r}, \ldots, \delta_{l1}\delta_{l2}\ldots\delta_{lr}$. Then a minset normal form of S is given by

$$S = \bigcup_{i=1}^{l}\left(\bigcap_{j=1}^{r} A_{ij}\right)$$

where

$$A_{ij} = \begin{cases} A_i' & \text{if } \delta_{ij} = 0 \\ A_i & \text{if } \delta_{ij} = 1 \end{cases}$$ □

Example 1-3

From Table 1-2 it is seen that the column labeled $(A \cap B) \cup (A' \cap C)$ has 1s in rows 001, 011, 110, and 111. Hence, a minset normal form of $(A \cap B) \cup (A' \cap C)$ is given by

$$(A' \cap B' \cap C) \cup (A' \cap B \cap C) \cup (A \cap B \cap C') \cup (A \cap B \cap C)$$

(See Table 1-5 for detailed verification.) □

PROBLEMS

1. Find minset normal forms of these sets, generated by A, B:
   (a) U (b) A (c) A' (d) $A \cup B$

2. Find minset normal forms of:
(a) $(A \cap B') \cup (B \cap (A \cup C'))$ (generated by A, B, C)
(b) $((A \cup D') \cap (B' \cup C)') \cup (A \cap B \cap D)$ (generated by A, B, C, D)

1-9. THE MAXSET NORMAL FORM

Let $\{A_i\}_{i=1}^r$ be a set of subsets in a universal set U. A set of the form $\bigcup_{i=1}^r \hat{A}_i$, where each $\hat{A}_i$ may be either A_i or A_i', is called a *maxset generated by* $A_1, A_2, \ldots, A_r$. There are 2^r such maxsets. Unlike the set of minsets, the set of maxsets does *not* constitute a partition of U.

It is convenient to denote a maxset $\bigcup_{i=1}^r \hat{A}_i$ by $\tilde{M}_{\delta_1\delta_2\ldots\delta_r}$, where each δ_i is either 0 or 1, according to the following rule: $\delta_i = 0$ if $\hat{A}_i = A_i$, and $\delta_i = 1$ if $\hat{A}_i = A_i'$. In this manner, the subscript of $\tilde{M}_{\delta_1\delta_2\ldots\delta_r}$ uniquely describes the maxset that it intends to represent. For example, the maxset $A \cup B \cup C'$ generated by A, B, C is denoted by $\tilde{M}_{001}$.

Envision an arbitrary maxset $\tilde{M}_{\delta_1\delta_2\ldots\delta_r} = \bigcup_{i=1}^r \hat{A}_i$ and its membership table. By definition of union, $u \notin \tilde{M}_{\delta_1\delta_2\ldots\delta_r}$ if and only if $u \notin A_1$, $u \notin A_2$, $\ldots$, $u \notin A_r$. Hence the column labeled $\tilde{M}_{\delta_1\delta_2\ldots\delta_r}$ has exactly one 0. This 0 appears in the row that contains $0, 0, \ldots, 0$ in columns $\hat{A}_1, \hat{A}_2, \ldots, \hat{A}_r$, respectively, and, hence, in the row that contains $\delta_1, \delta_2, \ldots, \delta_r$ in the columns labeled $A_1, A_2, \ldots, A_r$, respectively. Thus, column $\tilde{M}_{\delta_1\delta_2\ldots\delta_r}$ has 0 in row $\delta_1, \delta_2 \ldots \delta_r$, and 1s in all other rows. For example, the column labeled $\tilde{M}_{001}$ has 0 in row 001, and 1s elsewhere.

Now, consider the intersection $\tilde{M}_{\delta_1\delta_2\ldots\delta_r} \cap M_{\delta_1'\delta_2'\ldots\delta_r'}$ of two distinct maxsets generated by $A_1, A_2, \ldots, A_r$, and envision a membership table whose columns are labeled $A_1, A_2, \ldots, A_r, \tilde{M}_{\delta_1\delta_2\ldots\delta_r}, \tilde{M}_{\delta_1'\delta_2'\ldots\delta_r'}$, and $\tilde{M}_{\delta_1\delta_2\ldots\delta_r} \cap \tilde{M}_{\delta_1'\delta_2'\ldots\delta_r'}$. The last column can be constructed from the two preceding columns by direct invocation of Table 1-1(c), which defines the $\cap$ operation. From Table 1-1(c) it follows that the column labeled $\tilde{M}_{\delta_1\delta_2\ldots\delta_r} \cap M_{\delta_1'\delta_2'\ldots\delta_r'}$ would contain 0 only in those rows where columns $\tilde{M}_{\delta_1\delta_2\ldots\delta_r}$ or $\tilde{M}_{\delta_1'\delta_2'\ldots\delta_r'}$ contain 0: hence, only in rows $\delta_1\delta_2 \ldots \delta_r$ and $\delta_1'\delta_2' \ldots \delta_r'$.

More generally, consider the l distinct maxsets $\tilde{M}_{\delta_{11}\delta_{12}\ldots\delta_{1r}}, \tilde{M}_{\delta_{21}\delta_{22}\ldots\delta_{2r}}, \ldots, \tilde{M}_{\delta_{l1}\delta_{l2}\ldots\delta_{lr}}$, and the membership table for set

$$S = \tilde{M}_{\delta_{11}\delta_{12}\ldots\delta_{1r}} \cap \tilde{M}_{\delta_{21}\delta_{22}\ldots\delta_{2r}} \cap \ldots \cap \tilde{M}_{\delta_{l1}\delta_{l2}\ldots\delta_{lr}}$$

The column labeled S would have 0 in rows $\delta_{11}\delta_{12} \ldots \delta_{1r}, \delta_{21}\delta_{22} \ldots \delta_{2r}, \ldots, \delta_{l1}\delta_{l2} \ldots \delta_{lr}$, and 1 in all other rows. Table 1-6 shows how the membership table for set

$$S = (A \cup B \cup C) \cap (A \cup B' \cup C) \cap (A' \cup B \cup C) \cap (A' \cup B \cup C')$$

can be constructed from the membership tables of maxsets $A \cup B \cup C$, $A \cup B' \cup C$, $A' \cup B \cup C$, and $A' \cup B \cup C'$.

Table 1-6 MEMBERSHIP TABLE FOR $(A \cup B \cup C) \cap (A \cup B' \cup C) \cap (A' \cup B \cup C) \cap (A' \cup B \cup C')$

A B C	$A \cup B \cup C = \tilde{M}_{000}$	$A \cup B' \cup C = \tilde{M}_{010}$	$A' \cup B \cup C = \tilde{M}_{100}$	$A' \cup B \cup C' = \tilde{M}_{101}$	$\tilde{M}_{000} \cap \tilde{M}_{010} \cap \tilde{M}_{100} \cap \tilde{M}_{101}$
0 0 0	0	1	1	1	0
0 0 1	1	1	1	1	1
0 1 0	1	0	1	1	0
0 1 1	1	1	1	1	1
1 0 0	1	1	0	1	0
1 0 1	1	1	1	0	0
1 1 0	1	1	1	1	1
1 1 1	1	1	1	1	1

THEOREM 1-5

Every set S generated by $A_1, A_2, \ldots, A_r$ is equal to U or to an intersection of distinct maxsets generated by $A_1, A_2, \ldots, A_r$.

Proof If $S \neq U$, consider a set of maxsets selected as follows: $\tilde{M}_{\delta_{i1}\delta_{i2}\ldots\delta_{ir}}$ is in this set if and only if column S in the membership table of S contains 0 in row $\delta_{i1}\delta_{i2}\ldots\delta_{ir}$. Denote the intersection of all these maxsets by T. If a column labeled T is added to the membership table, then, by the preceding discussion, it would be identical to the column labeled S. Hence, $S = T$. □

When a set is expressed as U or as an intersection of distinct maxsets, it is said to be in a *maxset normal form* or *maxset canonical form*. The proof of Theorem 1-5 suggests the following procedure for deriving this form:

ALGORITHM 1-2

To find a maxset normal form of a given set S generated by $A_1, A_2, \ldots, A_r$:

(i) Construct the membership table of S. If the column labeled S contains no 0s, then $S = U$. Otherwise:

(ii) Suppose column S contains 0s in rows $\delta_{11}\delta_{12}\ldots\delta_{1r}$, $\delta_{21}\delta_{22}\ldots\delta_{2r}$, $\ldots$, $\delta_{l1}\delta_{l2}\ldots\delta_{lr}$. Then a maxset normal form of S is given by

$$S = \bigcap_{i=1}^{l} \left(\bigcup_{j=1}^{r} A_{ij}\right)$$

where

$$A_{ij} = \begin{cases} A_i & \text{if} \quad \delta_{ij} = 0 \\ A_i' & \text{if} \quad \delta_{ij} = 1 \end{cases}$$ □

Example 1-4

From Table 1-2 it is seen that the column labeled $(A \cap B) \cup (A' \cap C)$ has 0s in rows 000, 010, 100, and 101. Hence, a maxset normal form of $(A \cap B)$

$\cup (A' \cap C)$ is given by

$$(A \cup B \cup C) \cap (A \cup B' \cup C) \cap (A' \cup B \cup C) \cap (A' \cup B \cup C')$$

(See Table 1-6 for detailed verification.) □

PROBLEMS

1. Find maxset normal forms of these sets, generated by A, B:
   (a) $\varnothing$ (b) A (c) A' (d) $A \cap B$
2. Find maxset normal forms of the sets given in Problem 2, Sec. 1-8.
3. Show that the set of all maxsets generated by $A_1, A_2, \ldots, A_r$ does not form a partition of U.

1-10. MORE ON MINSET AND MAXSET NORMAL FORMS

Let the minset and maxset normal forms of S, as constructed by Algorithms 1-1 and 1-2, be given by

$$S = M_{\delta_{11}\delta_{12}\ldots\delta_{1r}} \cup M_{\delta_{21}\delta_{22}\ldots\delta_{2r}} \cup \ldots \cup M_{\delta_{l1}\delta_{l2}\ldots\delta_{lr}}$$
$$S = \tilde{M}_{\tilde{\delta}_{11}\tilde{\delta}_{12}\ldots\tilde{\delta}_{1r}} \cap \tilde{M}_{\tilde{\delta}_{21}\tilde{\delta}_{22}\ldots\tilde{\delta}_{2r}} \cap \ldots \cap \tilde{M}_{\tilde{\delta}_{m1}\tilde{\delta}_{m2}\ldots\tilde{\delta}_{mr}}$$

From these algorithms it follows that the sets of rows

$$\{\delta_{11}\delta_{12}\ldots\delta_{1r},\quad \delta_{21}\delta_{22}\ldots\delta_{2r},\quad \ldots,\quad \delta_{l1}\delta_{l2}\ldots\delta_{lr}\}$$
$$\{\tilde{\delta}_{11}\tilde{\delta}_{12}\ldots\tilde{\delta}_{1r},\quad \tilde{\delta}_{21}\tilde{\delta}_{22}\ldots\tilde{\delta}_{2r},\quad \ldots,\quad \tilde{\delta}_{m1}\tilde{\delta}_{m2}\ldots\tilde{\delta}_{mr}\}$$

are disjoint, and that their union equals the set of all 2^r rows in the membership table of S. Hence, if the minset normal form (as obtained by Algorithm 1-1) is the union of l minsets, then the maxset normal form (as obtained by Algorithm 1-2) is the intersection of $2^r - l$ maxsets. Moreover, if one of the forms is given, the other form can be readily constructed, as illustrated by this example:

Example 1-5

Suppose a minset normal form of $S = B \cap (A \cup (B \cap C'))$ has been determined (via Algorithm 1-1) to be

$$S = (A' \cap B \cap C') \cup (A \cap B \cap C') \cup (A \cap B \cap C)$$
$$= M_{010} \cup M_{110} \cup M_{111}$$

Then the maxset normal form is given by:

$$\begin{aligned} S &= \tilde{M}_{000} \cap \tilde{M}_{001} \cap \tilde{M}_{011} \cap \tilde{M}_{100} \cap \tilde{M}_{101} \\ &= (A \cup B \cup C) \cap (A \cup B \cup C') \cap (A \cup B' \cup C') \\ &\quad \cap (A' \cup B \cup C) \cap (A' \cup B \cup C') \end{aligned}$$ □

In what follows we shall outline a procedure that facilitates construction of a minset (or maxset) normal form without resorting to membership tables. This procedure is often faster than that given by Algorithm 1-1 (or 1-2).

ALGORITHM 1-3 (OR 1-4).[1]

To find a minset (or maxset) normal form of a given set S generated by $A_1, A_2, \ldots, A_r$:

(i) use the laws in Tables 1-3 and 1-4 to express S as $\varnothing$ (or U), or as a union (or intersection) of distinct intersections (or unions) of the form $\hat{A}_{i_1} \cap \hat{A}_{i_2} \cap \ldots \cap \hat{A}_{i_k}$ (or $\hat{A}_{i_1} \cup \hat{A}_{i_2} \cup \ldots \cup \hat{A}_{i_k}$), where $i_1 < i_2 < \ldots < i_k$. If $S = \varnothing$ (or $S = U$), S is in the desired form. Otherwise:

(ii) if every intersection (or union) is a minset (or maxset), then S is in the desired form. Otherwise:

(iii) from the expression last obtained, select an intersection (or union) of the form $\hat{A}_{i_1} \cap \hat{A}_{i_2} \cap \ldots \cap \hat{A}_{i_k}$ (or $\hat{A}_{i_1} \cup \hat{A}_{i_2} \cup \ldots \cup \hat{A}_{i_k}$), where $k < r$ and where, for some h, $\hat{A}_{i_h}$ does not appear, and replace it with

$$\hat{A}_{i_1} \cap \hat{A}_{i_2} \cap \ldots \cap \hat{A}_{i_k} \cap A_{i_h}) \cup (\hat{A}_{i_1} \cap \hat{A}_{i_2} \cap \ldots \cap \hat{A}_{i_k} \cap A'_{i_h})$$
$$[\text{or } \hat{A}_{i_1} \cup \hat{A}_{i_2} \cup \ldots \cup \hat{A}_{i_k} \cup A_{i_h}) \cap (\hat{A}_{i_1} \cup \hat{A}_{i_2} \cup \ldots \cup \hat{A}_{i_k} \cup A'_{i_h})]$$

which is permitted by virtue of the commutative, distributive, and complement laws in Table 1-3. Rearrange the $\hat{A}_{i_j}$ in the new intersections (or unions) in order of ascending subscript and delete all duplicate intersections (or unions). Return to step (ii). □

Example 1-6

The use of Algorithm 1-3 for constructing a minset normal form of $S = B \cap (A \cup (B \cap C'))$ is illustrated by

$$\begin{aligned} S &= B \cap (A \cup (B \cap C')) = (A \cap B) \cup (B \cap C') \\ &= (A \cap B \cap C) \cup (A \cap B \cap C') \cup (B \cap C') \\ &= (A \cap B \cap C) \cup (A \cap B \cap C') \cup (A' \cap B \cap C') \end{aligned}$$

[1] All parenthetically cited alternatives (or . . .) should be understood to belong together.

Using Algorithm 1-4 for constructing a maxset normal form of the same set S, we have

$$\begin{aligned} S &= B \cap (A \cup (B \cap C')) = B \cap (A \cup B) \cap (A \cup C') \\ &= (A \cup B) \cap (A' \cup B) \cap (A \cup C') \\ &= (A \cup B \cup C) \cap (A \cup B \cup C') \cap (A' \cup B) \cap (A \cup C') \\ &= (A \cup B \cup C) \cap (A \cup B \cup C') \cap (A' \cup B \cup C) \cap \\ &\qquad (A' \cup B \cup C') \cap (A \cup C') \\ &= (A \cup B \cup C) \cap (A \cup B \cup C') \cap (A' \cup B \cup C) \cap \\ &\qquad (A' \cup B \cup C') \cap (A \cup B' \cup C') \end{aligned}$$ □

In this and the preceding two sections we outlined procedures for expressing a given set S generated by $A_1, A_2, \ldots, A_r$ as a union of intersections (a minset normal form) or as an intersection of unions (a maxset normal form) of $\hat{A}_1, \hat{A}_2, \ldots, \hat{A}_r$, where each $\hat{A}_i$ is either A_i or A_i'. The curious thing about these forms is that, regardless of the particular procedure used for obtaining them, they always come out to be the same except, possibly, for the order in which the minsets or maxsets are arranged. This fact permits us to talk about *the* minset normal form and *the* maxset normal form of S because, essentially, only one of each kind exists. This conclusion will be formalized by:

THEOREM 1-6
Given a set S generated by $A_1, A_2, \ldots, A_r$, the minset (or maxset) normal form of S is unique up to minset (or maxset) ordering.

Proof Suppose a given minset (or maxset) normal form of S contains the minset (or maxset) $M_{\delta_{i1}\delta_{i2}\ldots\delta_{ir}}$ (or $\tilde{M}_{\delta_{i1}\delta_{i2}\ldots\delta_{ir}}$). Hence, column S of the membership table of S must contain 1 (or 0) in row $\delta_{i1}\delta_{i2}\ldots\delta_{ir}$. Because distinct minsets (or maxsets) contribute 1s (or 0s) to distinct rows of column S, it follows that *any* minset (or maxset) normal form of S must contain the minset (or maxset) $M_{\delta_{i1}\delta_{i2}\ldots\delta_{ir}}$ (or $\tilde{M}_{\delta_{i1}\delta_{i2}\ldots\delta_{ir}}$). Thus, except for ordering, any two minset (or maxset) normal forms of S must be identical. □

A direct corollary of Theorem 1-6 is:

THEOREM 1-7
Two sets generated by A_1 $A_2, \ldots, A_r$ are equal if and only if their minset (or maxset) normal forms are equal up to minset (or maxset) ordering.

Thus, the minset and maxset normal forms constitute standard forms into which every set generated by $A_1, A_2, \ldots, A_r$ can be reduced, and by means of which we can establish whether any two such sets are equal or not. (In fact, the adjectives "normal" and "canonical" are merely mathematical synonyms for the adjective "standard.")

PROBLEMS

1. Using Algorithms 1-3 and 1-4, find the minset and maxset normal forms of the following sets generated by A, B, C:
(a) U (b) A (c) $B \cap C'$ (d) $A \cup (B' \cap C)$
(e) $(A \cap B') \cup (B \cap (A \cup C'))$

2. S is a set generated by $A_1, A_2, \ldots, A_r$. How many minsets are there in the minset normal forms of the sets
(a) $A_{i_1} \cap A_{i_2} \cap \ldots \cap A_{i_k}$
(b) $A_{i_1} \cup A_{i_2} \cup \ldots \cup A_{i_k}$
(where $i_1, i_2, \ldots, i_k$ are distinct subscripts)?

3. S is a set whose minset normal form is given by

$$S = M_{\delta_{11}\delta_{12}\ldots\delta_{1r}} \cup M_{\delta_{21}\delta_{22}\ldots\delta_{2r}} \cup \ldots \cup M_{\delta_{l1}\delta_{l2}\ldots\delta_{lr}}$$

Show that the maxset normal form of S' is given by

$$S' = \tilde{M}_{\delta_{11}\delta_{12}\ldots\delta_{1r}} \cap \tilde{M}_{\delta_{21}\delta_{22}\ldots\delta_{2r}} \cap \ldots \cap \tilde{M}_{\delta_{l1}\delta_{l2}\ldots\delta_{lr}}$$

4. Use the minset and maxset normal forms to prove that the complement of $(A \cap B') \cup (A' \cap (B \cup C'))$ is $(A' \cup B) \cap (A \cup B') \cap (A \cup C)$.

REFERENCES

HALMOS, P. R., *Naive Set Theory*. Princeton, NJ: Van Nostrand, 1960.

HERSTEIN, I., *Topics in Algebra*. Waltham, MA: Xerox, 1964.

MACLANE, S., and G. BIRKHOFF, *Algebra*. New York: Macmillan, 1967.

PALEY, H., and P. M. WEICHSEL, *A First Course in Abstract Algebra*. New York: Holt, Rinehart and Winston, 1966.

VAN DER WAERDEN, B. L., *Modern Algebra*. Ungar, 1931.

2 RELATIONS

This chapter introduces the concepts of *r-tuples* and *Cartesian products* from which emanates the basic idea of a *relation*. It is shown how relations can be conveniently displayed by means of matrices and graphs. The operation of *composition* of relations is defined, and procedures are described for obtaining matrices and graphs of composite relations from those of their components. Several important properties of relations are enumerated, some of which define *equivalence relations*. It is demonstrated how equivalence relations give rise to partitions, and the notion of a *quotient set* defined by such partitions is introduced. Finally, relations called *partial orderings* (with *total orderings* and *well orderings* as special cases) are discussed in detail.

Like the notion of a set, the notion of a relation is of fundamental importance in computer science. It appears constantly in the theory of finite-state machines and formal languages and in applied areas such as compiler design and information retrieval. The related concept of a graph is indispensable in almost every phase of computer activity—from circuit and logical design to the description of programs and data structures.

2-1. CARTESIAN PRODUCTS

An ordered sequence of r elements (not necessarily distinct and not necessarily belonging to the same set), written in the form $(a_1, a_2, \ldots, a_r)$, is called an *r-tuple*.[1] An often-encountered special case is $r = 2$, where the

[1] Some authors call this sequence an *ordered r-tuple*, to differentiate it from an *unordered r-tuple* that is simply a collection of r elements without any imposed order.

ordered sequence is referred to as a *pair*. The ith element in $(a_1, a_2, \ldots, a_r)$ is sometimes called the ith *coordinate* of the r-tuple.

Let $A_1, A_2, \ldots, A_r$ be arbitrary sets. The set of all r-tuples $(a_1, a_2, \ldots, a_r)$, where $a_1 \in A_1, a_2 \in A_2, \ldots, a_r \in A_r$, is called the *Cartesian product* of $A_1, A_2, \ldots, A_r$, and is denoted by $A_1 \times A_2 \times \ldots \times A_r$. That is,

$$A_1 \times A_2 \times \ldots \times A_r = \{(a_1, a_2, \ldots, a_r) \mid a_i \in A_i, i = 1, 2, \ldots, r\}$$

If all the A_i are finite, then

$$\#(A_1 \times A_2 \times \ldots \times A_r) = (\#A_1)(\#A_2) \ldots (\#A_r)$$

When all the A_i are identical and equal to A, then $A_1 \times A_2 \times \ldots \times A_r$ can be denoted by A^r.

The set of all ith coordinates of the elements of $A_1 \times A_2 \times \ldots \times A_r$ (namely, the set A_i) is called the *projection of* $A_1 \times A_2 \times \ldots \times A_r$ *onto its* ith *coordinate*.

Example 2-1

If $A = \{0, 1\}$ and $B = \{1, 2, 3\}$, then

$$A \times B = \{(0, 1), (0, 2), (0, 3), (1, 1), (1, 2), (1, 3)\}$$
$$B \times A = \{(1, 0), (1, 1), (2, 0), (2, 1), (3, 0), (3, 1)\}$$
$$A^2 = \{(0, 0), (0, 1), (1, 0), (1, 1)\}$$

□

PROBLEMS

1. If $A = \{0, 1\}$ and $B = \{1, 2\}$, determine the sets
   (a) $A \times \{1\} \times B$ (b) $A^2 \times B$ (c) $(B \times A)^2$
2. Considering the usual Cartesian coordinate system with x and y axes, give an interpretation to the set $X \times Y$, where

$$X = \{x \mid x \in \mathbb{R}, -3 \leq x \leq 2\}$$
$$Y = \{y \mid y \in \mathbb{R}, -2 \leq y \leq 0\}$$

3. Prove that for any sets A, B, and C:
   (a) $A \times (B \cup C) = (A \times B) \cup (A \times C)$
   (b) $A \times (B \cap C) = (A \times B) \cap (A \times C)$

2-2. RELATIONS

A subset of the Cartesian product $A_1 \times A_2 \times \ldots \times A_r$ is called an *r-ary relation over* $A_1, A_2, \ldots, A_r$. A most important special case is $r = 2$; in this case the relation is a subset of $A_1 \times A_2$ (that is, a set of pairs, the first coordinate of which is from A_1 and the second from A_2), and is referred to as a *binary relation from* A_1 *into* A_2. Henceforth, whenever the term *relation* is used without qualifications, it means a *binary relation*.

If ρ is a relation from A into B and $(a, b) \in \rho$, we say that *a is related to b under ρ*, and write $a \rho b$ [if $(a, b) \notin \rho$, we write $a \rho' b$]. The *domain* of ρ is defined as the set of all $a \in A$ such that $a \rho b$ for some $b \in B$. The *range* of ρ is the set of all $b \in B$ such that $a \rho b$ for some $a \in A$.

Example 2-2

Consider the sets $A = \{2, 3, 4\}$ and $B = \{2, 3, 4, 5, 6\}$, and the relation ρ from A into B specified as follows: $a \rho b$ if and only if a divides b. Hence, ρ is a subset of $A \times B$ given by:

$$\{(2, 2), (2, 4), (2, 6), (3, 3), (3, 6), (4, 4)\} \tag{2-1}$$

The domain of ρ is $\{2, 3, 4\}$, and its range is $\{2, 3, 4, 6\}$. □

The *converse* of a relation ρ, denoted by $\breve{\rho}$, is defined as follows: If ρ is a relation from A into B, then $\breve{\rho}$ is a relation from B into A such that $b \breve{\rho} a$ if and only if $a \rho b$. That is, the pairs of $\breve{\rho}$ are the pairs of ρ, with the order of the elements reversed. For example, if ρ is specified by (2-1), then

$$\breve{\rho} = \{(2, 2), (4, 2), (6, 2), (3, 3), (6, 3), (4, 4)\}$$

Figure 2-1 shows a diagrammatic representation of a relation ρ from A into B. In this figure, the small circles labeled (generically) a_i and b_j represent elements from A and B, respectively, and an arrow points from a_i to b_j if and only if $a_i \rho b_j$. (The relation $\breve{\rho}$ is represented by the same diagram, with the arrows reversed.)

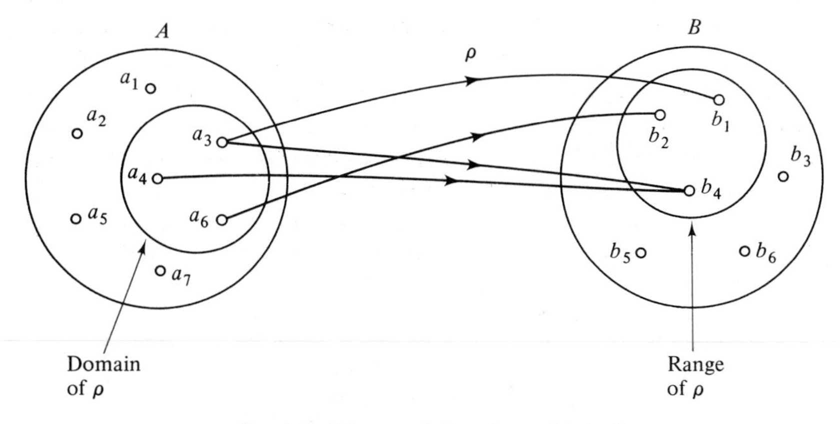

Fig. 2-1 Binary relation from A into B

When A and B are finite sets, a relation ρ from A into B can be conveniently represented by a $(\#A) \times (\#B)$ matrix[1] called a *relation matrix* and denoted by $\boldsymbol{\rho}$. If $A = \{a_1, a_2, \ldots, a_{\#A}\}$ and $B = \{b_1, b_2, \ldots, b_{\#B}\}$, then the (i, j) entry of $\boldsymbol{\rho}$ is 1 if $a_i \; \rho \; a_j$, and 0 otherwise. For example, the relation (2-1) can be specified by the relation matrix

$$\boldsymbol{\rho} = \begin{array}{c} \\ 2 \\ 3 \\ 4 \end{array} \begin{array}{c} \begin{array}{ccccc} 2 & 3 & 4 & 5 & 6 \end{array} \\ \begin{bmatrix} 1 & 0 & 1 & 0 & 1 \\ 0 & 1 & 0 & 0 & 1 \\ 0 & 0 & 1 & 0 & 0 \end{bmatrix} \end{array}$$

A relation from a set A into itself (that is, a subset of A^2) is called a *relation on A*. For example,

$$\rho = \{(0, 0), (0, 3), (2, 0), (2, 1), (2, 3), (3, 2)\} \qquad (2\text{-}2)$$

is a relation on $\{0, 1, 2, 3\}$. When $\rho = A^2$, it is called the *universal relation* on A, and is denoted by U_A; thus,

$$U_A = \{(a_i, a_j) \mid a_i, a_j \in A\}$$

The *identity relation* on A, denoted by I_A, is defined by

$$I_A = \{(a_i, a_i) \mid a_i \in A\}$$

For example, for $A = \{0, 1, 2\}$,

$$\begin{aligned} U_A &= \{(0, 0), (0, 1), (0, 2), (1, 0), (1, 1), (1, 2), (2, 0), (2, 1), (2, 2)\} \\ I_A &= \{(0, 0), (1, 1), (2, 2)\} \end{aligned}$$

A finite relation ρ on A can be represented by a diagram that we call the *graph* of ρ.[2] This graph consists of as many *vertices* as there are elements in A, each vertex drawn as a circle that bears the label of a different element. The vertices are interconnected by means of *edges* in this manner: an edge is an arrow that points from vertex a_i to vertex a_j if and only if $a_i \; \rho \; a_j$. (The graph of $\breve{\rho}$ is the same as that of ρ, with all arrows reversed.) As an example, Figure 2-2 shows the graph of ρ specified by (2-2).

Vertex a_{i_l} in a graph is said to be *reachable* from vertex a_{i_0} via a *path of length* $l \geq 1$, if there exist $l - 1$ vertices $a_{i_1}, a_{i_2}, \ldots, a_{i_{l-1}}$ such that $a_{i_0} \; \rho \; a_{i_1}$,

[1]An $m \times n$ *matrix* $\mathbf{M}$ is an array of mn elements arranged in m rows and n columns. The element located in the ith row and jth column is called the (i, j) *entry* of $\mathbf{M}$. In this book, symbols denoting matrices are printed in boldface.

[2]In Chapter 12 such a graph is called a *directed graph*.

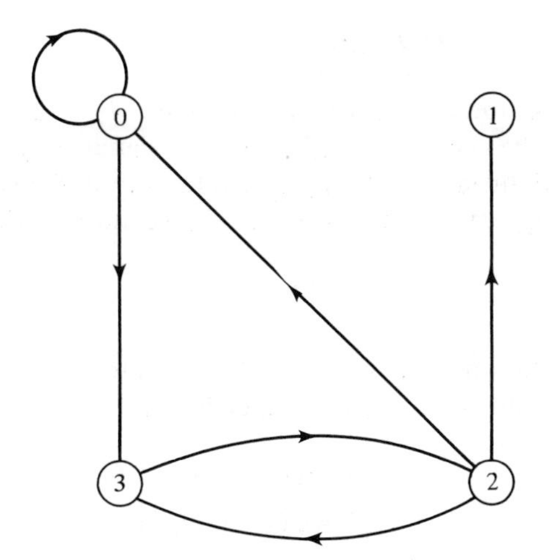

Fig. 2-2 Graph of relation (2-2)

$a_{i_1} \rho a_{i_2}, \ldots, a_{i_{l-1}} \rho a_{i_l}$. If $a_{i_0} = a_{i_l}$, the path is called a *loop*. Any vertex reachable from a_{i_0}, via a path of any length, is called a *descendent* of a_{i_0} (a *direct descendent* if the length is 1).

PROBLEMS

1. For each of the following cases, list the elements of relation ρ from A into B, determine the domain and range of ρ, and construct the relation matrix of ρ.

(a) $A = \{0, 1, 2\}$, $B = \{0, 2, 4\}$, $\rho = \{(a, b) \mid ab \in A \cap B\}$

(b) $A = \{1, 2, 3, 4, 5\}$, $B = \{1, 2, 3\}$, $\rho = \{(a, b) \mid a = b^2\}$

(c) $A = \mathbf{2}^{\{0,1\}}$, $B = \mathbf{2}^{\{0,1,2\}} - \mathbf{2}^{\{0\}}$, $\rho = \{(a, b) \mid a - b = \varnothing\}$

2. Let $A = \{1, 2, 3, 4, 5, 6\}$. For each of the following cases construct the graph of the relation ρ on A, and determine the domain and range of ρ.

(a) $\rho = \{(i, j) \mid i = j\}$

(b) $\rho = \{(i, j) \mid i \text{ divides } j\}$

(c) $\rho = \{(i, j) \mid i \text{ is a multiple of } j\}$

(d) $\rho = \{(i, j) \mid i > j\}$

(e) $\rho = \{(i, j) \mid i < j\}$

(f) $\rho = \{(i, j) \mid i \neq j, ij < 10\}$

(g) $\rho = \{(i, j) \mid (i - j)^2 \in A\}$

(h) $\rho = \{(i, j) \mid i/j \text{ is prime}\}$

3. A_1 and A_2 are finite sets with cardinalities n_1 and n_2, respectively. How many distinct relations are there from A_1 into A_2?

4. Characterize the relation matrices and graphs of the identity and universal relations on $A = \{a_1, a_2, \ldots, a_n\}$.

2-3. COMPOSITION OF RELATIONS

Let ρ_1 be a relation from a set A_1 into a set A_2, and ρ_2 a relation from A_2 into a set A_3. The *composition of* ρ_1 *and* ρ_2, denoted by $\rho_1\rho_2$, is a relation from A_1 into A_3 defined as: $a_i(\rho_1\rho_2)a_j$ if and only if, for some $a_k \in A_2$, we have $a_i\rho_1a_k$ and $a_k\rho_2a_j$. In this context, the relation $\rho_1\rho_2$ is said to be a *composite* relation.

Example 2-3

Let ρ_1 be a relation from $A_1 = \{1, 2, 3, 4\}$ into $A_2 = \{2, 3, 4\}$, and ρ_2 a relation from A_2 into $A_3 = \{1, 2, 3\}$, specified as

$$\begin{aligned}\rho_1 &= \{(a_i, a_k) \mid a_i + a_k = 6\}\\ &= \{(2, 4), (3, 3), (4, 2)\}\\ \rho_2 &= \{(a_k, a_j) \mid a_k - a_j = 1\}\\ &= \{(2, 1), (3, 2), (4, 3)\}\end{aligned}$$

The composite relation $\rho_1\rho_2$ (from A_1 into A_3) consists of all pairs (a_i, a_j) such that, for some $a_k \in A_2$, $a_i + a_k = 6$ and $a_k - a_j = 1$; hence,

$$\rho_1\rho_2 = \{(2, 3), (3, 2), (4, 1)\}$$

Note that the same result can also be obtained from the observation that

$$\rho_1\rho_2 = \{(a_i, a_j) \mid a_i + a_j = 5\}$$ □

From the definition of an identity relation it follows that, when ρ is a relation from A into B, then

$$I_A\rho = \rho I_B = \rho$$

Let ρ_1 be a relation from A_1 into A_2, ρ_2 a relation from A_2 into A_3, and ρ_3 a relation from A_3 into A_4. From the definition of composition it follows that the composite relations $\rho_1(\rho_2\rho_3)$ and $(\rho_1\rho_2)\rho_3$ (from A_1 into A_4) are equal and, hence, can be written unambiguously as $\rho_1\rho_2\rho_3$. More generally, if ρ_1 is a relation from A_1 into A_2, ρ_2 a relation from A_2 into $A_3, \ldots, \rho_r$ a relation from A_r into A_{r+1}, then the parentheses-free expression $\rho_1\rho_2 \ldots \rho_r$ uniquely represents a relation from A_1 into A_{r+1}. This relation, in fact, consists of all pairs (a_i, a_j) such that, for some $k_1, k_2, \ldots, k_{r-1}$, we have $a_i\rho_1a_{k_1}$, $a_{k_1}\rho_2a_{k_2}$, $\ldots$, $a_{k_{r-1}}\rho_ra_j$.

When $A_1 = A_2 = \ldots = A_r = A$, and $\rho_1 = \rho_2 = \ldots = \rho_r = \rho$ (that is, when all the ρ_i are the same relation ρ on A), the composite relation $\rho_1\rho_2 \ldots \rho_r$ (on A) can be denoted by ρ^r. In Example 2-3, if we let $\rho_2 = \rho$, we have:

$$\rho = \{(2, 1), (3, 2), (4, 3)\}$$
$$\rho^2 = \{(3, 1), (4, 2)\}$$
$$\rho^3 = \{(4, 1)\}$$

PROBLEMS

1. The following are relations on $A = \{0, 1, 2, 3\}$:
$\rho_1 = \{(i, j) \mid j = i + 1 \text{ or } j = i/2\}$
$\rho_2 = \{(i, j) \mid i = j + 2\}$
Determine the composite relations:
(a) $\rho_1\rho_2$ (b) $\rho_2\rho_1$ (c) $\rho_1\rho_2\rho_1$ (d) ρ_1^3

2. ρ_1, ρ_2 and ρ_3 are relations on a set A. Show that $\rho_1 \subset \rho_2$ implies:
(a) $\rho_1\rho_3 \subset \rho_2\rho_3$ (b) $\rho_3\rho_1 \subset \rho_3\rho_2$ (c) $\breve{\rho}_1 \subset \breve{\rho}_2$

3. Given:
$\rho_1 = \{(0, 1), (1, 2), (3, 4)\}$
$\rho_1\rho_2 = \{(1, 3), (1, 4), (3, 3)\}$
Find a relation of least cardinality that satisfies ρ_2. In general, given ρ_1 and $\rho_1\rho_2$, is ρ_2 uniquely determinable? Is ρ_2 of least cardinality uniquely determinable?

2-4. RELATION MATRICES OF COMPOSITE RELATIONS

For the purpose of this section, it is convenient to introduce *Boolean arithmetic* (which will be used extensively in Chapter 7). This arithmetic deals only with the digits 0 and 1, and defines addition (*Boolean addition*) and multiplication (*Boolean multiplication*) of these digits in the following manner:

$$0 + 0 = 0 \qquad 0 + 1 = 1 + 0 = 1 + 1 = 1$$
$$1 \cdot 1 = 1 \qquad 1 \cdot 0 = 0 \cdot 1 = 0 \cdot 0 = 0$$

(These operations are the same as ordinary addition and multiplication, with the exception that $1 + 1 = 1$.) For example, the expression

$$(1 \cdot 1) + (0 \cdot 0 \cdot 1) + (1 \cdot 1 \cdot 1) + 1 + (1 \cdot 0)$$

where all operations are Boolean, has the value 1. (Generally, a sum of products equals 1 if and only if at least one of the products is of the form $1 \cdot 1 \cdot \ldots \cdot 1$.)

We shall use Boolean arithmetic to define the *product* of two relation matrices as follows: Let $\boldsymbol{\rho}_1$ be an $l \times m$ relation matrix with the (i, j) entry

$r_{i,j}^{(1)}$, and $\boldsymbol{\rho}_2$ an $m \times n$ relation matrix with the (i, j) entry $r_{i,j}^{(2)}$. Then $\boldsymbol{\rho}_1\boldsymbol{\rho}_2$ is defined as the $l \times n$ relation matrix whose (i, j) entry is

$$\sum_{k=1}^{m} r_{i,k}^{(1)} \cdot r_{k,j}^{(2)} \qquad (i = 1, 2, \ldots, l, j = 1, 2, \ldots, n)$$

where all additions and multiplications are Boolean. Note that, for $\boldsymbol{\rho}_1\boldsymbol{\rho}_2$ to be defined, the number of columns in $\boldsymbol{\rho}_1$ must equal the number of rows in $\boldsymbol{\rho}_2$.

On the basis of the preceding definition it can be readily shown that the products $\boldsymbol{\rho}_1(\boldsymbol{\rho}_2\boldsymbol{\rho}_3)$ and $(\boldsymbol{\rho}_1\boldsymbol{\rho}_2)\boldsymbol{\rho}_3$, when defined, are identical and, hence, can be written unambiguously as $\boldsymbol{\rho}_1\boldsymbol{\rho}_2\boldsymbol{\rho}_3$. More generally, if the products $\boldsymbol{\rho}_1\boldsymbol{\rho}_2$, $\boldsymbol{\rho}_2\boldsymbol{\rho}_3, \ldots, \boldsymbol{\rho}_{r-1}\boldsymbol{\rho}_r$ are defined, then the parentheses-free expression $\boldsymbol{\rho}_1\boldsymbol{\rho}_2 \ldots \boldsymbol{\rho}_r$ uniquely represents the product of $\boldsymbol{\rho}_1, \boldsymbol{\rho}_2, \ldots, \boldsymbol{\rho}_r$. When $\boldsymbol{\rho}_1 = \boldsymbol{\rho}_2 = \boldsymbol{\rho}_r = \boldsymbol{\rho}$, this product can be denoted by $\boldsymbol{\rho}^r$.

Now, suppose ρ_1 is a relation from A_1 into A_2, and ρ_2 is a relation from A_2 into A_3, where A_1, A_2, and A_3 are finite sets. Let the matrices $\boldsymbol{\rho}_1$ and $\boldsymbol{\rho}_2$ of the preceding paragraph be the relation matrices of ρ_1 and ρ_2, respectively, and let $\boldsymbol{\rho}$ denote the relation matrix of the composite relation $\rho_1\rho_2$. From the definition of $\rho_1\rho_2$ it follows that the (i, j) entry of $\boldsymbol{\rho}$ is 1 if and only if, for some k, $r_{i,k}^{(1)} = r_{k,j}^{(2)} = 1$. Hence, the (i, j) entry of $\boldsymbol{\rho}$ is precisely

$$\sum_{k=1}^{\#A_2} r_{i,k}^{(1)} \cdot r_{k,j}^{(2)}$$

where all additions and multiplications are Boolean. Thus, we have:

$$\boldsymbol{\rho} = \boldsymbol{\rho}_1\boldsymbol{\rho}_2$$

More generally, we can arrive at this result:

THEOREM 2-1

Let ρ_1 be a relation from A_1 into A_2, ρ_2 a relation from A_2 into A_3, $\ldots$, ρ_r a relation from A_r into A_{r+1} (where $A_1, A_2, \ldots, A_{r+1}$ are finite sets), whose relation matrices are $\boldsymbol{\rho}_1, \boldsymbol{\rho}_2, \ldots, \boldsymbol{\rho}_r$, respectively. Then the relation matrix of the composite relation $\rho_1\rho_2 \ldots \rho_r$ (from A_1 into A_{r+1}) is given by $\boldsymbol{\rho}_1\boldsymbol{\rho}_2 \ldots \boldsymbol{\rho}_r$.

Example 2-4

Using the relations ρ_1 and ρ_2 of Example 2-3, we have

$$\boldsymbol{\rho}_1 = \begin{array}{c} \\ 1 \\ 2 \\ 3 \\ 4 \end{array} \begin{array}{c} \begin{array}{ccc} 2 & 3 & 4 \end{array} \\ \begin{bmatrix} 0 & 0 & 0 \\ 0 & 0 & 1 \\ 0 & 1 & 0 \\ 1 & 0 & 0 \end{bmatrix} \end{array}$$

$$\boldsymbol{\rho}_2 = \begin{array}{c} \\ 2 \\ 3 \\ 4 \end{array} \begin{array}{c} \begin{matrix} 1 & 2 & 3 \end{matrix} \\ \begin{bmatrix} 1 & 0 & 0 \\ 0 & 1 & 0 \\ 0 & 0 & 1 \end{bmatrix} \end{array}$$

Hence, the relation matrix of $\rho_1 \rho_2$ is given by

$$\boldsymbol{\rho}_1\boldsymbol{\rho}_2 = \begin{array}{c} \\ 1 \\ 2 \\ 3 \\ 4 \end{array} \begin{array}{c} \begin{matrix} 2 & 3 & 4 \end{matrix} \\ \begin{bmatrix} 0 & 0 & 0 \\ 0 & 0 & 1 \\ 0 & 1 & 0 \\ 1 & 0 & 0 \end{bmatrix} \end{array} \begin{array}{c} \\ 2 \\ 3 \\ 4 \end{array} \begin{array}{c} \begin{matrix} 1 & 2 & 3 \end{matrix} \\ \begin{bmatrix} 1 & 0 & 0 \\ 0 & 1 & 0 \\ 0 & 0 & 1 \end{bmatrix} \end{array}$$

$$= \begin{array}{c} \\ 1 \\ 2 \\ 3 \\ 4 \end{array} \begin{array}{c} \begin{matrix} 1 & 2 & 3 \end{matrix} \\ \begin{bmatrix} 0 & 0 & 0 \\ 0 & 0 & 1 \\ 0 & 1 & 0 \\ 1 & 0 & 0 \end{bmatrix} \end{array}$$

(which agrees with the result obtained in Example 2-3). □

When $\rho_1 = \rho_2 = \ldots = \rho_r = \rho$ and $A_1 = A_2 = \ldots = A_r = A$, Theorem 2-1 simplifies to:

THEOREM 2-2

Let ρ be a relation on a finite set A, with the relation matrix $\boldsymbol{\rho}$. Then the relation matrix of ρ^r is given by $\boldsymbol{\rho}^r$.

The composition of a relation ρ on a set A has an especially simple interpretation in terms of the graph of ρ. By definition, $a_i \rho^r a_j$ if, for some $k_1, k_2, \ldots, k_{r-1}$, we have $a_i \rho a_k, a_{k_1} \rho a_{k_2}, \ldots, a_{k_{r-1}} \rho a_j$. Hence, in the graph of ρ^r, an edge points from vertex a_i to vertex a_j if, in the graph of ρ, there are vertices $a_{k_1}, a_{k_2}, \ldots, a_{k_{r-1}}$, with edges pointing from a_i to a_{k_1}, from a_{k_1} to $a_{k_2}, \ldots$, from $a_{k_{r-1}}$ to a_j. This suggests the following procedure for constructing the graph of ρ^r from the graph of ρ: For each vertex a_i in the graph of ρ, determine the vertices that are reachable from a_i via a path of length r; these vertices are those towards which, in the graph of ρ^r, edges must point from vertex a_i. As an example, Fig. 2-3 shows the graphs of ρ^2 and ρ^3, constructed from the graph of ρ shown in Figure 2-2.

Let ρ be a relation on a set A. The *transitive closure* of ρ, denoted by ρ^+,

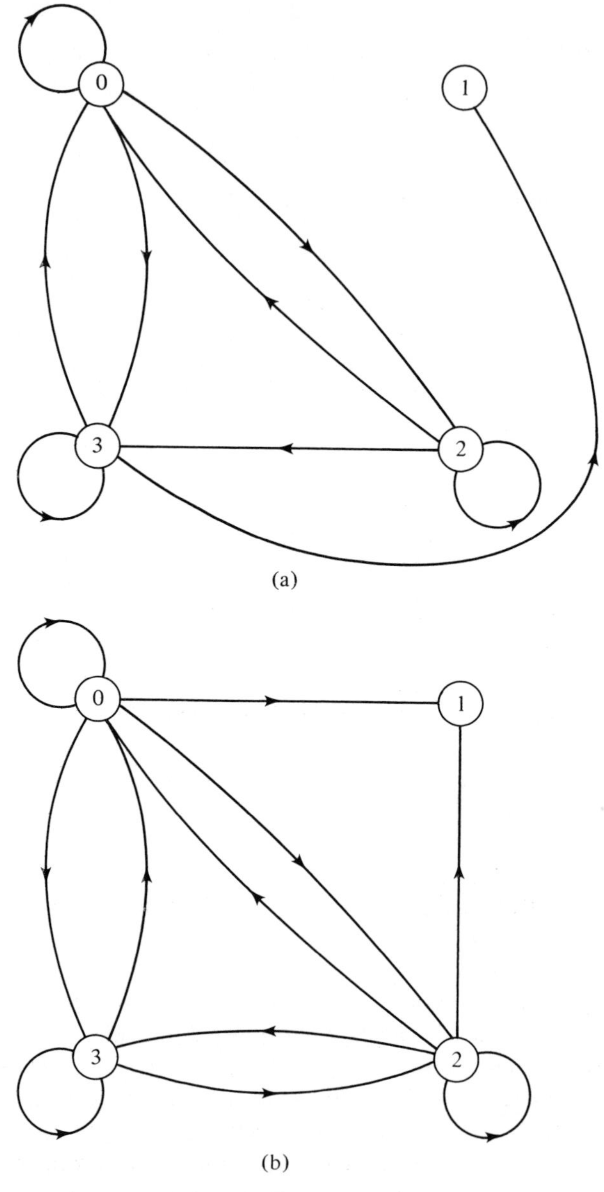

Fig. 2-3 Graph for: (a) ρ^2, (b) ρ^3 (for ρ of Fig. 2-2)

is a relation on A given by

$$\rho^+ = \bigcup_{i=1}^{\infty} \rho^i$$

When A is finite, there is only a finite number of distinct relations on A; hence, for some integers h and $k > h$, we must have $\rho^h = \rho^k$, consequently

$\rho^+ = \bigcup_{i=1}^{k-1} \rho^i$. Thus, the process of constructing the transitive closure of a relation on a finite set is finite. In terms of the graph of ρ, $a_i \rho^+ a_j$ if and only if a_j is reachable from a_i via a path of any finite length. From Theorem 2-2 it follows that the relation matrix $\boldsymbol{\rho}^+$ of ρ^+ can be obtained by adding (Boolean addition) corresponding entries of $\boldsymbol{\rho}, \boldsymbol{\rho}^2, \ldots, \boldsymbol{\rho}^{k-1}$. Symbolically, we can write

$$\boldsymbol{\rho}^+ = \sum_{i=1}^{k-1} \boldsymbol{\rho}^i$$

PROBLEMS

1. For the relations ρ_1 and ρ_2 of Problem 1, Sec. 2-3, construct the relation matrices:

(a) $\boldsymbol{\rho}_1$ (b) $\boldsymbol{\rho}_2$ (c) $\boldsymbol{\rho}_1\boldsymbol{\rho}_2$ (d) $\boldsymbol{\rho}_2\boldsymbol{\rho}_1$
(e) $\boldsymbol{\rho}_1\boldsymbol{\rho}_2\boldsymbol{\rho}_1$ (f) $\boldsymbol{\rho}_1^3$

2. Given:

$$\rho = \{(i, j) \mid i, j \in \mathbb{I}, j - i = 1\}$$

What is ρ^r?

3. Figure 2-A shows the graph of a relation ρ over $\{1, 2, 3, 4, 5, 6\}$. Draw the graphs of the relations ρ^5 and ρ^8.

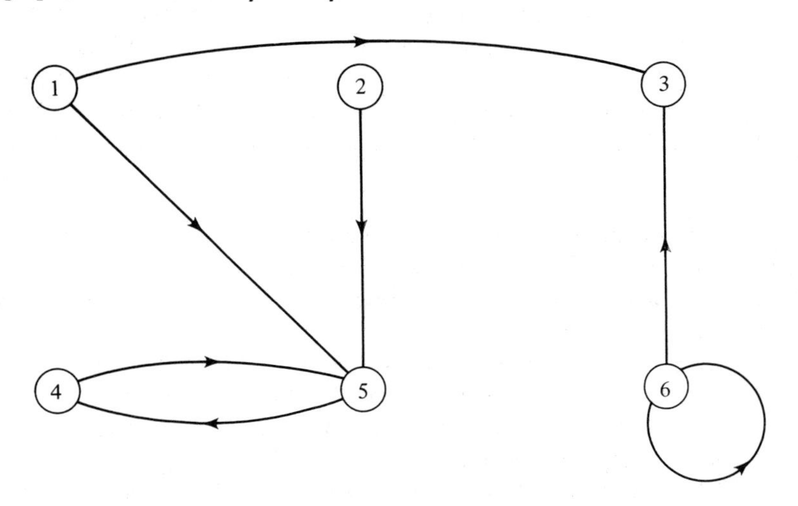

Fig. 2-A

4. Find the transitive closure of the relation ρ of Figure 2.

5. Show that if ρ is a relation on a set of cardinality n, then the transitive closure of ρ is given by $\rho^+ = \bigcup_{i=1}^{n} \rho^i$. (*Hint*: Show that if, in the graph of ρ, a_j is reachable from a_i, then it is reachable via a path of length at most n.)

6. Describe a procedure for computing ρ^r from ρ, using less than $2 \log_2 r$ matrix multiplications. (*Hint*: Represent r as a binary number.)

2-5. PROPERTIES OF RELATIONS

A relation ρ on a set A may exhibit various useful properties, some of which will be listed here.

Relation ρ is said to be *reflexive* if, for all $a_i \in A$, $a_i \rho a_i$. Otherwise, ρ is *irreflexive.*

Relation ρ is said to be *symmetric* if, for all $a_i, a_j \in A$, $a_i \rho a_j$ implies $a_j \rho a_i$ (that is, if ρ and its converse $\breve{\rho}$ are equal). Otherwise, ρ is *asymmetric.* If $a_i \rho a_j$ and $a_j \rho a_i$ imply $a_i = a_j$ (that is, if no distinct a_i and a_j exist such that both $a_i \rho a_j$ and $a_j \rho a_i$), then ρ is said to be *antisymmetric.*

Relation ρ is said to be *transitive* if, for all $a_i, a_j, a_k \in A$, $a_i \rho a_j$ and $a_j \rho a_k$ imply $a_i \rho a_k$. Otherwise, ρ is *nontransitive.*

When ρ is reflexive, then every vertex in the graph of ρ originates a single-edge loop. If ρ is symmetric, then in the graph for every edge pointing from vertex a_i to vertex a_j there is an edge pointing in the opposite direction. In fact, graphs of symmetric relations may be drawn with a single arrowless edge replacing each pair of arrowed edges that connect two vertices. When ρ is transitive, then for every pair of edges in the graph, one pointing from vertex a_i to vertex a_j and the other from vertex a_j to vertex a_k, there is an edge pointing from vertex a_i to vertex a_k.

When set A is finite and the relation ρ on A is specified by its relation matrix $\boldsymbol{\rho}$, the reflexivity of ρ can be readily detected by the fact that all entries along the principal diagonal[1] are 1. The symmetry of ρ can be detected by the fact that the matrix is symmetric relative to the principal diagonal [that is, by the fact that, for all i and j, the (i, j) and (j, i) entries are equal].

Example 2-5

We shall illustrate the properties described above with various relations on 2^S (the power set of S). (Assume that $S \neq \varnothing$.)

(a) The relation ρ, where $S_i \rho S_j$ if and only if $S_i \cap S_j = \varnothing$, is irreflexive, symmetric, and nontransitive.

(b) The relation ρ, where $S_i \rho S_j$ if and only if $\varnothing \rho \varnothing$ and, for all nonempty S_i and S_j, $S_i \cap S_j \neq \varnothing$, is reflexive, symmetric, and nontransitive.

(c) The relation ρ, where $S_i \rho S_j$ if and only if S_i is a proper subset of S_j, is irreflexive, asymmetric, and transitive.

(d) The relation ρ, where $S_i \rho S_j$ if and only if $S_i \subset S_j$, is reflexive, antisymmetric, and transitive.

(e) The relation ρ, where $S_i \rho S_j$ if and only if $S_i = S_j$ is reflexive, symmetric, and transitive.

Verification of these statements is left to the reader. □

[1]In an $n \times n$ (*square*) matrix, the *principal diagonal* consists of all (i, i) entries $(i = 1, 2, \ldots, n)$.

THEOREM 2-3

Let ρ be a relation on a set A. Then the transitive closure ρ^+ of ρ is transitive. Moreover, ρ^+ is included in every transitive relation that includes ρ.

Proof $a_i\rho^+a_j$ and $a_j\rho^+a_k$ imply that, for some integers $p \geq 1$ and $q \geq 1$, $a_i\rho^p a_j$ and $a_j\rho^q a_k$; and hence that, for some $p + q \geq 1$, $a_i\rho^{p+q}a_k$. Thus, $a_i\rho^+a_k$ and ρ^+ is transitive. Now, let $\bar{\rho}$ be any transitive relation on A that includes ρ. By definition of ρ^+, $a_i\rho^+a_j$ implies that, for some $b_1, b_2, \ldots, b_l \in A$, $a_i\rho b_1, b_1\rho b_2, \ldots, b_l\rho a_j$. Since $\rho \subset \bar{\rho}$, we have $a_i\bar{\rho}b_1, b_1\bar{\rho}b_2, \ldots, b_l\bar{\rho}a_j$. By transitivity of $\bar{\rho}$, $a_i\bar{\rho}b_1$ and $b_1\bar{\rho}b_2$ imply $a_i\bar{\rho}b_2$; $a_i\bar{\rho}b_2$ and $b_2\bar{\rho}b_3$ imply $a_i\bar{\rho}b_3$; . . . ; $a_i\bar{\rho}b_l$ and $b_l\bar{\rho}a_j$ imply $a_i\bar{\rho}a_j$. Hence, $a_i\rho^+a_j$ implies $a_i\bar{\rho}a_j$, which implies $\rho^+ \subset \bar{\rho}$. □

By Theorem 2-3, transitive closure of ρ can be viewed as the "smallest" transitive relation containing ρ. It is clear that the "smallest" transitive *and reflexive* relation containing ρ is given by $I_A \cup \rho^+$. This relation is sometimes called the *reflexive transitive closure* of ρ.

PROBLEMS

1. Prove the statements made in Example 2-5.

2. Give an example of a relation on $\mathbf{2}^S$ $(S \neq \varnothing)$ which is:
(a) Reflexive, asymmetric, and nontransitive
(b) Irreflexive, symmetric, and transitive
(c) Irreflexive, asymmetric, and nontransitive

3. Figure 2-B shows the graphs of twelve relations over $\{1, 2, 3\}$. For each of these graphs, decide whether the represented relation is reflexive or irreflexive; symmetric, asymmetric, or antisymmetric; transitive or nontransitive.

4. Which of the following relations ρ are reflexive, symmetric, or transitive?
(a) $i_1\rho i_2$ if and only if $|i_1 - i_2| \leq 10$ $(i_1, i_2 \in \mathbb{I})$.
(b) $n_1\rho n_2$ if and only if $n_1n_2 > 8$ $(n_1, n_2 \in \mathbb{N})$.
(c) $r_1\rho r_2$ if and only if $r_1 \leq r_2$ $(r_1, r_2 \in \mathbb{R})$.
(d) $r_1\rho r_2$ if and only if $r_1 \leq |r_2|$ $(r_1, r_2 \in \mathbb{R})$.

5. In the graph for a relation ρ on A, a path exists from a_i to a_j but not from a_j to a_i. Show that ρ is asymmetric.

6. Relation ρ on a set A is known to be symmetric and transitive. Suppose $a_i\rho a_j$. By symmetry, $a_j\rho a_i$. By transitivity, $a_i\rho a_j$ and $a_j\rho a_i$ imply $a_i\rho a_i$. Hence, ρ is reflexive. Thus, every relation that is symmetric and transitive is also reflexive. What is the fallacy in this argument?

7. When ρ is a transitive relation, show that, for every path leading from vertex a_i to vertex a_j in the graph of ρ, there is an edge pointing from a_i to a_j.

8. Show that when relation ρ is symmetric, so is ρ^k (for any $k \geq 1$).

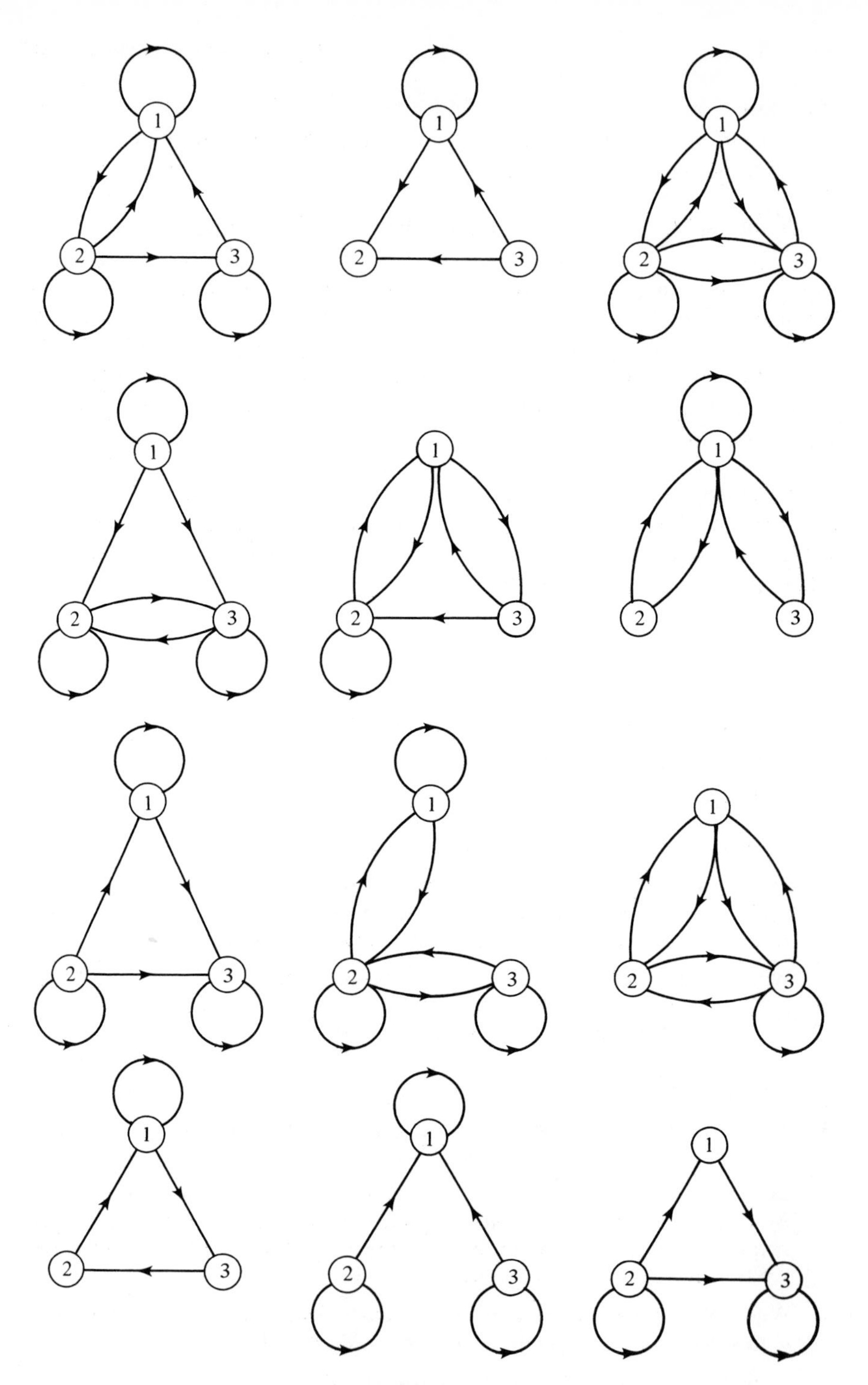

Fig. 2-B

9. Let ρ_1 and ρ_2 be symmetric relations on A. Show that $\rho_1\rho_2 \subset \rho_2\rho_1$ implies $\rho_1\rho_2 = \rho_2\rho_1$.

2-6. EQUIVALENCE RELATIONS

A relation ρ on a set A is called an *equivalence relation* (on A) if it is reflexive, symmetric, and transitive: that is, if

(a) $a_i\rho a_i$ for all $a_i \in A$ (*reflexivity*)
(b) $a_i\rho a_j$ implies $a_j\rho a_i$ (*symmetry*)
(c) $a_i\rho a_j$ and $a_j\rho a_k$ imply $a_i\rho a_k$ (*transitivity*)

The most familiar equivalence relation is that of equality among elements of a set (where ρ is usually denoted by $=$). Other examples of equivalence relations include the relation of parallelism among lines in plane geometry, the relation of similarity among triangles, the relation "lives in the same street as" among inhabitants of a given town.

Let ρ be an equivalence relation on A. We say that a_i is *equivalent to* a_j *(under* ρ*)* if $a_i\rho a_j$. If a_i is equivalent to a_j, then, since ρ is symmetric, a_j is equivalent to a_i. We can thus simply say that, if $a_i\rho a_j$, then a_i and a_j are *equivalent* (under ρ). The *equivalence class containing* a_i (under ρ), denoted by $\mathcal{E}_\rho(a_i)$, is the set of all elements in A that are equivalent to a_i; that is,

$$\mathcal{E}_\rho(a_i) = \{a_j \mid a_i\rho a_j\}$$

THEOREM 2-4

Let ρ be an equivalence relation on a set A. Then the set of equivalence classes $\{\mathcal{E}_\rho(a_i) \mid a_i \in A\}$ constitutes a partition of A.

Proof Since ρ is reflexive, for all $a_i \in A$ we have $a_i\rho a_i$ and, hence, no $\mathcal{E}_\rho(a_i)$ can be empty. Since the $\mathcal{E}_\rho(a_i)$ are constructed for every $a_i \in A$, every element of A is in at least one $\mathcal{E}_\rho(a_i)$. Suppose an element $a \in A$ is in both $\mathcal{E}_\rho(a_i)$ and $\mathcal{E}_\rho(a_j)$. Then $a_i\rho a$ and $a_j\rho a$. Since ρ is symmetric, $a_j\rho a$ implies $a\rho a_j$. If a_{jj} is *any* element of $\mathcal{E}_\rho(a_j)$, then $a_j\rho a_{jj}$. Since ρ is transitive, $a\rho a_j$ and $a_j\rho a_{jj}$ imply $a\rho a_{jj}$. Finally, $a_i\rho a$ and $a\rho a_{jj}$ imply $a_i\rho a_{jj}$ which, in turn, implies that any element of $\mathcal{E}_\rho(a_j)$ is also in $\mathcal{E}_\rho(a_i)$: that is, $\mathcal{E}_\rho(a_j) \subset \mathcal{E}_\rho(a_i)$. Analogously, it can be proved that $\mathcal{E}_\rho(a_i) \subset \mathcal{E}_\rho(a_j)$ and, hence, that $\mathcal{E}_\rho(a_i) = \mathcal{E}_\rho(a_j)$. Thus, $a \in A$ cannot belong to two distinct $\mathcal{E}_\rho(a_i)$. In conclusion, the sets $\mathcal{E}_\rho(a_i)$ constitute a partition of A. $\square$

The partition of A that consists of the equivalence classes under ρ is said to be the *equivalence partition induced on A by* ρ, and is denoted by π_ρ^A. Given π_ρ^A, ρ can always be constructed by forming all possible pairs of elements belonging to the same equivalence class. Hence, π_ρ^A is just as adequate a representation of ρ as ρ itself. Specifying an equivalence relation ρ by means of π_ρ^A is often less cumbersome than specifying ρ by listing all of its pairs. For example, consider the set of cities

$$A = \{\text{Chicago (CH), Mexico City (MC), Montreal (MO), New York (NY), San Francisco (SF), Toronto (TO)}\}$$

and a relation ρ on A, where $a_i \rho a_j$ if and only if a_i and a_j are in the same country. Thus,

$$\rho = \{(\text{CH, CH}), (\text{MC, MC}), (\text{MO, MO}), (\text{NY, NY}), (\text{SF, SF}), (\text{TO, TO}), (\text{CH, NY}), (\text{NY, CH}), (\text{CH, SF}), (\text{SF, CH}), (\text{NY, SF}), (\text{SF, NY}), (\text{MO, TO}), \text{TO, MO})\} \tag{2-3}$$

Clearly, ρ is an equivalence relation that induces on A the equivalence partition

$$\pi_\rho^A = \{\{\text{CH, NY, SF}\}, \{\text{MO, TO}\}, \{\text{MC}\}\} \tag{2-4}$$

The partition (2-4) imparts the same information about ρ as does the explicit listing (2-3), and is considerably more compact.

Note that, for any set A, the identity relation I_A and the universal relation U_A are both equivalence relations. In the equivalence partition induced by I_A, each equivalence class consists of a single element. In the equivalence partition induced by U_A, there is only one equivalence class that contains all elements of A. These partitions are sometimes referred to as the *trivial* partitions of A.

The *quotient set of A relative to ρ*, denoted by A/ρ, is the set whose elements are the equivalence classes of A under the equivalence relation ρ. The cardinality of A/ρ (that is, the number of distinct equivalence classes of A under ρ) is called the *rank* of ρ.

Example 2-6

Consider the following relation on $A = \{0, 1, 2, 3, 4, 5\}$:

$$\rho = \{(0, 0), (1, 1), (1, 2), (1, 3), (2, 1), (2, 2), (2, 3), (3, 1), (3, 2), (3, 3), (4, 4), (4, 5), (5, 4), (5, 5)\}$$

ρ is seen to be reflexive, symmetric, and transitive; hence, ρ is an equivalence relation. It induces on A the following equivalence partition:

$$\pi_\rho^A = \{\{0\}, \{1, 2, 3\}, \{4, 5\}\}$$

Using the notation $C_0 = \{0\}$, $C_1 = \{1, 2, 3\}$, $C_4 = \{4, 5\}$, the quotient set of A relative to ρ is given by

$$A/\rho = \{C_0, C_1, C_4\}$$ □

Example 2-7

For any integer i and any nonnegative integer m, $\text{res}_m(i)$ will be defined as the nonnegative remainder obtained upon division of i by m [hence, $0 \leq \text{res}_m(i) < m$]. We say that i_1 and i_2 are *equal modulo m*, written $i_1 = i_2$ (mod m), if $\text{res}_m(i_1) = \text{res}_m(i_2)$. If the quotients obtained upon division of i_1 and i_2 by m are q_1 and q_2, respectively, then $\text{res}_m(i_1) = \text{res}_m(i_2)$ if and only if $i_1 - q_1 m = i_2 - q_2 m$, or $i_1 - i_2 = (q_1 - q_2)m$. Thus, $i_1 = i_2 \pmod{m}$ if and only if $i_1 - i_2$ is a multiple of m.

Now, let ρ be a relation on the set of integers $\mathbb{I}$, defined as $i_1 \rho i_2$ if and only if $i_1 = i_2 \pmod{m}$. Since $i_1 - i_1 = 0 = 0 \cdot m$, $i_1 = i_1 \pmod{m}$; hence, ρ is reflexive. Since $i_1 - i_2 = qm$ implies $i_2 - i_1 = (-q)m$, $i_1 = i_2 \pmod{m}$ implies $i_2 = i_1 \pmod{m}$; hence, ρ is symmetric. Finally, since $i_1 - i_2 = q_1 m$ and $i_2 - i_3 = q_2 m$ imply $i_1 - i_3 = (q_1 + q_2)m$, $i_1 = i_2 \pmod{m}$ and $i_2 = i_3 \pmod{m}$ imply $i_1 = i_3 \pmod{m}$; hence, ρ is transitive. Therefore, ρ is an equivalence relation. Each integer is equivalent to exactly one of the integers $0, 1, \ldots, m - 1$; hence, the equivalence partition induced by ρ on $\mathbb{I}$ is

$$\pi_\rho^{\mathbb{I}} = \{\mathcal{E}_\rho(0), \mathcal{E}_\rho(1), \ldots, \mathcal{E}_\rho(m-1)\}$$

[The $\mathcal{E}_\rho(i)$, in this case, are called *residue classes modulo m*]. Denoting the set $\mathcal{E}_\rho(i)$ by C_i $(i = 1, 2, \ldots, m - 1)$, we have

$$\mathbb{I}/\rho = \{C_0, C_1, \ldots, C_{m-1}\}$$ □

While Theorem 2-4 states that every equivalence relation defines a partition, Theorem 2-5 states the converse—that every partition defines an equivalence relation:

THEOREM 2-5

Let $\pi = \{A_i\}_{i \in K}$ be a partition of a set A. Then there exists a relation ρ on A such that π is the equivalence partition induced on A by ρ.

Proof Simply define ρ as $a_i \rho a_j$ if and only if a_i and a_j belong to the same block A_i of π. □

From Theorems 2-4 and 2-5 we can conclude that the concepts of "partition" and "equivalence relation" are essentially the same, and, indeed, many authors use these terms interchangeably.

PROBLEMS

1. Which of the following are equivalence relations?

Set	*Relation on Set*
(a) People	Is the brother of
(b) People	Has the same parents as
(c) Points on a map	Is connected by a road to
(d) Lines in plane geometry	Is perpendicular to
(e) Positive integers	For some integer k, equals 10^k times

2. Figure 2-C shows the graphs of two relations on $\{1, 2, 3\}$. Are these relations equivalence relations?

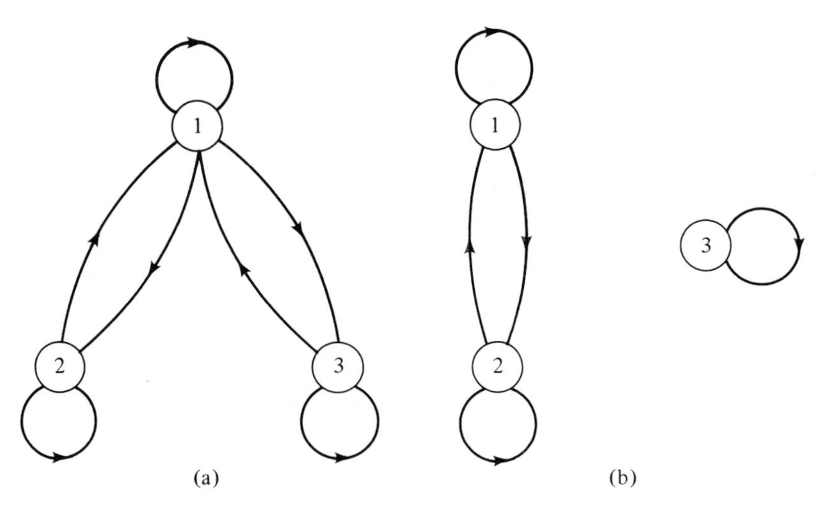

Fig. 2-C

3. How can equivalence classes be identified in the graph of an equivalence relation?

4. The relation ρ on $\mathbb{N}$ is defined as $n_i \rho n_j$ if and only if n_i/n_j can be expressed in the form 2^m, where m is an arbitrary integer
(a) Show that ρ is an equivalence relation.
(b) What are the equivalence classes under ρ?

5. A relation ρ on a set A is said to be *circular* if, for all $a_i, a_j, a_k \in A$, $a_i \rho a_j$, and $a_j \rho a_k$ imply $a_k \rho a_i$. Show that ρ is reflexive and circular if and only if it is an equivalence relation.

6. Let ρ_1 and ρ_2 be equivalence relations on A. Prove that $\rho_1 \subset \rho_2$ if and only if every equivalence class in $\pi^A_{\rho_1}$ is included in some equivalence class in $\pi^A_{\rho_2}$.

7. ρ_1 and ρ_2 are equivalence relations on A of ranks r_1 and r_2, respectively. Show that $\rho_1 \cap \rho_2$ is an equivalence relation on A of rank at most $r_1 r_2$. Also show that $\rho_1 \cup \rho_2$ is not necessarily an equivalence relation on A.

2-7. PARTIAL ORDERINGS

A relation on a set A is called a *partial ordering* (on A) if it is reflexive, antisymmetric, and transitive. A partial ordering is commonly denoted by $\leq$. Thus, we have

(a) $a_i \leq a_i$ for all $a_i \in A$.
(b) $a_i \leq a_j$ and $a_j \leq a_i$ imply $a_i = a_j$.
(c) $a_i \leq a_j$ and $a_j \leq a_k$ imply $a_i \leq a_k$.[1]

Clearly, the converse of a partial ordering is also a partial ordering. When a partial ordering is denoted by $\leq$, its converse is usually denoted by $\geq$.

A partial ordering on A is called a *total ordering* (on A) if, for all $a_i, a_j \in A$, either $a_i \leq a_j$ or $a_j \leq a_i$. The elements of A on which a total ordering is imposed can always be arranged in a linear sequence $a_1, a_2, a_3, \ldots$, where $a_i \leq a_j$ if and only if $i \leq j$. (For that reason, a total ordering is sometimes referred to as a *linear ordering*.)

A partial ordering on A is called a *well ordering* (on A) if, for every subset $S \subset A$, the following holds: There exists an element a_S in S (a *least element* of S) such that, for all $s \in S$, $a_S \leq s$. Clearly, every well ordering is also a total ordering.

The most familiar example of partial ordering (as well as total and well ordering) is the relation "less than or equal to" defined over the set of integers or real numbers. In fact, the generic symbol $\leq$, denoting partial orderings, was borrowed from this special case. The following is another familiar example.

Example 2-8

Let L denote an alphabet (say, the 26 Latin letters) on which a well ordering ρ is defined (say, the conventional alphabetic ordering $A, B, \ldots, Z$). If $\bar{L}$ denotes all of the words that can be constructed from elements of L, a relation $\bar{\rho}$ can be defined on L in this manner: $(l_1 \, l_2 \ldots l_p)\bar{\rho}(l'_1 \, l'_2 \ldots l'_q)$ (where the l_i and l'_i are elements of L) if and only if: (a) for some $k < p$ we have $l_1 = l'_1, l_2 = l'_2, \ldots, l_k = l'_k$ and $l_{k+1}\rho l'_{k+1}$, or (b) $p < q$ and $l_1 = l'_1, l_2 = l'_2, \ldots, l_p = l'_p$. The relation $\bar{\rho}$ is a well ordering on L and is known as a *lexicographic ordering*; it is precisely the ordering in which entries are arranged in dictionaries, telephone books, and similar listings. □

When A is finite, a partial ordering on A can be conveniently depicted by means of an *ordering diagram* (or a *Hasse diagram*). This diagram consists of $\#A$ vertices, labeled as the elements of A; a vertex a_i appears below every vertex a_j such that $a_j \neq a_i$ and $a_i \leq a_j$; an edge connects vertex a_i

[1]The statement "$a_i \leq a_j$ and $a_j \leq a_k$" is often abbreviated to "$a_i \leq a_j \leq a_k$."

to every vertex $a_j \neq a_i$ that satisfies the following condition: $a_i \leq a_j$, and there is no a_k (other than a_i and a_j) such that $a_i \leq a_k \leq a_j$. Thus, $a_i \leq a_j$ if and only if $a_i = a_j$ or if one can reach vertex a_i from vertex a_j via a *descending* path. The ordering diagram of the converse relation $\geq$ is simply that of $\leq$ turned upside down. (Note that the ordering diagram of a total ordering consists simply of a vertical sequence of edges.)

Example 2-9

The relation $|$ defined over the set $\mathbb{N}$ of positive integers, where $i \mid j$ if and only if i is a divisor of j, is a partial ordering. Figure 2-4 shows the ordering diagram for the relation $|$ on $J = \{2, 3, 4, 6, 8, 12, 36, 60\}$. In this diagram there is a descending path from every integer $j \in J$ to every other integer $i \in J$ such that $i \mid j$. □

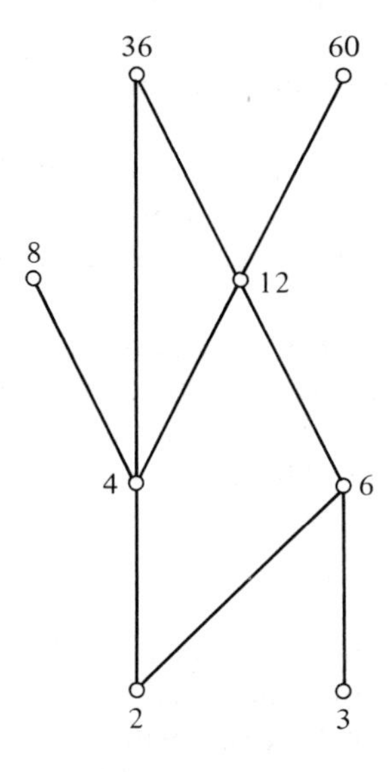

Fig. 2-4 Ordering diagram for $|$ on $\{2, 3, 4, 6, 8, 12, 36, 60\}$

Example 2-10

The relation $\subset$ defined over the power set of a universal set U, where $A \subset B$ if and only if A is included in B, is a partial ordering. Figure 2-5 shows the ordering diagram for the relation $\subset$ on $\mathbf{2}^U$, where $U = \{a, b, c\}$. In this diagram, there is a descending path from every set $S \in \mathbf{2}^U$ to every one of the proper subsets of S. □

Given a partial ordering ρ, one may wish to find a total ordering ρ_T such that $\rho \subset \rho_T$. This process of "embedding" a partial ordering in a total one is sometimes called *topological sorting*. In terms of the ordering diagram, this process is equivalent to "stretching" the diagram in such a way that all

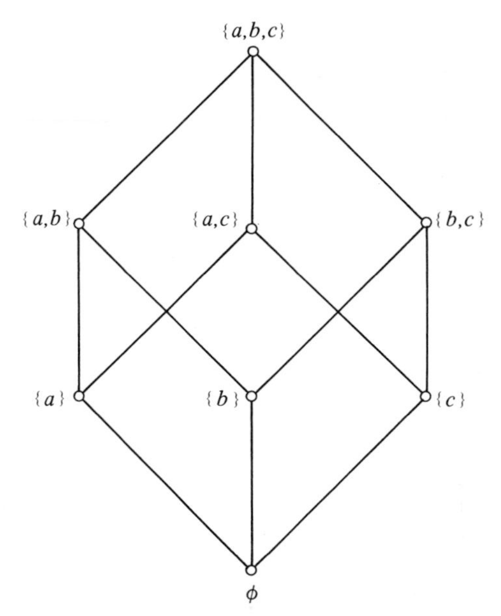

Fig. 2-5 Ordering diagram for $\subset$ on $\mathbf{2}^{\{a,b,c\}}$

vertices are aligned in a single column, with all descending paths preserved. (In general, given a partial ordering, the total ordering that embeds it is not unique.)

Example 2-11

Consider the partial ordering $\subset$ on $\mathbf{2}^U$, where $U = \{a, b, c\}$ (see Example 2-10), and the total ordering $\subset_T$, specified by the sequence

$$\varnothing, \{c\}, \{b\}, \{a\}, \{b, c\}, \{a, c\}, \{a, b\}, \{a, b, c\}$$

(where $A \subset_T B$ if and only if $A = B$ or A precedes B). Clearly, $A \subset B$ implies $A \subset_T B$; hence, $\subset$ is embedded in $\subset_T$. The "stretched" version of the ordering diagram in this example is shown in Figure 2-6. □

Generally, topological sorting of a partial ordering on a finite set can be effected as follows:

ALGORITHM 2-1

Given the ordering diagram of a partial ordering on $A = \{a_1, a_2, \ldots, a_n\}$, to find a total ordering on A:

(i) set $k = 1$ and let D_1 denote the ordering diagram,

(ii) let a_{i_k} be any vertex in D_k such that there is no descending path to a_{i_k} from any other vertex,

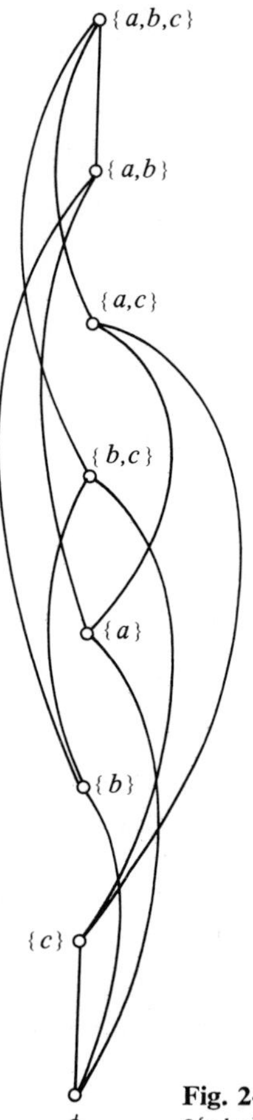

Fig. 2-6 Topological sorting for $\subset$ on $2^{\{a,b,c\}}$

(iii) if $k = n$, the desired total ordering is given by $a_{i_n}, a_{i_{n-1}}, \ldots, a_{i_1}$. Otherwise, let D_{k+1} be D_k with the vertex a_{i_k} (and all edges originated by a_{i_k}) removed; increment k by 1 and return to step (ii). □

PROBLEMS

1. Verify that the following relations are partial orderings on $\mathbb{N}$:
(a) Is a multiple of
(b) Equals 10^k (for some nonnegative integer k) times
Which of these is a well ordering?

2. Prove that a lexicographic ordering on $\bar{L}$ is a well ordering (see Example 2-8).

3. Draw the ordering diagram for the relation | on the sets
   (a) $\{1, 2, 3, 4, 6, 12\}$
   (b) $\{1, 2, 3, 4, 6, 8, 12, 24\}$
   (c) $\{1, 2, 3, \ldots, 12\}$

4. Figure 2-D shows the graphs of four partial orderings on $\{1, 2, 3, 4\}$. Draw the ordering diagram for each of these. Which of the orderings are total and which are well orderings?

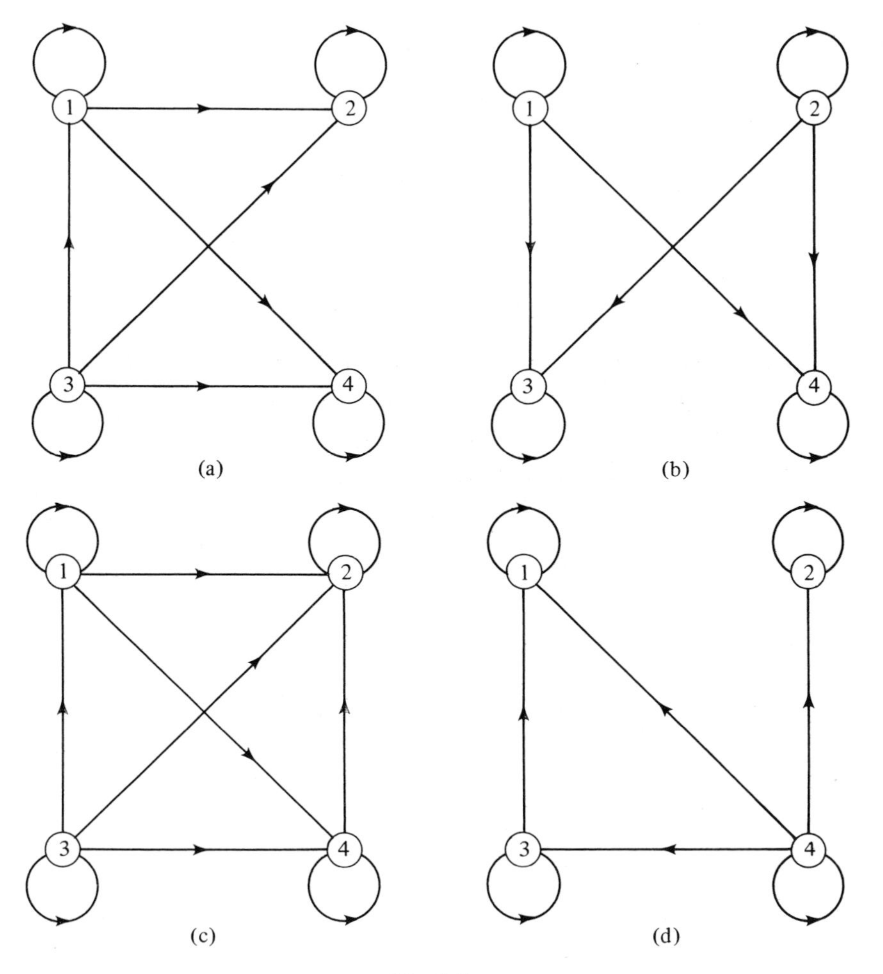

Fig. 2-D

5. Figure 2-E shows the ordering diagram of a partial ordering. Perform a topological sorting on this ordering.

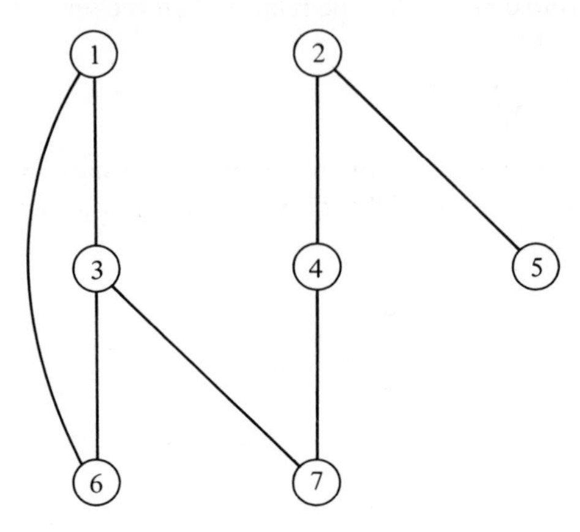

Fig. 2-E

REFERENCES

MacLane, S., and G. Birkhoff, *Algebra.* New York: Macmillan, 1967.

Paley, H., and P. M. Weichsel, *A First Course in Abstract Algebra.* New York: Holt, Rinehart and Winston, 1966.

Van der Waerden, B. L., *Modern Algebra.* Ungar, 1931.

3 FUNCTIONS

In this chapter, the concept of a *function* is introduced and the special functions known as *injections*, *surjections*, and *bijections* are characterized. The *composition* of functions (analogous to that of relations) is described, and the *identity* and *inverse* functions are defined. The important *characteristic* and *permutation* functions are introduced and discussed in detail. Our knowledge of functions at this point facilitates a more rigorous treatment of the concept of *cardinality*, especially as it relates to infinite sets. Next, the *Peano's successor function* is defined, which leads to the characterization of natural numbers (via the *Peano postulates*) and the fundamental *principle of finite induction*. The power of this principle is illustrated by a variety of theorem proofs and recursive definitions. The chapter closes with a discussion of some useful results relating to integers, such as the *division theorem*, the *Euclidean algorithm*, the *prime factorization theorem*, and others.

Like the notions of a set and a relation, the notion of a function is indispensable to the computer scientist. It appears constantly in areas such as switching theory, automata theory, and computability. The permutation function is fundamental to many algorithmic processes involving enumeration. Finite induction is an invaluable tool in all theoretical investigations in computer science; the related concept of recursion is essential in the characterization, design, and implementation of programming languages and data structures. The properties of integers introduced at the end of this chapter are important in the study of number representations and arithmetic operations within a computer; as we shall see later, they are also relevant to schemes for error detection and correction in digital systems.

3-1. FUNCTIONS

A *function f from a set A into a set B* is a rule that assigns to each element $a \in A$ a *unique* element $b \in B$. Symbolically, this statement is written as

$$f: \quad A \longrightarrow B$$

The element assigned to $a \in A$ is denoted by $f(a)$ and is referred to as the *value of f for the argument a*. The set A is called the *domain* of f, and B the *codomain* of f. The set of all elements of B assigned to elements of A by f is called the *range* of f, and is denoted by $f(A)$. Thus,

$$f(A) = \{b \in B \mid b = f(a) \text{ for some } a \in A\}$$

When the argument of f is an r-tuple, say, $(a_1, a_2, \ldots, a_r)$, the expression $f((a_1, a_2, \ldots, a_r))$ is usually simplified to $f(a_1, a_2, \ldots, a_r)$.

The term *mapping* (or *transformation*) is synonymous with function. The element $f(a)$ is sometimes referred to as the *image of a* (under the mapping f), and the set $f(A)$ is called the *image of A* (under the mapping f). We also say that *f maps a into $f(a)$ and A into $f(A)$*. If $b = f(a)$, then a is said to be the *inverse image* of b (under the mapping f).

The fact that every element in the domain of a function has exactly one image in the range, is sometimes emphasized by stating that the function is *single-valued* or *well-defined.*

Example 3-1

The function

$$f: \mathbb{I} \longrightarrow \mathbb{N}$$

where $f(i) = |2i| + 1$ is a function from the set of integers into the set of positive integers. Its range is the set of all odd integers. □

Example 3-2

The function

$$f: \quad \mathbf{2}^U \times \mathbf{2}^U \longrightarrow \mathbf{2}^U$$

where $f(S_1, S_2) = S_1 \cap S_2$ is a function from the set of all pairs of subsets of U into the power set of U. In this case, the codomain and range of f are equal. (Why?) □

The functions $f: A \longrightarrow B$ and $g: C \longrightarrow D$ are said to be *equal* (written $f = g$) if $A = C$, $B = D$, and for all $x \in A$ (or $x \in C$) we have $f(x) = g(x)$.

Consider the function f: $A \longrightarrow B$. The function g: $\tilde{A} \longrightarrow B$ is called a *restriction of f to A* if $\tilde{A} \subset A$ and if, for all $a \in \tilde{A}$, $g(a) = f(a)$. (When this is the case, f is called an *extension of g to A.*) For example, the function g: $\mathbb{Z} \longrightarrow \mathbb{Z}$, where $g(z) = 2z + 1$, is the restriction of f of Example 3-1 to the set of nonnegative integers $\mathbb{Z}$.

A function

$$f: \quad A_1 \times A_2 \times \ldots \times A_r \longrightarrow A_i \quad (0 \leq i \leq r)$$

such that $f(a_1, a_2, \ldots, a_r) = a_i$ is called a *projection of* $A_1 \times A_2 \times \ldots \times A_r$ *onto* A_i. For example, if $X \times Y$ represents the set of all points on the Cartesian x-y plane, then a projection of $X \times Y$ onto X is a function whose value, for each point, is the x-coordinate of this point.

The set of all distinct functions from A into B is denoted by B^A; that is,

$$B^A = \{f \mid f: \quad A \longrightarrow B\}$$

If A and B are finite, then the number of possible images for each $a \in A$ is $\#B$; hence, the number of distinct functions f: $A \longrightarrow B$ is $(\#B)^{\#A}$. Thus,

$$\#(B^A) = (\#B)^{\#A} \tag{3-1}$$

Figure 3-1 shows a diagrammatic representation of a function, illustrating some of the terms introduced above. Comparing Fig. 3-1 with Fig. 2-1, it is seen that every function from A into B defines a relation from A into B that consists of all pairs $(a, f(a))$. Conversely, a relation ρ from A into B constitutes a function from A into B, provided that:

(a) ρ has the domain A;
(b) ρ has the range B;
(c) $a\rho b_i$ and $a\rho b_j$ imply $b_i = b_j$.

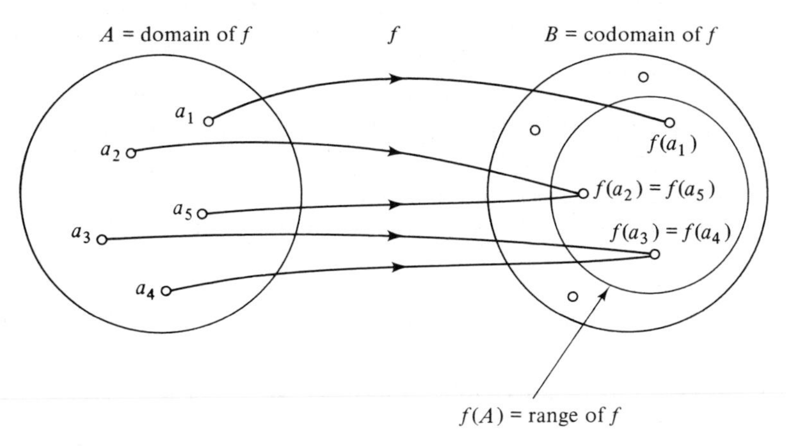

Fig. 3-1 f: $A \longrightarrow B$

PROBLEMS

1. Prove that the codomain and range of f of Example 3-2 are equal.

2. Which of the following relations constitute functions?
(a) $\{(n_1, n_2) \mid n_1, n_2 \in \mathbb{N}, n_1 + n_2 < 10\}$
(b) $\{(r_1, r_2) \mid r_1, r_2 \in \mathbb{R}, r_2 = r_1^2\}$
(c) $\{(r_1, r_2) \mid r_1, r_2 \in \mathbb{R}, r_2^2 = r_1\}$
(d) $\{(n_1, n_2) \mid n_1, n_2 \in \mathbb{N}, n_2 = \text{number of prime numbers less than } n_1\}$

3. Consider the function $f: A \longrightarrow B$, where $A = \{-1, 0, 1\}^2$ and

$$f(a_1, a_2) = \begin{cases} 0 & \text{if } a_1 a_2 > 0 \\ a_1 - a_2 & \text{otherwise} \end{cases}$$

(a) What relation is defined by f?
(b) What is the range of f?
(c) Define the restriction of f to $\{0, 1\}^2$.
(d) How many distinct functions are there with the same domain and range as f?

4. Let ρ_1 and ρ_2 be the relations defined by the functions f_1 and f_2, respectively. Show that there always exists a function that defines the relation $\rho_1 \cap \rho_2$, but that there does not always exist a function that defines the relation $\rho_1 \cup \rho_2$.

3-2. INJECTIONS, SURJECTIONS, AND BIJECTIONS

Let f be a function from A into B. When $a_i \neq a_j$ implies $f(a_i) \neq f(a_j)$ (that is, when distinct elements of A have distinct images under f), then f is called an *injection* (or a *one-to-one mapping*) *from A into B*. For example, the function $f: \mathbb{Z} \longrightarrow \mathbb{Z}$, where $f(z) = z^2$, is an injection from the set of non-negative integers into itself.

When $f(A) = B$ (that is, when every element of B is the image of at least one element of A under f), then f is called a *surjection* (or an *onto mapping*) *from A onto B*. For example, the function $f: \mathbb{I} \longrightarrow \mathbb{Z}_m$, where $f(i) = i \,(\text{mod}\, m)$ (see Example 2-7), is a surjection from the set of integers onto the set $\{1, 2, \ldots, m - 1\}$. (Note, however, that f is not an injection.)

When a function f is both an injection from A into B and a surjection from A onto B, it is called a *bijection from A onto B*. For example, the function $f: \mathbf{2}^U \longrightarrow \mathbf{2}^U$, where $f(S) = S'$, is a bijection from the power set of U onto itself.

Figure 3-2 shows diagrammatic representations of an injection, a surjection, and a bijection.

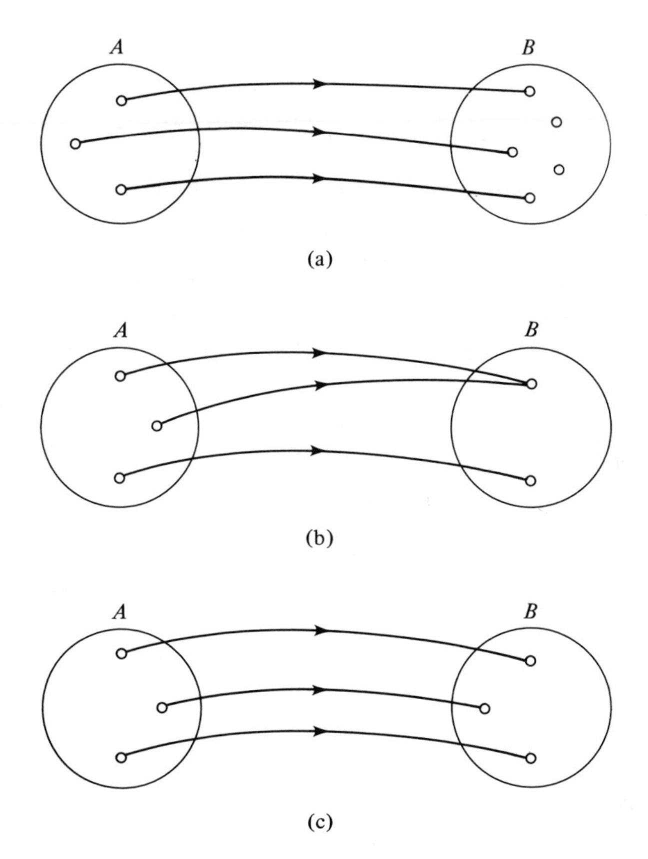

Fig. 3-2 (a) Injection (b) Surjection (c) Bijection

PROBLEMS

1. Which of the following functions are injections, surjections, or bijections?

	Domain	*Codomain*	*Function*
(a)	$\mathbb{N}$	$\mathbb{R}$	$f(n) = \log_{10} n$
(b)	$\mathbb{N}$	$\mathbb{Z}$	$f(n) =$ least integer which equals or exceeds $\log_{10} n$
(c)	$\mathbb{N}$	$\mathbb{Z}$	$f(n) =$ number of perfect squares less than n
(d)	$\mathbb{R}$	$\mathbb{R}$	$f(r) = 2r - 15$
(e)	$\mathbb{R}$	$\mathbb{R}$	$f(r) = r^2 + 2r - 15$
(f)	$\mathbb{N}$	$\mathbb{N}$	$f(n_1, n_2) = n_1^{n_2}$
(g)	$(\mathbf{2}^U)^2$	$(\mathbf{2}^U)^2$	$f(S_1, S_2) = (S_1 \cup S_2, S_1 \cap S_2)$

2. A and B are finite sets. How many distinct injections $f: A \longrightarrow B$ are there (in terms of $\#A$ and $\#B$)? How many distinct bijections?

3-3. COMPOSITION OF FUNCTIONS

Given the functions

$$f: A \longrightarrow B, \qquad g: B \longrightarrow C$$

the *composition of f and g*, denoted by gf, is the function

$$gf: A \longrightarrow C$$

where $gf(a) = g(f(a))$. Thus, if $b \in B$ is the image of $a \in A$ under f, and $c \in C$ is the image of b under g, then c is the image of a under gf (see Fig. 3-3). In this context, the function gf is said to be a *composite* function.

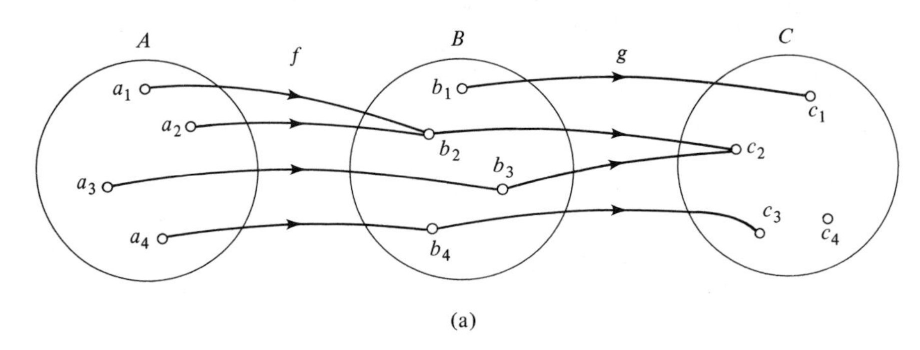

(a)

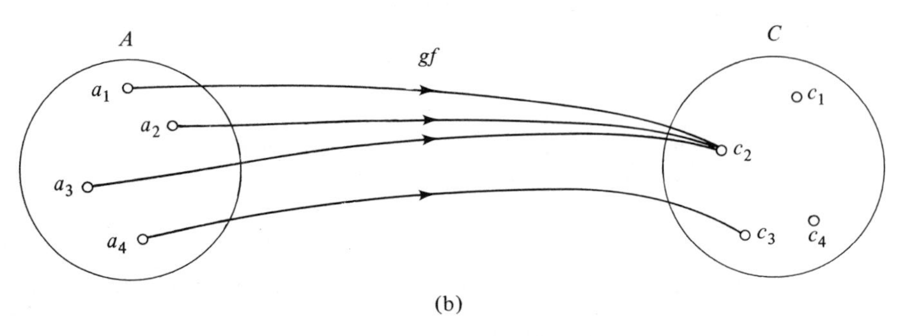

(b)

Fig. 3-3 (a) f and g (b) gf

Example 3-3

Consider the function

$$f: \mathbf{2}^U \longrightarrow \mathbb{Z}$$

where U is a finite set and where $f(S) = \#S$, and the function

$$g: \mathbb{Z} \longrightarrow \mathbb{R}$$

where $g(z) = (z - 5)/2$. Then the composition of f and g is the function

$$gf: \mathbf{2}^U \longrightarrow \mathbb{R}$$

where $$gf(S) = g(f(S)) = g(\#S) = [(\#S) - 5]/2. \qquad \square$$

Consider these three functions:

$$f\colon\ A \longrightarrow B, \qquad g\colon\ B \longrightarrow C, \qquad h\colon\ C \longrightarrow D$$

From the definition of composition it follows that

$$\begin{aligned}[(hg)f](a) = (hg)(f(a)) = h(g(f(a))) &= h((gf)(a)) \\ &= [h(gf)](a)\end{aligned}$$

Thus, the composite functions $(hg)f$ and $h(gf)$ (from A into D) are equal and, hence, can be written unambiguously as hgf. More generally, given the r functions

$$f_1\colon\ A_1 \longrightarrow A_2, \qquad f_2\colon\ A_2 \longrightarrow A_3, \ldots, \qquad f_r\colon\ A_r \longrightarrow A_{r+1}$$

the parentheses-free expression $f_r f_{r-1} \ldots f_1$ uniquely represents a function from A_1 into A_{r+1}. (The same conclusion could be reached by recognizing that functions define relations, and then using the results of Sec. 2-3.)

When $A_1 = A_2 = \ldots = A_r = A$ and $f_1 = f_2 = \ldots = f_r = f$, the composite function $f_r f_{r-1} \ldots f_1$ (from A into A) can be denoted by f^r.

Example 3-4

Consider the function

$$f\colon\ ▯ \longrightarrow ▯$$

where $f(i) = 2i + 1$. Then the function f^3 is given by

$$f^3\colon\ ▯ \longrightarrow ▯$$

where $$\begin{aligned}f^3(i) &= 2f^2(i) + 1 = 2(2f(i) + 1) + 1 \\ &= 2(2(2i + 1) + 1) + 1 = 8i + 7 \qquad \square\end{aligned}$$

If f is a function from A into A, and if $f^2 = f$, then f is said to be *idempotent*. For example, the function $f\colon ▯ \longrightarrow \mathbb{Z}_m$, where $f(i) = i \pmod{m}$, is idempotent. When f is idempotent, then

$$\begin{aligned}f^3 &= f(f^2) = ff = f \\ f^4 &= f(f^3) = ff = f \\ f^5 &= f(f^4) = ff = f \\ &\vdots\end{aligned}$$

Thus, for all $k \geq 1$, $f^k = f$.

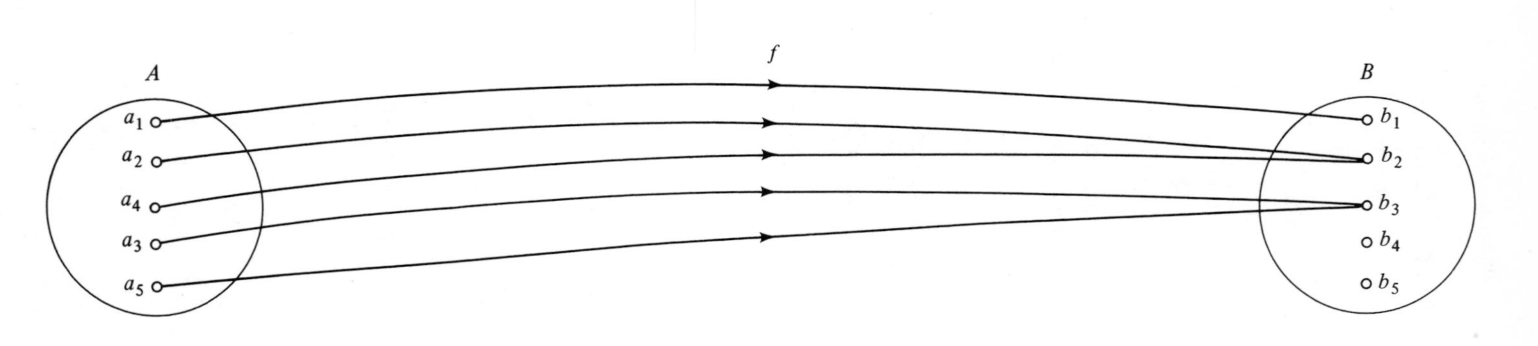

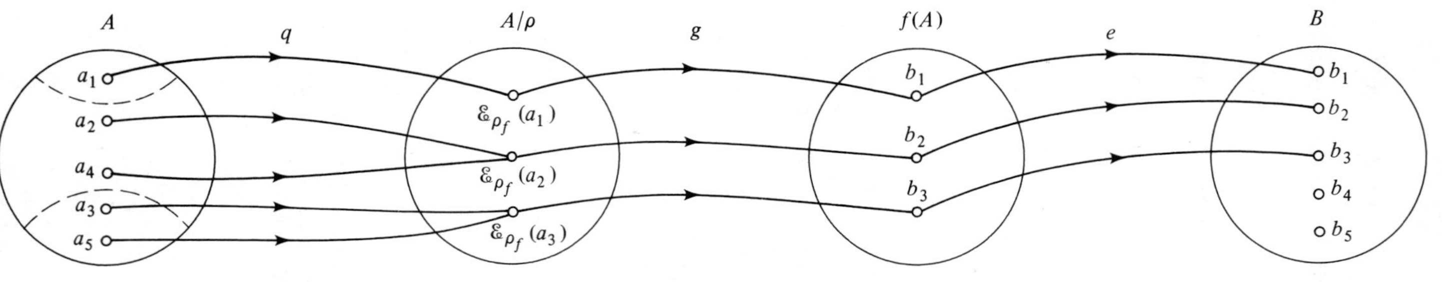

Fig. 3-4. Canonical factorization of $f: A \longrightarrow B$

Given a function f: $A \longrightarrow B$, the *equivalence kernel* of f is a relation ρ_f on A, defined as $a_i \rho_f a_j$ if and only if $f(a_i) = f(a_j)$. Clearly, ρ_f is an equivalence relation and, hence, induces on A an equivalence partition $\pi^A_{\rho_f}$, where each equivalence class consists of all elements of A whose image is a given element in the range of f. Thus, there is a bijection from the elements of the quotient set A/ρ_f onto the range of f.

This observation can be formalized by stating that every function f: $A \longrightarrow B$ is expressible in the form

$$f = egq \tag{3-2}$$

where q, g, and e are functions defined as

$$q: \quad A \longrightarrow A/\rho_f$$

where $q(a_i) = \mathcal{E}_{\rho_f}(a_i)$. That is, the image of $a_i \in A$ under q is the equivalence class in $\pi^A_{\rho_f}$ which contains a_i. Next,

$$g: \quad A/\rho_f \longrightarrow f(A)$$

where $g(\mathcal{E}_{\rho_f}(a_i)) = f(a_i)$. That is, the image of $\mathcal{E}_{\rho_f} \in A/\rho_f$ under g is the element of B whose inverse image is a_i. [The function g is well defined, because all $a_i \in A$ whose image under q is $\mathcal{E}_{\rho_f}(a_i)$ are the inverse images of the same element of B.] Finally,

$$e: \quad f(A) \longrightarrow B$$

where $e(b_j) = b_j$. That is, e is the same as g, except that the codomain is now B rather than $f(A)$.

The form (3-2) of f (illustrated in Fig. 3-4) is called the *canonical factorization* of f. It shows that, except for the bijection g which simply renames elements, and the injection e, which eliminates the discrepancy between B and $f(A)$, f is completely determined by q and, hence, by ρ_f.

PROBLEMS

1. Consider the functions

f: $\mathbb{N}_{1000} \longrightarrow \mathbb{N}_{1000}$

where $f(n)$ = number of integers less than or equal to n which are of the form 2^k $(k \geq 0)$

and g: $\mathbb{N}_{1000} \longrightarrow \mathbb{N}_{1000}$

where $g(n)$ = least integer which exceeds $100 \log_2 n$

Determine the functions:

(a) gf (b) fg (c) fgf (d) f^2 (e) f^3
(f) g^2 (g) g^3

2. Consider the functions f: $A \longrightarrow B$ and g: $B \longrightarrow C$, and prove or disprove these statements:
(a) gf is an injection if and only if g and f are injections.
(b) gf is a surjection if and only if g and f are surjections.
(c) gf is a bijection if and only if g and f are bijections.

3. Consider the function f: $\mathbf{2}^{\mathbb{N}} \longrightarrow \mathbf{2}^{\mathbb{N}}$, where:
(a) $f(S) = \{n \mid n \in S \cap \mathbb{P}\}$
(b) $f(S) = \{n \mid n$ is the largest element in $S\}$
(c) $f(S) = \{n \mid n \in S \cap \mathbb{P}$, n is not the largest element in $S\}$
Which of these functions is idempotent? For which of these functions does there exist an integer $i \geq 1$ such that f^i is idempotent?

4. Find the canonical factorization of the function

$$f\colon\ \mathbb{N}_{20} \longrightarrow \mathbb{N}_{10}$$

where $f(n) =$ number of perfect squares which equal or exceed n

3-4. IDENTITY AND INVERSE FUNCTIONS

The *identity function of a set A*, denoted by $\mathbf{1}_A$, is defined as the function

$$\mathbf{1}_A\colon\ A \longrightarrow A$$

where

$$\mathbf{1}_A(a) = a$$

That is, $\mathbf{1}_A$ is a bijection from A onto A, such that the image (and inverse image) of every element in A is the element itself. From this definition it follows that, for any function f: $A \longrightarrow B$, we have

$$\mathbf{1}_B f = f \mathbf{1}_A = f$$

Consider the functions f: $A \longrightarrow B$ and g: $B \longrightarrow A$. If

$$gf = \mathbf{1}_A$$

then g is called a *left inverse* of f, and f is called a *right inverse* of g. If, in addition,

$$fg = \mathbf{1}_B$$

then g is called a *two-sided inverse* of f, and f is called a *two-sided inverse* of g.

THEOREM 3-1

(a) The function $f: A \longrightarrow B$ has a left inverse if and only if f is an injection.
(b) The function $g: B \longrightarrow A$ has a right inverse if and only if g is a surjection.

Proof (a) If f has a left inverse, then there exists a function $g: B \longrightarrow A$ such that $gf = \mathbf{1}_A$. If $f(a) = f(a')$, then

$$a = \mathbf{1}_A(a) = gf(a) = g(f(a)) = g(f(a')) = gf(a') = \mathbf{1}_A(a') = a'$$

and, hence, f is an injection. Conversely, if f is an injection, we can define a function $g: B \longrightarrow A$, where

$$g(b) = \begin{cases} a \text{ for all } b \in B \text{ such that } f(a) = b \\ \text{some fixed element } a_0 \in A \text{ otherwise} \end{cases}$$

Since f is an injection, there is exactly one $a \in A$ for every $b \in B$ such that $f(a) = b$ and, hence, g is well defined. Also, since for every $a \in A$ we have $gf(a) = g(f(a)) = a$, it follows that $gf = \mathbf{1}_A$ and, hence, that f has a left inverse.

(b) If g has a right inverse, then there exists a function $f: A \longrightarrow B$ such that $gf = \mathbf{1}_A$. Hence, for every $a \in A$, $gf(a) = g(f(a)) = a$, which implies that every $a \in A$ is the image of some $b \in B$ under g and, hence, that g is a surjection. Conversely, if g is a surjection, we can define a function $f: A \longrightarrow B$, where $f(a)$ is any $b \in B$ such that $g(b) = a$. Since g is a surjection, there is at least one $b \in B$ for each $a \in A$ such that $g(b) = a$ and, hence, f is well-defined. Also, since for every $a \in A$, $gf(a) = g(f(a)) = a$, we have $gf = \mathbf{1}_A$; hence, g has a right inverse. □

THEOREM 3-2

Consider the two functions $f: A \longrightarrow B$ and $g: B \longrightarrow C$. (a) If f and g are injections, then gf is an injection. (b) If f and g are surjections, then gf is a surjection. (c) If f and g are bijections, then gf is a bijection.

Proof (a) By Theorem 3-1(a), if f and g are injections, then they have left inverses, say, f_l and g_l, respectively. Hence,

$$(f_l g_l)(gf) = f_l(g_l g)f = f_l \mathbf{1}_B f = f_l(\mathbf{1}_B f) = f_l f = \mathbf{1}_A$$

Thus, gf has a left inverse (namely, $f_l g_l$) and, again by Theorem 3-1(a), gf is an injection.

(b) By Theorem 3-1(b), if f and g are surjections, then they have right inverses, say, f_r and g_r, respectively. Hence,

$$(gf)(f_r g_r) = g(ff_r)g_r = g\mathbf{1}_B g_r = (g\mathbf{1}_B)g_r = gg_r = \mathbf{1}_C$$

Thus, gf has a right inverse (namely, $f_r g_r$) and, again by Theorem 3-1(b), gf is a surjection.

(c) The last assertion follows immediately from parts (a) and (b) and from the definition of bijection. □

THEOREM 3-3

A function $f: A \longrightarrow B$ is a bijection if and only if f has both a left inverse $g_l: B \longrightarrow A$ and a right inverse $g_r: B \longrightarrow A$. When this is the case, $g_l = g_r$ and f has a unique two-sided inverse which is itself a bijection.

Proof The first statement follows immediately from Theorem 3-1 and the definition of bijection. If g_l and g_r exist, then

$$g_l = g_l \mathbf{1}_B = g_l(f g_r) = (g_l f) g_r = \mathbf{1}_A g_r = g_r$$

Thus, any left inverse must equal any right inverse and, hence, the two-sided inverse $g = g_l = g_r$ of f is unique. Since $g_l f = gf = \mathbf{1}_A$ and $fg_r = fg = \mathbf{1}_B$, g has a left inverse and a right inverse (namely, f); hence, g must be a bijection. □

The unique two-sided inverse of a bijection $f: A \longrightarrow B$ is called simply the *inverse* of f and denoted by f^{-1}. Thus,

$$f^{-1}f = \mathbf{1}_A, \qquad ff^{-1} = \mathbf{1}_B$$

If f is a bijection from A onto A, then

$$f^{-1}f = ff^{-1} = \mathbf{1}_A$$

PROBLEMS

1. Consider the following functions f and the relations ρ defined by them:
   (a) $f\colon \mathbb{N}_3 \longrightarrow \mathbb{N}_5$, $\rho = \{(1, 3), (2, 4), (3, 2)\}$
   (b) $f\colon \mathbb{N}_4 \longrightarrow \mathbb{N}_3$, $\rho = \{(1, 3), (2, 3), (3, 2), (4, 1)\}$
   (c) $f\colon \mathbb{N}_4 \longrightarrow \mathbb{N}_4$, $\rho = \{(1, 2), (2, 4), (3, 1), (4, 3)\}$
   (d) $f\colon \mathbb{N}_4 \longrightarrow \mathbb{N}_4$, $\rho = \{(1, 2), (2, 2), (3, 4), (4, 3)\}$
   For each of these functions, find (whenever possible) a left inverse, a right inverse, and a two-sided inverse.

2. Consider the functions $f: A \longrightarrow B$ and $g: B \longrightarrow C$, and prove the following:
   (a) If gf is an injection, then f is an injection.
   (b) If gf is a surjection, then g is a surjection.
   (c) If gf is a bijection, then f is an injection and g is a surjection.

3. Show that, if f_1 and f_2 are bijections, then

$$(f_1 f_2)^{-1} = f_2^{-1} f_1^{-1}$$

3-5. PERMUTATIONS

A bijection from a finite set $A = \{a_1, a_2, \ldots, a_n\}$ onto itself is called a *permutation on A*. The integer n is called the *order* of the permutation. A permutation $p: A \longrightarrow A$ of order n is often specified in the form

$$p = \begin{pmatrix} a_1 & a_2 & \ldots & a_n \\ p(a_1) & p(a_2) & \ldots & p(a_n) \end{pmatrix}$$

(where the order of the n columns is arbitrary). The number of distinct permutations of order n (on any fixed set A) is $n \cdot (n-1) \cdot (n-2) \cdot \ldots \cdot 3 \cdot 2 = n!$. A systematic way of constructing all these permutations is described in Sec. 3-10.

The *identity permutation* on $A = \{a_1, a_2, \ldots, a_n\}$ is the identity function $\mathbf{1}_A: A \longrightarrow A$, and thus the permutation

$$\begin{pmatrix} a_1 & a_2 & \ldots & a_n \\ a_1 & a_2 & \ldots & a_n \end{pmatrix}$$

Given any permutation on A,

$$p = \begin{pmatrix} a_1 & a_2 & \ldots & a_n \\ p(a_1) & p(a_2) & \ldots & p(a_n) \end{pmatrix}$$

the inverse of p is, clearly, the permutation given by

$$p^{-1} = \begin{pmatrix} p(a_1) & p(a_2) & \ldots & p(a_n) \\ a_1 & a_2 & \ldots & a_n \end{pmatrix}$$

Example 3-5

The set of all six permutations of order 3 (on a, b, c) is given by

$$\left\{ \begin{pmatrix} a & b & c \\ a & b & c \end{pmatrix}, \begin{pmatrix} a & b & c \\ a & c & b \end{pmatrix}, \begin{pmatrix} a & b & c \\ b & a & c \end{pmatrix}, \begin{pmatrix} a & b & c \\ b & c & a \end{pmatrix}, \begin{pmatrix} a & b & c \\ c & a & b \end{pmatrix}, \begin{pmatrix} a & b & c \\ c & b & a \end{pmatrix} \right\}$$

The first element is the identity permutation. The inverse of $\begin{pmatrix} a & b & c \\ c & a & b \end{pmatrix}$, for example, is $\begin{pmatrix} a & b & c \\ b & c & a \end{pmatrix}$.

□

Let p be a permutation on $A = \{a_1, a_2, \ldots, a_n\}$. For any $a_i \in A$, consider the sequence of elements

$$a_i, p(a_i), p^2(a_i), p^3(a_i), \ldots$$

Since A is finite, the sequence must have repetitions; that is, there must be a *least* integer l such that $0 \leq k < l \leq n$ and such that $p^k(a_i) = p^l(a_i)$ (with p^0 denoting the identity permutation). Writing $(p^{-1})^k$ as p^{-k}, we have

$$p^{-k}(p^k(a_i)) = p^{-k}(p^l(a_i))$$

which, letting $r_i = l - k$, implies $p^{r_i}(a_i) = a_i$ $(1 \leq r_i \leq n)$. Hence, the above sequence of elements can be written in the "periodic" form

$$a_i, p(a_i), p^2(a_i), \ldots, p^{r_i-1}(a_i), a_i, p(a_i), p^2(a_i), \ldots, p^{r_i-1}(a_i), \ldots$$

The first r elements of this sequence, which must be distinct, are usually displayed in the form of an r_i-tuple

$$(a_i, p(a_i), p^2(a_i), \ldots, p^{r_i-1}(a_i)) \tag{3-3}$$

and referred to as a *cycle of order* r_i. We can see that the permutation p, when applied r_i times, "sends" a_i (or any other element in the cycle containing a_i) back to itself. If $r_i < n$, then there is at least one element $a_j \in A$ not included in the cycle (3-3). Repeating the same process, we can determine the cycle which contains a_j, say,

$$(a_j, p(a_j), p^2(a_j), \ldots, p^{r_j-1}(a_j)) \tag{3-4}$$

Cycles (3-3) and (3-4) cannot have any elements in common. For suppose $p^k(a_i) = p^l(a_j)$ for some k and l; then

$$p^{r_j-l}(p^k(a_i)) = p^{r_j-l}(p^l(a_j))$$

or

$$p^{r_j+k-l}(a_i) = p^{r_j}(a_j) = a_j$$

But $p^{r_j+k-l}(a_i)$ is in cycle (3-3), while a_j is not. By contradiction, then, the two cycles are disjoint.

Proceeding in this manner, A can finally be partitioned into subsets of elements, each subset constituting a different cycle. This partition is usually expressed as a "product" of disjoint cycles, in the form

$$(a_i, p(a_i), \ldots, p^{r_i-1}(a_i))(a_j, p(a_j), \ldots, p^{r_j-1}(a_j))(a_m, p(a_m), \ldots, p^{r_m-1}(a_m))$$

(where the order in which the cycles are written is arbitrary, and where cycles of order 1 are commonly deleted). This form completely specifies the permutation p.

Example 3-6

Consider the permutation

$$p = \begin{pmatrix} a & b & c & d & e & f & g & h \\ h & g & b & f & e & d & a & c \end{pmatrix}$$

The cycle containing the element a, that is, $(a, p(a), p^2(a), \ldots)$, is

$$(a, h, c, b, g)$$

(Any element in this cycle is sent back to itself after five applications of p.) The cycle containing d, that is, $(d, p(d), p^2(d), \ldots)$, is

$$(d, f)$$

Finally, the cycle containing e is simply

$$(e)$$

Thus, p can be specified in the form

$$(a, h, c, b, g)(d, f)(e)$$

[from which (e) can be deleted]. □

PROBLEMS

1. Express the following permutations as products of disjoint cycles. Also, find the inverses of these permutations:

(a) $\begin{pmatrix} a & b & c & d & e & f & g \\ g & f & e & d & c & b & a \end{pmatrix}$

(b) $\begin{pmatrix} a & b & c & d & e & f & g \\ e & g & f & b & a & c & d \end{pmatrix}$

(c) $\begin{pmatrix} a & b & c & d & e & f & g & h & i & j \\ h & d & b & a & c & f & j & i & g & e \end{pmatrix}$

2. Denoting the permutations of Problems 1(a) and 1(b) by p_1 and p_2, respectively, evaluate these permutations:

(a) $p_1 p_2$ (b) $p_2 p_1$ (c) $(p_1 p_2 p_1)^2$

3. Find the permutations corresponding to the following products of cycles:

(a) $(a, c, b, f)(d, h, i, j)(e)(g)$

(b) $(a, g)(f, e)(b, i)(h, c)(d, j)$

(c) $(a, h, c, d, j, f, g, i, b, e)$

4. A *transposition* is a permutation that effects the interchange of two elements, leaving all other elements intact. For example,

$$\begin{pmatrix} a & b & c & d & e \\ d & b & c & a & e \end{pmatrix}$$

is a transposition on $\{a, b, c, d, e\}$. Show that every permutation can be expressed as the composition of a finite number of transpositions. Express the permutation of Example 3-6 in this manner.

5. A typist is given an unusual typewriter: whenever a character is pressed on the keyboard, the wrong character appears on the paper. (For example, whenever A is pressed, H appears, whenever 7 is pressed, $ appears, etc.) However, every character marked on the keyboard can be typed by pressing *some* key on the keyboard. Now, the typist is presented with an English text that she types in the normal manner, obtaining, of course, a gibberish copy. She takes the copy and types *it* in the normal manner, obtaining (possibly) another gibberish copy. She takes the second copy and types it to obtain still another version, and so forth. If the typist continues in this manner, will she eventually produce a copy of the original English text? Justify your answer.

3-6. CARDINALITY

Now that we have learned certain facts about relations and functions, we are in a position to discuss the concept of cardinality more rigorously. We start by proving:

THEOREM 3-4

Let S denote a set of sets, and let ρ be a relation on S defined as $A\rho B$ if and only if a bijection exists from A onto B. Then ρ is an equivalence relation.

Proof The function $\mathbf{1}_A$ is clearly a bijection from A onto A, hence, ρ is reflexive. If a bijection f exists from A onto B, then, by Theorem 3-3, f^{-1} is a bijection from B onto A. Hence, ρ is symmetric. If a bijection f exists from A onto B and a bijection g exists from B onto C, then, by Theorem 3-2, gf is a bijection from A onto C. Hence, ρ is transitive. In conclusion, ρ is an equivalence relation. □

Thus, the relation ρ defined in Theorem 3-4 induces an equivalence partition on S. The equivalence classes of this partition are called *cardinality classes*, and sets that belong to the same class in this partition are said to have the same *cardinality*.

A set is said to be *finite*, and of cardinality m, if it belongs to the same cardinality class as the set $\mathbb{N}_m = \{1, 2, \ldots, m\}$ (for some positive integer m). Otherwise, the set is said to be *infinite*. (The cardinality of $\varnothing$ is defined as 0.)

Thus, the cardinality of any finite set can be determined by establishing a bijection from this set onto some set $\mathbb{N}_m$. A basic property of finite sets is that no bijection exists from a finite set A onto a proper subset of A. (This property is known as the *pigeonhole principle*: we cannot place more than k pigeons in k pigeonholes without forcing at least two pigeons into the same pigeonhole.)

A set is said to be *denumerable*, and of cardinality $\aleph_0$ ("aleph-null"), if it belongs to the same cardinality class as the set $\mathbb{N}$ of positive integers.

To establish the fact that a set A is finite or denumerable, it is sufficient to devise a procedure for arranging the elements of A in some linear order; once this task is accomplished, the first element in this arrangement can be associated with 1, the second with 2, and so forth, to form a bijection from A onto $\mathbb{N}_m$ or $\mathbb{N}$. Since this procedure is essentially that for "counting" the elements of A, sets that are either finite or denumerable are said to be *countable*. Otherwise, they are said to be *uncountable*.

Example 3-7

The Latin alphabet {A, B, . . . , Z} has cardinality 26, since we can establish the following bijection from this alphabet onto $\mathbb{N}_{26}$:

$$\begin{array}{ccccc} \text{A} & \text{B} & \text{C} & \ldots & \text{Z} \\ \updownarrow & \updownarrow & \updownarrow & & \updownarrow \\ 1 & 2 & 3 & \ldots & 26 \end{array}$$

□

Example 3-8

The set of integers $\mathbb{I}$ is denumerable, because we can establish the following bijection from $\mathbb{I}$ onto $\mathbb{N}$:

$$\begin{array}{cccccccc} 0 & 1 & -1 & 2 & -2 & 3 & -3 & \ldots \\ \updownarrow & \updownarrow & \updownarrow & \updownarrow & \updownarrow & \updownarrow & \updownarrow & \\ 1 & 2 & 3 & 4 & 5 & 6 & 7 & \ldots \end{array}$$

□

Example 3-9

The set $\mathbb{Q}$ of rational numbers is denumerable. To verify this fact, first list all positive rational numbers in this manner:

$$\begin{array}{cccccc} 1 & 2 & 3 & 4 & 5 & \ldots \\ \frac{1}{2} & \frac{2}{2} & \frac{3}{2} & \frac{4}{2} & \frac{5}{2} & \ldots \\ \frac{1}{3} & \frac{2}{3} & \frac{3}{3} & \frac{4}{3} & \frac{5}{3} & \ldots \\ \frac{1}{4} & \frac{2}{4} & \frac{3}{4} & \frac{4}{4} & \frac{5}{4} & \ldots \\ \frac{1}{5} & \frac{2}{5} & \frac{3}{5} & \frac{4}{5} & \frac{5}{5} & \ldots \\ \vdots & \vdots & \vdots & \vdots & \vdots & \end{array}$$

Next, delete all rational numbers not represented in their lowest terms (for example, $\frac{2}{2}$, $\frac{2}{4}$, etc.), and scan the remaining numbers in the order indicated by this arrowed path:

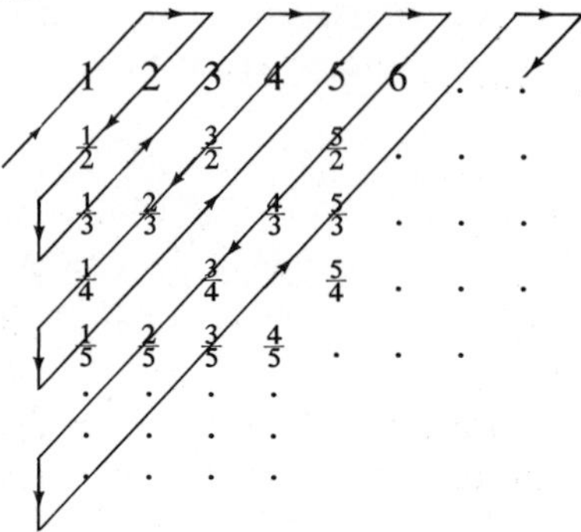

The positive rational numbers can now be listed linearly as

$$1, 2, \tfrac{1}{2}, \tfrac{1}{3}, 3, 4, \tfrac{3}{2}, \tfrac{2}{3}, \tfrac{1}{4}, \tfrac{1}{5}, 5, \ldots$$

Finally, by starting the list with 0 and inserting after every rational number q the number $-q$, the following bijection can be established from $\mathbb{Q}$ onto $\mathbb{N}$:

$$\begin{array}{cccccccccccccc} 0 & 1 & -1 & 2 & -2 & \frac{1}{2} & -\frac{1}{2} & \frac{1}{3} & -\frac{1}{3} & 3 & -3 & 4 & -4 & \ldots \\ \updownarrow & \updownarrow & \updownarrow & \updownarrow & \updownarrow & \updownarrow & \updownarrow & \updownarrow & \updownarrow & \updownarrow & \updownarrow & \updownarrow & \updownarrow & \\ 1 & 2 & 3 & 4 & 5 & 6 & 7 & 8 & 9 & 10 & 11 & 12 & 13 & \ldots \end{array}$$ □

Note that, in the last two examples, bijections were established from given sets ($\mathbb{I}$ and $\mathbb{Q}$) onto a proper subset (namely, $\mathbb{N}$) of these sets. Such bijections are possible only among infinite sets.

We have already shown (see Theorem 1-2) that the cardinality of $\mathbf{2}^A$, where A is a finite set, is $2^{\#A}$. Hence, the cardinality of any finite set A must be less than that of $\mathbf{2}^A$. When A is infinite, we can arrive at the same conclusion by noting that the function $f\colon A \longrightarrow \mathbf{2}^A$, where $f(a) = \{a\}$, cannot be a surjection. Thus, for every *infinite* set A there exists at least one set (namely, $\mathbf{2}^A$) that is "more infinite" than A. We can conceive of a hierarchy of cardinalities, where for each cardinality there exists one that "exceeds" it. In particular, there exist sets whose cardinality "exceeds" that of $\mathbb{N}$; hence, there exist sets that are uncountable. The following theorem gives an example of such a set.

THEOREM 3-5

The set $R = \{r \mid r \in \mathbb{R}, 0 < r < 1\}$ is uncountable.

Proof Every element of R can be written as a decimal fraction $0.\delta_{i1}\delta_{i2}\delta_{i3}\ldots$, where $0 \leq \delta_{ij} \leq 9$. Assuming that R (which is infinite) is denumerable, some bijection f must exist from R onto $\mathbb{N}$. For every $n \geq 1$, let the real number whose image under f is n be $0.\delta_{n1}\delta_{n2}\delta_{n3}\ldots$. Thus, f is given by

$$\begin{array}{cccc} 0.\delta_{11}\delta_{12}\delta_{13}\ldots & 0.\delta_{21}\delta_{22}\delta_{23}\ldots & 0.\delta_{31}\delta_{32}\delta_{33}\ldots & \ldots \\ \updownarrow & \updownarrow & \updownarrow & \\ 1 & 2 & 3 & \ldots \end{array}$$

Now, consider the real number $0.\bar{\delta}_{11}\bar{\delta}_{22}\bar{\delta}_{33}\ldots$, where $\bar{\delta}_{ii}$ is any decimal digit that differs from δ_{ii} (for example, $\bar{\delta}_{ii} = 9 - \delta_{ii}$). Thus, $0.\bar{\delta}_{11}\bar{\delta}_{22}\bar{\delta}_{33}\ldots$ differs from *any* $0.\delta_{i1}\delta_{i2}\delta_{i3}\ldots$ and cannot belong to the domain of f. Hence, a bijection from R onto $\mathbb{N}$ cannot exist. □

Theorem 3-5 implies that the sets R and $\mathbb{N}$ belong to different cardinality classes. All sets that belong to the same cardinality class as R are said to be of cardinality $\aleph_1$. That $\aleph_1$ is also the cardinality of $\mathbb{R}$ can be demonstrated as follows: Represent $\mathbb{R}$ by an infinite coordinate axis (a line on which each point represents a different real number) and R by a finite line segment (on which each point represents a different real number between 0 and 1). Now, bend the R segment into a semicircle, and let the $\mathbb{R}$ axis be tangent to this semicircle at the midpoint of the latter (see Fig. 3-5). If lines are drawn from the center of the semicircle, intersecting both the semicircle and the axis, the points of intersection can be paired off to form a bijection from R onto $\mathbb{R}$. Hence, R and $\mathbb{R}$ have the same cardinality—namely, $\aleph_1$. (Note the existence of a bijection from $\mathbb{R}$ onto a proper subset of $\mathbb{R}$.)

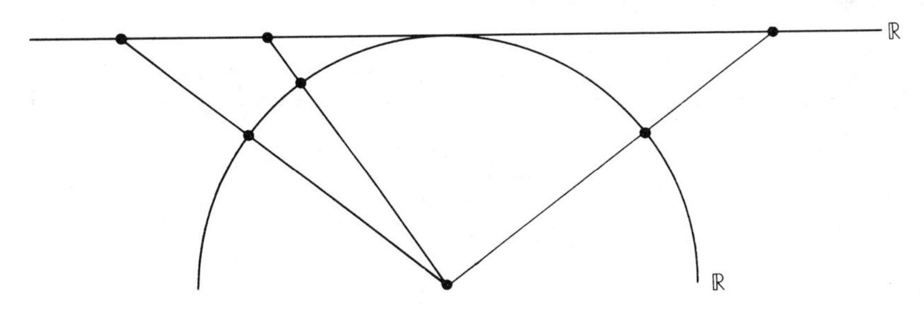

Fig 3-5 Bijection from R onto $\mathbb{R}$

The proof employed in Theorem 3-5 is referred to as a *diagonalization argument*. This term is derived from the fact that, if the fractions $0.\delta_{i1}\delta_{i2}\delta_{i3}\ldots$ are written one underneath the other, then $0.\bar{\delta}_{11}\bar{\delta}_{22}\bar{\delta}_{33}\ldots$ is constructed by scanning the resulting array of digits along the diagonal that originates in the upper left corner of the array. In general, a diagonalization argument refers to a procedure which produces an element differing from each element of a given set in at least one attribute (in Theorem 3-5, this attribute is a digit to the right of the decimal point). For example, to prove that the set F of functions of the form $f: \mathbb{N} \longrightarrow \mathbb{N}$ is uncountable, we can employ this diagonalization argument: Suppose that F is countable. We can list its elements as $f_1, f_2, f_3, \ldots$; we can now construct a function $g: \mathbb{N} \longrightarrow \mathbb{N}$ such that, for every positive integer i, $g(i) = f_i(i) + 1$. Clearly, g cannot equal any of the f_i; hence, the assumption that F is countable must be false.

PROBLEMS

1. Prove that if the sets A and B are denumerable, so is the set $A \times B$.

2. Show that these sets are denumerable:
(a) $\{k \mid k = 3n - 2, n \in \mathbb{N}\}$
(b) $\{k \mid k = n^2, n \in \mathbb{N}\}$
(c) The union of the sets of parts (a) and (b)
(d) $\{q_1 + q_2\sqrt{-1} \mid q_1, q_2 \in \mathbb{Q}\}$

3. Is the set of all polynomials with coefficients from $\{0, 1\}$ countable? Prove your answer.

4. Through a technique exemplified in Fig. 3-5, establish a bijection:
(a) from the set of all points on a line A onto the set of all points on a line B;
(b) from the set of all points on a circle C onto the set of all points on a triangle T.

5. Show that the union of a countable number of countable sets is itself countable.

3-7. THE CHARACTERISTIC FUNCTION

A function often used in set-theoretic studies is the *characteristic function* of a subset A of a universal set U. This function, denoted by e_A, is defined by

$$e_A: \ U \longrightarrow \{0, 1\}$$

where

$$e_A(u) = \begin{cases} 1 & \text{if } u \in A \\ 0 & \text{if } u \notin A \end{cases}$$

Thus, if A_1 and A_2 are subsets of U, then $A_1 = A_2$ if and only if $e_{A_1} = e_{A_2}$.

The characteristic function may be used to give a convenient representation to functions with a finite range. Consider the function $f: A \longrightarrow B$, where $B = \{b_1, b_2, \ldots, b_k\}$. Let

$$\pi^A_{\rho_f} = \{A_1, A_2, \ldots, A_k\}$$

be the equivalence kernel of f (see Sec. 3-3), where

$$A_i = \{a \mid a \in A, f(a) = b_i\}$$

Since, for any $a \in A_i$, exactly one of the k values $e_{A_1}(a), e_{A_2}(a), \ldots, e_{A_k}(a)$ is 1 [namely, $e_{A_i}(a)$], while all other values are 0, we can write

$$f(a) = \textstyle\sum_{i=1}^{k} b_i e_{A_i}(a)$$

Example 3-10

Consider the function $f\colon \mathbb{I} \longrightarrow \mathbb{Z}_m$, where $f(i) = i \pmod{m}$, and let

$$C_j = \{i \mid i \in \mathbb{I}, i = j \pmod{m}\} \qquad (j = 0, 1, \ldots, m-1)$$

Then we can write

$$f(i) = \textstyle\sum_{j=0}^{m-1} j e_{C_j}(i)$$

For example, when $m = 3$,

$$f(i) = e_{C_1}(i) + 2e_{C_2}(i)$$ □

The characteristic function enables us to express various set relations in a concise numerical manner. For example, it can be readily verified that

$$e_{A'}(u) = 1 - e_A(u) \tag{3-5}$$

$$e_{A \cup B}(u) = e_A(u) + e_B(u) - e_A(u)e_B(u) \tag{3-6}$$

$$e_{A \cap B}(u) = e_A(u)e_B(u) \tag{3-7}$$

Using these identities, more complex ones can be derived. For example, the identity

$$A \cap (B \cup C) = (A \cap B) \cup (A \cap C)$$

can be verified by

$$\begin{aligned} e_{A \cap (B \cup C)}(u) &= e_A(u)e_{B \cup C}(u) \\ &= e_A(u)[e_B(u) + e_C(u) - e_B(u)e_C(u)] \\ &= e_A(u)e_B(u) + e_A(u)e_C(u) - e_A(u)e_B(u)e_A(u)e_C(u) \\ &= e_{A \cap B}(u) + e_{A \cap C}(u) - e_{A \cap B}(u)e_{A \cap C}(u) \\ &= e_{(A \cap B) \cup (A \cap C)}(u) \end{aligned}$$

From (3-5), (3-6), and (3-7) it can be noted that, when values of $e_{A'}(u)$, $e_{A \cup B}(u)$, and $e_{A \cap B}(u)$ are tabulated for all possible combinations of $e_A(u)$ and $e_B(u)$ values, the result is precisely the membership tables for A', $A \cup B$, and $A \cap B$ shown in Tables 1-1(a), (b), and (c). More generally, if S is a set generated by $A_1, A_2, \ldots, A_r$, then the values of $e_S(u)$, tabulated for all value combinations of $e_{A_1}(u), e_{A_2}(u), \ldots, e_{A_r}(u)$, constitute the membership table for S as described in Sec. 1-5. In conclusion, then, membership tables are merely means for representing characteristic functions of sets.

Consider the function

$$f: \ \{0, 1)^U \longrightarrow \mathbf{2}^U$$

(from the set of all characteristic functions with domain U into the power set of U), where

$$f(e_A) = A$$

Because $A_1 = A_2$ if and only if $e_{A_1} = e_{A_2}$, f is a bijection; hence, its domain and range must have the same cardinality; that is,

$$\#(\{0, 1\}^U) = \#(\mathbf{2}^U)$$

In particular, when U is finite, by (3-1) we have:

$$\#(\{0, 1\}^U) = (\#\{0, 1\})^{\#U} = 2^{\#U}$$

Hence,

$$\#(\mathbf{2}^U) = 2^{\#U}$$

which agrees with Theorem 1-2.

PROBLEMS

1. Let $A_1, A_2, \ldots, A_r$ be subsets of U, and S a set generated by $A_1, A_2, \ldots, A_r$. Suppose the minset normal form of S is given by

$$S = \bigcup_{i=1}^{l} M_i$$

(where the M_i are minterms generated by $A_1, A_2, \ldots, A_r$). Show that

$$e_S(u) = \sum_{i=1}^{l} e_{M_i}(u)$$

2. Consider the function f: $A \longrightarrow B$, where

$$f(a) = 0e_{A_1}(a) + 1e_{A_2}(a)$$

($\{A_1, A_2\}$ being the equivalence kernel of f), and the function g: $A \longrightarrow C$, where

$$g(a) = 0e_{A_3}(a) + 1e_{A_4}(a) + 3e_{A_5}(a)$$

($\{A_3, A_4, A_5\}$ being the equivalence kernel of g). Show that

$$f(a) + g(a) = 0e_{A_1 \cap A_3}(a) + e_{A_1 \cap A_4}(a) + e_{A_2 \cap A_3}(a) + 2e_{A_2 \cap A_4}(a) + 3e_{A_1 \cap A_5}(a) + 4e_{A_2 \cap A_5}$$

3. Prove: For all $u \in U$,
(a) $e_A(u) \leq e_B(u)$ if and only if $A \subset B$.
(b) $e_{A \cap B}(u) =$ the smaller of $e_A(u)$ and $e_B(u)$.
(c) $e_{A \cup B}(u) =$ the larger of $e_A(u)$ and $e_B(u)$.
(d) $e_{A-B}(u) = e_A(u) - e_{A \cap B}(u)$.

4. Let $S = (A \cap B) \cup (A' \cap C) \cup (B \cap C)$, where A, B, and C are subsets of U. Tabulate the values of $e_S(u)$ for all value combinations of $e_A(u)$, $e_B(u)$, and $e_C(u)$. (Compare with Table 1-2.)

5. With every subset A of $U = \{u_1, u_2, \ldots, u_n\}$, associate an n-digit binary number $\beta_1 \beta_2 \ldots \beta_n$, where $\beta_i = e_A(u_i)$. Based on this association, prove (once more) that $\#(\mathbf{2}^U) = 2^{\#U}$.

3-8. THE PEANO POSTULATES AND FINITE INDUCTION

Perhaps the most basic mathematical function is the so-called *Peano's successor function* (named after the Italian mathematician Giuseppe Peano, 1858–1932). This function, denoted by s, assigns to every positive integer the next larger integer. Thus,

$$s\colon\ \mathbb{N} \to \mathbb{N}$$

where

$$s(n) = n + 1$$

We can think of s as a function which performs the rudimentary operation of counting.

In terms of the function s, a number of basic properties can be formulated which characterize the set of positive integers $\mathbb{N}$. These properties, known as the *Peano postulates*, can be summarized as:

(a) $1 \in \mathbb{N}$;
(b) if $n \in \mathbb{N}$, then $s(n) \in \mathbb{N}$;
(c) for no $n \in \mathbb{N}$ is $s(n) = 1$;
(d) if $s(n) = s(m)$, then $n = m$;
(e) any subset of $\mathbb{N}$ that contains 1, and that contains $s(n)$ whenever it contains n, must equal $\mathbb{N}$.

Postulate (e) leads to one of the most important principles of mathematics:

THEOREM 3-6 (*The principle of finite induction*)

Let $P(i)$ be a statement which, for each integer i, may be either true or false. To prove that $P(i)$ is true for all integers $i \geq i_0$, it suffices to prove that:

(i) (*Basis*). $P(i_0)$ is true.

(ii) (*Induction step*). For all $k \geq i_0$, the assumption that $P(k)$ is true (called the *induction hypothesis*) implies that $P(k + 1)$ is true.

Proof Define $P'(i) = P(i - i_0 + 1)$. Thus, to prove that $P(i)$ is true for all integers $i \geq i_0$, it suffices to prove that $P'(i)$ is true for all $i \in \mathbb{N}$. Now, define

$$S = \{i \mid i \in \mathbb{N}, P'(i) \text{ is true}\}$$

If condition (i) in the theorem holds, then $P(i_0) = P'(1)$ is true; hence, $1 \in S$. If condition (ii) holds, then, for all $k \geq i_0$, the truth of $P(k)$ implies the truth of $P(k+1)$; hence, for all $k \geq 1$, the truth of $P'(k)$ implies the truth of $P'(k+1)$. Thus, S contains $s(k) = k + 1$ whenever S contains k, and by Peano postulate (e) we must have $S = \mathbb{N}$. In conclusion, if conditions (i) and (ii) hold, $P'(i)$ is true for all $i \in \mathbb{N}$; hence, $P(i)$ is true for all $i \geq i_0$. □

The principle of finite induction provides a powerful tool for proving statements in every branch of mathematics. The variable i that appears in Theorem 3-6 is sometimes referred to as the *induction variable*, and the proving procedure as a *proof by induction on i*. We illustrate the application of proof by induction in the following important result.

Theorem 3-7

Let $I = \{i \mid i \in \text{II}, i \geqslant i_0\}$ and let J be any subset of I. Then J contains a unique integer j (called the *least integer*) such that, for all $i \in J$, $j \leqslant i$.[1]

Proof It can be readily shown that j, if exists, must be unique. If $i_0 \in J$ then, clearly, $j = i_0$. Otherwise, let K be the set of all integers smaller than *every* integer in J. Thus, K must contain i_0. If, for some $k \in K$ there is $j_0 \in J$ such that $k + 1 = j_0$, then $j = j_0$. Otherwise $K = \{k \mid k \geqslant i_0\}$ and hence $K = I$, which implies that J is empty.

In some cases, the proof of $P(k + 1)$, needed in the induction step, is greatly facilitated if the induction hypothesis is the truth of $P(i_0)$, $P(i_0 + 1)$, $P(i_0 + 2)$, . . . , $P(k)$, rather than the truth of $P(k)$ alone. In these cases, it is convenient to invoke this principle:

THEOREM 3-8 (*The second principle of finite induction*)

Let $P(i)$ be a statement which, for each integer i, may be either true or false. To prove that $P(i)$ is true for all integers $i \geq i_0$, it suffices to prove that:

(i) (*Basis*). $P(i_0)$ is true.

[1] A corollary of this theorem is that the set of positive integers $\mathbb{N}$ is a well-ordering with respect to the relation "less than or equal to" (see Sec. 2-7).

(ii) (*Induction step*). For all $k \geq i_0$, the assumption that $P(i)$ is true for $i = i_0, i_0 + 1, i_0 + 2, \ldots, k$ (called the *induction hypothesis*) implies that $P(k + 1)$ is true.

Proof Define

$$I = \{i \mid i \in \mathbb{I}, i \geq i_0\}$$
$$J = \{i \mid i \in I, P(i) \text{ is false}\}$$

If $J \neq \varnothing$, then, by Theorem 3-7, it contains a least integer, say, j. If condition (i) in the theorem holds, $i_0 \notin J$; hence, $j > i_0$. Thus, $P(i)$ is true for $i = i_0$, $i_0 + 1, \ldots, j - 1$, where $j - 1 \geq i_0$. If condition (ii) holds, then the truth of $P(i)$ for $i = i_0, i_0 + 1, \ldots, j - 1$ implies the truth of $P(j)$, which contradicts the fact that $j \in J$ and, hence, that $J \neq \varnothing$. In conclusion, $J = \varnothing$, and $P(i)$ is true for all $i \geq i_0$. □

PROBLEMS

1. Consider the following "theorem" and its "proof":

THEOREM
All citizens are treated equally under the law.

Proof (by induction on the number i of citizens) (*Basis*). When $i = 1$, there is only one citizen, and the theorem is trivially true. (*Induction step*). Suppose the theorem is true for any k citizens, and consider any group of $k + 1$ citizens, denoted by $c_1, c_2, \ldots, c_k, c_{k+1}$. By induction hypothesis, $c_1, c_2, \ldots, c_k$ are treated equally under the law, and so are $c_2, c_3, \ldots, c_{k+1}$. Hence, the treatment of $c_1, c_2, \ldots, c_k, c_{k+1}$ under the law is the same as that of c_2 and, hence, equal. Thus, the theorem is true for $i = k + 1$. □

Avoiding political arguments, point out the fallacy in this proof.

2. Consider the following "theorem" and its "proof":

THEOREM
For all $i \geq 0$,

$$\sqrt{1 + i\sqrt{1 + (i + 1)\sqrt{1 + (i + 2)\sqrt{1 + (i + 3)\sqrt{\ldots}}}}} = i + 1$$

Proof (by induction on i) (*Basis*). For $i = 0$, the theorem reduces to $\sqrt{1 + 0} = 0 + 1$ which, clearly, is true. (*Induction step*). The induction hypothesis is:

$$\sqrt{1 + k\sqrt{1 + (k + 1)\sqrt{1 + (k + 2)\sqrt{1 + (k + 3)\sqrt{\ldots}}}}} = k + 1$$

from which we have (by squaring, subtracting 1, and dividing by k):

$$\sqrt{1 + (k + 1)\sqrt{1 + (k + 2)\sqrt{1 + (k + 3)\sqrt{\ldots}}}}$$
$$= \frac{(k + 1)^2 - 1}{k} = \frac{k^2 + 2k}{k} = k + 2 = (k + 1) + 1$$

Thus, the theorem is true for $i = k + 1$. □

There is a fallacy in this proof. What is it?

3. Prove the following *principle of double induction:*
   Let $P(i, j)$ be a statement which, for each integer i and each integer j, may be either true or false. To prove that $P(i, j)$ is true for all integers $i \geq i_0$ and $j \geq j_0$, it suffices to prove:
   (i) (*Basis*). $P(i_0, j_0)$ is true.
   (ii) (*Induction step*). For all $k \geq i_0$ and $l \geq j_0$, the assumption that $P(k, l)$ is true (called the *induction hypothesis*) implies that $P(k + 1, l)$ and $P(k, l + 1)$ are true.

3-9. EXAMPLES OF PROOF BY INDUCTION

In this section we shall reinforce our discussion of finite induction with a number of examples.

Example 3-11

THEOREM

For all integers $i \geq 1$,

$$\textstyle\sum_{m=1}^{i} m = \frac{1}{2}i(i + 1)$$

Proof (by induction on i) (*Basis*). For $i = 1$, the theorem is true, because it reduces to

$$\textstyle\sum_{m=1}^{1} m = \frac{1}{2}\cdot 1 \cdot 2 = 1$$

(*Induction step*). The induction hypothesis is that, for $k \geq 1$,

$$\textstyle\sum_{m=1}^{k} m = \frac{1}{2}k(k + 1)$$

Hence: $\textstyle\sum_{m=1}^{k+1} m = (\sum_{m=1}^{k} m) + (k + 1)$

$$= [\tfrac{1}{2}k(k + 1)] + (k + 1) = \tfrac{1}{2}(k + 1)(k + 2)$$

which proves the theorem for $i = k + 1$. □

Example 3-12

THEOREM

Let r be a positive integer, and let $a_0, a_1, \ldots, a_j$ be nonnegative integers less than r. Then, for all $j \geq 0$,

$$a_0 + a_1 r + a_2 r^2 + \ldots + a_j r^j < r^{j+1}$$

Proof (by induction on j) (*Basis*). For $j = 0$, the theorem is true since $a_0 < r$. (*Induction step*). The induction hypothesis is that, for $k \geq 0$,

$$a_0 + a_1 r + a_2 r^2 + \ldots + a_k r^k < r^{k+1}$$

Hence,

$$\begin{aligned} &a_0 + a_1 r + a_2 r^2 + \ldots + a_{k+1} r^{k+1} \\ &= (a_0 + a_1 r + a_2 r^2 + \ldots + a_k r^k) + a_{k+1} r^{k+1} \\ &< r^{k+1} + a_{k+1} r^{k+1} = (1 + a_{k+1}) r^{k+1} \\ &\leq r \cdot r^{k+1} = r^{k+2} \end{aligned}$$

which proves the theorem for $j = k + 1$. □

Example 3-13

THEOREM

For all integers $r \geq 1$,

$$(A_1 \cup A_2 \cup \ldots \cup A_r)' = A_1' \cap A_2' \cap \ldots \cap A_r'$$
$$(A_1 \cap A_2 \cap \ldots \cap A_r)' = A_1' \cup A_2' \cup \ldots \cup A_r'$$

Proof (by induction on r) (*Basis*). For $r = 1$, the theorem is trivially true. (*Induction step*). The induction hypothesis is that, for $k \geq 1$,

$$(A_1 \cup A_2 \cup \ldots \cup A_k)' = A_1' \cap A_2' \cap \ldots \cap A_k'$$
$$(A_1 \cap A_2 \cap \ldots \cap A_k)' = A_1' \cup A_2' \cup \ldots \cup A_k'$$

Hence, using De Morgan's laws cited in Table 1-4, we have

$$\begin{aligned} &(A_1 \cup A_2 \cup \ldots \cup A_{k+1})' = [(A_1 \cup A_2 \cup \ldots \cup A_k) \cup A_{k+1}]' \\ &= (A_1 \cup A_2 \cup \ldots \cup A_k)' \cap A_{k+1}' \\ &= (A_1' \cap A_2' \cap \ldots \cap A_k') \cap A_{k+1}' \\ &= A_1' \cap A_2' \cap \ldots \cap A_{k+1}' \end{aligned}$$

$$(A_1 \cap A_2 \cap \ldots \cap A_{k+1})' = [(A_1 \cap A_2 \cap \ldots \cap A_k) \cap A_{k+1}]'$$
$$= (A_1 \cap A_2 \cap \ldots \cap A_k)' \cup A'_{k+1}$$
$$= (A'_1 \cup A'_2 \cup \ldots \cup A'_k) \cup A'_{k+1}$$
$$= A'_1 \cup A'_2 \cup \ldots \cup A'_{k+1}$$

which proves the theorem for $r = k + 1$. □

The following illustrates the use of the second principle of finite induction.

Example 3-14

THEOREM

Every positive integer $n \geq 2$ can be written as a product of prime numbers (called a *prime factorization* of n).

Proof (by induction on n) (*Basis*) For $n = 2$, the theorem is true, since 2 is a prime number. (*Induction step*). The induction hypothesis is that, for $k \geq 2$, the integers $2, 3, 4, \ldots, k$ can be written as products of prime numbers. If $k + 1$ is a prime number, the proof is complete. Otherwise, $k + 1$ is divisible by some integer other than 1 or $k + 1$, say, i. Hence, $k + 1 = ij$, where $2 \leq i \leq k$ and $2 \leq j \leq k$. By induction hypothesis, both i and j can be written as products of prime numbers; hence, ij can be written as a product of prime numbers. This proves the theorem for $n = k + 1$. □

PROBLEMS

1. Prove the following statements, by induction on n:
   (a) $\sum_{i=1}^{n} (2i - 1) = n^2 \quad (n \geq 1)$
   (b) $\sum_{i=1}^{n} i^2 = \frac{1}{6}n(n + 1)(2n + 1) \quad (n \geq 1)$
   (c) $\sum_{i=1}^{n} i^3 = [\frac{1}{2}n(n + 1)]^2 \quad (n \geq 1)$
   (d) $\sum_{i=1}^{n} i(i!) = (n + 1)! - 1 \quad (n \geq 1)$
   (e) The sum of the digits of any positive integer $9n$ is divisible by 9 $(n \geq 1)$.
   (f) If f is an idempotent function, then $f^n = f \quad (n \geq 1)$.
   (g) If A is a denumerable set, so is $A^n \quad (n \geq 1)$. (See Problem 1, Sec. 3-6)
   (h) For any a such that $0 < a < 1$, $(1 - a)^n \geq 1 - na \; (n \geq 1)$.
   (i) $2^n > n^3 \quad (n \geq 10)$

   $\left[\textit{Hint}: \; (n + 1)^3 = \left(1 + \frac{1}{n}\right)^3 n^3\right]$.

2. Use the principle of double induction (Problem 3, Sec. 3-8) to prove:

$$2^{mn} > m^n \qquad (m \geq 1, n \geq 1)$$

3. Consider the function $f: \mathbb{I} \longrightarrow \mathbb{I}$, where $f(i) = 2i + 1$. Obtain an expression for f^r (in terms of r and i), and prove (by induction on r) the validity of this expression.

4. Consider the following game: There are n upright pins; players A and B take turns (with A playing first) knocking down any number of pins from 1 to m (where m is fixed and less than n); the player who knocks down the last pin is the winner. For example, with $n = 9$ and $m = 3$ we may have the following sequence of events: A knocks down 2 pins, B knocks down 3, A knocks down 1, B knocks down 1, A knocks down 2 and wins. Now, prove the following theorem: Player A, by a suitable strategy, can always guarantee victory to himself if and only if n is not a multiple of $m + 1$. [*Hint*: Let $n = i(m + 1) + r$, where $0 \leq r \leq m$, and prove the theorem by induction on i].

5. Prove by induction that, in a group of any size, the total number of people who shake the hands of an odd number of other people is even.

3-10. RECURSIVE DEFINITIONS

In this section we see how the principles of finite induction can be used for "defining"—that is, for computing or constructing—various entities whose values or structures are functions of integer arguments.

From the second principle of finite induction we can immediately conclude:

THEOREM 3-9

Let $E(i)$ be an entity which, for each integer i, may be either defined or undefined. Then $E(i)$ is defined for all integers $i \geq i_0$ if:

(i) (*Basis*). $E(i_0)$ is defined.

(ii) (*Induction step*). For every $k \geq i_0$, the assumption that $E(i)$ is defined for $i = i_0, i_0 + 1, i_0 + 2, \ldots, k$ implies that $E(k + 1)$ is defined.

Theorem 3-9 suggests the following method for defining an entity $E(i)$ for all integer $i \geq i_0$:

(i) (*Basis*). Define $E(i_0)$.

(ii) (*Induction step*). For every $i \geq i_0$, define $E(i + 1)$ in terms of $E(i_0)$, $E(i_0 + 1), E(i_0 + 2), \ldots, E(i)$.

A definition expressed in this manner is called a *recursive definition*; $E(i)$ is said to be *recursively defined.*

Example 3-15

The factorial function $i!$ $(i \geq 0)$ is defined recursively as

(*Basis*). $0! = 1$

(*Induction step*). $(i + 1)! = (i + 1)i! \qquad (i = 0, 1, 2, \ldots)$ □

Example 3-16

The *Fibonacci numbers* F_i $(i \geq 0)$ are defined recursively by

(*Basis*). $F_0 = 0,\ F_1 = 1$

(*Induction step*). $F_{i+1} = F_{i-1} + F_i \qquad (i = 1, 2, 3, \ldots)$

For example, the first ten Fibonacci numbers are 0, 1, 1, 2, 3, 5, 8, 13, 21, 34. □

Example 3-17

The set P_n of all permutations of order $n \geq 1$ on the set $\{a_1, a_2, \ldots, a_n\}$ (see Sec. 3-5) is defined recursively by

(*Basis*). $P_1 = \left\{\begin{pmatrix} a_1 \\ a_1 \end{pmatrix}\right\}$

(*Induction step*). For each element

$$\begin{pmatrix} a_1 & a_2 & \ldots & a_{i-1} \\ a'_1 & a'_2 & \ldots & a'_{i-1} \end{pmatrix}$$

in P_{i-1} $(i = 2, 3, \ldots, n)$, include in P_i the i elements

$$\begin{pmatrix} a_1 & a_2 & \ldots & a_{i-1} & a_i \\ a_i & a'_1 & \ldots & a'_{i-2} & a'_{i-1} \end{pmatrix}, \begin{pmatrix} a_1 & a_2 & \ldots & a_{i-1} & a_i \\ a'_1 & a_i & \ldots & a'_{i-2} & a'_{i-1} \end{pmatrix}, \ldots,$$

$$\begin{pmatrix} a_1 & a_2 & \ldots & a_{i-1} & a_i \\ a'_1 & a'_2 & \ldots & a_i & a'_{i-1} \end{pmatrix}, \begin{pmatrix} a_1 & a_2 & \ldots & a_{i-1} & a_i \\ a'_1 & a'_2 & \ldots & a'_{i-1} & a_i \end{pmatrix}$$

For example, P_3 can be constructed as

$$P_1 = \left\{\begin{pmatrix} a_1 \\ a_1 \end{pmatrix}\right\}$$

$$P_2 = \left\{\begin{pmatrix} a_1 & a_2 \\ a_2 & a_1 \end{pmatrix}, \begin{pmatrix} a_1 & a_2 \\ a_1 & a_2 \end{pmatrix}\right\}$$

$$P_3 = \left\{\begin{pmatrix} a_1 & a_2 & a_3 \\ a_3 & a_2 & a_1 \end{pmatrix}, \begin{pmatrix} a_1 & a_2 & a_3 \\ a_2 & a_3 & a_1 \end{pmatrix}, \begin{pmatrix} a_1 & a_2 & a_3 \\ a_2 & a_1 & a_3 \end{pmatrix},\right.$$
$$\left.\begin{pmatrix} a_1 & a_2 & a_3 \\ a_3 & a_1 & a_2 \end{pmatrix}, \begin{pmatrix} a_1 & a_2 & a_3 \\ a_1 & a_3 & a_2 \end{pmatrix}, \begin{pmatrix} a_1 & a_2 & a_3 \\ a_1 & a_2 & a_3 \end{pmatrix}\right\}$$

□

Example 3-18

A *rooted tree* of *height* l $(l \geq 0)$ is a relation graph (see Sec. 2-2) whose set of vertices is partitionable recursively into l subsets, denoted by $V_0, V_1, \ldots, V_l$, in this manner:

(*Basis*). V_0 consists of a single vertex, called the *root* of the rooted tree.

(*Induction step*). V_{i+1} ($i = 0, 1, \ldots, l - 1$) is a finite nonempty set of vertices such that each vertex in V_{i+1} is pointed to by an edge originating from exactly one vertex in V_i.

From this definition it follows that every vertex in a rooted tree is reachable from the root via some path of length at most l, and that a tree has no loops. Figure 3-6 shows a rooted tree of height 4. (Since arrows associated with edges of a rooted tree always point *away* from the root, they are normally omitted from the diagram.) □

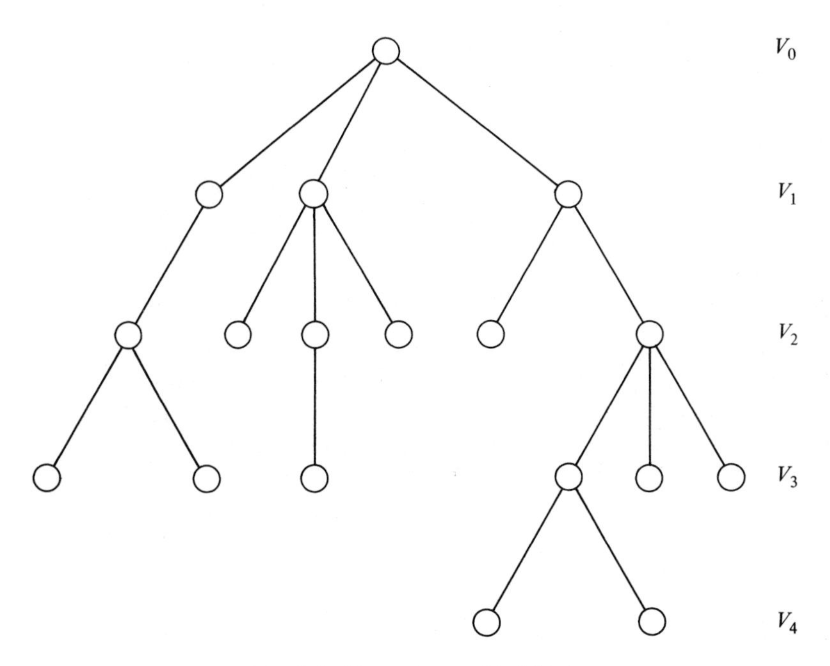

Fig. 3-6 Rooted tree of height 4

Example 3-19

A number of discs stacked in decreasing size on a spindle A are to be moved to spindle C (stacked in the original order), using, if necessary, a spindle B for temporary storage (see Fig. 3-7). In the moving process, these two rules should be obeyed: (a) Only one disc may be moved at a time (from any spindle to any other spindle). (b) At no time may a disc rest on top of a smaller one.

The procedure $P(i)$ which moves $i \geq 1$ discs from spindle A to spindle C can be defined recursively as follows:

(*Basis*). $P(1)$ consists of moving the single disc from A directly to C.

(*Induction step*). $P(i + 1)$ consists of the following three steps: (a) Using $P(i)$ (with A, C, and B as initial, temporary, and final spindles, respectively),

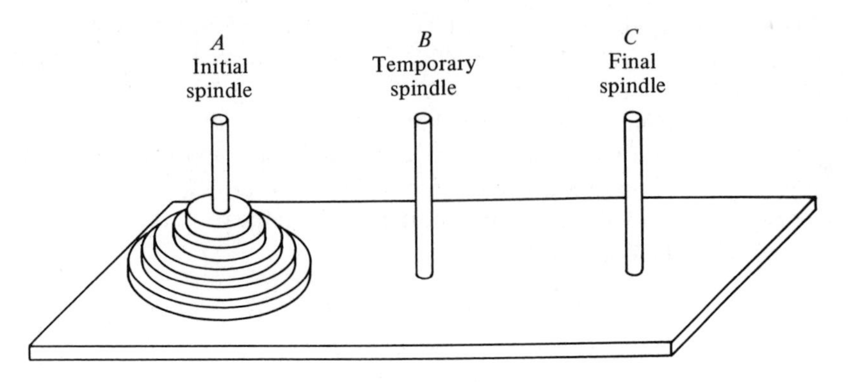

Fig. 3-7 For Example 3-19

move the topmost i discs of A from A to B. (b) Move the (new) topmost disc of A from A to C. (c) Using $P(i)$ (with B, A, and C as initial, temporary, and final spindles, respectively), move the i discs of B from B to C.

Figure 3-8 indicates how $P(5)$, for example, is executed. The discs, from top to bottom, are denoted by the integers 1, 2, 3, 4, 5. In the figure, the expression of the form $\{1, 2, \ldots, j\}\alpha \xrightarrow{\beta} \gamma$ stands for the step "move the set of discs $\{1, 2, \ldots, j\}$ from spindle α to spindle γ, using spindle β for temporary storage" (with the symbol β deleted when no temporary storage is needed). When $j > 1$, each such move is carried out by performing the three substeps which constitute the induction step of $P(i)$:

$$\{1, 2, \ldots, j-1\}\,\alpha \xrightarrow{\gamma} \beta$$
$$\{j\}\,\alpha \longrightarrow \gamma$$
$$\{1, 2, \ldots, j-1\}\beta \xrightarrow{\alpha} \gamma$$

When $j = 1$, the move is carried out simply by $\{j\}\alpha \longrightarrow \gamma$, which constitutes the basis of $P(i)$. From the figure it follows that $P(5)$ consists of this sequence of steps:

$$\begin{aligned}
&\{1\}A \longrightarrow C,\ \{2\}A \longrightarrow B,\ \{1\}C \longrightarrow B,\ \{3\}A \longrightarrow C,\ \{1\}B \longrightarrow A,\\
&\{2\}B \longrightarrow C,\ \{1\}A \longrightarrow C,\ \{4\}A \longrightarrow B,\ \{1\}C \longrightarrow B,\ \{2\}C \longrightarrow A,\\
&\{1\}B \longrightarrow A,\ \{3\}C \longrightarrow B,\ \{1\}A \longrightarrow C,\ \{2\}A \longrightarrow B,\ \{1\}C \longrightarrow B,\\
&\{5\}A \longrightarrow C,\ \{1\}B \longrightarrow A,\ \{2\}B \longrightarrow C,\ \{1\}A \longrightarrow C,\ \{3\}B \longrightarrow A,\\
&\{1\}C \longrightarrow B,\ \{2\}C \longrightarrow A,\ \{1\}B \longrightarrow A,\ \{4\}B \longrightarrow C,\ \{1\}A \longrightarrow C,\\
&\{2\}A \longrightarrow B,\ \{1\}C \longrightarrow B,\ \{3\}A \longrightarrow C,\ \{1\}B \longrightarrow A,\ \{2\}B \longrightarrow C,\\
&\{1\}A \longrightarrow C
\end{aligned}$$

□

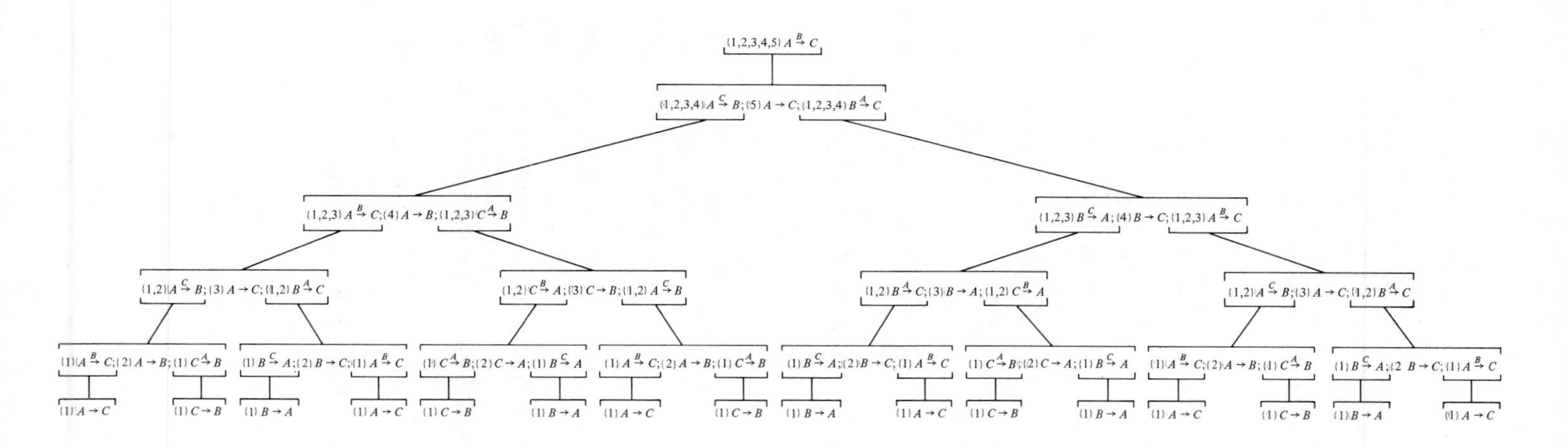

Gill 3-8

Fig. 3-8 Execution of $P(5)$

Recursive definitions are often used in defining one-dimensional, two-dimensional, or generally multi-dimensional patterns of symbols. In these definitions the induction variable is the "size" (or the so-called *shape*) of the pattern. The basis consists of the definition of the individual symbols, and the induction step defines any pattern that is not a symbol in terms of "smaller" patterns. The following example illustrates such a definition.

Example 3-20

A set generated by $A_1, A_2, \ldots, A_r$ (which are subsets of a universal set U) is defined recursively as

(*Basis*). $\varnothing, U, A_1, A_2, \ldots, A_r$ are sets generated by $A_1, A_2, \ldots, A_r$.

(*Induction step*). If S and T are sets generated by $A_1, A_2, \ldots, A_r$, so are: (a) (S), (b) S', (c) $S \cup T$, and (d) $S \cap T$.

The preceding definition can be used to decide whether or not any expression consisting of the symbols $\varnothing, U, A_1, A_2, \ldots, A_r, ', \cup, \cap,$ (, and) is indeed a set generated by $A_1, A_2, \ldots, A_r$. For example, the expression

$$((A \cup B') \cap C)' \cup A'$$

is a set generated by $A, B, C,$ since each one of the following is such a set:

A	by Basis
B	by Basis
B'	by Induction Step, (b)
$A \cup B'$	by Induction Step, (c)
$(A \cup B')$	by Induction Step, (a)
C	by Basis
$(A \cup B') \cap C$	by Induction Step, (d)
$((A \cup B') \cap C)$	by Induction Step, (a)
$((A \cup B') \cap C)'$	by Induction Step, (b)
A'	by Induction Step, (b)
$((A \cup B') \cap C)' \cup A'$	by Induction Step, (c)

Note that, by the above definition, an expression such as $A \cup B \cap C \cap B' \cup C'$ is also a set generated by A, B, C. In the absence of parentheses, it is understood that $'$ has precedence over $\cup$ and $\cap$, and $\cap$ has precedence over $\cup$. □

PROBLEMS

1. Define the following entities $E(i)$ recursively:

(a) $E(i) = \binom{n}{i} \qquad (i \geq 0)$

(b) $E(i)$ = ith smallest prime number [$E(1) = 2$, $E(2) = 3$, $E(3) = 5$, $E(4) = 7$, $E(5) = 11$, etc.]
(c) $E(i)$ = number of distinct partitions of i elements $(i \geq 1)$
(d) $E(i)$ = number of distinct graphs of a relation $A = \{a_1, a_2, \ldots, a_i\}$ $(i \geq 1)$
(e) $E(i)$ = a procedure which rearranges the elements of an i-tuple $(k_1, k_2, \ldots, k_i)$ (where all elements are positive integers) in order of ascending magnitude $(i \geq 1)$. [For example, $E(5)$ transforms (3, 2, 4, 1, 4) into (1, 2, 3, 4, 4)]

2. How many steps are needed to execute the procedure $P(i)$ of Example 3-19?

3. *Ackerman's function* $A(m, n)$ $(m \geq 0, n \geq 0)$ can be defined recursively (with respect to both its integer arguments) in this manner:

$$A(0, n) = n + 1 \qquad (n \geq 0)$$
$$A(m, 0) = A(m - 1, 1) \qquad (m > 0)$$
$$A(m, n) = A(m - 1, A(m, n - 1)) \qquad (m > 0, n > 0)$$

Compute $A(2, 3)$.

4. In a computer program, a *recursive function* (or *recursive procedure*) is one which "calls" itself. In ALGOL programs, recursive functions are permitted. For example, the function FACTORIAL(i), which computes $i!$ for any integer $i \geq 0$, can be written in ALGOL as

```
integer procedure FACTORIAL(i); integer i;
FACTORIAL: = if i = 0 then 1 else i*FACTORIAL(i - 1);
```

Another example: The function FIBONACCI(i) which computes F_i for any integer $i \geq 0$, can be written in ALGOL as

```
integer procedure FIBONACCI(i); integer i;
FIBONACCI: = if i ≤ 1 then 1 else FIBONACCI(i - 1)
                                  + FIBONACCI(i - 2);
```

In FORTRAN, recursive functions are illegal. Write FORTRAN functions which compute $i!$ and F_i (for any integer $i \geq 0$) *nonrecursively.*

5. Define recursively an arithmetic expression composable of the symbols X, +, −, *, /, **, (, and). (You may assume that the expressions are subject to the conventional FORTRAN precedence rules with respect to arithmetic operations.)

3-11. SOME PROPERTIES OF INTEGERS

We conclude this chapter by deriving some useful results relating to integers.

THEOREM 3-10 (*The division theorem*)

For any integers $n \geq 0$ and $m > 0$ there exist unique integers $q \geq 0$ and

r, such that

$$n = mq + r \qquad \text{where } 0 \leq r < m$$

Proof Let J be the set of nonnegative integers obtainable by subtracting multiples of m from n. That is,

$$J = \{j \mid n = mv + j, j \geq 0, v \geq 0\}$$

Since J is nonempty (it must contain at least n), it has a least integer, say, r. Thus, for some $r \geq 0$ and $q \geq 0$, we have $n = mq + r$. Now, hypothesize that $r \geq m$. Then, for some a such that $0 \leq a < r$, we have $r = m + a$. Hence,

$$n = mq + r = mq + m + a = m(q + 1) + a$$

which implies that $a \in J$. But this contradicts the fact that $r > a$ is the least integer in J. Thus, we must have $r < m$.

To prove uniqueness, suppose we can write

$$n = qm + r = q'm + r' \qquad \text{where } 0 \leq r < m, 0 \leq r' < m$$

Hence, $(q - q')m = r' - r$. Without loss in generality we can assume that $q \geq q'$. If $q > q'$, $q - q' > 0$ and $r' - r \geq m$. This implies that $r' \geq m$, a contradiction. Hence, $q = q'$ and $r = r'$. □

The quantities q and r are the familiar *quotient* and *remainder*, respectively, obtained upon division of n by m.

Given nonnegative integers n and m, m is said to *divide* n (or be a *divisor* or a *factor* of n) if, for some integer q, $n = qm$. In this case we write $m \mid n$, and say that n is a *multiple* of m.

If m is a divisor of every integer in some nonempty set $N = \{n_1, n_2, \ldots, n_k\}$ of positive integers, it is said to be a *common divisor* of N. Since, for every such set there is at least one common divisor (namely, 1), N must have a *greatest common divisor*, which will be denoted by $\gcd(n_1, n_2, \ldots, n_k)$. If any of the n_i is 1, then, clearly, $\gcd(n_1, n_2, \ldots, n_k) = 1$. Otherwise, each n_i can be expressed in its prime factorization form (see Example 3-14); $\gcd(n_1, n_2, \ldots, n_k)$ is then a product of powers of prime numbers such that each factor in the product is the largest power of a prime number which appears in every one of the k factorizations. For example, if $N = \{n_1, n_2, n_3\}$, where

$$n_1 = 180 = 2^2 \cdot 3^2 \cdot 5$$
$$n_2 = 200 = 2^3 \cdot 5^2$$
$$n_3 = 600 = 2^3 \cdot 3 \cdot 5^2$$

then $$\gcd(n_1, n_2, n_3) = 2^2 \cdot 5 = 20$$

The computation of gcd (n_1, n_2) for any positive integers n_1 and n_2 whose prime factorization is not readily available, is facilitated by the so-called *Euclidean algorithm*. It consists of forming, with the aid of the division theorem, this series of identities:

$$\begin{aligned}
n_1 &= q_1 n_2 + n_3 && (0 \leq n_3 < n_2) \\
n_2 &= q_2 n_3 + n_4 && (0 \leq n_4 < n_3) \\
n_3 &= q_3 n_4 + n_5 && (0 \leq n_5 < n_4) \\
&\vdots \\
n_{h-2} &= q_{h-2} n_{h-1} + n_h && (0 \leq n_h < n_{h-1}) \\
n_{h-1} &= q_{h-1} n_h + 0
\end{aligned} \tag{3-8}$$

Since $n_i < n_{i-1}$ for $i = 3, 4, 5, \ldots$, it is guaranteed that eventualy a zero remainder will be obtained, as indicated in the last identity. Now, any divisor of n_2 and n_3 is also a divisor of n_1, and any divisor of n_1 and n_2 is also a divisor of n_3. Thus, the common divisors of $\{n_1, n_2\}$ are the same as those of $\{n_2, n_3\}$; hence, gcd $(n_1, n_2) =$ gcd (n_2, n_3). Proceeding down the series of identities (3-8), we have by the same argument:

$$\begin{aligned}
\gcd(n_1, n_2) = \gcd(n_2, n_3) = \ldots &= \gcd(n_{h-2}, n_{h-1}) \\
&= \gcd(n_{h-1}, n_h) = n_h
\end{aligned}$$

In conclusion, gcd (n_1, n_2) equals the last nonzero remainder in the series (3-8).

Example 3-21

The series of identities (3-8) for $n_1 = 6099$ and $n_2 = 2166$ is given by

$$\begin{aligned}
6099 &= 2 \cdot 2166 + 1767 \\
2166 &= 1 \cdot 1767 + 399 \\
1767 &= 4 \cdot 399 + 171 \\
399 &= 2 \cdot 171 + 57 \\
171 &= 3 \cdot 57 + 0
\end{aligned}$$

Hence, $$\gcd(6099, 2166) = 57$$

A convenient way of computing the preceding quotients and remainders is the "divide and invert" method illustrated here:

```
              2
     2166 | 6099
            4332     1
            ----
            1767 | 2166
                   1767     4
                   ----
                    399 | 1767
                          1596    2
                          ----
                           171 | 399
                                 342    3
                                 ---
                                  57 | 171
                                       171
                                       ---
                                         0
```

□

THEOREM 3-11

Let m and n be positive integers, with $\gcd(m, n) = d$. Then there exist integers a and b such that

$$d = am + bn$$

Proof Let J be the set of all positive integers j expressible in the form $j = am + bn$. Since J is nonempty (it must contain m and n), it has a least element, say, k. Thus, if $m = t_1 d$ and $n = t_2 d$, we have

$$k = am + bn = at_1 d + bt_2 d = d(at_1 + bt_2)$$

which implies that $d \leq k$. By the division theorem, we can write $m = kq + r$, where $r < k$; hence,

$$r = m - kq = m - (am + bn)q = (1 - a)m + (-bq)n$$

which implies either $r = 0$ or $r \in J$. Since $r < k$ and k is the least element in J, we must have $r = 0$; hence, k is a divisor of m. Analogously it can be shown that k is a divisor of n; hence, k is a common divisor of m and n. Thus, $k \geq d$. Since we have already shown that $d \leq k$, we can conclude that $d = k$ and, hence, $d = am + bn$. □

Example 3-22

We illustrate the computation of the constants a and b of Theorem 3-11 with $m = 6099$ and $n = 2166$. By Example 3-21, we can write

$$\begin{aligned} 57 &= 399 - 2 \cdot 171 \\ &= 399 - 2 \cdot (1767 - 4 \cdot 399) = -2 \cdot 1767 + 9 \cdot 399 \end{aligned}$$

$$= -2 \cdot 1767 + 9 \cdot (2166 - 1 \cdot 1767) = 9 \cdot 2166 - 11 \cdot 1767$$
$$= 9 \cdot 2166 - 11 \cdot (6099 - 2 \cdot 2166) = -11 \cdot 6099 + 31 \cdot 2166$$

Hence, $\gcd(6099, 2166) = -11 \cdot 6099 + 31 \cdot 2166$ □

The positive integers $n_1, n_2, \ldots, n_k$ $(k \geq 2)$ are said to be *relatively prime* if their only common divisor is 1, that is, if $\gcd(n_1, n_2, \ldots, n_k) = 1$. Thus, if the positive integers m and n are relatively prime, there always exist integers a and b such that $am + bn = 1$. The integers $n_1, n_2, \ldots, n_k$ are said to be *relatively prime in pairs* if, for all i and $j \neq i$, n_i and n_j are relatively prime. For example, the integers 6, 9, and 11 are relatively prime, but not relatively prime in pairs. On the other hand, integers 5, 9, and 11 are relatively prime in pairs.

THEOREM 3-12

Let p be a prime number which divides mn $(m > 0, n > 0)$. Then p must either divide m or divide n.

Proof Hypothesizing that p does not divide m, we have $\gcd(p, m) = 1$; hence, $1 = ap + bm$. Consequently we can write, with some integer q,

$$n = apn + bmn = apn + bqp = p(an + bq)$$

which implies that p divides n. Thus, p must either divide m or divide n. □

THEOREM 3-13 (*The prime factorization theorem*)

Every integer $n \geq 2$ equals a product of prime numbers which is unique except for the ordering of the factors.

Proof (by induction on n) (*Basis*). For $n = 2$, the theorem is true since 2 is a prime number. (*Induction step*). Suppose n has two prime factorizations:

$$n = p_1 p_2 \ldots p_r = q_1 q_2 \ldots q_s$$

where the p_i and q_i are prime. By Theorem 3-12, since p_1 divides $q_1 q_2 \ldots q_s$, it must divide one of the q_i, say, q_1. Since q_1 is prime, $p_1 = q_1$ and we can write

$$n' = \frac{n}{p_1} = p_2 p_3 \ldots p_r = q_2 q_3 \ldots q_s \tag{3-9}$$

If the induction hypothesis (using the second principle of finite induction) is that the theorem holds for $n' = 2, 3, \ldots, n - 1$, then n' of (3-9) has unique prime factorization. That is, every p_i in (3-9) equals some q_j, and every q_j equals some p_i. Hence, n must have a unique prime factorization also. □

If m is a multiple of every integer in some nonempty set $N = \{n_1, n_2, \ldots, n_k\}$ of positive integers, it is said to be a *common multiple* of N. Since, for every such set there is at least one common multiple (namely, $n_1 n_2 \ldots n_k$), N must have a *least common multiple*, which is denoted by lcm $(n_1, n_2, \ldots, n_k)$. If each $n_i \geq 2$ is expressed in its prime factorization form, then lcm $(n_1, n_2, \ldots, n_k)$ is the product of powers of prime numbers such that each factor in the product is the greatest power of a prime number which appears in at least one of the k factorizations. For example, if $N = \{n_1, n_2, n_3\}$, where

$$n_1 = 180 = 2^2 \cdot 3^2 \cdot 5$$

$$n_1 = 200 = 2^3 \cdot 5^2$$

$$n_3 = 600 = 2^3 \cdot 3 \cdot 5^2$$

then

$$\text{lcm}\,(n_1, n_2, n_3) = 2^3 \cdot 3^2 \cdot 5^2 = 1800$$

THEOREM 3-14

Given any integer $n \geq 1$ and $b \geq 2$, then n can be expressed uniquely in the form

$$n = a_0 + a_1 b + a_2 b^2 + \ldots + a_j b^j \tag{3-10}$$

where $0 \leq a_i < b$ $(i = 1, 2, \ldots, j)$ and $a_j \neq 0$.

Proof (by induction on n) (*Basis*). When $n = 1$, we can write n in form (3-10) by setting $j = 0$ and $a_0 = 1$. Since $b \geq 2$, no other way of writing n in form (3-10) (with some $j > 0$) is possible. (*Induction step*). Hypothesize that the theorem is true for $n = 1, 2, \ldots, i$. By the division theorem, we can write uniquely,

$$i + 1 = bq + r \qquad \text{where } 0 \leq r < b$$

Since $b \geq 2$, we have $q < i + 1$ and, by induction hypothesis, we can write uniquely,

$$q = a'_0 + a'_1 b + a'_2 b^2 + \ldots + a'_j b^j$$

where $0 \leq a'_i < b$ $(i = 1, 2, \ldots, j)$ and $a_j \neq 0$. Then, we can write uniquely,

$$\begin{aligned} i + 1 &= r + a'_0 b + a'_1 b^2 + \ldots + a'_j b^{j+1} \\ &= a_0 + a_1 b + a_2 b^2 + \ldots + a_{j+1} b^{j+1} \end{aligned}$$

where $0 \leq a_i < b$ $(i = 1, 2, \ldots, j + 1)$ and $a_{j+1} \neq 0$. □

The integer b in (3-10) is referred to as the *base* (or *radix*) of the expression. When the radix is known in advance (say, $b = 10$ in the decimal system, or $b = 2$ in the binary system), the form (3-10) is abbreviated to the usual form $a_j a_{j-1} \ldots a_1 a_0$ in which positive integers are represented.

Example 3-23

To express the decimal number 427 in the base-8 (octal) system, we must find the a_i in the expression

$$427 = a_0 + a_1 8 + a_2 8^2 + \ldots + a_j 8^j$$

Noting that $$427 = a_0 + 8(a_1 + a_2 8 + \ldots + a_j 8^{j-1})$$

we see that a_0 is the remainder of 427 upon division by 8:

$$427 = 53 \cdot 8 + 3$$

Hence, $a_0 = 3$. Now, we can write

$$\begin{aligned} 53 &= a_1 + a_2 8 + \ldots + a_j 8^{j-1} \\ &= a_1 + 8(a_2 + a_3 8 + \ldots + a_j 8^{j-2}) \end{aligned}$$

Thus, a_1 is the remainder of 53 upon division by 8:

$$53 = 6 \cdot 8 + 5$$

Hence, $a_1 = 5$. Finally, we can write

$$6 = a_2 + a_3 8 + \ldots + a_j 8^{j-2}$$

which implies $a_2 = 6$. In conclusion, the decimal number 427 is the octal number 653. (Generalization of this method of conversion from one base to another is left to the reader.) □

PROBLEMS

1. For the following values of m and n, compute $\gcd(m, n)$ and express it in the form $am + bn$:
   (a) $m = 14, n = 35$ (b) $m = 180, n = 252$
   (c) $m = 4148, n = 7684$
2. Produce the prime factorization forms of 144 and 162. Use these forms to compute gcd(144, 162) and lcm (144, 162).
3. If i, j, and k are positive integers, prove:
   (a) $i \mid j$ and $i \mid k$ imply $i \mid (j + k)$
   (b) $\gcd(ij, ik) = i \cdot \gcd(j, k)$
   (c) $\gcd(\gcd(i, j), k) = \gcd(\gcd(i, k), j) = \gcd(\gcd(j, k), i)$

(d) $\gcd(i, j) = 1$ and $i \mid jk$ imply $i \mid k$
(e) $\gcd(i, j) = \gcd(i, k) = 1$ implies $\gcd(i, jk) = 1$
(f) $ij = \gcd(i, j) \cdot \text{lcm}(i, j)$
(g) $ij = k^2$ (for some k) and $\gcd(i, j) = 1$ imply $i = k_1^2$ and $j = k_2^2$ (for some k_1, k_2).

4. Prove that the number of prime numbers is infinite. (*Hint*: Show that if $p_1, p_2, \ldots, p_r$ are prime numbers, then $1 + p_1 p_2 \ldots p_r$ is not divisible by any of the p_i).

5. Express the decimal numbers 374 and 3046 in binary and octal forms.

6. What are the decimal values of the binary number 100101 and the ternary (base-3) number 100101? Convert these numbers to octal numbers.

7. Using the method illustrated in Example 3-23, write a recursive procedure for converting a base-b_1 number into a base-b_2 number.

8. Consider the binary number

$$\beta_{k3}\beta_{k2}\beta_{k1} \ldots \beta_{33}\beta_{32}\beta_{31}\beta_{23}\beta_{22}\beta_{21}\beta_{13}\beta_{12}\beta_{11}$$

(where the digits, with leading zeros if necessary, are separated into groups of three). Show that the octal representation of this number is given by

$$\gamma_k \ldots \gamma_3\gamma_2\gamma_1$$

where γ_i denotes the octal value of the binary number $\beta_{i1}\beta_{i2}\beta_{i3}$. (For example, the octal representation of the binary number 011010110 is 326.)

9. At a drugstore, a pharmacist is instructed to place pills in 10 vials in this manner: If a certain number of pills from 1 to 1023 were requested, the pharmacist could quickly pull out one or more vials and deliver the exact amount without opening them.
(a) How many pills should be placed in each of the 10 vials?
(b) What is the maximum number n of pills that can be placed in 10 vials, so that it is still possible to deliver any amount 1 to n of pills without opening the vials?

REFERENCES

KEMENY, J. G., H. MIRKIL, J. L. SNELL, and G. L. THOMPSON, *Finite Mathematical Structures*. Englewood Cliffs, N.J: Prentice-Hall, 1959.

MACLANE, S., and G. BIRKHOFF, *Algebra*. New York: Macmillan, 1967.

VAN DER WAERDEN, B. L., *Modern Algebra*, Ungar, 1931.

4 ALGEBRAIC SYSTEMS

In this chapter we introduce the concept of an *algebraic system*—a mathematical structure consisting of *relations, operations, postulates, theorems, definitions,* and *algorithms.* After enumerating some typical postulates (*commutativity, associativity, identity, inverse, distributivity*), the *integral domain* is studied as an example of an algebraic system. Also discussed is the algebra of sets (whose properties are already familiar to the reader from Chapter 1); the principle of *duality* is stated and illustrated with respect to this system. A *homomorphism* from one algebraic system into another is defined (with *isomorphism* being an important special case), and the relation between homomorphisms and *congruence relations* is studied. The chapter concludes with a discussion of the *direct product* of algebraic systems.

The notion of an algebraic system is indispensable in almost every theoretical area of computer science. The algebra of sets postulated in this chapter is important in switching theory, while the algebra of propositions and the algebraic systems known as groups, rings, and fields (introduced in later chapters) are essential in areas such as program verification, the design of arithmetic and logical units, and error detection and correction. The concepts of homomorphism and direct product are important in the theoretical study as well as the practical design of finite-state machines.

4-1. OPERATIONS

Given a set A, an *m-ary operation over* A is a function from A^m into A. In this context, m is called the *order* of the operation. The most commonly encountered operations are of order 1 (*unary* operations) and of order 2 (*binary* operations). It is customary to designate unary and binary operations with special symbols; in the unary case, the symbol is placed before, above, or after $a_i \in A$ to denote the image of a_i under the operation; in the binary case, the symbol is usually inserted between $a_i \in A$ and $a_j \in A$ to denote the image of (a_i, a_j) under the operation.[1]

For example, the unary operation of *negation* (denoted by $-$) can be defined over the set of integers $\mathbb{I}$, with $-i$ denoting the image of $i \in \mathbb{I}$ under negation. The unary operation of *complementation* (denoted by $'$) can be defined over a power set $\mathbf{2}^U$, with S' denoting the image of $S \in \mathbf{2}^U$ under complementation. The binary operation of *addition* (denoted by $+$) can be defined over $\mathbb{I}$, with $i_1 + i_2$ denoting the image of $(i_1, i_2) \in \mathbb{I}^2$ under addition. The binary operation of *union* (denoted by $\cup$) can be defined over $\mathbf{2}^U$, with $S_1 \cup S_2$ denoting the image of $(S_1, S_2) \in (\mathbf{2}^U)^2$ under union.

When A is finite, say, $A = \{a_1, a_2, \ldots, a_n\}$, a unary or a binary operation o over A is often defined by means of an *operation table* of the form:

a_i	$o(a_i)$
a_1	$o(a_1)$
a_2	$o(a_2)$
.	.
.	.
.	.
a_n	$o(a_n)$

or of the form:

o	a_1	a_2	$\ldots$	a_n
a_1	$o(a_1, a_1)$	$o(a_1, a_2)$	$\ldots$	$o(a_1, a_n)$
a_2	$o(a_2, a_1)$	$o(a_2, a_2)$	$\ldots$	$o(a_2, a_n)$
.	.	.		.
.	.	.		.
.	.	.		.
a_n	$o(a_n, a_1)$	$o(a_n, a_2)$	$\ldots$	$o(a_n, a_n)$

respectively. A familiar operation table is the multiplication table for the set of integers $\{1, 2, \ldots, 10\}$.

Suppose an m-ary operation o is defined over A, and let S be any subset

[1]In the so-called *Polish suffix notation*, a binary operation symbol is placed *after*, rather than between, the two operands. For example $ij+$ denotes $i + j$, $ijk+-l+$ denotes $i - (j + k) + l$, etc. This notation obviates the need for parentheses and is quite convenient in computer processing of arithmetic expressions. The *Polish prefix notation* is an analogous method where the binary operation symbol is placed before the two operands.

of A. If the image of every $(a_1, a_2, \ldots, a_m) \in S^m$ under o is an element of S, we say that S is *closed under* o (A itself, by definition, is always closed under o). For example, the set of even numbers is closed under both addition and multiplication of positive integers; the set of odd numbers, however, is closed only under multiplication.

THEOREM 4-1

Let o be an operation defined on A, and let S_1 and S_2 be subsets of A closed under o. Then $S_1 \cap S_2$ is also closed under o.

Proof Suppose o is an m-ary operation. Since S_1 and S_2 are closed under o, for every $a_1, a_2, \ldots, a_m \in S_1$, $o(a_1, a_2, \ldots, a_m) \in S_1$, and for every $a_1, a_2, \ldots, a_m \in S_2$, $o(a_1, a_2, \ldots, a_m) \in S_2$. Hence, for every $a_1, a_2, \ldots, a_m \in S_1 \cap S_2$, $o(a_1, a_2, \ldots, a_m) \in S_1 \cap S_2$. □

By a simple induction argument it can be shown that if $\{S_i\}_{i \in K}$ is any set of subsets of A closed under o, then $\bigcap_{i \in K} S_i$ is also closed under o.

PROBLEMS

1. Is the set $\{1, 2, \ldots, 10\}$ closed under the binary operations $*$ defined below?

(a) $i_1 * i_2 = \gcd(i_1, i_2)$
(b) $i_1 * i_2 = \operatorname{lcm}(i_1, i_2)$
(c) $i_1 * i_2 =$ least integer which equals or exceeds $i_1 i_2$
(d) $i_1 * i_2 =$ number of prime numbers p such that $i_1 \leq p \leq i_2$
(e) $i_1 * i_2 = |2^{i_1+i_2} - 10^6|$

2. Which of the following subsets of $\mathbb{N}$ are closed under addition? Justify your answers.

(a) $\{n \mid \text{some power of } n \text{ is divisible by } 16\}$
(b) $\{n \mid n \text{ and } 5 \text{ are relatively prime}\}$
(c) $\{n \mid n \text{ divides } 30\}$
(d) $\{n \mid 6 \text{ divides } n, \text{ and } 24 \text{ divides } n^2\}$
(e) $\{n \mid 9 \text{ divides } 21n\}$

3. Show that the set of integers closed under subtraction must also be closed under addition.

4. An $n \times n$ matrix, with the (i, j) entry m_{ij}, is called a *stochastic matrix*, if

$$0 \leq m_{ij} \leq 1 \qquad (i = 1, 2, \ldots, n; j = 1, 2, \ldots, n)$$
$$\sum_{j=1}^{n} m_{ij} = 1 \qquad (i = 1, 2, \ldots, n)$$

Show that the set of all $n \times n$ stochastic matrices is closed under matrix multiplication.

4-2. ALGEBRAIC SYSTEMS

Roughly speaking, an *algebraic system* (or an *abstract algebra*) is a mathematical structure where various results are derived from certain basic premises. These premises, as a rule, are of sufficiently general nature so as to render the results useful in a number of abstract and/or applied areas which are seemingly unrelated. Thus, an algebraic system often provides a unifying framework for diverse disciplines—a framework that facilitates their cross-fertilization and growth. Numerous examples of algebraic systems appear throughout the remainder of this book.

More precisely, an algebraic system consists of:

(a) A nonempty set V, called the *domain* of the algebra.
(b) A set of *relations* on V. (One indispensable relation is that of equality, usually denoted by $=$, where $v_i = v_j$ if and only if v_i and v_j are the same element.)
(c) A set of *operations* over V.
(d) A set of *postulates* (or *axioms*) that are basic rules to which the operations are subjected.
(e) A set of *theorems* that are statements concerning V and can be proved from the postulates by means of accepted rules of logic. (In the development of an algebraic system, the usual goal is to derive the theorems from the smallest possible set of postulates; for pedagogic reasons, however, it is common to classify certain theorems as postulates.)
(f) A set of *definitions* that constitute agreements concerning symbolism and terminology. As theorems increase in number and complexity, definitions are introduced to facilitate the derivation and presentation of new theorems, to unify ideas, and to establish concepts.
(g) A set of *algorithms* that constitute step-by-step procedures (programs) for the construction and computation of entities incorporated in the definitions.

An algebraic system $\mathcal{V}$, whose domain is V and whose set of operations (or operation symbols) is $\{o_1, o_2, \ldots, o_l\}$, is denoted by

$$\mathcal{V} = \langle V; o_1, o_2, \ldots, o_l \rangle$$

The *cardinality* of $\mathcal{V}$ is synonymous with the cardinality of V. When V is finite, $\mathcal{V}$ is said to be a *finite* algebraic system.

Example 4-1

The simplest nontrivial algebraic system is $\langle \mathbb{N}; s \rangle$, where $\mathbb{N}$ is the set of positive integers and s is a unary operation over $\mathbb{N}$, specified by the Peano successor function of Sec. 3-8. The set of postulates for this system consists of Peano's postulates listed in that section. One of the most important theorems of this algebraic system is the principle of finite induction. □

Some of the material developed in Chapters 2 and 3 could alternatively be developed via the study of various algebraic systems. For example, denoting the set of all relations on A by R_A, and the composition of two relations ρ_1 and ρ_2 by $\rho_1 \cdot \rho_2$ (abbreviated $\rho_1 \rho_2$), the algebraic system $\langle R_A; \cdot \rangle$ could serve as a vehicle for developing properties of relations on A. Similarly, denoting the composition of the functions $f\colon A \longrightarrow A$ and $g\colon A \longrightarrow A$ by $g \cdot f$(abbreviated gf), the algebraic system $\langle A^A; \cdot \rangle$ could be used as a vehicle for developing properties of functions from A into A.

Let $\mathcal{V} = \langle V; o_1, o_2, \ldots, o_l \rangle$ be an algebraic system, and let $\tilde{V}$ be a nonempty subset of V which is closed under every operation of $\mathcal{V}$. That is, for every operation o_i of order k_i, and for every $(v_1, v_2, \ldots, v_{k_i}) \in V^{k_i}$,

$$o_i(v_1, v_2, \ldots, v_{k_i}) \in \tilde{V}$$

We can now define a new algebraic system, namely, $\tilde{\mathcal{V}} = \langle \tilde{V}; \tilde{o}_1, \tilde{o}_2, \ldots, \tilde{o}_l \rangle$, where the operation $\tilde{o}_i$ $(i = 1, 2, \ldots, l)$ is the restriction of the original operation o_i to the new domain $\tilde{V}$. (In practice, the symbols used for the o_i are retained for the $\tilde{o}_i$.) The algebraic system $\tilde{\mathcal{V}}$ thus defined is called a *subsystem* of $\mathcal{V}$ (and $\mathcal{V}$ is called an *extension* of $\tilde{\mathcal{V}}$). If $\tilde{V}$ is a proper subset of V, then $\tilde{\mathcal{V}}$ is called a *proper subsystem* of $\mathcal{V}$. For example, the algebraic system $\langle E; +, \cdot \rangle$, where E is the set of all integral multiples of 2, and $+$ and $\cdot$ are ordinary addition and multiplication, is a proper subsystem of $\langle \mathbb{I}; +, \cdot \rangle$.

PROBLEMS

1. Enumerate some algebraic systems you are familiar with.
2. Suggest a set of postulates, and derive a theorem or two, for these algebraic systems:
   (a) $\langle \{0, 1\}; +, \cdot \rangle$, where $+$ and $\cdot$ are Boolean addition and multiplication (see Sec. 2-4).
   (b) $\langle 2^U; \cup, \cap \rangle$, where $U = \{a, b\}$.
   (c) $\langle \tilde{R}_A; \cdot \rangle$, where $\tilde{R}_A$ is the set of all relation matrices for relations on A, and $\cdot$ denotes multiplication of relation matrices (see Sec. 2-4).

4-3. SOME TYPICAL POSTULATES

In this section we state some postulates which are frequently encountered in the study of algebraic systems.

The algebraic system $\langle \mathcal{V} = V; *, \ldots \rangle$ is said to obey the *commutative law* with respect to the binary operation $*$ (or $*$ is said to be a *commutative operation*) if, for all $v_i, v_j \in V$,

$$v_i * v_j = v_j * v_i$$

$\mathcal{V}$ is said to obey the *associative law* with respect to $*$ (or $*$ is said to be an *associative operation*) if, for all $v_i, v_j, v_k \in V$,

$$v_i * (v_j * v_k) = (v_i * v_j) * v_k$$

If the algebraic system $\langle V; *, \ldots \rangle$ obeys the associative law with respect to $*$, the elements of V represented by

$$(\ldots (((v_1 * v_2) * v_3) * v_4) \ldots * v_r)$$

can be represented unambiguously by the parentheses-free expression

$$v_1 * v_2 * \ldots * v_r$$

(since the order in which the $*$ operations are performed is immaterial). In some such systems, $v * v * \ldots * v$ (r times) is written as v^r and referred to as the rth *power* of v. In this context, r is called the *exponent* of v. Formally, v^r can be defined recursively as

$$v^1 = v$$

$$v^{r+1} = v^r * v \qquad (r = 1, 2, \ldots)$$

This definition leads to the following well-known "laws of exponents":

THEOREM 4-2

For any positive integers p and r,

(a) $v^p * v^r = v^{p+r}$

(b) $(v^p)^r = v^{pr}$

Proof (by induction on r) (*Basis*). When $r = 1$, the theorem follows, since, by definition, $v^p * v = v^{p+1}$ and $(v^p)^1 = v^p$. (*Induction step*). The induction hypothesis is that, for all $k \geq 1$, $v^p * v^k = v^{p+k}$ and $(v^p)^k = v^{pk}$. Hence,

$$v^p * v^{k+1} = v^p * (v^k * v) = (v^p * v^k) * v = (v^{p+k}) * v = v^{p+k+1}$$

$$(v^p)^{k+1} = (v^p)^k * v^p = v^{pk} * v^p = v^{pk+p} = v^{p(k+1)}$$

which proves the theorem for $r = k + 1$. □

If $\langle V; *, \ldots\rangle$ obeys both the associative and commutative laws with respect to $*$, the order of the v_i in $v_1 * v_2 * \ldots * v_r$ can be changed arbitrarily without altering the represented element.

The algebraic system $\mathcal{V} = \langle V; *, \ldots\rangle$ is said to have an *identity* e with respect to the binary operation $*$, if there exists an element $e \in V$ such that, for all $v_i \in V$,

$$v_i * e = e * v_i = v_i$$

An element $v_i \in V$ is said to be *invertible* with respect to $*$, if there exists an element v_j (called the *inverse* of v_i) such that

$$v_i * v_j = v_j * v_i = e$$

THEOREM 4-3

If $\mathcal{V} = \langle V; *, \ldots\rangle$ has an identity with respect to $*$, then this identity is unique.

Proof Suppose e_1 and e_2 are identities of $\mathcal{V}$ with respect to $*$. Then $e_1 * e_2 = e_1$ and $e_1 * e_2 = e_2$, which implies $e_1 = e_2$. □

THEOREM 4-4

Let $\mathcal{V} = \langle V; *, \ldots\rangle$ be an algebraic system which obeys the associative law and has an identity e with respect to $*$. If $v_i \in V$ is invertible with respect to $*$, then its inverse is unique.

Proof Suppose v_j and v'_j are inverses of v_i with respect to $*$. Using the associative law, we have

$$\begin{aligned} v_j = e * v_j = (v'_j * v_i) * v_j &= v'_j * (v_i * v_j) \\ &= v'_j * e = v'_j \end{aligned}$$

□

The algebraic system $\mathcal{V} = \langle V; *, \square, \ldots\rangle$ is said to obey the *distributive law* with respect to the binary operations $*$ and $\square$ (or $*$ is said to be a *distributive operation* over $\square$) if, for all $v_i, v_j, v_k \in V$,

$$v_i * (v_j \square v_k) = (v_i * v_j) \square (v_i * v_k)$$
$$(v_j \square v_k) * v_i = (v_j * v_i) \square (v_k * v_i)$$

(Only one of these identities is needed if $*$ is commutative.)

PROBLEMS

1. The following are various definitions of the binary operation $*$ over the set of real numbers $\mathbb{R}$. In each case, decide whether $*$ is commutative, whether $*$ is associative, whether $\mathbb{R}$ contains an identity with respect to $*$, and—if it does

contain an identity—whether every element in $\mathbb{R}$ is invertible with respect to $*$.

(a) $r_1 * r_2 = |r_1 - r_2|$
(b) $r_1 * r_2 = r_1 + 2r_2$
(c) $r_1 * r_2 = \frac{1}{2}(r_1 + r_2)$
(d) $r_1 * r_2 = (r_1^2 + r_2^2)^{1/2}$
(e) $r_1 * r_2 = r_1/r_2$

2. The following are various operation tables of the binary operation $*$ over the set $A = \{a, b, c\}$. In each case decide whether $*$ is commutative or associative, whether A has an identity with respect to *, and—in case it does contain an identity—whether every element in A is invertible with respect to $*$.

(a)

$*$	a	b	c
a	a	b	c
b	b	c	a
c	c	a	b

(b)

$*$	a	b	c
a	a	b	c
b	b	a	c
c	c	c	c

(c)

$*$	a	b	c
a	a	b	c
b	a	b	c
c	a	b	c

3. The following are operation tables of the binary operations $\cdot$ and $+$ over $\{a, b\}$. In each case decide whether the operation is commutative or associative, and whether $\{a, b\}$ contains an identity with respect to the operation. Also, decide whether $\cdot$ is distributive over $+$ and conversely.

$\cdot$	a	b
a	a	a
b	a	b

$+$	a	b
a	a	b
b	b	a

4. Define the binary operation $*$ over a set of positive integers J as follows: For all $j_1, j_2 \in J$,

$$j_1 * j_2 = j_1^{j_2}$$

Determine whether * is associative or commutative when:

(a) $J = \mathbb{N}$

(b) J is the set of all integers of the form

$$2^{2^{2^{\cdot^{\cdot^{\cdot^{2}}}}}}$$

5. How can a commutative binary operation be recognized from its operation table? How can the identity and inverses (if they exist) be recognized from the table?

6. $\langle V; *\rangle$ is an algebraic system where $*$ is an associative binary operation and where, for all v_i, $v_j \in V$, $v_i * v_j = v_j * v_i$ implies $v_i = v_j$. Show that, for any $v \in V$, $v * v = v$. (In this case, v is said to be *idempotent* with respect to $*$.)

7. Prove that, for any positive integer p,

$$(v_1 * v_2)^p = v_1^p * v_2^p$$

4-4. INTEGRAL DOMAINS

Some of the most familiar algebraic systems are special cases of a system known as "integral domain." Formally, an *integral domain* $\langle J; +, \cdot\rangle$ is a two-operation algebraic system, with these postulates:

(a) *Commutativity*. For all $i, j \in J$,

$$i + j = j + i, \qquad i \cdot j = j \cdot i$$

(b) *Associativity*. For all $i, j, k \in J$,

$$i + (j + k) = (i + j) + k, \qquad i \cdot (j \cdot k) = (i \cdot j) \cdot k$$

(c) *Distributivity*. For all $i, j, k \in J$,

$$i \cdot (j + k) = (i \cdot j) + (i \cdot k)$$

(d) *Identities*. J contains the distinct elements 0 (called *zero*) and 1 (called *one* or *unity*) such that, for all $i \in J$,

$$i + 0 = 0 + i = i, \qquad i \cdot 1 = 1 \cdot i = i$$

(e) *Inverse*. For every element $i \in J$, there is an element $j \in J$ (commonly denoted by $-i$), such that

$$i + j = j + i = 0$$

(f) *Cancellation*. If $i \neq 0$, then, for all $j, k \in J$,

$$i \cdot j = i \cdot k \text{ implies } j = k$$

Thus, $\langle J; +, \cdot\rangle$ obeys the commutative and associative laws with respect to $+$ and $\cdot$, and the distributive law with respect to $\cdot$ over $+$; it has the identity 0 with respect to $+$ and the identity 1 with respect to $\cdot$, and every element in J is invertible with respect to $+$. In addition, nonzero i can be "cancelled" in every equation of the form $i \cdot j = i \cdot k$ (or $j \cdot i = k \cdot i$).

Example 4-2

The most familiar integral domain is the algebraic system $\mathcal{I} = \langle \mathbb{I}; +, \cdot \rangle$, where $\mathbb{I}$ is the set of integers, and $+$ and $\cdot$ are the ordinary addition and multiplication operations. □

Example 4-3

Let $\mathbb{Q}$ be the set of all pairs (i, j), where $i, j \in \mathbb{I}$ and where $j \neq 0$. An element (i, j) of $\mathbb{Q}$ is called a *rational number*, and is commonly written as i/j. Define equality, addition, and multiplication of rational numbers as follows:

$$(i, j) = (i', j') \text{ if and only if } i \cdot j' = i' \cdot j$$

$$(i, j) + (i', j') = (i \cdot j' + i' \cdot j, j \cdot j')$$

$$(i, j) \cdot (i', j') = (i \cdot i', j \cdot j')$$

(where the $+$ and $\cdot$ operations appearing on the right are those of $\langle \mathbb{I}; +, \cdot \rangle$). With these defintions, the algebraic system $\langle \mathbb{Q}; +, \cdot \rangle$ can be shown to be an integral domain. One of the most important properties of $\langle \mathbb{Q}; +, \cdot \rangle$ is that every equation $i \cdot x = 1$, where $i \in \mathbb{Q}$ and $i \neq 0$, has a solution in $\mathbb{Q}$. □

PROBLEMS

1. If $\langle J; +, \cdot \rangle$ is an integral domain, prove that:
   (a) For all $i, j, k \in J$, $i + j = i + k$ implies $j = k$.
   (b) For all $i, j \in J$, the equation $i + x = j$ has a unique solution in J.
   (c) For all $i \in J$, $i \cdot 0 = 0 \cdot i = 0$.
   (d) For all $i, j \in J$, $i \cdot j = 0$ implies $i = 0$ or $j = 0$.
   (e) For all $i \in J$, $-(-i) = i$.
   (f) For all $i \in J$, $-i = (-1) \cdot i$.
   (g) For all $i, j \in J$, $-(i + j) = (-i) + (-j)$.
   (h) For all $i, j \in J$, $(-i) \cdot j = i \cdot (-j) = -(ij)$.
   (i) For all $i, j \in J$, $(-i) \cdot (-j) = i \cdot j$.
2. Is $\langle \{a, b\}; +, \cdot \rangle$ of Problem 3, Sec. 4-3, an integral domain?
3. Show that the equality of rational numbers, as defined in Example 4-2, is an equivalence relation.
4. Prove that $\langle \mathbb{Q}; +, \cdot \rangle$ is an integral domain.
5. Prove that $\langle \mathbb{C}; +, \cdot \rangle$, where $\mathbb{C}$ is the set of complex numbers and $+$ and $\cdot$ are complex addition and multiplication, is an integral domain.
6. Is $\langle \{2i \mid i \in \mathbb{I}\}; +, \cdot \rangle$ (where $+$ and $\cdot$ are ordinary addition and multiplication) an integral domain?

4-5. THE ALGEBRA OF SETS

One of the most important algebraic systems encountered in mathematics is the *algebra of sets* which incorporates most of the results presented in Chapter 1. This system can be denoted by $\langle \mathbf{2}^U; ', \cup, \cap \rangle$, where $\mathbf{2}^U$ is the power set of a universal set U, and $'$, $\cup$, and $\cap$ are the complementation, union, and intersection operations introduced in Sec. 1-3. The basic relations of the system are those of equality ($=$) and inclusion ($\subset$). The postulates of the algebra of sets consist of the "set laws" listed in Table 1-3, together with the stipulation that U and $\varnothing$ are distinct. From this table it is apparent that the system obeys the associative and commutative laws with respect to both $\cup$ and $\cap$; that it obeys the distributive law with respect to $\cup$ over $\cap$, and with respect to $\cap$ over $\cup$; and that it has the identity $\varnothing$ with respect to $\cup$, and the identity U with respect to $\cap$. In addition, it obeys the "complement laws," namely, $A \cup A' = U$ and $A \cap A' = \varnothing$, for any $A \in \mathbf{2}^U$.

The following shows that the complement laws are consistent with the assumption that $'$ is an operation over $\mathbf{2}^U$, that is, with the assumption that every set has exactly one complement.

THEOREM 4-5
For every $S \in \mathbf{2}^U$, S' is unique.

Proof Suppose S has two complements, S'_1 and S'_2; then

$$S \cup S'_1 = U, \qquad S \cup S'_2 = U$$
$$S \cap S'_1 = \varnothing, \qquad S \cap S'_2 = \varnothing$$

Hence,

$$\begin{aligned}
S'_2 &= U \cap S'_2 && \text{(Identity law)}\\
&= (S \cup S'_1) \cap S'_2 && \text{(Assumption)}\\
&= S'_2 \cap (S \cup S'_1) && \text{(Commutative law)}\\
&= (S'_2 \cap S) \cup (S'_2 \cap S'_1) && \text{(Distributive law)}\\
&= (S \cap S'_2) \cup (S'_1 \cap S'_2) && \text{(Commutative law)}\\
&= \varnothing \cup (S'_1 \cap S'_2) && \text{(Assumption)}\\
&= (S \cap S'_1) \cup (S'_1 \cap S'_2) && \text{(Assumption)}\\
&= (S'_1 \cap S) \cup (S'_1 \cap S'_2) && \text{(Commutative law)}\\
&= S'_1 \cap (S \cup S'_2) && \text{(Distributive law)}\\
&= S'_1 \cap U && \text{(Assumption)}\\
&= S'_1 && \text{(Identity law)}
\end{aligned}$$

Thus, the two complements must be identical. □

Various set identities familiar to the reader from Chapter 1 can be derived as consequences of the postulates of the algebra of sets. In what follows we illustrate such derivations.

THEOREM 4-6 (*Involution law*)
For every $A \in \mathbf{2}^U$,

$$(A')' = A$$

Proof Applying Theorem 4-5 to the case $S = A'$, it follows that $(A')'$ is unique. Using the complement and commutative laws, we have

$$A' \cup A = U, \qquad A' \cap A = \varnothing$$

which implies that $(A')' = A$. □

THEOREM 4-7 (*Idempotent laws*)
For all $A \in \mathbf{2}^U$,

(a) $A \cup A = A$ (b) $A \cap A = A$

Proof

(a) $A \cup A$	(b) $A \cap A$	
$= (A \cup A) \cap U$	$= (A \cap A) \cup \varnothing$	(Identity)
$= (A \cup A) \cap (A \cup A')$	$= (A \cap A) \cup (A \cap A')$	(Complement)
$= A \cup (A \cap A')$	$= A \cap (A \cup A')$	(Distributive)
$= A \cup \varnothing$	$= A \cap U$	(Complement)
$= A$	$= A$	(Identity)

[First and second columns of identities pertain to parts (a) and (b), respectively, of the theorem; the last column indicates the postulates invoked in the succesive steps.] □

PROBLEMS

1. Consider the algebra of sets $\langle \mathbf{2}^U; ', \cup, \cap \rangle$, where $U = \{a, b\}$. Define this algebraic system by constructing the operation tables for the operations $'$, $\cup$, and $\cap$ over $\mathbf{2}^U$.

2. Let $A = \{a, b, c\}$ be a subset of $U = \{a, b, c, d, e\}$. Construct the $'$, $\cup$, and $\cap$ operation tables for the algebraic system $\langle \{\varnothing, A, A', U\}; ', \cup, \cap \rangle$. Compare these tables with those constructed in Problem 1.

4-6. THE DUALITY PRINCIPLE

The *dual* of a set or of an identity involving elements of $\mathbf{2}^U$ and the operations $'$, $\cup$, and $\cap$, is the same set or identity with $\varnothing$ replacing U, U replacing $\varnothing$, $\cup$ replacing $\cap$, and $\cap$ replacing $\cup$. For example, the dual of the identity

$$(A \cup B) \cup (A' \cap B') = U$$

is the identity

$$(A \cap B) \cap (A' \cup B') = \varnothing$$

The dual of a set or an identity P is denoted by P^D. Clearly, if P^D is the dual of P, then P is the dual of P^D. Thus, we can simply say that P and P^D are *dual* (to each other).

The reader has undoubtedly observed that laws 1′, 2′, 3′, 4′, and 5′ in Table 1-3 are the duals of laws 1, 2, 3, 4, and 5, respectively. Thus, every postulate of the algebra of sets is matched by a dual postulate. Now, suppose we are given a valid sequence of set identities,

$$S_1 = S_2 = S_3 = \ldots = S_q$$

(such as those appearing in the proofs of Theorems 4-5 and 4-7), where S_i follows from S_{i-1} $(i = 2, 3, \ldots, q)$ by virtue of postulate p_i. Then,

$$S_1^D = S_2^D = S_3^D = \ldots = S_q^D$$

where S_i^D follows from $S_{i-1}^D (i = 2, 3, \ldots, q)$ by virtue of postulate p_i^D, must also be valid. Hence, if $S_1 = S_q$ is valid, so must be $S_1^D = S_q^D$. In conclusion, *given any valid identity in the algebra of sets, then the dual of this identity must also be valid.* This so-called *duality principle* is illustrated in the parallel proofs of parts (a) and (b) of Theorem 4-7.

The duality principle is extremely useful inasmuch as it obviates the proof of a theorem, once the dual of this theorem has been proved. For example, the identity $A \cap A = A$ could be deduced immediately from the identity $A \cup A = A$ without the burden of a separate proof. By the same token, in Table 1-4, the "additional set laws" 2′ through 5′ could be deduced immediately from laws 2 through 5, respectively. Each of the theorems appearing below contains a pair of dual identities, only one of which requires a proof.

THEOREM 4-8 (*Null laws*)
For all $A \in \mathbf{2}^U$,

$$\text{(a) } A \cup U = U \qquad \text{(b) } A \cup \varnothing = \varnothing$$

Proof (a)

$$\begin{aligned} A \cup U &= (A \cup U) \cap U && \text{(Identity)}\\ &= (A \cup U) \cap (A \cup A') && \text{(Complement)}\\ &= A \cup (U \cap A') && \text{(Distributive)}\\ &= A \cup A' && \text{(Identity)}\\ &= U && \text{(Complement)} \end{aligned}$$

□

THEOREM 4-9 (*Absorption laws*)
For all $A, B \in \mathbf{2}^U$,

(a) $A \cup (A \cap B) = A$ (b) $A \cap (A \cup B) = A$

Proof (a) $A \cup (A \cap B)$

$$\begin{aligned} &= (A \cap U) \cup (A \cap B) && \text{(Identity)}\\ &= A \cap (U \cup B) && \text{(Distributive)}\\ &= A \cap (B \cup U) && \text{(Commutative)}\\ &= A \cap U && \text{(Theorem 4-8)}\\ &= A && \text{(Identity)} \end{aligned}$$

□

THEOREM 4-10 (*De Morgan's laws*)
For all $A, B \in \mathbf{2}^U$,

(a) $(A \cup B)' = A' \cap B'$ (b) $(A \cap B)' = A' \cup B'$

Proof (a) By Theorem 4-5, it will suffice to show that

$$(A \cup B) \cup (A' \cap B') = U$$
$$(A \cup B) \cap (A' \cap B') = \varnothing$$

This can be done as follows:

$$\begin{aligned} &(A \cup B) \cup (A' \cap B')\\ &\quad = ((A \cup B) \cup A') \cap ((A \cup B) \cup B') && \text{(Distributive)}\\ &\quad = (B \cup (A \cup A')) \cap (A \cup (B \cup B')) && \text{(Commutative and associative)}\\ &\quad = (B \cup U) \cap (A \cup U) && \text{(Complement)}\\ &\quad = U \cap U && \text{(Theorem 4-8)}\\ &\quad = U && \text{(Identity)} \end{aligned}$$

$$
\begin{aligned}
(A \cup B) \cap (A' \cap B') & \\
&= (A' \cap B') \cap (A \cup B) && \text{(Commutative)}\\
&= ((A' \cap B') \cap A) \cup ((A' \cap B') \cap B) && \text{(Distributive)}\\
&= ((A \cap A') \cap B') \cup ((B \cap B') \cap A') && \text{(Commutative and associative)}\\
&= (\varnothing \cap B') \cup (\varnothing \cap A') && \text{(Complement)}\\
&= \varnothing \cup \varnothing && \text{(Theorem 4-8)}\\
&= \varnothing && \text{(Identity)} \quad \square
\end{aligned}
$$

A generalization of De Morgan's laws (named after the English mathematician Augustus De Morgan 1806–1871) is proved by induction in Example 3-13.

PROBLEMS

1. Example 3-20 defines recursively a set generated by $A_1, A_2, \ldots, A_r$. Extend this definition to cover the *dual* of such a set.

2. A set S generated by $A_1, A_2, \ldots, A_r$ is said to be *self-dual* if $S = S^D$. Determine all self-dual sets generated by A, B.

3. Let S_1 and S_2 be sets generated by $A_1, A_2, \ldots, A_r$. Show that if $S_1 \subset S_2$, then $S_1^D \supset S_2^D$.

4. A and B are subsets of U. Using the postulates of the algebra of sets and their corollaries (Tables 1-3 and 1-4), prove these identities and construct their duals:

(a) $(A' \cup B')' \cup (A' \cup B)' = A$

(b) $(A \cup B') \cap (A \cup B) \cap (A' \cup B') = (A' \cup B)'$

(c) $B \cup ((A' \cup B) \cap A)' = U$

5. The minset normal form of a set S generated by A, B, C is given by

$$S = (A \cap B \cap C') \cup (A \cap B' \cap C) \cup (A' \cap B \cap C')$$

Using De Morgan's laws, find the maxset normal form of S'.

4-7. HOMOMORPHISM AND ISOMORPHISM

Consider these two algebraic systems:

$$\mathcal{V}_1 = \langle V_1;\ o_{11}, o_{12}, \ldots, o_{1l} \rangle$$
$$\mathcal{V}_2 = \langle V_2;\ o_{21}, o_{22}, \ldots, o_{2l} \rangle$$

and let h be a function from V_1 into V_2. Suppose a bijection exists from $\{o_{11}, o_{12}, \ldots, o_{1l}\}$ onto $\{o_{21}, o_{22}, \ldots, o_{2l}\}$ such that, if o_{2i} is the image of o_{1i} $(i = 1, 2, \ldots, l)$ under this bijection, then:

(a) o_{1i} and o_{2i} are operations of the same order, say, k_i.
(b) For every k_i-tuple $(v_1, v_2, \ldots, v_{k_i}) \in V_1^{k_i}$,

$$h(o_{1i}(v_1, v_2, \ldots, v_{k_i}) = o_{2i}(h(v_1), h(v_2), \ldots, h(v_{k_i}))$$

(See Fig. 4-1.) If this is the case, h is said to be a *homomorphism from* $\mathcal{V}_1$ *into* $\mathcal{V}_2$. The algebraic system $\mathcal{V}_2$ is sometimes called the *homomorphic image* of $\mathcal{V}_1$ (under h). We also say that h *carries* the operation o_{1i} into the operation o_{2i} $(i = 1, 2, \ldots, l)$.

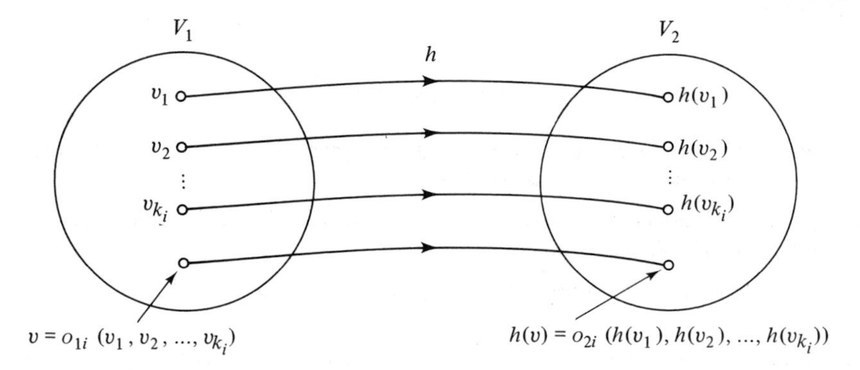

Fig. 4-1 Homomorphism

The above definition may be clarified by specializing it to the case $l = 2$, where both operations are binary. For example, consider the algebraic systems,

$$\mathcal{V}_1 = \langle V_1; *, \square \rangle$$
$$\mathcal{V}_2 = \langle V_2; \star, \bullet \rangle$$

and the function $h: V_1 \longrightarrow V_2$. Then h is a homomorphism from $\mathcal{V}_1$ into $\mathcal{V}_2$ (carrying $*$ into $\star$ and $\square$ into $\bullet$) if, for all $(v_1, v_2) \in V_1^2$,

$$h(v_1 * v_2) = h(v_1) \star h(v_2)$$
$$h(v_1 \square v_2) = h(v_1) \bullet h(v_2)$$

It is seen that a homomorphism is an "operation-preserving" mapping, in this sense: The image of the "$*$-product" of any two elements v_1 and v_2 in V_1 is the "$\star$-product" of the images of v_1 and v_2 in V_2; similarly, the image of the "$\square$-product" of v_1 and v_2 is the "$\bullet$-product" of the images of v_1 and v_2.

Example 4-4

Consider the algebraic system $\mathcal{I} = \langle \mathbb{I}; +, \cdot \rangle$ (see Example 4-2). Also, consider the algebraic system $\mathcal{Z}_m = \langle \mathbb{Z}_m; \oplus, \odot \rangle$, where $\mathbb{Z}_m = \{0, 1, \ldots, m - 1\}$ and $\oplus$ (*addition modulo m*) and $\odot$ (*multiplication modulo m*) are defined by

$$z_1 \oplus z_2 = \text{res}_m(z_1 + z_2)$$
$$z_1 \odot z_2 = \text{res}_m(z_1 \cdot z_2)$$

(see Example 2-7). The function $h: \mathbb{I} \longrightarrow \mathbb{Z}_m$, where

$$h(i) = \text{res}_m(i)$$

is a homomorphism from $\mathcal{I}$ into $\mathcal{Z}_m$ (carrying $+$ into $\oplus$ and $\cdot$ into $\odot$). That is, for all $(i_1, i_2) \in \mathbb{I}^2$,

$$\text{res}_m(i_1 + i_2) = \text{res}_m(i_1) \oplus \text{res}_m(i_2)$$
$$\text{res}_m(i_1 \cdot i_2) = \text{res}_m(i_1) \cdot \text{res}_m(i_2)$$

For example, if $m = 3$:

$$\text{res}_3(4 + 8) = \text{res}_3(12) = 0$$
$$\text{res}_3(4) \oplus \text{res}_3(8) = 1 \oplus 2 = 0$$
$$\text{res}_3(4 \cdot 8) = \text{res}_3(32) = 2$$
$$\text{res}_3(4) \odot \text{res}_3(8) = 1 \odot 2 = 2$$

□

When the homomorphism h is an injection from V_1 into V_2, it is called a *monomorphism from* $\mathcal{V}_1$ *into* $\mathcal{V}_2$. When h is a surjection from V_1 onto V_2, it is called an *epimorphism from* $\mathcal{V}_1$ *onto* $\mathcal{V}_2$. When h is a bijection from V_1 onto V_2, it is called an *isomorphism from* $\mathcal{V}_1$ *onto* $\mathcal{V}_2$. In the latter case, an isomorphism also exists from $\mathcal{V}_2$ onto $\mathcal{V}_1$ (namely, the function h^{-1}), and we can simply say that $\mathcal{V}_1$ and $\mathcal{V}_2$ are *isomorphic* (to each other).

Except (possibly) for the names of the elements and the symbols for the operations, isomorphic algebraic systems are identical. (Such systems are often referred to as "identical up to isomorphism.") Thus, isomorphic algebraic systems can be recognized by the fact that their operation tables —except, possibly, for the order and labels of elements and operations—are the same. If $\mathcal{V}_1$ and $\mathcal{V}_2$ are as defined in the first paragraph of this section, and if h is an isomorphism, then any statement made with respect to $\mathcal{V}_1$ can be made with respect to $\mathcal{V}_2$ simply by replacing every $v \in V_1$ with $h(v) \in V_2$, and every o_{1i} with o_{2i}. Thus, a theory developed for $\mathcal{V}_1$ is directly applicable to any algebraic system isomorphic to $\mathcal{V}_1$. When exploring properties of a new algebraic system, a great deal of effort can be saved by establishing that this system is isomorphic to another system whose properties have already been discovered.

Example 4-5

Consider the algebraic system

$$\mathcal{V}_1 = \langle\{\varnothing, A, A', U\}; ', \cup, \cap\rangle$$

where A is a fixed subset of a universal set U, and $'$, $\cup$, and $\cap$ are the usual set operations. Also, consider the algebraic system

$$\mathcal{V}_2 = \langle\{1, 2, 5, 10\}; \tilde{}, \vee, \wedge\rangle$$

where $i_1 \vee i_2$ denotes the least common multiple of i_1 and i_2, $i_1 \wedge i_2$ the greatest common divisor of i_1 and i_2, and $\tilde{i}$ the quotient upon division of 10 by i. Thus, $\mathcal{V}_1$ and $\mathcal{V}_2$ can be defined by these operation tables:

S	S'
$\varnothing$	U
A	A'
A'	A
U	$\varnothing$

$\cup$	$\varnothing$	A	A'	U
$\varnothing$	$\varnothing$	A	A'	U
A	A	A	U	U
A'	A'	U	A'	U
U	U	U	U	U

$\cap$	$\varnothing$	A	A'	U
$\varnothing$	$\varnothing$	$\varnothing$	$\varnothing$	$\varnothing$
A	$\varnothing$	A	$\varnothing$	A
A'	$\varnothing$	$\varnothing$	A'	A'
U	$\varnothing$	A	A'	U

i	$\tilde{i}$
1	10
2	5
5	2
10	1

$\vee$	1	2	5	10
1	1	2	5	10
2	2	2	10	10
5	5	10	5	10
10	10	10	10	10

$\wedge$	1	2	5	10
1	1	1	1	1
2	1	2	1	2
5	1	1	5	5
10	1	2	5	10

It is apparent that the three lower tables can be obtained from the three upper ones simply by replacing $\varnothing$, A, A', and U with 1, 2, 5, and 10, respectively; and $'$, $\cup$, and $\cap$ with $\tilde{}$, $\vee$, and $\wedge$, respectively. Hence, $\mathcal{V}_1$ and $\mathcal{V}_2$ are isomorphic, with the isomorphism

$$h: \{\varnothing, A, A', U\} \longrightarrow \{1, 2, 5, 10\}$$

where $\quad h(\varnothing) = 1, \quad h(A) = 2, \quad h(A') = 5, \quad h(U) = 10$

Thus, any properties developed for $\mathcal{V}_1$ can be immediately applied to $\mathcal{V}_2$, simply by performing the above replacements. □

THEOREM 4-11

Let h be an epimorphism from the algebraic system $\mathcal{V}_1 = \langle V_1; o_{11}, o_{12}, \ldots, o_{1l}\rangle$ onto the algebraic system $\mathcal{V}_2 = \langle V_2; o_{21}, o_{22}, \ldots, o_{2l}\rangle$, where the operation o_{1i} is carried into the operation o_{2i} $(i = 1, 2, \ldots, l)$. If o_{1i}

is a binary commutative and/or associative operation, so is o_{2i}. If $\mathcal{V}_1$ has an identity e_i with respect to o_{1i}, then $\mathcal{V}_2$ has the identity $h(e_i)$ with respect to o_{2i}. If every element $v \in V_1$ has an inverse v^{-1} with respect to o_{1i}, then every element $h(v) \in V_2$ has an inverse $h(v^{-1})$ with respect to o_{2i}. If o_{1i} is distributive over the binary operation o_{1j}, then o_{2i} is distributive over o_{2j}.

Proof Since h is an epimorphism, it is also a surjection; hence, every element in V_2 can be written in the form $h(v)$, where $v \in V_1$. If o_{1i} is commutative, then for all $h(v_1), h(v_2) \in V_2$,

$$\begin{aligned} o_{2i}(h(v_1), h(v_2)) &= h(o_{1i}(v_1, v_2)) = h(o_{1i}(v_2, v_1)) \\ &= o_{2i}(h(v_2), h(v_1)) \end{aligned}$$

which proves that o_{2i} is commutative.

If e_i is an identity with respect to o_{1i}, then for all $h(v) \in V_2$,

$$o_{2i}(h(v), h(e_i)) = h(o_{1i}(v, e_i)) = h(v)$$

which proves that $h(e_i)$ is the identity with respect to o_{2i}.

Proofs to the remaining statements in the theorem are analogous and left to the reader. □

Theorem 4-11 demonstrates that some important postulates (such as the commutative, associative, distributive, identity, and inverse laws) associated with an algebraic system $\mathcal{V}_1$ are "preserved" in any epimorphic (and, in particular, isomorphic) image of $\mathcal{V}_1$. These postulates are sometimes referred to as the "structure" of $\mathcal{V}_1$, and it is common to say that epimorphism is a "structure-preserving" mapping.

Clearly, every algebraic system is isomorphic to itself. Also, if $\mathcal{V}_1$ is isomorphic to $\mathcal{V}_2$, then $\mathcal{V}_2$ is isomorphic to $\mathcal{V}_1$. Finally, if $\mathcal{V}_1$ is isomorphic to $\mathcal{V}_2$, and $\mathcal{V}_2$ isomorphic to $\mathcal{V}_3$, then $\mathcal{V}_1$ is isomorphic to $\mathcal{V}_3$. Thus, the relation of isomorphism over the set of all algebraic systems is an equivalence relation, and we can partition this set into equivalence classes where each class consists of isomorphic algebraic systems which share the same "structure."

When $\mathcal{V}_1$ and $\mathcal{V}_2$ are the same algebraic system $\mathcal{V}$, then a homomorphism from $\mathcal{V}_1$ into $\mathcal{V}_2$ is called an *endomorphism* of $\mathcal{V}$, and an isomorphism from $\mathcal{V}_1$ onto $\mathcal{V}_2$ is called an *automorphism* of $\mathcal{V}$.

PROBLEMS

1. Consider the algebraic systems $\langle \mathbb{N}; \cdot \rangle$ and $\langle \{0, 1\}; \cdot \rangle$, where $\cdot$ is ordinary multiplication. Show that the function $h: \mathbb{N} \to \{0, 1\}$, where

$$h(n) = \begin{cases} 1 \text{ if } n = 2^k \text{ for some } k \geq 0 \\ 0 \text{ otherwise} \end{cases}$$

is a homomorphism from $\langle \mathbb{N}; \cdot \rangle$ into $\langle \{0, 1\}; \cdot \rangle$.

2. The following tables postulate the algebraic systems $\langle\{a, b, c, d\}; *\rangle$ and $\langle\{\alpha, \beta, \gamma, \delta\}; \cdot\rangle$. Verify that these systems are isomorphic.

$*$	a	b	c	d
a	d	a	b	d
b	d	b	c	d
c	a	d	c	c
d	a	b	a	a

$\cdot$	α	β	γ	δ
α	β	β	β	δ
β	α	α	δ	β
γ	γ	β	γ	α
δ	α	α	γ	δ

3. Consider the algebraic systems $\langle\mathbb{C}; +, \cdot\rangle$, where $\mathbb{C}$ is the set of complex numbers and $+$ and $\cdot$ are complex addition and multiplication, and $\langle M; +, \cdot\rangle$, where M is the set of all 2×2 matrices of the form

$$\begin{bmatrix} r_1 & r_2 \\ -r_2 & r_1 \end{bmatrix} \qquad (r_1, r_2 \in \mathbb{R})$$

and $+$ and $\cdot$ are matrix addition and multiplication. Show that these systems are isomorphic.

4. Is the algebraic system $\langle\{0, 1\}; +, \cdot\rangle$, where $+$ and $\cdot$ denote Boolean addition and multiplication (see Sec. 2-4), isomorphic to the algebraic system $\langle\{-1, 1\}; \wedge, \vee\rangle$, where $i \wedge j$ and $i \vee j$ denote the largest and smallest, respectively, of the elements i and j? Is $\mathcal{Z}_2 = \langle\mathbb{Z}_2; \oplus, \odot\rangle$ (as defined in Example 4-4) isomorphic to either of these two systems? Justify your answers.

5. Let $U = \{a, b, c\}$. Are the algebraic systems $\langle\{\varnothing, U\}; \cup, \cap\rangle$ and $\langle\{\{a, b\}, \{a, b, c\}\}; \cup, \cap\rangle$ isomorphic? Compare these with the systems specified in Problem 4.

6. Complete the proof of Theorem 4-11.

4-8. CONGRUENCE RELATIONS

Consider the algebraic system

$$\mathcal{V} = \langle V; o_1, o_2, \ldots, o_l\rangle$$

where all the o_i are unary operations; that is,

$$o_i\colon V \longrightarrow V \qquad (i = 1, 2, \ldots, l)$$

Let ρ be an equivalence relation on V. We say that ρ satisfies the *substitution property with respect to* o_i if, for any $v_1, v_2 \in V$,

$$v_1 \rho v_2 \text{ implies } (o_i(v_1))\rho(o_i(v_2))$$

That is: ρ satisfies the substitution property with respect to o_i if the fact that v_1 and v_2 are equivalent implies that their images (under o_i) are also equivalent. When this is the case for *all* the operations o_i, ρ is called a *congruence relation* on $\mathcal{V}$.

Example 4-6

Consider the algebraic system $\langle \mathbb{I}; o \rangle$, where o is a unary operation defined as

$$o(i) = \text{res}_m(i^2) \qquad (\text{for some } m > 0)$$

[For example, if $m = 3$, then $o(4) = 1$, $o(12) = 0$, etc.] Let ρ be a relation on $\mathbb{I}$ defined as $i_1 \rho i_2$ if and only if $\text{res}_m(i_1) = \text{res}_m(i_2)$. By Example 2-7, ρ is an equivalence relation. Now, let i_1 and i_2 be any two elements such that $i_1 \rho i_2$. Hence, $\text{res}_m(i_1) = \text{res}_m(i_2)$ and, by the division theorem, we can write $i_1 = q_1 m + r$ and $i_2 = q_2 m + r$, where $0 \leq r \leq m - 1$. Consequently,

$$\text{res}_m(i_1^2) = \text{res}_m((q_1 m + r)^2) = \text{res}_m(q_1^2 m^2 + 2q_1 mr + r^2) = \text{res}_m(r^2)$$
$$\text{res}_m(i_2^2) = \text{res}_m((q_2 m + r)^2) = \text{res}_m(q_2^2 m^2 + 2q_2 mr + r^2) = \text{res}_m(r^2)$$

which implies that $(o(i_1))\rho(o(i_2))$. Hence, ρ is a congruence relation on $\langle \mathbb{I}; o \rangle$. □

THEOREM 4-12

Let h be a homomorphism from the algebraic system

$$\mathcal{V} = \langle V; o_1, o_2, \ldots, o_l \rangle$$

where all the o_i are unary operations, into the algebraic system

$$\mathcal{V}^* = \langle V^*; o_1^*, o_2^*, \ldots, o_l^* \rangle$$

where o_i^* $(i = 1, 2, \ldots, l)$ is the operation into which h carries o_i. Define a relation ρ on V, such that $v_1 \rho v_2$ if and only if $h(v) = h(v_1)$.[1] Then ρ is a congruence relation on $\mathcal{V}$.

Proof Clearly, ρ is an equivalence relation. Now, suppose $v_1 \rho v_2$. Then $h(v_1) = h(v_2)$ and, hence,

$$o_i^*(h(v_1)) = o_i^*(h(v_2)) \qquad (i = 1, 2, \ldots, l)$$

Since h is a homomorphism, this implies

$$h(o_i(v_1)) = h(o_i(v_2)) \qquad (i = 1, 2, \ldots, l)$$

[1] In Sec. 3-3, this relation was called the *equivalence kernel* of h.

and, hence, $(o_i(v_1))\rho(o_i(v_2))$. Thus, ρ satisfies the substitution property with respect to every o_i. □

Consider the algebraic system

$$\mathcal{V} = \langle V; o_1, o_2, \ldots, o_l \rangle$$

where all the o_i are unary operations, and let ρ be a congruence relation on $\mathcal{V}$. We can construct a new algebraic system

$$\tilde{\mathcal{V}} = \langle \tilde{V}; \tilde{o}_1, \tilde{o}_2, \ldots, \tilde{o}_l \rangle$$

where

$$\tilde{V} = V/\rho = \{\mathcal{E}_\rho(v) \mid v \in V\} \tag{4-1}$$

and where each $\tilde{o}_i$ is a unary operation defined by:

$$\tilde{o}_i(\mathcal{E}_\rho(v)) = \mathcal{E}_\rho(o_i(v)) \qquad (i = 1, 2, \ldots, l) \tag{4-2}$$

That is, $\tilde{V}$ consists of the equivalence classes of V under ρ (see Sec. 2-6), and the image of the class $\mathcal{E}_\rho(v) \in \tilde{V}$ under $\tilde{o}_i$ is the class to which the image of $v \in V$ under o_i belongs. Since ρ satisfies the substitution property with respect to o_i, the equivalence class to which $o_i(v)$ belongs is the same for all $v \in \mathcal{E}_\rho(v)$; hence, $\tilde{o}_i$ is well defined. The new algebraic system $\tilde{\mathcal{V}}$ is called the *quotient algebra of* $\mathcal{V}$ *modulo* ρ, and is denoted by $\mathcal{V}/\rho$.

THEOREM 4-13

Let h be an epimorphism from

$$\mathcal{V} = \langle V; o_1, o_2, \ldots, o_l \rangle$$

where all the o_i are unary operations, onto the algebraic system

$$\mathcal{V}^* = \langle V^*; o_1^*, o_2^*, \ldots, o_l^* \rangle$$

where o_i^* $(i = 1, 2, \ldots, l)$ is the operation into which h carries o_i. Define a relation ρ on V, such that $v_1 \rho v_2$ if and only if $h(v_1) = h(v_2)$. Then $\mathcal{V}^*$ is isomorphic to $\mathcal{V}/\rho$.

Proof (See Fig. 4-2.) Because epimorphism is a special case of homomorphism, Theorem 4-12 implies that ρ is a congruence relation. Hence, the quotient algebra

$$\tilde{\mathcal{V}} = \mathcal{V}/\rho = \langle \tilde{V}; \tilde{o}_1, \tilde{o}_2, \ldots, \tilde{o}_l \rangle$$

is constructible as indicated by (4-1) and (4-2). Note that, since h is a surjection from V onto V^*, every $v^* \in V^*$ can be written as $h(v)$ for some $v \in V$.

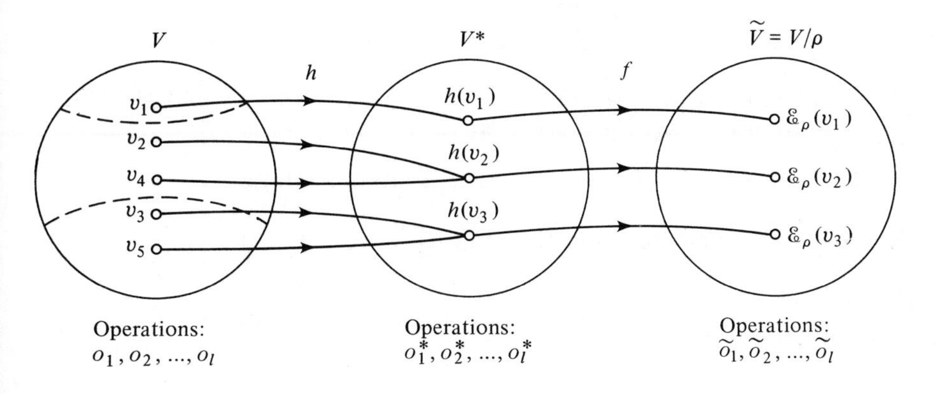

Fig. 4-2 For Theorem 4-13

Now, define the function

$$f\colon\ V^* \longrightarrow \tilde{V}$$

where

$$f(h(v)) = \mathcal{E}_\rho(v)$$

This function is an injection, since $\mathcal{E}_\rho(v_1) = \mathcal{E}_\rho(v_2)$ implies $v_1 \rho v_2$; hence, $h(v_1) = h(v_2)$. Also, f is a surjection, since for every $\mathcal{E}_\rho(v)$ there is some $h(v)$ such that $f(h(v)) = \mathcal{E}_\rho(v)$. Thus, f is a bijection. Since h is a homomorphism, for every o_i^* we have

$$f(o_i^*(h(v))) = f(h(o_i(v))) = \mathcal{E}_\rho(o_i(v)) = \tilde{o}_i(\mathcal{E}_\rho(v)) = \tilde{o}_i(f(h(v)))$$

which implies that f is an isomorphism from $\mathcal{V}^*$ onto $\tilde{\mathcal{V}}$. Thus, $\mathcal{V}^*$ is isomorphic to $\mathcal{V}/\rho$. □

THEOREM 4-14
Let

$$\mathcal{V} = \langle V; o_1, o_2, \ldots, o_l \rangle$$

be an algebraic system where all the o_i are unary operations, and let ρ be a congruence relation on $\mathcal{V}$. Then there exists an epimorphism from $\mathcal{V}$ onto $\mathcal{V}/\rho$.

Proof Following (4-1) and (4-2), construct

$$\tilde{\mathcal{V}} = \mathcal{V}/\rho = \langle \tilde{V}; \tilde{o}_1, \tilde{o}_2, \ldots, \tilde{o}_l \rangle$$

and define the function

$$h\colon V \longrightarrow \tilde{V}$$

where $$h(v) = \mathcal{E}_\rho(v)$$

This function is a surjection, since for every $\mathcal{E}_\rho(v)$ there is some v such that $h(v) = \mathcal{E}_\rho(v)$. Moreover, for every o_i we have

$$h(o_i(v)) = \mathcal{E}_\rho(o_i(v)) = \tilde{o}_i(\mathcal{E}_\rho(v)) = \tilde{o}_i(h(v))$$

Hence, h is an epimorphism from $\mathcal{V}$ onto $\tilde{\mathcal{V}}$. □

Example 4-7

Consider the algebraic system

$$\mathcal{I} = \langle \mathbb{I}; ' \rangle$$

where $\mathbb{I}$ is the set of integers and

$$i' = i + 1$$

Also, consider the algebraic system

$$\mathcal{B} = \langle B; * \rangle$$

where $B = \{0, 1\}$ and

$$b^* = \text{res}_2(b + 1)$$

(that is, $0^* = 1$ and $1^* = 0$). Define the surjection

$$h: \mathbb{I} \longrightarrow B$$

where $$h(i) = \text{res}_2(i)$$

Since

$$h(i') = h(i + 1) = \text{res}_2(i + 1) = \text{res}_2((\text{res}_2(i)) + 1)$$
$$= (\text{res}_2(i))^* = (h(i))^*$$

h is an epimorphism.

Now, define a relation ρ on $\mathbb{I}$, such that $i_1 \rho i_2$ if and only if $h(i_1) = h(i_2)$; that is, if and only if $\text{res}_2(i_1) = \text{res}_2(i_2)$. By Example 2-7, ρ is an equivalence relation. Moreover, if $i_1 \rho i_2$, then $\text{res}_2(i_1) = \text{res}_2(i_2)$, and we have

$$\text{res}_2(i'_1) = \text{res}_2(i_1 + 1) = (\text{res}_2(i_1))^* = (\text{res}_2(i_2))^*$$
$$= \text{res}_2(i_2 + 1) = \text{res}_2(i'_2)$$

Thus, $i'_1 \rho i'_2$, and ρ is indeed a congruence relation, as asserted by Theorem 4-12. The equivalence partition induced by ρ on $\mathbb{I}$ is

$$\pi_\rho^{\mathbb{I}} = \{\mathcal{E}_\rho(0), \mathcal{E}_\rho(1)\}$$

Denoting $\mathcal{E}_\rho(i)$ by C_i $(i = 0, 1)$, we have:

$$\mathbb{I}/\rho = \{C_0, C_1\}$$

The quotient algebra of $\mathcal{I}$ modulo ρ is

$$\tilde{\mathcal{I}} = \mathcal{I}/\rho = \langle \tilde{I}; \star \rangle$$

where $\tilde{I} = \mathbb{I}/\rho$ and

$$C_i^\star = C_{i^*}$$

(that is, $C_0^\star = C_1$ and $C_1^\star = C_0$). Define the function

$$f\colon\ B \longrightarrow \tilde{I}$$

where $$f(b) = C_b$$

Clearly, f is a bijection. Moreover,

$$f(b^*) = C_{b^*} = (C_b)^\star = (f(b))^\star$$

and, hence, f is an isomorphism from $\mathcal{B}$ onto $\tilde{\mathcal{I}}$. Thus, $\mathcal{B}$ is indeed isomorphic to $\mathcal{I}/\rho$, as asserted by Theorem 4-13.

Finally, given $\mathcal{I}$ and ρ as defined above, it is possible to define the surjection

$$\tilde{h}\colon\ \mathbb{I} \longrightarrow \tilde{I}$$

where $$\tilde{h}(i) = C_{\text{res}_2(i)}$$

Since

$$\tilde{h}(i') = \tilde{h}(i+1) = C_{\text{res}_2(i+1)} = C_{(\text{res}_2(i))^*} = C^\star_{\text{res}_2(i)} = (\tilde{h}(i))^\star$$

$\tilde{h}$ is an epimorphism from $\mathcal{I}$ onto $\mathcal{I}/\rho$, as asserted by Theorem 4-14. □

When $\mathcal{V} = \langle V; o_1, o_2, \ldots, o_l \rangle$ is an algebraic system where the o_i are of arbitrary order (not necessarily unary), the substitution property can be generalized to: The equivalence relation ρ on V satisfies the *substitution property with respect to the operation* o_i of order k_i, if for all $(v_{11}, v_{12}, \ldots, v_{1k_i}), (v_{21}, v_{22}, \ldots, v_{2k_i}) \in V^{k_i}$,

$$v_{11}\rho v_{21}, v_{12}\rho v_{22}, \ldots, v_{1k_i}\rho v_{2k_i} \text{ imply}$$
$$(o_i(v_{11}, v_{12}, \ldots, v_{1k_i}))\rho(o_i(v_{21}, v_{22}, \ldots, v_{2k_i}))$$

The definition of congruence relation can now be generalized as: ρ is a *congruence relation* on $\mathcal{V}$ if it satisfies the (generalized) substitution property with respect to *all* the operations o_i.

With these extended definitions, Theorems 4-12, 4-13, and 4-14 can be proved for any algebraic system $\langle V; o_1, o_2, \ldots, o_l \rangle$, with no restrictions imposed on the orders of the o_i. (The proofs are straightforward modifications of the ones presented for the unary case, and are left to the reader.) In summary, we have these results:

THEOREM 4-15

For any algebraic system $\mathcal{V}$:

(a) If h is a homomorphism from $\mathcal{V}$ into $\mathcal{V}^*$, then the equivalence kernel of h is a congruence relation on $\mathcal{V}$.

(b) If h is an epimorphism from $\mathcal{V}$ onto $\mathcal{V}^*$, and ρ is the equivalence kernel of h, then $\mathcal{V}^*$ is isomorphic to $\mathcal{V}/\rho$.

(c) If ρ is a congruence relation on $\mathcal{V}$, then there exists an epimorphism from $\mathcal{V}$ onto $\mathcal{V}/\rho$.

Let $\mathcal{V} = \langle V; * \rangle$ be an algebraic system, where $*$ is a binary operation. An equivalence relation ρ on V is said to be a *right congruence relation on* $\mathcal{V}$ if, for all $v_1, v_2, v_3 \in V$,

$$v_1 \rho v_2 \text{ implies } (v_1 * v_3)\rho(v_2 * v_3)$$

The equivalence relation ρ on V is a *left congruence relation on* $\mathcal{V}$ if, for all $v_1, v_2, v_3 \in V$,

$$v_1 \rho v_2 \text{ implies } (v_3 * v_1)\rho(v_3 * v_2)$$

THEOREM 4-16

Let ρ be both a right congruence relation and a left congruence relation on $\mathcal{V} = \langle V; * \rangle$. Then ρ is a congruence relation on $\mathcal{V}$.

Proof For all $v_1, v_2, v_3, v_4 \in V$, $v_1 \rho v_2$ and $v_3 \rho v_4$ imply $(v_1 * v_3)\rho(v_2 * v_3)$ and $(v_2 * v_3)\rho(v_2 * v_4)$. Since ρ is an equivalence relation, transitivity holds and $(v_1 * v_3)\rho(v_2 * v_4)$. Thus, ρ is a congruence relation on $\mathcal{V}$. □

PROBLEMS

1. Consider the algebraic system $\langle \mathbb{I}; o \rangle$, where o is a unary operation defined as

$$o(i) = \text{res}_m(i^k) \qquad (\text{for some } m > 0,\ k > 0)$$

The relation ρ on $\mathbb{I}$ is defined by: $i_1 \rho i_2$ if and only if $\text{res}_m(i_1) = \text{res}_m(i_2)$. Is ρ a congruence relation on $\langle \mathbb{I}; o \rangle$?

2. Consider the algebraic system $\langle \mathbb{I}; +, \cdot \rangle$ (where $+$ and $\cdot$ are ordinary addition and multiplication) and the relation ρ, where $i_1 \rho i_2$ if and only if $|i_1| = |i_2|$. Does ρ satisfy the (generalized) substitution property with respect to $+$? With respect to $\cdot$?

3. The algebraic system $\mathfrak{A} = \langle A; o_1, o_2\rangle$, where $A = \{a_1, a_2, a_3, a_4, a_5\}$, is specified by the table shown below. The relation ρ on A induces on A the partition $\{\{a_1, a_3\}, \{a_2, a_5\}, \{a_4\}\}$. Show that ρ is a congruence relation on $\mathfrak{A}$. Define the quotient algebra $\mathfrak{A}/\rho$ (by constructing its operation tables) and the epimorphism from $\mathfrak{A}$ onto $\mathfrak{A}/\rho$.

a_i	$o_1(a_i)$	$o_2(a_i)$
a_1	a_4	a_3
a_2	a_3	a_2
a_3	a_4	a_1
a_4	a_2	a_3
a_5	a_1	a_5

4. Let I_V and U_V be the identity and universal relations on a set V (see Sec. 2-2). Show that I_V and U_V are congruence relations on $\langle V; o_1, o_2, \ldots, o_l\rangle$, where the o_i are unary operations.

5. Consider the algebraic system $\mathcal{Z}_3 = \langle \mathbb{Z}_3; \oplus, \odot\rangle$ (see Example 4-4), and any equivalence relation ρ on $\mathbb{Z}_3$.
(a) Show that if ρ satisfies the (generalized) substitution property with respect to $\oplus$, it must also satisfy it with respect to $\odot$.
(b) Find an equivalence relation on $\mathbb{Z}_3$ which satisfies the (generalized) substitution property with respect to $\odot$ but not with respect to $\oplus$.

6. Prove Theorem 4-15 by generalizing the proofs of Theorems 4-12, 4-13, and 4-14.

7. Illustrate Theorem 4-15 with the algebraic systems $\mathcal{I} = \langle \mathbb{I}; +, \cdot\rangle$ (where $+$ and $\cdot$ are ordinary addition and multiplication) and $\mathcal{Z}_3 = \langle \mathbb{Z}_3; \oplus, \odot\rangle$ (where $\oplus$ and $\odot$ are as defined in Example 4-4).

8. Let $\mathcal{V} = \langle V; o_1, o_2, \ldots, o_l\rangle$ be an algebraic system where all the o_i are unary operations. Let ρ_1 and ρ_2 be congruence relations on $\mathcal{V}$ such that $\rho_1 \subset \rho_2$. Show that there exists a homomorphism from $\mathcal{V}/\rho_1$ into $\mathcal{V}/\rho_2$. (*Hint*: Make use of Problem 6, Sec. 2-6.)

9. The operation table of the algebraic system $\mathcal{V} = \langle V; *\rangle$, where $V = \{a, b, c, d\}$, is given by:

*	a	b	c	d
a	b	d	b	c
b	b	a	d	d
c	b	a	a	d
d	c	a	b	b

Find all right congruence relations on $\mathcal{V}$.

4-9. DIRECT PRODUCT OF ALGEBRAIC SYSTEMS

Consider the algebraic systems

$$\begin{aligned} \mathcal{V}_1 &= \langle V_1;\, o_{11}, o_{12}, \ldots, o_{1l}\rangle \\ \mathcal{V}_2 &= \langle V_2;\, o_{21}, o_{22}, \ldots, o_{2l}\rangle \\ &\vdots \\ \mathcal{V}_r &= \langle V_r;\, o_{r1}, o_{r2}, \ldots, o_{rl}\rangle \end{aligned} \tag{4-3}$$

where $o_{1i}, o_{2i}, \ldots, o_{ri}$ $(i = 1, 2, \ldots, l)$ are operations of the same order k_i. The *direct product* of $\mathcal{V}_1, \mathcal{V}_2, \ldots, \mathcal{V}_r$ is the algebraic system

$$\mathcal{V} = \langle V;\, o_1, o_2, \ldots, o_l\rangle \tag{4-4}$$

where $V = V_1 \times V_2 \times \ldots \times V_r$ and where o_i $(i = 1, 2, \ldots, l)$ is a k_i-ary operation defined as follows: For every $(v_1^{(1)}, v_2^{(1)}, \ldots, v_r^{(1)}), (v_1^{(2)}, v_2^{(2)}, \ldots, v_r^{(2)}), \ldots, (v_1^{(k_i)}, v_2^{(k_i)}, \ldots, v_r^{(k_i)}) \in V$,

$$\begin{aligned} &o_i((v_1^{(1)}, \ldots, v_r^{(1)}), (v_1^{(2)}, \ldots, v_r^{(2)}), \ldots, (v_1^{(k_i)}, \ldots, v_r^{(k_i)})) \\ &\quad = (o_{1i}(v_1^{(1)}, \ldots, v_1^{(k_i)}), o_{2i}(v_2^{(1)}, \ldots, v_2^{(k_i)}), \ldots, o_{ri}(v_r^{(1)} \ldots, v_r^{(k_i)})) \end{aligned}$$

$\mathcal{V}$ is commonly denoted by $\mathcal{V}_1 \times \mathcal{V}_2 \times \ldots \times \mathcal{V}_r$.

Example 4-8

Consider the algebraic systems

$$\begin{aligned} \mathcal{A} &= \langle A; *\rangle = \langle\{a_1, a_2\}; *\rangle \\ \mathcal{B} &= \langle B; \circ\rangle = \langle\{b_1, b_2, b_3\}; \circ\rangle \end{aligned}$$

where $*$ and $\circ$ are the binary operations defined in Table 4-1. The direct product of $\mathcal{A}$ and $\mathcal{B}$ is the algebraic system

$$\begin{aligned} \mathcal{A} \times \mathcal{B} &= \langle A \times B; \square\rangle \\ &= \langle\{(a_1, b_1), (a_1, b_2), (a_1, b_3), (a_2, b_1), (a_2, b_2), (a_2, b_3)\}; \square\rangle \end{aligned}$$

Table 4-1 THE $*$ AND $\circ$ OPERATIONS OF $\mathcal{A}$ AND $\mathcal{B}$

$*$	a_1	a_2
a_1	a_1	a_2
a_2	a_2	a_1

$\circ$	b_1	b_2	b_3
b_1	b_1	b_1	b_3
b_2	b_2	b_2	b_3
b_3	b_1	b_3	b_3

Table 4-2 THE □ OPERATION OF $\mathcal{A} \times \mathcal{B}$

□	(a_1, b_1)	(a_1, b_2)	(a_1, b_3)	(a_2, b_1)	(a_2, b_2)	(a_2, b_3)
(a_1, b_1)	(a_1, b_1)	(a_1, b_1)	(a_1, b_3)	(a_2, b_1)	(a_2, b_1)	(a_2, b_3)
(a_1, b_2)	(a_1, b_2)	(a_1, b_2)	(a_1, b_3)	(a_2, b_2)	(a_2, b_2)	(a_2, b_3)
(a_1, b_3)	(a_1, b_1)	(a_1, b_3)	(a_1, b_3)	(a_2, b_1)	(a_2, b_3)	(a_2, b_3)
(a_2, b_1)	(a_2, b_1)	(a_2, b_1)	(a_2, b_3)	(a_1, b_1)	(a_1, b_1)	(a_1, b_3)
(a_2, b_2)	(a_2, b_2)	(a_2, b_2)	(a_2, b_3)	(a_1, b_2)	(a_1, b_2)	(a_1, b_3)
(a_2, b_3)	(a_2, b_1)	(a_2, b_3)	(a_2, b_3)	(a_1, b_1)	(a_1, b_3)	(a_1, b_3)

where, for all $(a_i, b_j), (a_i', b_j') \in A \times B$,

$$(a_i, b_j) \,\square\, (a_i', b_j') = (a_i * a_i', b_j \circ b_j')$$

For example,

$$(a_1, b_2) \,\square\, (a_2, b_1) = (a_1 * a_2, b_2 \circ b_1) = (a_2, b_2)$$

Table 4-2 shows the operation table of □. □

THEOREM 4-17

Let $\mathcal{V}_1, \mathcal{V}_2, \ldots, \mathcal{V}_r$ and $\mathcal{V}$ be the algebraic systems defined in (4-3) and (4-4), where the o_{j1} $(j = 1, 2, \ldots, r)$ are binary operations. If the o_{j1} are commutative and/or associative operations, so is o_1. If $\mathcal{V}_j$ has an identity e_j with respect to o_{j1} $(j = 1, 2, \ldots, r)$, then $\mathcal{V}$ has an identity $(e_1, e_2, \ldots, e_r)$ with respect to o_1. If every element $v_j \in V_j$ has an inverse v_j^{-1} with respect to o_{j1} $(j = 1, 2, \ldots, r)$, then every element $(v_1, v_2, \ldots, v_r) \in V$ has an inverse $(v_1^{-1}, v_2^{-1}, \ldots, v_r^{-1})$ with respect to o_1. If the o_{j1} are distributive over the binary operations o_{j2}, then o_1 is distributive over o_2.

Proof If the o_{j1} are commutative, then, for all $(v_1, v_2, \ldots, v_r), (v_1', v_2', \ldots, v_r') \in V$,

$$\begin{aligned} o_1((v_1, \ldots, v_r), (v_1', \ldots, v_r')) &= (o_{11}(v_1, v_1'), \ldots, o_{r1}(v_r, v_r')) \\ &= (o_{11}(v_1', v_1), \ldots, o_{r1}(v_r', v_r)) \\ &= o_1((v_1', \ldots, v_r'), (v_1, \ldots, v_r)) \end{aligned}$$

which proves that o_1 is commutative.

If $e_1, e_2, \ldots, e_r$ are the identities with respect to $o_{11}, o_{21}, \ldots, o_{r1}$, respectively, then, for all $(v_1, v_2, \ldots, v_r) \in V$,

$$\begin{aligned} o_1((v_1, \ldots, v_r), (e_1, \ldots, e_r)) &= (o_{11}(v_1, e_1), \ldots, o_{r1}(v_r, e_r) \\ &= (v_1, \ldots, v_r) \end{aligned}$$

which proves that $(e_1, \ldots, e_r)$ is the identity with respect to o_1.

Proofs to the remaining statements in the theorem are analogous and left to the reader. □

Theorem 4-17 implies that some important postulates associated with algebraic systems (such as the commutative, associative, distributive, identity, and inverse laws) are "preserved" in the direct product of such systems.

PROBLEMS

1. Complete the proof of Theorem 4-17.

2. The algebraic systems $\mathbf{Z}_2 = \langle \mathbb{Z}_2; \oplus, \odot \rangle$ and $\mathbf{Z}_3 = \langle \mathbb{Z}_3; \oplus, \odot \rangle$ are as defined in Example 4-4. Construct the operation tables for $\mathbf{Z}_2 \times \mathbf{Z}_3$ and $\mathbf{Z}_3 \times \mathbf{Z}_2$. Use these tables to confirm Theorem 4-17.

3. Consider the algebraic systems $\mathcal{R} = \langle \mathbb{R}; * \rangle$ where, for all real numbers r_1 and r_2, $r_1 * r_2 = (r_1^2 - r_2^2)^{1/2}$, and $\mathcal{I} = \langle \mathbb{I}; \cdot \rangle$ where, for all integers i_1 and i_2, $i_1 \cdot i_2 = |i_1 - i_2|$. What is the identity of the algebraic system $\mathcal{R} \times \mathcal{I} = \langle A; \square \rangle$? Compute the value of $(5, 3) \square (3, 5)$ and find the inverse of this element with respect to $\square$.

REFERENCES

FRALEIGH, J. B., *A First Course in Abstract Algebra.* Reading, MA: Addison-Wesley, 1969.

KEMENY, J. G., H. MIRKIL, J. L. SNELL, and G. L. THOMPSON, *Finite Mathematical Structures.* Englewood Cliffs, NJ: Prentice-Hall, 1959.

KUROSH, A. G., *Lectures on General Algebra.* New York: Chelsea, 1963.

MACLANE, S., and G. BIRKHOFF, *Algebra.* New York: Macmillan, 1967.

VAN DER WAERDEN, B. L., *Modern Algebra.* Ungar, 1931.

5 PROPOSITIONS

In this chapter we introduce the concept of a *proposition* and the related concepts of *truth set*, *tautology*, and *contradiction*. It is shown how the operations of *negation*, *disjunction*, *conjunction*, *implication*, and *equivalence* can be employed to generate new propositions from given ones. *Truth tables* are introduced as means for computing *truth values* of given propositions. Postulates are formulated for an "algebra of propositions," which is shown to be isomorphic to the algebra of sets postulated in Chapter 4. Many results concerning propositions (for example, duality, normal forms) can thus be obtained directly from the set-theoretic results already available. Using various properties of propositions, it is shown how the validity of "logical" arguments can be checked systematically, and how theorems can be proved by contradiction. The chapter concludes with a discussion of *quantifiers* and of schemes for proving quantified propositions.

To the computer theorist, the notion of a proposition is important because it appears wherever a formal argument is developed. It also has applications in more practical activities such as program verification and computer-aided theorem proving.

5-1. PROPOSITIONS

A *proposition* is a declarative statement that can be either true or false. For example, "For every integer $n > 1$, the integer $2n$ is not a prime" is a true proposition; "There exists an integer i such that $3i + 4 = 5$" is a false proposition.

Of special interest to mathematicians are propositions that describe properties of elements of a given universal set U. Such propositions are referred to as *propositions over U*. For example, if the universal set is the set of integers $\mathbb{I}$, then "i is not divisible by 2 and not a power of 3," and "$i^2 - 6i + 8 = 0$" are propositions over $\mathbb{I}$.

Generally, if p is a proposition over U, describing some property of the generic element $u \in U$, then p is either true or false, depending on the particular $u \in U$ substituted in the proposition. Thus, p induces on U a partition consisting of the subsets T_p and T'_p, where T_p consists of all elements $u \in U$ for which p is true, while $T'_p = U - T_p$ consists of all elements $u \in U$ for which p is false; that is,

$$T_p = \{u \mid p \text{ is true}\}$$

$$T'_p = \{u \mid p \text{ is false}\}$$

T_p is called the *truth set* (or *solution set*) of p; T'_p is called the *falsity set* of p. For example, the truth set of the proposition "$i^2 - 6i + 8 = 0$" (over $\mathbb{I}$) is $\{2, 4\}$; the falsehood set of this propositions is the set $\mathbb{I} - \{2, 4\}$.

A proposition p which is true for all $u \in U$, that is, such that $T_p = U$, is called a *tautology*. A proposition p that is false for all $u \in U$, that is, such that $T_p = \varnothing$, is called a *contradiction*. For example, if U is the set $\mathbb{P}$ of prime numbers, then the proposition "u is not divisible by 6" is a tautology; the proposition "u is a multiple of 10" is a contradiction.

PROBLEMS

1. Find the truth sets of the following propositions over the set of positive integers $\mathbb{N}$:

(a) $n^2(n^2 - 1) = 0$

(b) $n^3 - 2n = 0$

(c) $n^3 - 2n - 21 = 0$

(d) $n! \leq n$

(e) There are $n - 2$ prime numbers less than n.

(f) If $n > 10$ and $n^2 < 200$, then n is a multiple of 3.

(g) The binary representation of n has at most $\log_2 (n + 1)$ 1s.

(h) The proposition "2^n exceeds $10n$" is false.

(i) The proposition "$2^n > n$ is a tautology" is a contradiction.

Which of these propositions is a tautology? A contradiction?

5-2. NEGATION, DISJUNCTION, AND CONJUNCTION

Given a proposition p, the proposition "not p" is true if and only if p is false. This new proposition is called the *negation* of p, and is denoted by $\sim p$.

Given two propositions p and q, new propositions can be formed by the so-called "disjunction" and "conjunction" of p and q. The *disjunction* of p

and q is the proposition "p or q." This proposition, denoted by $p \vee q$, is true if and only if p is true or q is true. The *conjunction* of p and q is the proposition "p and q." This proposition, denoted by $p \wedge q$, is true if and only if both p and q are true.

Let p and q be propositions over a universal set U, with the truth sets T_p and T_q, respectively. From the preceding definitions of the $\sim$, $\vee$, and $\wedge$ operations, it follows that

$$T_{\sim p} = T'_p \tag{5-1}$$

$$T_{p \vee q} = T_p \cup T_q \tag{5-2}$$

$$T_{p \wedge q} = T_p \cap T_q \tag{5-3}$$

Example 5-1

Let $U = \{0, 1, 2, 3, 4, 5\}$. If p is the proposition "u is odd" and q the proposition "$u > 2$," then $\sim p$, $p \vee q$, and $p \wedge q$ are the propositions "u is not odd," "u is odd or $u > 2$," and "u is odd and $u > 2$," respectively. Since $T_p = \{1, 3, 5\}$ and $T_q = \{3, 4, 5\}$, we have

$$T_{\sim p} = \{1, 3, 5\}' = \{0, 2, 4\}$$

$$T_{p \vee q} = \{1, 3, 5\} \cup \{3, 4, 5\} = \{1, 3, 4, 5\}$$

$$T_{p \wedge q} = \{1, 3, 5\} \cap \{3, 4, 5\} = \{3, 5\} \qquad \square$$

Propositions that are constructed by taking negations, disjunctions, and conjunctions of "elementary" propositions (that is, of propositions that do not involve the symbols $\sim$, $\vee$, and $\wedge$) are sometimes called *compound propositions.* The truth sets of compound propositions can be determined by successive application of rules (5-1), (5-2), and (5-3). For example, if r is the proposition "u is a prime number less than 6" ($T_r = \{2, 3, 5\}$) and p and q are as defined in Example 5-1, then the proposition "u is odd and $u \leq 2$, or u is prime" (over $U = \{0, 1, 2, 3, 4, 5\}$) is, symbolically, the proposition $(p \wedge (\sim q)) \vee r$ and, hence, has the truth set

$$(\{1, 3, 5\} \cap \{3, 4, 5\}') \cup \{2, 3, 5\} = \{1, 2, 3, 5\}$$

If $T_p \cap T_q = \varnothing$, p and q are said to be *mutually exclusive.* In this case the truth of p always implies the falsity of q (and conversely).

It is customary to associate the numerical value 1 with a true proposition, and the value 0 with a false one. In this context, 0 and 1 are referred to as *truth values.*[1]

The truth values of compund propositions can be determined from the truth values of the elementary propositions of which they are composed,

[1]Some authors use the letters T and F for the truth values 1 and 0, respectively.

Table 5-1

(a) TRUTH TABLE FOR p

p	$\sim p$
0	1
1	0

(b) TRUTH TABLE FOR $p \vee q$

p	q	$p \vee q$
0	0	0
0	1	1
1	0	1
1	1	1

(c) TRUTH TABLE FOR $p \wedge q$

p	q	$p \wedge q$
0	0	0
0	1	0
1	0	0
1	1	1

Table 5-2 TRUTH TABLE FOR $((p \wedge q) \vee (\sim p) \wedge r)) \vee (q \wedge r)$

p	q	r	$\sim p$	$p \wedge q$	$(\sim p) \wedge r$	$q \wedge r$	$(p \wedge q) \vee ((\sim p) \wedge r)$	$((p \wedge q) \vee ((\sim p) \wedge r)) \vee (q \wedge r)$
0	0	0	1	0	0	0	0	0
0	0	1	1	0	1	0	1	1
0	1	0	1	0	0	0	0	0
0	1	1	1	0	1	1	1	1
1	0	0	0	0	0	0	0	0
1	0	1	0	0	0	0	0	0
1	1	0	0	1	0	0	1	1
1	1	1	0	1	0	1	1	1

using the definitions of negation, disjunction, and conjunction. This evaluation can be systematized by constructing *truth tables* such as shown in Tables 5-1 and 5-2.

Table 5-1 shows the truth tables for the propositions $\sim p$, $p \vee q$, and $p \wedge q$ (the entries follow immediately from the definitions). Table 5-2 illustrates the construction of the truth table for the proposition

$$s = ((p \wedge q) \vee ((\sim p) \wedge r)) \vee (q \wedge r)$$

In this table, the first three columns list all possible truth values of p, q, and r. The remaining columns show how the truth values of s are determined by successively determining those of $\sim p$, $p \wedge q$, $(\sim p) \wedge r$, $q \wedge r$, $(p \wedge q) \vee ((\sim p) \wedge r)$, and, finally, s. Each column is constructed from the preceding ones by direct invocation of the truth tables in Table 5-1.

If a compund proposition s is a tautology, then it must always be true—regardless of the truth values of the elementary propositions. Thus, s is a tautology if and only if column s in the truth table contains only 1s. Similarly, s is a contradiction if and only if column s in the truth table contains only 0s.

PROBLEMS

1. Consider the propositions:

p: John is watching television
q: John is inside
r: John is doing his homework

Translate the following compound propositions into symbolic notation, using the symbols $p, q, r, \sim, \vee, \wedge$, and parentheses only.
(a) Either John is watching television or he is inside.
(b) Neither is John watching television, nor is he doing his homework.
(c) John is watching television and not doing his homework.
(d) John is inside doing his homework, not watching television.

2. Using the specifications of p, q, and r of Problem 1, translate the following propositions into acceptable English:
(a) $(\sim p) \wedge (\sim q)$ (b) $p \vee (q \wedge r)$
(c) $\sim((\sim q) \wedge r)$ (d) $((\sim p) \wedge q) \vee (\sim r)$

3. The following are propositions over $U = \{0, 1, 2, 3, 4, 5, 6, 7\}$:

p: $n^2 < 4$
q: $n \geq 3$
r: n is a multiple of 2

Find the truth sets of:
(a) $p \vee q$ (b) $p \wedge q$
(c) $(\sim p) \vee q$ (d) $p \wedge (\sim q)$
(e) $\sim((\sim p) \wedge (\sim q))$ (f) $(\sim p) \vee (q \wedge (\sim r))$
(g) $(p \wedge (\sim q)) \vee (r \wedge q)$

4. Using the definitions of p, q, and r of Problem 3, find a compound proposition whose truth set is $\{0, 1, 2, 3, 5, 7\}$.

5. If p and q are propositions over U, prove that the truth set of $\sim(p \wedge q)$ and the truth set of $(\sim p) \vee (\sim q)$ are the same.

6. Let p and q be propositions over U. Express the truth set of the proposition "p or q but not both p and q" in terms of T_p and T_q.

7. Construct the truth tables of the following compound propositions, composed of p, q, and r:
(a) $(p \wedge q) \vee (\sim r)$ (b) $(p \vee q) \wedge ((\sim p) \vee (\sim r))$
(c) $(((\sim p) \wedge (\sim q)) \vee r) \vee ((p \wedge (\sim q)) \vee ((\sim p) \wedge (\sim q)))$

8. Prove that the following propositions are tautologies:
(a) $(p \wedge q) \vee ((\sim p) \wedge q) \vee (p \wedge (\sim q)) \vee ((\sim p) \wedge (\sim q))$
(b) $(p \vee q) \wedge ((\sim p) \vee q) \wedge (p \vee (\sim q)) \wedge ((\sim p) \vee (q))$
(c) $\sim(p \vee q) \vee ((\sim p) \wedge q) \vee p$

5-3. IMPLICATION

If p and q are propositions, a proposition of the form "if p, then q" is called an *implication* and is denoted by $p \Rightarrow q$. In the implication $p \Rightarrow q$, p

is called the *hypothesis* (or *antecedent*) and q the *conclusion* (or *consequent*). By definition, the proposition $p \Rightarrow q$ is true in all cases except where a true hypothesis p appears with a false conclusion q. For example, the propositions "If $2 + 2 = 4$, then $6 > 0$"; "If $2 + 2 = 5$, then $6 > 0$"; and "If $2 + 2 = 5$, then $6 < 0$" are all true; but the proposition "If $2 + 2 = 4$, then $6 < 0$" is false. The truth table of $p \Rightarrow q$ is shown in Table 5-3.

Table 5-3 TRUTH TABLE FOR $p \Rightarrow q$

p	q	$p \Rightarrow q$
0	0	1
0	1	1
1	0	0
1	1	1

If p and q are propositions over a universal set U, the implication $p \Rightarrow q$ is true for all $u \in U$, except for those u which are in both the truth set of p and the falsity set of q. That is,

$$T_{p \Rightarrow q} = (T_p \cap T'_q)'$$

or, using (1-3),

$$T_{p \Rightarrow q} = (T_p - T_q)'$$

Implication is illustrated by the Venn diagram in Fig. 5-1, where the cross-hatched region represents $T_{p \Rightarrow q}$ (and the clear region $T_p - T_q$). Using this figure, or De Morgan's law, we can also write

$$T_{p \Rightarrow q} = T'_p \cup T_q$$

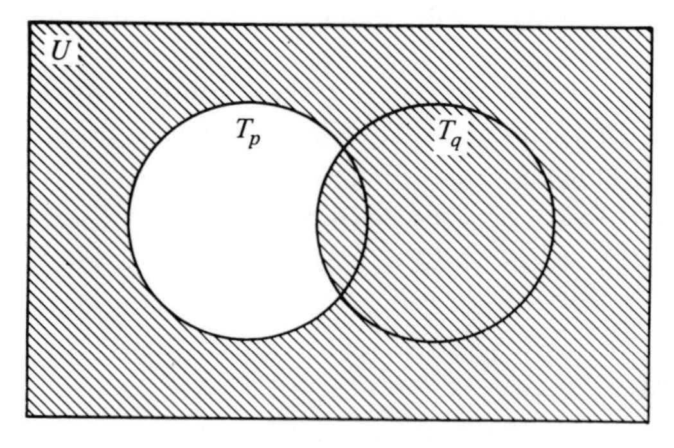

Fig. 5-1 Venn diagram displaying $T_{p \Rightarrow q}$

Example 5-2

Let U be the set $\mathbb{N}$ of positive integers. If p is the proposition "n is a prime number" and q is the proposition "n is an odd number," then T_p is the set $\mathbb{P}$ of prime numbers, and T_q the set O of odd numbers. The implication $p \Rightarrow q$, that is, the proposition "If n is a prime number, then n is an odd number," has the truth set $\mathbb{P}' \cup O$, which consists of all positive integers except 2. □

It is clear that $T_{p \Rightarrow q} = (T_p - T_q)' = U$ if and only if $T_p - T_q = \varnothing$ (in which case the clear region in Fig. 5-1 vanishes) and, hence, if and only if $T_p \subset T_q$. Consequently, $p \Rightarrow q$ is a tautology if and only if $T_p \subset T_q$. Thus, the task of proving that $p \Rightarrow q$ is a tautology can be accomplished by proving that T_p is a subset of T_q. When $p \Rightarrow q$ is a tautology, we say that *p implies q*, or *p is sufficient for q*, or *q is necessary for p*.

Example 5-3

Consider the following theorem: "For all $n \in \mathbb{N}$, if $2^n < 100$, then $n^2 < 100$." This theorem can be proved by proving that the proposition $p \Rightarrow q$, where p and q are the propositions "$2^n < 100$" and "$n^2 < 100$," respectively, is a tautology. This, in turn, follows from the fact that $T_p = \{1, 2, 3, 4, 5, 6\}$ and $T_q = \{1, 2, 3, 4, 5, 6, 7, 8, 9\}$; hence, that $T_p \subset T_q$. □

PROBLEMS

1. Consider the propositions p, q, and r of Problem 1, Sec. 5-2, and translate the following compound propositions into symbolic notation, using the symbols $p, q, r, \sim, \vee, \wedge, \Rightarrow$, and parentheses only.
   (a) If John is inside and not watching television, then he is doing his homework.
   (b) If John is not inside, then neither does he watch television nor is he doing his homework.
   (c) If John is either outside or doing his homework, then he is not watching television.

2. Find the truth sets of these propositions over $\mathbb{N}$:
   (a) $(n^2 - 5n - 6 \geq 0) \Rightarrow (n \leq 10)$
   (b) $((n < 2) \vee (n > 4)) \Rightarrow ((n \geq 3) \wedge (n \leq 7))$
   (c) $(n! = m) \Rightarrow$ (m is a multiple of 10)
   (d) $(2^n < 50) \Rightarrow$ (2^n is divisible by 6)
   (e) $((n > 10) \Rightarrow (n \leq 5)) \Rightarrow (8 \leq n \leq 15)$
   Are there any tautologies among these propositions? Any contradictions?

3. Construct the truth tables of these compound propositions, composed of p, q, and r:
   (a) $(p \wedge (\sim q)) \Rightarrow ((\sim p) \vee (\sim r))$

(b) $((p \vee q) \Rightarrow (q \wedge r)) \Rightarrow (p \wedge (\sim r))$
(c) $(((\sim p) \Rightarrow (p \wedge (\sim q)) \Rightarrow r) \wedge q) \vee (\sim r)$

4. Let p, q, and r be propositions over U. Express the truth sets of these propositions in terms of T_p, T_q, and T_r:
(a) $p \Rightarrow (\sim q)$ (b) $((\sim p) \wedge q) \Rightarrow (\sim r)$
(c) $((\sim p) \vee q) \Rightarrow (p \wedge (\sim q))$

5. Prove that these propositions are tautologies:
(a) $p \Rightarrow (p \vee q)$ (b) $((p \Rightarrow q) \wedge p) \Rightarrow q$
(c) $((p \Rightarrow q) \Rightarrow ((\sim q) \Rightarrow (\sim p))) \Rightarrow (((\sim q) \Rightarrow (\sim p)) \Rightarrow (p \Rightarrow q))$

6. Prove these theorems (following the technique used in Example 5-3):
(a) For all $n \in \mathbb{N}$, $n^3 > 200$ implies $\log_2 n > 2$.
(b) For all $n \in \mathbb{N}$, n is a divisor of 30 other than 1, implies that n is not a power of 4.
(c) For all $i \in \mathbb{I}$, i is a solution of $i^2 + 3i - 18 = 0$ implies that i is divisible by 3.

5-4. EQUIVALENCE

If p and q are propositions, the compound proposition $(p \Rightarrow q) \wedge (q \Rightarrow p)$ is called an *equivalence* and is denoted by $p \Leftrightarrow q$. The truth table of $p \Leftrightarrow q$, constructible from those of Tables 5-1 and 5-3, is shown in Table 5-4. It can be observed that $p \Leftrightarrow q$ is true only when both p and q are true, or when both p and q are false. When this is the case, p and q are said to be *equivalent*. For example, the propositions "Triangle A is equilateral" and "Triangle A is equiangular" are equivalent because they are either both true or both false.

Table 5-4 TRUTH TABLE FOR $p \Leftrightarrow q$

p	q	$p \Leftrightarrow q$
0	0	1
0	1	0
1	0	0
1	1	1

If p and q are propositions over a universal set U, then the equivalence $p \Leftrightarrow q$ is true only for those u which are in both the truth sets of p and q, or are in both the falsity sets of p and q. That is,

$$T_{p \Leftrightarrow q} = (T_p \cap T_q) \cup (T'_p \cap T'_q)$$

This condition is illustrated by the Venn diagram in Fig. 5-2, where the cross-hatched region represents $T_{p \Leftrightarrow q}$. Using this figure, or the distributive and De

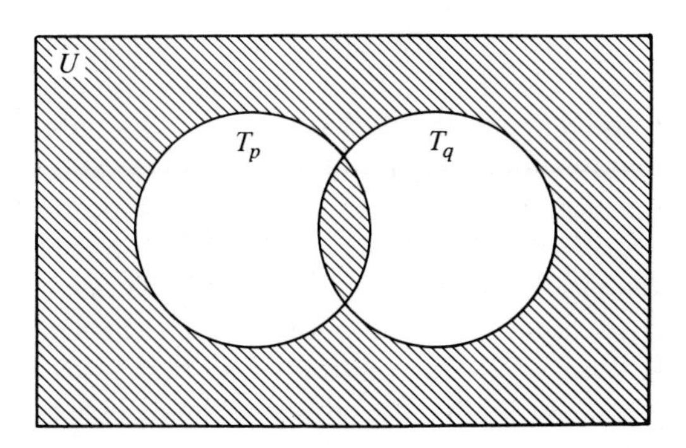

Fig 5-2. Venn diagram displaying $T_{p \Leftrightarrow q}$

Morgan's laws, we can also write

$$T_{p \Leftrightarrow q} = ((T_p - T_q) \cup (T_q - T_p))'$$

Example 5-4

Let $U = \{n \in \mathbb{N} \mid 1 \leq n \leq 9\}$. If p is the proposition "n is an odd number" and q is the proposition "$(3 \leq n \leq 8) \wedge (n \neq 7)$," then $T_p = \{1, 3, 5, 7, 9\}$ and $T_q = \{3, 4, 5, 6, 8\}$. The equivalence $p \Leftrightarrow q$ has the truth set

$$(T_p \cap T_q) \cup (T'_p \cap T'_q) = \{3, 5\} \cup \{2\} = \{2, 3, 5\}$$

Indeed, "2 is an odd number" and "$(3 \leq 2 \leq 8) \wedge (2 \neq 7)$" are both false; "3 is an odd number" and "$(3 \leq 3 \leq 8) \wedge (3 \neq 7)$" are both true; "5 is an odd number" and "$(3 \leq 5 \leq 8) \wedge (5 \neq 7)$" are both true. □

It is clear that $T_{p \Leftrightarrow q} = ((T_p - T_q) \cup (T_q - T_p))' = U$ if and only if $(T_p - T_q) \cup (T_q - T_p) = \varnothing$ (in which case the clear region in Fig. 5-2 vanishes), hence, if and only if $T_p = T_q$. Thus, $p \Leftrightarrow q$ is a tautology if and only if $T_p = T_q$. The task of proving that $p \Leftrightarrow q$ is a tautology can be accomplished by proving that T_p and T_q are identical. When $p \Leftrightarrow q$ is a tautology, we say that *p if and only if q* (and conversely),[1] or *p is necessary and sufficient for q* (and conversely).

Example 5-5

Consider this theorem: "For all $n \in \mathbb{N}$, n is a power of 3 less than 10 if and only if $n^2 - 12n + 27 = 0$." This theorem can be verified by proving that the proposition $p \Leftrightarrow q$, where p and q are the propositions "n is a power of 3 less than 10" and "$n^2 - 12n + 27 = 0$," respectively, is a tautology. This, in turn, follows from the fact that $T_p = T_q = \{3, 9\}$. □

[1] "If and only if" is sometimes abbreviated into "iff."

Let s_1 and s_2 be compound propositions composed of the same elementary propositions, and consider a truth table containing columns s_1 and s_2. From the definition of equivalence it follows that s_1 and s_2 are equivalent (and, hence, $s_1 \Leftrightarrow s_2$ is a tautology) if and only if columns s_1 and s_2 are identical.

Given a proposition s of the form $p \Rightarrow q$, the proposition $(\sim q) \Rightarrow (\sim p)$ is called the *contrapositive* of s, the proposition $q \Rightarrow p$ is called the *converse* of s, and the proposition $(\sim p) \Rightarrow (\sim q)$ is called the *inverse* of s. Table 5-5 shows that s and its contrapositive are equivalent; hence, s can be proved either by proving that $p \Rightarrow q$ is a tautology or that $(\sim q) \Rightarrow (\sim p)$ is a tautology. The table also shows that the converse and inverse of s are equivalent.

Table 5-5 TRUTH TABLE FOR $p \Rightarrow q$, $(\sim q) \Rightarrow (\sim p)$, $q \Rightarrow p$, AND $(\sim p) \Rightarrow (\sim q)$

p	q	$\sim p$	$\sim q$	$p \Rightarrow q$	$(\sim q) \Rightarrow (\sim p)$	$q \Rightarrow p$	$(\sim p) \Rightarrow (\sim q)$
0	0	1	1	1	1	1	1
0	1	1	0	1	1	0	0
1	0	0	1	0	0	1	1
1	1	0	0	1	1	1	1

PROBLEMS

1. Use the set laws of Tables 1-3 and 1-4 to prove that

$$T_{p \Leftrightarrow q} = ((T_p - T_q) \cup (T_q - T_p))'$$

2. Find the truth sets of these propositions over $\mathbb{N}$:
(a) $(n^2 - 5n - 6 \geq 0) \Leftrightarrow (5 \leq n \leq 10)$
(b) $((n < 2) \vee (n > 4)) \Leftrightarrow ((n \geq 3) \wedge (n \leq 7))$
(c) $(n^2 - n = 0) \Leftrightarrow ((n^2 - 4n + 3 = 0) \wedge (n \leq 2))$
(d) $(2^n < 100) \Leftrightarrow (n^2 < 100)$
(e) $(n \geq 10) \Leftrightarrow (2^n \leq 1000)$
Are there any tautologies among these propositions? Any contradictions?

3. Construct the truth tables of these propositions, composed of p, q, and r:
(a) $(p \wedge (\sim q)) \Leftrightarrow ((\sim p) \vee (\sim r))$
(b) $((p \vee q) \Leftrightarrow (q \wedge r)) \Leftrightarrow (p \wedge (\sim r))$
(c) $((\sim p) \Leftrightarrow ((p \wedge (\sim q)) \Leftrightarrow r) \vee q) \wedge (\sim r)$

4. Let p, q, and r be propositions over U. Express the truth sets of the following propositions in terms of T_p, T_q, and T_r:
(a) $(\sim p) \Leftrightarrow q$
(b) $(p \wedge q) \Leftrightarrow \sim(\sim q)$
(c) $((\sim q) \vee r) \Leftrightarrow ((\sim p) \wedge q)$

5. Prove that these propositions are tautologies:
(a) $(p \Rightarrow q) \Leftrightarrow ((\sim p) \vee q)$
(b) $(p \Leftrightarrow q) \Leftrightarrow ((p \wedge q) \vee ((\sim p) \wedge (\sim q)))$
(c) $(p \Leftrightarrow q) \Leftrightarrow (((r \wedge p) \Leftrightarrow (r \wedge q)) \wedge ((r \vee p) \Leftrightarrow (r \vee q)))$

6. Prove these theorems (following the technique used in Example 5-5):
   (a) For all $n \in \mathbb{N}$, $2^n \leq 100$ if and only if $n^2 < 40$.
   (b) For all $n \in \mathbb{N}$, n is a prime number and a divisor of 30 if and only if n is neither greater than 5 nor a power of 4.
   (c) For all $i \in \mathbb{I}$, i is a solution of $i^3 - i = 0$ if and only if $|i^2| = |i|$.

7. Show that $p \Leftrightarrow (q \Leftrightarrow r)$ is equivalent to $(p \Leftrightarrow q) \Leftrightarrow r$. Is $p \Rightarrow (q \Rightarrow r)$ equivalent to $(p \Rightarrow q) \Rightarrow r$?

5-5. THE ALGEBRA OF PROPOSITIONS

Using truth tables, a number of useful equivalences involving negation, disjunction, and conjunction of propositions can be derived. Some of these equivalences, referred to as propositional laws, are shown in Table 5-6, where p, q, and r denote arbitrary propositions, and where 1 and 0 denote tautology and contradiction, respectively. The propositional laws, together with the stipulation that 1 and 0 be distinct, can serve as postulates for an algebraic system $\langle P_U; \sim, \vee, \wedge \rangle$, whose domain P_U is the set of all propositions over U, and where element equality is denoted by the equivalence symbol $\Leftrightarrow$. This algebraic system is called the *algebra of propositions* (or *propositional calculus*).

Table 5-6 PROPOSITIONAL LAWS

Commutative Laws	1. $(p \vee q) \Leftrightarrow (q \vee p)$	1′. $(p \wedge q) \Leftrightarrow (q \wedge p)$
Associative Laws	2. $(p \vee (q \vee r)) \Leftrightarrow ((p \vee q) \vee r)$	2′. $(p \wedge (q \wedge r)) \Leftrightarrow ((p \wedge q) \wedge r)$
Distributive Laws	3. $(p \wedge (q \vee r)) \Leftrightarrow ((p \wedge q) \vee (p \wedge r))$	3′. $(p \vee (q \wedge r)) \Leftrightarrow ((p \vee q) \wedge (p \vee r))$
Identity Laws	4. $(p \vee 0) \Leftrightarrow (0 \vee p) \Leftrightarrow p$	4′. $(p \wedge 1) \Leftrightarrow (1 \wedge p) \Leftrightarrow p$
Negation Laws	5. $(p \vee (\sim p)) \Leftrightarrow 1$	5′. $(p \wedge (\sim p)) \Leftrightarrow 0$

Comparison of the propositional laws of Table 5-6 with the set laws of Table 1-3 reveals that the former can be obtained from the latter simply by replacing sets with propositions, equalities with equivalences, negations with complements, unions with disjunctions, intersections with conjunctions, U with 1, and $\varnothing$ with 0. Since the set laws of Table 1-3 constitute the postulates of the algebra of sets $\langle \mathbf{2}^U; ', \cup, \cap \rangle$ (see Sec. 4-5), this observation immediately raises the possibility that the algebra of sets and the algebra of propositions might be isomorphic. This, indeed is the case, and can be verified as follows: Let h be the function

$$h: \mathbf{2}^U \longrightarrow P_U$$

where $\qquad h(S) = \text{proposition whose truth set is } S$

Since propositions are equivalent if and only if their truth sets are equal, and since every proposition over U has a truth set in $\mathbf{2}^U$, h is a bijection. If $h(S) = p$ (and, hence, $S = T_p$), we have, by (5-1),

$$h(S') = h(T'_p) = h(T_{\sim p}) = \sim p = \sim(h(S))$$

If $h(S_1) = p_1$ and $h(S_2) = p_2$ (and, hence, $S_1 = T_{p_1}$ and $S_2 = T_{p_2}$), we have, by (5-2) and (5-3).

$$\begin{aligned} h(S_1 \cup S_2) &= h(T_{p_1} \cup T_{p_2}) = h(T_{p_1 \vee p_2}) \\ &= p_1 \vee p_2 = h(S_1) \vee h(S_2) \\ h(S_1 \cap S_2) &= h(T_{p_1} \cap T_{p_2}) = h(T_{p_1 \wedge p_2}) \\ &= p_1 \wedge p_2 = h(S_1) \wedge h(S_2) \end{aligned}$$

Thus, h is an isomorphism from $\langle \mathbf{2}^U; ', \cup, \cap \rangle$ onto $\langle P_U; \sim, \vee, \wedge \rangle$; hence, the two algebraic systems are isomorphic.

This isomorphism enables us to produce theorems in the algebra of propositions simply by taking theorems from the algebra of sets and replacing $=$, $'$, $\cup$, $\cap$, U, and $\varnothing$ with $\Leftrightarrow$, $\sim$, $\vee$, $\wedge$, 1, and 0, respectively. For example, the equivalences listed in Table 5-7 can be obtained directly from the identities of Table 1-4.

The *dual* of a compund proposition (without $\Rightarrow$ symbols) is the same proposition with 0 replacing 1, 1 replacing 0, $\vee$ replacing $\wedge$, and $\wedge$ replacing $\vee$. For example, the dual of the equivalence

$$((p \vee q) \vee ((\sim p) \wedge (\sim p))) \Longleftrightarrow 1$$

is

$$((p \wedge q) \wedge ((\sim p) \vee (\sim q))) \Longleftrightarrow 0$$

The isomorphism of the algebras of sets and propositions implies that the principle of duality which holds in the algebra of sets, also prevails in the

Table 5-7 ADDITIONAL PROPOSITIONAL LAWS

Involution Law	1. (and 1′.) $(\sim(\sim p)) \Leftrightarrow p$	
Idempotent Laws	2. $(p \vee p) \Leftrightarrow p$	2′. $(p \wedge p) \Leftrightarrow p$
Null Laws	3. $(p \vee 1) \Leftrightarrow 1$	3′. $(p \wedge 0) \Leftrightarrow 0$
Absorption Laws	4. $(p \vee (p \wedge q)) \Leftrightarrow p$	4′. $(p \wedge (p \vee q)) \Leftrightarrow p$
De Morgan's Laws	5. $(\sim(p \vee q)) \Leftrightarrow ((\sim p) \wedge (\sim q))$	5′. $(\sim(p \wedge q)) \Leftrightarrow ((\sim p) \vee (\sim q))$

algebra of propositions. Thus, given any true proposition s in P_U, then its dual s^D must also be true.

An element of $\langle P_U; \sim, \vee, \wedge \rangle$ which is formed by applying the operations $\sim$, $\vee$, and $\wedge$ to 0, 1, $p_1, p_2, \ldots, p_r$, is referred to as a *proposition generated by* $p_1, p_2, \ldots, p_r$. In the algebra of sets, this corresponds to a set generated by $A_1, A_2, \ldots, A_r$ which, as shown in Secs. 1-8 and 1-9, is always expressible in the minset and maxset normal forms. In the algebra of propositions, these forms are called the *disjunctive normal form* and *conjunctive normal form*, respectively. A procedure for obtaining these forms can be deduced directly from Algorithms 1-1 and 1-2, and is outlined in Algorithm 5-1 below. In this algorithm, "row $\delta_1 \delta_2 \ldots \delta_r$" (where δ_i is either 0 or 1) denotes the truth-table row that contains the digits $\delta_1, \delta_2, \ldots, \delta_r$ in the columns labeled "p_1," "p_2," ..., "p_r," respectively. The symbols $\bigvee_{k=1}^{h} q_k$ and $\bigwedge_{k=1}^{h} q_k$ stand for the propositions $q_1 \vee q_2 \vee \ldots \vee q_h$ and $q_1 \wedge q_2 \wedge \ldots \wedge q_h$, respectively.

ALGORITHM 5-1

To find the disjunctive and conjunctive normal forms of a given proposition s generated by $p_1, p_2, \ldots, p_r$:

(i) Construct the truth table of s. If the column labeled s contains no 1s, then $s \Leftrightarrow 0$; if this column contains no 0s, then $s \Leftrightarrow 1$. Otherwise:

(ii) If column s contains 1s in rows $\delta_{11} \delta_{12} \ldots \delta_{1r}$, $\delta_{21} \delta_{22} \ldots \delta_{2r}$, ..., $\delta_{l1} \delta_{l2} \ldots \delta_{lr}$, then the disjunctive normal form of s is given by

$$s \Longleftrightarrow \bigvee_{i=1}^{l} \left(\bigwedge_{j=1}^{r} p_{ij}\right)$$

where

$$p_{ij} = \begin{cases} \sim p_i & \text{if } \delta_{ij} = 0 \\ p_i & \text{if } \delta_{ij} = 1 \end{cases}$$

If column s contains 0s in these rows, then the conjunctive normal form of s is given by

$$s \Longleftrightarrow \bigwedge_{i=1}^{l} \left(\bigvee_{j=1}^{r} p_{ij}\right)$$

where

$$p_{ij} = \begin{cases} p_i & \text{if } \delta_{ij} = 0 \\ \sim p_i & \text{if } \delta_{ij} = 1 \end{cases}$$

□

Example 5-6

Consider the proposition

$$((p \wedge q) \vee ((\sim p) \wedge r)) \vee (q \wedge r)$$

generated by p, q, r. From the truth table shown in Table 5-2, the disjunctive normal form of this proposition is given by

$$((\sim p) \wedge (\sim q) \wedge r) \vee ((\sim p) \wedge q \wedge r) \vee (p \wedge q \wedge (\sim r)) \vee (p \wedge q \wedge r)$$

while the conjunctive normal form is given by

$$(p \vee q \vee r) \wedge (p \vee (\sim q) \vee r) \wedge ((\sim p) \vee q \vee r) \wedge ((\sim p) \vee q \vee (\sim r)) \quad \square$$

The normal forms of propositions generated by $p_1, p_2, \ldots, p_r$ can also be found through methods analogous to those described in Algorithms 1-3 and 1-4.

PROBLEMS

1. Show that if the truth set of p is T_p, then the truth set of p^D is T_p^D. (*Hint*: Prove by induction on the shape of p.)
2. Let p and q be propositions in P_U. Show that if $p \Longrightarrow q$, then $q^D \Longrightarrow p^D$. (See Problem 3, Sec. 4-6.)
3. Using the postulates of the algebra of propositions listed in Table 5-6 and their corollaries listed in Table 5-7, prove these tautologies and write their duals:
   (a) $(((\sim p) \vee (\sim q)) \vee (\sim((\sim p) \vee q))) \Leftrightarrow p$
   (b) $((p \vee (\sim q)) \wedge (p \vee q) \wedge ((\sim p) \vee (\sim q))) \Leftrightarrow (\sim((\sim p) \vee q))$
   (c) $(q \vee (\sim(((\sim p) \vee q) \wedge p))) \Leftrightarrow 1$
4. Find the disjunctive and conjunctive normal forms of these propositions generated by p and q:
   (a) 1 (b) 0 (c) p (d) $\sim p$
   (e) $p \vee q$ (f) $p \wedge q$
5. Find the disjunctive and conjunctive normal forms of:
   (a) $(p \wedge (\sim q)) \vee (q \wedge (p \vee)\sim r)))$ (generated by p, q, r)
   (b) $((p \vee (\sim s)) \wedge (\sim((\sim q) \vee r))) \vee (p \wedge q \wedge s)$ (generated by p, q, r, s)

5-6. THEOREM PROVING

A task often encountered in mathematics, as well as in daily life, is that of proving an alleged statement on the basis of a given set of assumptions. More precisely, we are given the *premises* $p_1, p_2, \ldots, p_r$, and a *consequence* c, and are called upon to prove that the proposition

$$(p_1 \wedge p_2 \wedge \ldots \wedge p_r) \Longrightarrow c$$

is *valid*—that is, a tautology. A proposition of this form is called an *argumental*. Mathematical theorems are often expressed in the form of argumentals, where the premises are postulates, defintions, or previously proven theorems.

Example 5-7

Premises: Businessmen are ambitious. Early risers do not like orange juice. Ambitious people are early risers.

Consequence: Businessmen do not like orange juice.

The "theorem" can be proved by defining the propositions:

p: An individual is a businessman
q: An individual is ambitious
r: An individual is an early riser
s: An individual likes orange juice

and then demonstrating that the argumental

$$((p \Longrightarrow q) \wedge (r \Longrightarrow (\sim s)) \wedge (q \Longrightarrow r)) \Longrightarrow (p \Longrightarrow (\sim s)) \qquad (5\text{-}4)$$

is a tautology. This is done in Table 5-8. □

Table 5-8 TRUTH TABLE FOR EXAMPLE 5-7

p	q	r	s	$\sim s$	u: $p \Longrightarrow q$	v: $r \Longrightarrow (\sim s)$	w: $q \Longrightarrow r$	x: $u \wedge v \wedge w$	y: $p \Longrightarrow (\sim s)$	$x \Longrightarrow y$
0	0	0	0	1	1	1	1	1	1	1
0	0	0	1	0	1	1	1	1	1	1
0	0	1	0	1	1	1	1	1	1	1
0	0	1	1	0	1	0	1	0	1	1
0	1	0	0	1	1	1	0	0	1	1
0	1	0	1	0	1	1	0	0	1	1
0	1	1	0	1	1	1	1	1	1	1
0	1	1	1	0	1	0	1	0	1	1
1	0	0	0	1	0	1	1	0	1	1
1	0	0	1	0	0	1	1	0	0	1
1	0	1	0	1	0	1	1	0	1	1
1	0	1	1	0	0	0	1	0	0	1
1	1	0	0	1	1	1	0	0	1	1
1	1	0	1	0	1	1	0	0	0	1
1	1	1	0	1	1	1	1	1	1	1
1	1	1	1	0	1	0	1	0	0	1

The truth-table method of proving theorems becomes increasingly cumbersome as the number of elementary propositions grows (when this number is k, the number of table rows is 2^k). An alternative procedure is to use various "standard" implications and equivalences to simplify the argumental in successive steps until its validity becomes evident. Although this procedure is not systematic, it often results in concise and elegant proofs of complex

Table 5-9 SOME IMPLICATIONS AND EQUIVALENCES

Detachment	$((p \Rightarrow q) \wedge p) \Rightarrow q$
Disjunctive addition	$p \Rightarrow (p \vee q)$
Conjunctive simplification	$(p \wedge q) \Rightarrow p$
Chain rule	$((p \Rightarrow q) \wedge (q \Rightarrow r)) \Rightarrow (p \Rightarrow r)$
Disjunctive simplification	$((p \vee q) \wedge (\sim p)) \Rightarrow q$
Conditional rules	(a) $(p \Rightarrow q) \Leftrightarrow ((\sim p) \vee q)$
	(b) $(p \Rightarrow q) \Leftrightarrow ((\sim q) \Rightarrow (\sim p))$
Equivalence rules	(a) $(p \Leftrightarrow q) \Leftrightarrow ((p \Rightarrow q) \wedge (q \Rightarrow p))$
	(b) $(p \Leftrightarrow q) \Leftrightarrow ((p \wedge q) \vee ((\sim p) \wedge (\sim q)))$
	(c) $(\sim (p \Leftrightarrow q)) \Leftrightarrow (p \Leftrightarrow (\sim q))$
Implication rules	(a) $(p \Rightarrow q) \Rightarrow ((p \wedge r) \Rightarrow (q \wedge r))$
	(b) $(p \Rightarrow q) \Rightarrow ((p \vee r) \Rightarrow (q \vee r))$
Contrapositive inference	$((p \Rightarrow q) \wedge (\sim q)) \Rightarrow (\sim p)$

theorems. Table 5-9 lists some of the implications and equivalences that can be used in such proofs.

Example 5-8

The argumental (5-4) can be proved by first applying the "chain rule" twice:

$$((r \Longrightarrow (\sim s)) \wedge (q \Longrightarrow r)) \Longrightarrow (q \Longrightarrow (\sim s))$$
$$((p \Longrightarrow q) \wedge (q \Longrightarrow (\sim s))) \Longrightarrow (p \Longrightarrow (\sim s))$$

Then, using the "implication rule (a)":

$$((p \Longrightarrow q) \wedge (r \Longrightarrow (\sim s)) \wedge (q \Longrightarrow r)) \Longrightarrow$$
$$((p \Longrightarrow q) \wedge (q \Longrightarrow (\sim s))) \Longrightarrow (p \Longrightarrow (\sim s)) \qquad \square$$

Example 5-9

The argumental

$$((p \vee q) \wedge (q \Longrightarrow (\sim r)) \wedge (\sim p)) \Longrightarrow (\sim r)$$

can be proved as follows:

$$(p \vee q) \Longrightarrow ((\sim p) \Longrightarrow q) \qquad \text{Conditional (a)}$$
$$(((\sim p) \Longrightarrow q) \wedge (q \Longrightarrow (\sim r))) \Longrightarrow ((\sim p) \Longrightarrow (\sim r)) \qquad \text{Chain}$$
$$(((\sim p) \Longrightarrow (\sim r)) \wedge (\sim p)) \Longrightarrow (\sim r) \qquad \text{Detachment}$$
$$((p \vee q) \wedge (q \Longrightarrow (\sim r)) \wedge (\sim p)) \Longrightarrow$$
$$(((\sim p) \Longrightarrow q) \wedge (q \Longrightarrow (\sim r)) \wedge (\sim p)) \Longrightarrow$$
$$(((\sim p) \Longrightarrow (\sim r)) \wedge (\sim p)) \Longrightarrow (\sim r) \qquad \text{Implication (a)} \quad \square$$

In Sec. 5-4 we saw that a theorem and its contrapositive are equivalent. Thus, we can prove that $p \Rightarrow q$ by proving that $(\sim q) \Rightarrow (\sim p)$. This procedure is an example of an *indirect proof*, which can be attempted when direct proof procedure, illustrated in Examples 5-7 to 5-9, is too difficult to carry out. Another example of indirect proof is the so-called *proof by contradiction* (or *reductio ad absurdum*), which exploits the fact that $p \Rightarrow q$ is true if and only if $p \wedge (\sim q)$ is false. Thus, proof that the argumental

$$(p_1 \wedge p_2 \wedge \ldots \wedge p_r) \Longrightarrow c$$

is a tautology can be accomplished by proving that

$$p_1 \wedge p_2 \wedge \ldots \wedge p_r \wedge (\sim c)$$

is a contradiction. This can be done by reducing $p_1 \wedge p_2 \wedge \ldots \wedge p_r \wedge (\sim c)$ into a conjunction that contains a contradiction of the form $s \wedge (\sim s)$. Often, s is one of the premises, and the proof culminates with the demonstration that this premise is false.

Example 5-10

Premises: If individual X does not live in France, then he does not speak French. X does not drive a Chevrolet. If X lives in France, then he rides a bicycle. Either X speaks French, or he drives a Chevrolet.

Consequence: X rides a bicycle.

Proof Define the propositions:

p: X lives in France
q: X speaks French
r: X drives a Chevrolet
s: X rides a bicycle

The argumental whose validity is to be proved is

$$(((\sim p) \Longrightarrow (\sim q)) \wedge (\sim r) \wedge (p \Longrightarrow s) \wedge (q \vee r)) \Longrightarrow s$$

Instead, let us prove that the proposition

$$((\sim p) \Longrightarrow (\sim q)) \wedge (\sim r) \wedge (p \Longrightarrow s) \wedge (q \vee r) \wedge (\sim s)$$

is a contradiction. Using the rules of Table 5-9, this can be done as follows:

$(q \vee r) \Longrightarrow ((\sim q) \Longrightarrow r)$ Conditional (a)

$(((\sim p) \Longrightarrow (\sim q)) \wedge ((\sim q) \Longrightarrow r))$
$\Longrightarrow ((\sim p) \Longrightarrow r)$ Chain

$(p \Longrightarrow s) \Longrightarrow ((\sim s) \Longrightarrow (\sim p))$	Conditional (b)
$(((\sim s) \Longrightarrow (\sim p)) \wedge (\sim s)) \Longrightarrow (\sim p)$	Detachment
$(((\sim p) \Longrightarrow r) \wedge (\sim p)) \Longrightarrow r$	Detachment

$$(((\sim p) \Longrightarrow (\sim q)) \wedge (\sim r) \wedge (p \Longrightarrow s) \wedge (q \vee r) \wedge (\sim s)) \Longrightarrow$$
$$(((\sim p) \Longrightarrow (\sim q)) \wedge (\sim r) \wedge (p \Longrightarrow s) \wedge ((\sim q) \Longrightarrow r) \wedge (\sim s)) \Longrightarrow$$
$$(((\sim p) \Longrightarrow r) \wedge (\sim r) \wedge (p \Longrightarrow s) \wedge (\sim s)) \Longrightarrow$$
$$(((\sim p) \Longrightarrow r) \wedge (\sim r) \wedge ((\sim s) \Longrightarrow (\sim p)) \wedge (\sim s)) \Longrightarrow$$
$$(((\sim p) \Longrightarrow r) \wedge (\sim r) \wedge (\sim p)) \Longrightarrow$$
$$(r \wedge (\sim r)) \Longrightarrow 0 \qquad \text{Implication (a)} \quad \square$$

An argumental that is not a tautology is said to be a *fallacy*. Thus, to show that $p_1 \wedge p_2 \wedge \ldots \wedge p_r \Rightarrow c$ is a fallacy, it is sufficient to show that for some particular $p_1, p_2, \ldots, p_r$ such that $p_1 \wedge p_2 \wedge \ldots \wedge p_r$ is true, c is false. That is, it is sufficient to produce one instance in which all the premises are true but the consequence is false. This instance is called a *counterexample*.

Example 5-11

The argumental

$$((p \Longrightarrow q) \wedge (r \Longrightarrow (\sim q)) \wedge (p \vee r)) \Longrightarrow r$$

is a fallacy. The counterexample corresponds to the instance where p is true, q is true, and r is false. In this case the premises $p \Rightarrow q$, $r \Rightarrow (\sim q)$, and $p \vee r$ are all true, while the consequence r is false. $\square$

PROBLEMS

1. Prove that $p \Rightarrow q$ is true if and only if $p \wedge (\sim q)$ is false.
2. Prove the implications and equivalences of Table 5-9.
3. Prove that these propositions (adapted from Lewis Carroll) are valid:
   (a) Babies are illogical; despised persons cannot manage crocodiles; illogical persons are despised. Therefore, babies cannot manage crocodiles.
   (b) No ducks waltz; no officers ever decline to waltz; all my poultry are ducks. Therefore, none of my poultry are officers.
   (c) Promise-breakers are untrustworthy; wine drinkers are very communicative; a man who keeps his promise is honest; all pawnbrokers are wine drinkers; we can always trust a very communicative person. Therefore, all pawnbrokers are honest.

4. Prove the following theorems by contradiction:
   (a) $((p \Rightarrow q) \wedge p) \Rightarrow q$
   (b) $((p \Rightarrow (\sim q)) \wedge (q \vee (\sim r)) \wedge (r \wedge (\sim s))) \Rightarrow (\sim p)$
   (c) $\sqrt{2}$ is not a rational number. (*Hint*: If $\sqrt{2}$ is a rational number, then it can be written in the form i/j, where i and j are positive integers that are not both even.)

5. By producing counterexamples, prove that these argumentals are fallacies:
   (a) $((p \Rightarrow q) \wedge r) \Rightarrow p$
   (b) $((p \Rightarrow q) \wedge (q \Rightarrow r)) \Rightarrow (r \Rightarrow p)$
   (c) $((p \wedge (\sim q)) \wedge (q \Rightarrow r) \wedge (r \vee (\sim s))) \Rightarrow (s \wedge q)$

5-7. QUANTIFIERS

Propositions over a universal set U frequently contain phrases such as "for all $u \in S$" and "there exists $u \in S$ such that," where S is a subset of U. These phrases, called *quantifiers*, appear sufficiently often in mathematical discussions to warrant special notation.

The expression "for all $u \in S$" (or "for every $u \in S$." or "for each $u \in S$") is represented by the expression $(\forall u)_S$ (the subscript S may be omitted when $S = U$). This expression is called a *universal quantifier*. If p is a proposition over U, then the *universal proposition*

$$(\forall u)_S(p)$$

is true if and only if p is true for all $u \in S$; that is, if and only if $S \subset T_p$, or $S \cap T'_p = \varnothing$. For example, if S is the set of prime numbers $\mathbb{P}$, then the proposition

$$(\forall n)_{\mathbb{P}}(n \neq 2 \text{ implies } n + 1 \text{ is even})$$

(over the set $\mathbb{N}$ of positive integers) is true. The proposition

$$(\forall n)_{\mathbb{P}}(n \text{ is odd})$$

(over $\mathbb{N}$) is false.

The expression "there exists $u \in S$ such that" (or "for some $u \in S$") is represented by the expression $(\exists u)_S$ (the subscript S can be omitted when $S = U$). This expression is called an *existential quantifier*. If p is a proposition over U, then the *existential proposition*

$$(\exists u)_S(p)$$

is true if and only if there exists at least one u for which p is true; that is, if and only if $S \not\subset T'_p$, or $S \cap T_p \neq \varnothing$. For example, the proposition

$$(\exists\, n)_{\mathbb{P}}(n^2 - 10n + 21 = 0)$$

(over $\mathbb{N}$) is true. The proposition

$$(\exists\, n)_{\mathbb{P}}((n^2 > 10) \wedge (n \text{ is even}))$$

(over $\mathbb{N}$) is false.

Table 5-10 lists some quantified propositions together with the conditions that make them true. Rows 1 and 2 of the table repeat the preceding definitions of universal and existential quantifiers. Rows 3 through 6 follow immediately from 1 and 2. Comparing row 4 with 5, and 3 with 6, these equivalences can be deduced:

$$(\sim((\forall u)_S(p))) \Longleftrightarrow ((\exists\, u)_S(\sim p))$$
$$(\sim((\exists\, u)_S(p))) \Longleftrightarrow ((\forall u)_S(\sim p))$$

Thus, to refute the assertion that p is true for all $u \in S$, it suffices to produce a single $u \in S$ such that p is false [this u is a counterexample to the proposition $(\forall u)_S(p)$]. On the other hand, to prove that no $u \in S$ exists such that p is true, it is required to prove that p is false for *every* $u \in S$.

Table 5-10 QUANTIFIED PROPOSITIONS

Proposition	True if and only if
$(\forall u)_S(p)$	$S \cap T'_p = \varnothing$
$(\exists\, u)_S(p)$	$S \cap T_p \neq \varnothing$
$(\forall u)_S(\sim p)$	$S \cap T_p = \varnothing$
$(\exists\, u)_S(\sim p)$	$S \cap T'_p \neq \varnothing$
$\sim((\forall u)_S(p))$	$S \cap T'_p \neq \varnothing$
$\sim((\exists\, u)_S(p))$	$S \cap T_p = \varnothing$

Example 5-12

Consider the proposition over $\mathbb{N}$, "For all $n \geq 2$, $n^3 - n - 6 > 0$." This can be written as

$$(\forall n)_J(n^3 - n - 6 > 0), \qquad J = \{n \mid n \in \mathbb{N}, n \geq 2\}$$

To prove the negation of this proposition, it suffices to prove $(\exists\, n)_J(n^3 - n - 6 \leq 0)$, hence, to produce a single $n \in J$ such that $n^3 - n - 6 \leq 0$. The existence of such an n (namely, $n = 2$) proves that the original proposition is false. □

Example 5-13

Consider the proposition, "There exist two positive integers, $n_1 \leq 2$ and $n_2 \leq 3$, such that $n_1^2 = n_2 + 4$." This can be written as

$$(\exists (n_1, n_2))_{J_1 \times J_2}(n_1^2 = n_2 + 4)$$
$$J_1 = \{1, 2\}, \qquad J_2 = \{1, 2, 3\}$$

To prove the negation of this proposition, it is necessary to prove the proposition $(\forall (n_1, n_2))_{J_1 \times J_2}(n_1^2 \neq n_2 + 4)$, hence, to demonstrate that $n_1^2 \neq n_2 + 4$ for *all* possible elements of $J_1 \times J_2$ [namely, (1, 1), (1, 2), (1, 3), (2, 1), (2, 2), (2, 3)]. Since this is indeed the case, the original proposition turns out to be false. □

PROBLEMS

1. Determine the truth or falsity of these propositions over the set of integers $\mathbb{I}$:
   (a) $(\forall i)(i^2 - 1 \geq 0)$
   (b) $(\forall i)(i^2 - i - 1 \neq 0)$
   (c) $(\exists i)(2i^2 - 3i + 1 = 0)$
   (d) $(\exists i)(i^2 - 3 = 0)$
   (e) $(\forall i)((\exists j)(i^2 = j))$
   (f) $(\exists i)((\forall j)(i^2 = j))$
   (g) $(\forall j)((\exists i)(i^2 = j))$
   (h) $(\exists j)((\forall i)(i^2 = j))$
2. Write the negations (as universal or existential propositions) of the propositions (a) through (d) of Problem 1.
3. Prove these equivalences over U:
   (a) $(\sim((\forall u)((\exists v)(p)))) \Leftrightarrow ((\exists u)((\forall v)(\sim p)))$
   (b) $(\sim((\exists u)((\forall v)(p)))) \Leftrightarrow ((\forall u)((\exists v)(\sim p)))$
   Use these equivalences to write the negations (as universal or existential propositions) of propositions (e) through (h) of Problem 1.

REFERENCES

KEMENY, J. G., H. MIRKIL, J. L. SNELL, and G. L. THOMPSON, *Finite Mathematical Structures*, Englewood Cliffs, NJ: Prentice-Hall, 1959.

STOLL, R. R., *Sets, Logic and Axiomatic Theories*. San Francisco: Freeman, 1961.

SUPPES, P., *Introduction to Logic*. Princeton, NJ: Van Nostrand, 1957.

6 LATTICES AND BOOLEAN ALGEBRAS

This chapter introduces the algebraic system called *lattice* (whose structure is based on the partial ordering relation introduced in Chapter 2). Properties of lattices are developed, and various examples of lattices (for example, the *lattice of partitions*) are presented. Reinforced with additional postulates, lattices become *Boolean algebras*—algebraic systems of utmost importance to computer scientists. Properties of subsystems and homomorphisms of Boolean algebras are derived, and it is shown that every finite Boolean algebra is isomorphic to some algebra of sets (and, hence, that the cardinality of every finite Boolean algebra is a power of 2). It is also shown that every Boolean algebra of cardinality 2^r is isomorphic to the direct product of r Boolean algebras of cardinality 2. The chapter concludes with a discussion of *Boolean functions* and their normal forms.

The concept of a lattice is important in many aspects of the theory of finite-state machines. Boolean algebras are of special significance to computer scientists because of their direct applicability to switching theory and logical design—as we show in Chapter 7.

6-1. POSETS

A set on which a partial ordering $\leq$ and its converse $\geq$ are defined (see Sec. 2-7) is called a *partially ordered set*, or *poset* for short. Thus, L is a poset if, for all $l_1, l_2, l_3 \in L$,

$$l_1 \leq l_1 \qquad (6\text{-}1)$$

(Reflexivity)

$$l_1 \geq l_1 \qquad (6\text{-}1')$$

$$(l_1 \leq l_2, l_2 \leq l_1) \Longrightarrow (l_1 = l_2) \qquad \text{(Antisymmetry)} \tag{6-2}$$
$$(l_1 \geq l_2, l_2 \geq l_1) \Longrightarrow (l_1 = l_2) \tag{6-2'}$$
$$(l_1 \leq l_2, l_2 \leq l_3) \Longrightarrow (l_1 \leq l_3) \qquad \text{(Transitivity)} \tag{6-3}$$
$$(l_1 \geq l_2, l_2 \geq l_3) \Longrightarrow (l_1 \geq l_3) \tag{6-3'}$$

Let l_1 and l_2 be elements in a poset L. An element $a \in L$ is called a *greatest lower bound* (abbreviated glb) of l_1 and l_2, if

$$a \leq l_1, \qquad a \leq l_2$$

and if, for any $a' \in L$,

$$(a' \leq l_1, a' \leq l_2) \Longrightarrow (a' \leq a)$$

An element $b \in L$ is called a *least upper bound* (abbreviated lub) of l_1 and l_2, if

$$b \geq l_1, \qquad b \geq l_2$$

and if, for any $b' \in L$,

$$(b' \geq l_1, b' \geq l_2) \Longrightarrow (b' \geq b)$$

THEOREM 6-1

Let l_1 and l_2 be elements in a poset L. If l_1 and l_2 have a glb, then this glb is unique. If l_1 and l_2 have a lub, then this lub is unique.

Proof Let a_1 and a_2 be glb's of l_1 and l_2. By definition:

$$a_1 \leq l_1, a_1 \leq l_2, a_2 \leq l_1, a_2 \leq l_2$$
$$a_1 \leq a_2, a_2 \leq a_1$$

hence, by antisymmetry, $a_1 = a_2$. Proof for the uniqueness of the lub is analogous. □

In the ordering diagram of L, the fact that l_1 and l_2 have a glb is reflected by the fact that there exists at least one vertex in the diagram that is reachable from both vertices l_1 and l_2 via a *descending* path. The uppermost such vertex represents the glb of l_1 and l_2. Similarly, the fact that l_1 and l_2 have a lub is reflected by the fact that there exists at least one vertex in the diagram that is reachable from both vertices l_1 and l_2 via an *ascending* path. The lowermost such vertex represents the lub of l_1 and l_2.

Example 6-1

The power set $\mathbf{2}^U$ of a universal set U, on which the partial ordering $\subset$ is defined, is a poset (see Example 2-10). The ordering diagram of $\mathbf{2}^U$, where

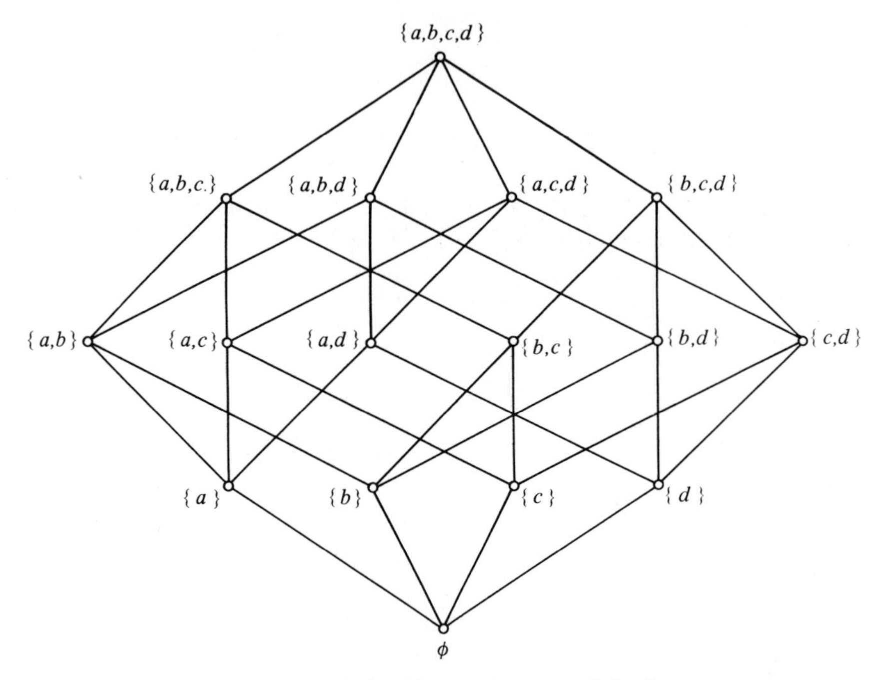

Fig. 6-1 Ordering diagram for poset $\mathbf{2}^{\{a,b,c,d\}}$

$U = \{a, b, c, d\}$, is shown in Fig. 6-1. It is seen, for example, that the glb of $\{a, c\}$ and $\{b, c, d\}$ is $\{c\}$, and that their lub is $\{a, b, c, d\}$. Since every vertex can reach the vertex $\varnothing$ via a descending path and the vertex $\{a, b, c, d\}$ via an ascending path, it can be concluded that every pair of elements of this poset has both a glb and a lub. □

An element a in a poset L is called a *least element* if, for all $l \in L$, $a \leq l$. An element $b \in L$ is called a *greatest element* if, for all $l \in L$, $b \geq l$. In Example 6-1, $\varnothing$ is a least element and $\{a, b, c, d\}$ is a greatest element.

THEOREM 6-2

If a poset L has a least element, then this least element is unique. If L has a greatest element, then this greatest element is unique.

Proof Suppose a_1 and a_2 are least elements, and b_1 and b_2 greatest elements in L. Then $a_1 \leq a_2$, $a_2 \leq a_1$, $b_1 \leq b_2$, and $b_2 \leq b_1$. By antisymmetry, $a_1 = a_2$ and $b_1 = b_2$. □

The least and greatest elements of a lattice L (when these exist) are often referred to as the *zero* of L (denoted by 0) and the *one* of L (denoted by 1), respectively.

PROBLEMS

1. Each of the following sets L is a poset with respect to the partial ordering $|$ ("is a divisor of"). In each set, determine the glb and lub of every pair of elements. Which of the posets has a least element? A greatest element?
   (a) $L = \{1, 2, 3, 4, 6, 12\}$
   (b) $L = \{1, 2, 3, 4, 6, 8, 12, 24\}$
   (c) $L = \{1, 2, 3, \ldots, 12\}$
2. Let L_1 and L_2 be posets. Define the relation $\leq$ on $L_1 \times L_2$ in this manner: For all $(l_1, l_2), (l'_1, l'_2) \in L_1 \times L_2$,

$$((l_1, l_2) \leq (l'_1, l'_2)) \Longleftrightarrow (l_1 \leq l'_1, l_2 \leq l'_2)$$

   Show that $L_1 \times L_2$ is a poset.
3. Let L be a set on which the relation $<$ is defined, with these properties:

$$l_1 < l_1 \text{ is false for all } l_1 \in L$$
$$(l_1 < l_2, l_2 < l_3) \Longrightarrow (l_1 < l_3) \text{ for all } l_1, l_2, l_3 \in L$$

   Show that L is a poset under the relation $\leq$, where

$$(l_1 \leq l_2) \Longleftrightarrow (l_1 < l_2 \text{ or } l_1 = l_2)$$

6-2. LATTICES

A *lattice* $\langle L; \vee, \wedge \rangle$ (*under* $\leq$) is an algebraic system where the domain L is a poset (under the partial ordering $\leq$) in which every pair of elements has both a lub and a glb, and where $\vee$ and $\wedge$ are defined as

$$l_1 \vee l_2 = \text{lub of } l_1 \text{ and } l_2$$
$$l_1 \wedge l_2 = \text{glb of } l_1 \text{ and } l_2$$

Theorem 6-1 guarantees the uniqueness of $l_1 \vee l_2$ and $l_1 \wedge l_2$; hence, it establishes $\vee$ and $\wedge$ as bona fide operations over L. The operations $\vee$ and $\wedge$ are sometimes called the *join* and *meet* operations, respectively.

Example 6-2

In the poset $\mathbf{2}^U$, on which the partial ordering $\subset$ is defined, every pair of elements S_1 and S_2 has both a lub and a glb, namely:

$$\text{lub of } S_1 \text{ and } S_2 = S_1 \cup S_2$$
$$\text{glb of } S_1 \text{ and } S_2 = S_1 \cap S_2$$

Hence, $\langle \mathbf{2}^U; \cup, \cap \rangle$ is a lattice (under $\subset$). □

Example 6-3

The relation | on the set of positive integers $\mathbb{N}$, where $n_1 | n_2$ if and only if n_1 is a divisor of n_2, is a partial ordering (see Example 2-9). Every pair of elements n_1 and n_2 in $\mathbb{N}$ has both a lub and a glb, namely:

$$\text{lub of } n_1 \text{ and } n_2 = \text{lcm}\,(n_1, n_2)$$

$$\text{glb of } n_1 \text{ and } n_2 = \text{gcd}\,(n_1, n_2)$$

Hence, $\langle \mathbb{N}; \text{lcm}, \text{gcd} \rangle$ is a lattice (under |). □

Notice that not every poset is a lattice. For example, the set $\{2, 3, 4, 6, 8, 12, 36, 60\}$ is not a lattice under | (the integers 2 and 3 have no common divisor, and the integers 4 and 36 have no common multiple in this set).

In terms of $\vee$ and $\wedge$, the definitions of lub and glb introduced in Sec. 6-1 can be reformulated as follows: For all $l_1, l_2, l_3 \in L$,

$$l_1 \vee l_2 \geq l_1, l_1 \vee l_2 \geq l_2 \tag{6-4}$$

$$(l_3 \geq l_1, l_3 \geq l_2) \Longrightarrow (l_3 \geq l_1 \vee l_2) \tag{6-5}$$

$$l_1 \wedge l_2 \leq l_1, l_1 \wedge l_2 \leq l_2 \tag{6-4'}$$

$$(l_3 \leq l_1, l_3 \leq l_2) \Longrightarrow (l_3 \leq l_1 \wedge l_2) \tag{6-5'}$$

These definitions, together with the poset properties (6-1), (6-2), (6-3), (6-1′), (6-2′), and (6-3′), can serve as a set of postulates for the algebraic system $\langle L; \vee, \wedge \rangle$, from which a mathematical theory of lattices can evolve. For example, these postulates lead to:

THEOREM 6-3

If l_1 and l_2 are elements of a lattice $\langle L; \vee, \wedge \rangle$, then

$$(l_1 \vee l_2 = l_1) \Longleftrightarrow (l_1 \wedge l_2 = l_2) \Longleftrightarrow (l_2 \leq l_1)$$

Proof Suppose $l_1 \vee l_2 = l_1$. By (6-4), $l_2 \leq l_1$, and, by reflexivity, $l_2 \leq l_2$. Thus, by (6-5′), $l_2 \leq l_1 \wedge l_2$. Also, by (6-4′), $l_2 \geq l_1 \wedge l_2$. Hence, by antisymmetry, $l_1 \wedge l_2 = l_2$. If $l_1 \wedge l_2 = l_2$, then, by (6-4′), $l_2 \leq l_1$. In conclusion:

$$(l_1 \vee l_2 = l_1) \Longrightarrow (l_1 \wedge l_2 = l_2) \Longrightarrow (l_2 \leq l_1) \tag{6-6}$$

Now, suppose $l_2 \leq l_1$. By reflexivity, $l_1 \leq l_1$, and, hence, by (6-5), $l_1 \geq l_1 \vee l_2$. Since, by (6-4), $l_1 \leq l_1 \vee l_2$, we have, by antisymmetry, $l_1 = l_1 \vee l_2$. Similarly, by reflexivity, $l_2 \leq l_2$, and, hence, by (6-5′), $l_2 \leq l_1 \wedge l_2$. Since, by

(6-4′), $l_2 \geq l_1 \wedge l_2$, we have, by antisymmetry, $l_2 = l_1 \wedge l_2$. In conclusion:

$$(l_2 \leq l_1) \Longrightarrow (l_1 \vee l_2 = l_1), (l_2 \leq l_1) \Longrightarrow (l_1 \wedge l_2 = l_2)$$

Hence, using (6-6):

$$(l_1 \wedge l_2 = l_2) \Longrightarrow (l_1 \vee l_2 = l_1) \qquad \square$$

PROBLEMS

1. Fig. 6-A shows the ordering diagrams of four posets. Which of these posets forms a lattice?

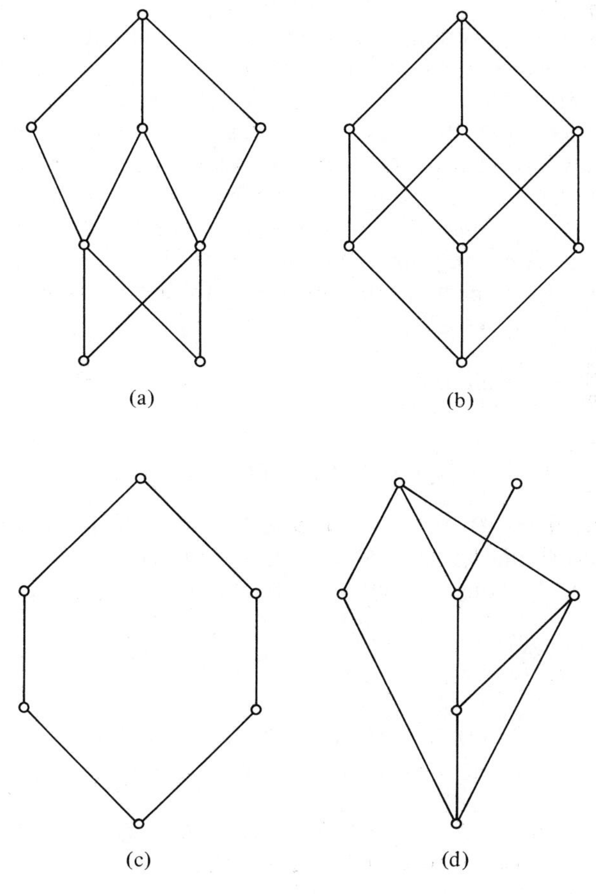

Fig. 6-A

2. Which of the posets of Problem 1, Sec. 6-1, forms a lattice?

3. Show that if $\langle L; \vee, \wedge \rangle$ is a finite lattice, then L must contain both a least element and a greatest element.

4. Let L be a poset with a least element and a greatest element. Show that L forms a lattice, if the following condition is satisfied: For any $a_1, a_2, b_1, b_2 \in L$, where $a_i \leq b_j$ $(i, j \in \{1, 2\})$, there is an element $c \in L$ such that $a_i \leq c \leq b_j$ $(i, j \in \{1, 2\})$.

6-3. BASIC LAWS OF LATTICES

The *dual* of any proposition involving elements of a lattice and the symbols $=, \leq, \geq, \vee$, and $\wedge$, is the same proposition with $\leq$ replacing $\geq$, $\geq$ replacing $\leq$, $\vee$ replacing $\wedge$, and $\wedge$ replacing $\vee$. If proposition P^D is the dual of proposition P, then, clearly, P is the dual of P^D; hence, we can simply say that propositions P and P^D are *dual* (to each other).

It is seen that the statements (6-1′), (6-2′), (6-3′), (6-4′), and (6-5′) are the duals of (6-1), (6-2), (6-3), (6-4), and (6-5), respectively; hence, that for every lattice postulate there is a dual postulate. Through reasoning analogous to that which led to the duality principle in the algebra of sets (Sec. 4-6), we can arrive at this *duality principle* for lattices: *Given any true proposition for $\langle L; \vee, \wedge \rangle$, then the dual of this proposition is also true.*

Each of the following theorems consists of dual identities. In each case, only one identity is proved; the other one follows directly by the duality principle.

THEOREM 6-4 (*Commutative laws*)
For all $l_1, l_2 \in L$,

$$\text{(a) } l_1 \vee l_2 = l_2 \vee l_1 \qquad \text{(b) } l_1 \wedge l_2 = l_2 \wedge l_1$$

Proof (a) By (6-4), $l_1 \vee l_2 \geq l_2$ and $l_1 \vee l_2 \geq l_1$. Hence, by (6-5), $l_1 \vee l_2 \geq l_2 \vee l_1$. Similarly, by (6-4), $l_2 \vee l_1 \geq l_1$ and $l_2 \vee l_1 \geq l_2$; hence, $l_2 \vee l_1 \geq l_1 \vee l_2$. Thus, by antisymmetry, $l_1 \vee l_2 = l_2 \vee l_1$. □

THEOREM 6-5 (*Associative laws*)
For all $l_1, l_2, l_3 \in L$,

$$\text{(a) } l_1 \vee (l_2 \vee l_3) = (l_1 \vee l_2) \vee l_3$$

$$\text{(b) } l_1 \wedge (l_2 \wedge l_3) = (l_1 \wedge l_2) \wedge l_3$$

Proof (a) Let $a = l_1 \vee (l_2 \vee l_3)$ and $a' = (l_1 \vee l_2) \vee l_3$. By (6-4), $a \geq l_1$ and $a \geq l_2 \vee l_3$; also, by (6-4) and transitivity, $a \geq l_2$ and $a \geq l_3$. Now, $a \geq l_1$ and $a \geq l_2$ imply, by (6-5), $a \geq l_1 \vee l_2$; $a \geq l_1 \vee l_2$ and $a \geq l_3$ imply

$$a \geq (l_1 \vee l_2) \vee l_3 = a'$$

Analogously, it can be shown that

$$a' \geq l_1 \vee (l_2 \vee l_3) = a$$

Hence, by antisymmetry, $a = a'$. □

The associative laws imply that one can interpret unambiguously the parentheses-free expressions $l_1 \vee l_2 \vee \ldots \vee l_r$ (abbreviated $\bigvee_{i=1}^{r} l_i$) and $l_1 \wedge l_2 \wedge \ldots \wedge l_r$ (abbreviated $\bigwedge_{i=1}^{r} l_i$). Consequently, we can talk about the lub (or join) and the glb (or meet) of any number of lattice elements (not just a pair of such elements).

THEOREM 6-6 (*Idempotent laws*)
For all $l \in L$,

$$\text{(a) } l \vee l = l \qquad \text{(b) } l \wedge l = l$$

Proof (a) By (6-4), $l \leq l \vee l$. By reflexivity, $l \geq l$; hence, by (6-5), $l \geq l \vee l$. Thus, by (6-2), $l = l \vee l$. □

THEOREM 6-7 (*Absorption laws*)
For all $l_1, l_2 \in L$,

$$\text{(a) } l_1 \vee (l_1 \wedge l_2) = l_1$$
$$\text{(b) } l_1 \wedge (l_1 \vee l_2) = l_1$$

Proof (a) By (6-4), $l_1 \leq l_1 \vee (l_1 \wedge l_2)$. By reflexivity, $l_1 \geq l_1$ and, by (6-4), $l_1 \geq l_1 \wedge l_2$. Hence, by (6-5), $l_1 \geq l_1 \vee (l_1 \wedge l_2)$. Thus, by antisymmetry, $l_1 = l_1 \vee (l_1 \wedge l_2)$. □

Instead of using (6-1), (6-2), (6-3), (6-4), (6-5), and their duals as lattice postulates, we can just as well use Theorems 6-4 through 6-7. This is confirmed by:

THEOREM 6-8
Let L be a set over which the binary operations $\vee$ and $\wedge$ are defined, and in which the commutative, associative, idempotent, and absorption laws (Theorems 6-4, 6-5, 6-6, and 6-7) hold. Then there exists a partial ordering on L such that, for every $l_1, l_2 \in L$, $l_1 \vee l_2$ is the lub and $l_1 \wedge l_2$ is the glb of l_1 and l_2 under this partial ordering.

Proof Suppose Theorems 6-4 through 6-7 hold. Let us define a relation $\leq$ on L as: For all $l_1, l_2 \in L$,

$$(l_2 \leq l_1) \Longleftrightarrow (l_1 \vee l_2 = l_1) \tag{6-7}$$

By the idempotent law, for any $l \in L$ we have $l \vee l = l$; hence, by (6-7), $l \leq l$. Thus, $\leq$ is reflexive.

Suppose $l_1 \leq l_2$ and $l_2 \leq l_1$. By (6-7), $l_2 \vee l_1 = l_2$ and $l_1 \vee l_2 = l_1$. By the commutative law, $l_1 = l_2$ and, hence, $\leq$ is antisymmetric.

Suppose $l_1 \leq l_2$ and $l_2 \leq l_3$. By (6-7), $l_2 \vee l_1 = l_2$ and $l_3 \vee l_2 = l_3$. Using the associative law, we can write

$$\begin{aligned} l_3 \vee l_1 = (l_3 \vee l_2) \vee l_1 &= l_3 \vee (l_2 \vee l_1) \\ &= l_3 \vee l_2 = l_3 \end{aligned}$$

which, by (6-7), implies $l_1 \leq l_3$. Hence, $\leq$ is transitive. In conclusion, $\leq$ is a partial ordering on L, which satisfies conditions (6-1), (6-2), and (6-3).

By (6-7), $l_1 \vee l_2 \geq l_1$ if and only if $(l_1 \vee l_2) \vee l_1 = l_1 \vee l_2$ which, by the commutative, associative, and idempotent laws is true for all $l_1, l_2 \in L$. Hence, for all $l_1, l_2 \in L, l_1 \vee l_2 \geq l_1$. Similarly, for all $l_1, l_2 \in L, l_1 \vee l_2 \geq l_2$. Now, by (6-7), we have

$$(l_3 \geq l_1, l_3 \geq l_2) \Longleftrightarrow (l_3 \vee l_1 = l_3, l_3 \vee l_2 = l_3)$$
$$(l_3 \geq l_1 \vee l_2) \Longleftrightarrow ((l_1 \vee l_2) \vee l_3 = l_3)$$

But, by the commutative, associative, and idempotent laws,

$$(l_3 \vee l_1 = l_3, l_3 \vee l_2 = l_3) \Longrightarrow ((l_1 \vee l_2) \vee l_3 = l_3)$$

and, hence,

$$(l_3 \geq l_1, l_3 \geq l_2) \Longrightarrow (l_3 \geq l_1 \vee l_2)$$

Thus, $\vee$ satisfies conditions (6-4) and (6-5).

Using the commutative and absorption laws, we have

$$(l_1 \vee l_2 = l_1) \Longrightarrow (l_2 \wedge (l_1 \vee l_2) = l_2 \wedge l_1) \Longrightarrow (l_1 \wedge l_2 = l_2)$$
$$(l_1 \wedge l_2 = l_2) \Longrightarrow (l_1 \vee (l_1 \wedge l_2) = l_1 \vee l_2) \Longrightarrow (l_1 \vee l_2 = l_1)$$

hence,

$$(l_1 \vee l_2 = l_1) \Longleftrightarrow (l_1 \wedge l_2 = l_2)$$

Thus, the relation $\leq$ can also be specified as: For all $l_1, l_2 \in L$,

$$(l_2 \leq l_1) \Longleftrightarrow (l_1 \wedge l_2 = l_2) \tag{6-8}$$

Using arguments dual to those produced above, we can show, with the aid of (6-8), that $\leq$ satisfies conditions (6-1′), (6-2′), and (6-3′), and that $\wedge$ satisfies conditions (6-4′) and (6-5′).

In conclusion, $\leq$ is a partial ordering under which, for every $l_1, l_2 \in L$, $l_1 \vee l_2$ is the lub and $l_1 \wedge l_2$ the glb of l_1 and l_2. □

PROBLEMS

1. Let $\langle L; \vee, \wedge \rangle$ be a lattice. Show that, for all $l_1, l_2, l_3 \in L$,

$$(l_2 \leq l_1) \Longrightarrow (l_2 \vee (l_1 \wedge l_3) \leq l_1 \wedge (l_2 \vee l_3))$$

2. A lattice $\langle L; \vee, \wedge \rangle$ is said to be *modular* if, for all $l_1, l_2, l_3 \in L$,

$$(l_2 \leq l_1) \Longrightarrow (l_2 \vee (l_1 \wedge l_3) = l_1 \wedge (l_2 \vee l_3))$$

Is the lattice whose ordering diagram is shown in Fig. 6-B modular? Prove your answer.

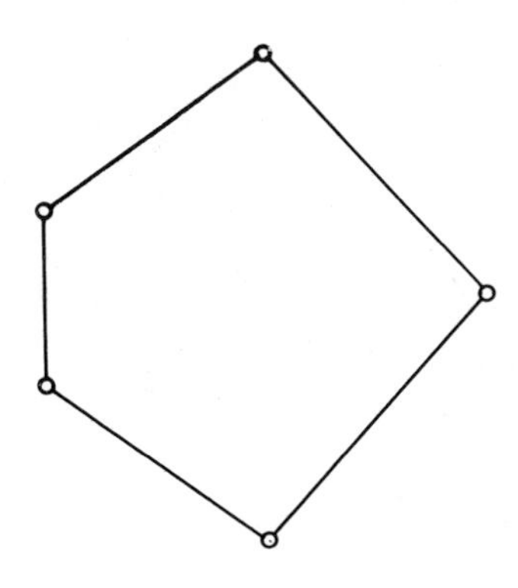

Fig. 6-B

3. Show that in a modular lattice (see Problem 2),

$$(l_2 \geq l_1) \Longrightarrow (l_2 \wedge (l_1 \vee l_3) = l_1 \vee (l_2 \wedge l_3))$$

(thereby verifying that the duality principle holds for modular lattices).

6-4. THE LATTICE OF PARTITIONS

Let A be a set, and $P(A)$ the set of all partitions of A. Define a relation $\leq$ on $P(A)$ as follows: For all $\pi_1, \pi_2 \in P(A)$,

$$(\pi_1 \leq \pi_2) \Longleftrightarrow (\pi_1 \text{ is a refinement of } \pi_2) \tag{6-9}$$

That $\leq$ is reflexive, antisymmetric, and transitive, follows immediately from the definition of "refinement" (see Sec. 1-7). Thus, $\leq$, as defined in (6-9), is a partial ordering. Figure 6-2 shows the ordering diagram for the poset $P(\{a, b, c, d\})$.

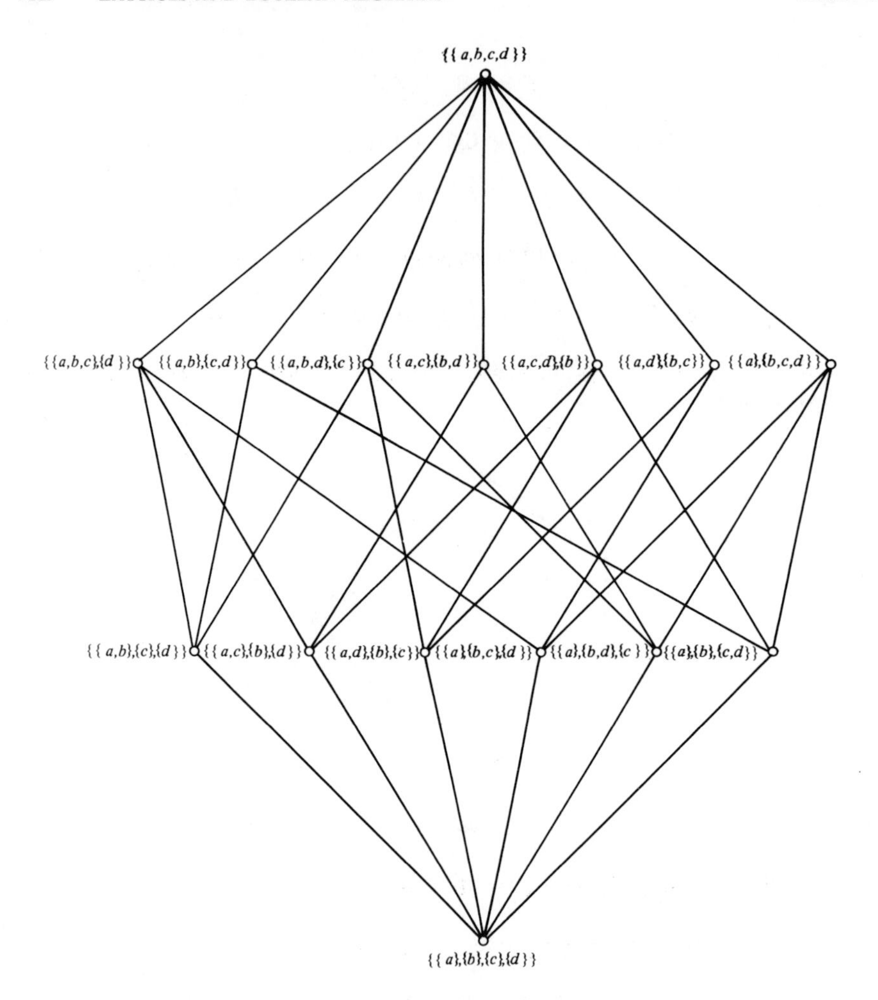

Fig. 6-2 Ordering diagram for poset $P(\{a, b, c, d\})$

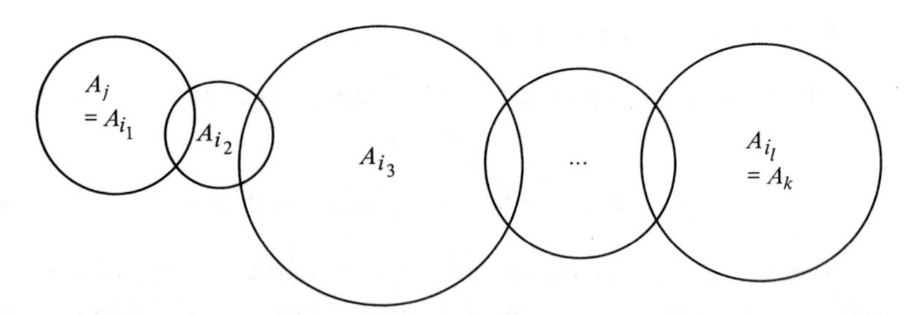

Fig. 6-3 A_j is connected to A_k.

Let $A_1, A_2, \ldots, A_m$ be subsets of A (not necessarily distinct or disjoint). We say that A_j is *directly connected to* A_k if $A_j \cap A_k \neq \varnothing$. We say that A_j is *connected to* A_k *in* $\{A_1, A_2, \ldots, A_m\}$ if there exists a finite sequence of A_i, say, $A_{i_1}, A_{i_2}, \ldots, A_{i_l}$, such that $A_{i_1} = A_j$, $A_{i_l} = A_k$, and A_{i_ν} is directly connected to $A_{i_{\nu+1}}$ $(\nu = 1, 2, \ldots, l-1)$. The latter concept is illustrated in Fig. 6-3.

THEOREM 6-9

Let

$$\pi_1 = \{S_1, S_2, \ldots, S_{r_1}\}$$
$$\pi_2 = \{T_1, T_2, \ldots, T_{r_2}\}$$

be partitions of a set A. Define

$$\pi_1 \vee \pi_2 = \{A^{(i)} \mid A^{(i)} = A_{j_1} \cup A_{j_2} \cup \ldots \cup A_{j_h}, \text{ where } A_{j_\nu} \ (\nu = 1, 2, \ldots, h) \text{ is connected to } S_i \text{ in } \pi_1 \cup \pi_2\}$$
$$\pi_1 \wedge \pi_2 = \{A^{(i,j)} \mid A^{(i,j)} = S_i \cap T_j, \text{ where } S_i \text{ is directly connected to } T_j\}$$

Then $\pi_1 \vee \pi_2$ and $\pi_1 \wedge \pi_2$ are the lub and glb, respectively, of π_1 and π_2 under the partial ordering $\leq$ defined by (6-9).

Before proving the theorem, we illustrate it by:

Example 6-4

Consider the set $A = \{a, b, c, d, e, f, g, h, i\}$ and its partitions

$$\begin{aligned}\pi_1 &= \{S_1, S_2, S_3, S_4\}\\ &= \{\{a, b\}, \{c, d\}, \{e, f\}, \{g, h, i\}\}\\ \pi_2 &= \{T_1, T_2, T_3, T_4, T_5\}\\ &= \{\{a, f\}, \{b, c\}, \{d, e\}, \{g, h\}, \{i\}\}\end{aligned}$$

(see Fig. 6-4). To evaluate $\pi_1 \vee \pi_2$, first construct

$$\begin{aligned}\pi_1 \cup \pi_2 = \{&\{a, b\}, \{c, d\}, \{e, f\}, \{g, h, i\},\\ &\{a, f\}, \{b, c\}, \{d, e\}, \{g, h\}, \{i\}\}\end{aligned}$$

(which, incidentally, is *not* a partition of A). Now,

$$\begin{aligned}A^{(1)} &= \{a, b\} \cup \{a, f\} \cup \{b, c\} \cup \{e, f\} \cup \{c, d\} \cup \{d, e\}\\ &= \{a, b, c, d, e, f\}\\ A^{(2)} &= \{c, d\} \cup \{b, c\} \cup \{d, e\} \cup \{a, b\} \cup \{e, f\} \cup \{a, f\}\\ &= \{a, b, c, d, e, f\}\end{aligned}$$

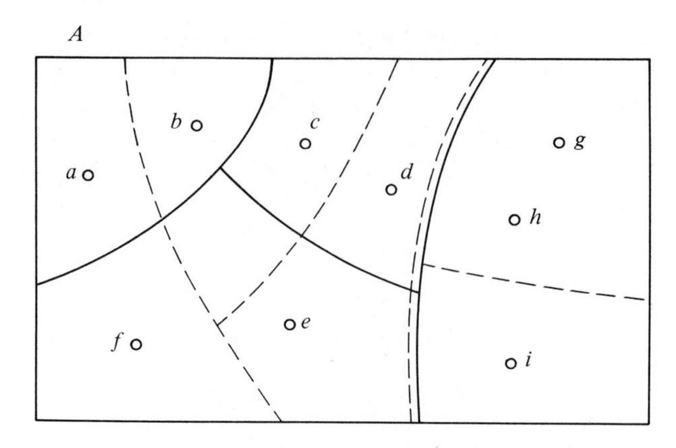

Fig. 6-4 Partitions π_1 (solid boundaries) and π_2 (dotted boundaries) of A

$$A^{(3)} = \{e, f\} \cup \{d, e\} \cup \{a, f\} \cup \{c, d\} \cup \{a, b\} \cup \{b, c\}$$
$$= \{a, b, c, d, e, f\}$$
$$A^{(4)} = \{g, h, i\} \cup \{g, h\} \cup \{i\} = \{g, h, i\}$$

Hence,

$$\pi_1 \vee \pi_2 = \{\{a, b, c, d, e, f\}, \{g, h, i\}\}$$

(Roughly speaking, $\pi_1 \vee \pi_2$ is the partition obtained by retaining only those "boundary lines" that are common to both π_1 and π_2.)

To evaluate $\pi_1 \wedge \pi_2$, we compute

$$A^{(1,1)} = \{a, b\} \cap \{a, f\} = \{a\}$$
$$A^{(1,2)} = \{a, b\} \cap \{b, c\} = \{b\}$$
$$A^{(2,2)} = \{c, d\} \cap \{b, c\} = \{c\}$$
$$A^{(2,3)} = \{c, d\} \cap \{d, e\} = \{d\}$$
$$A^{(3,1)} = \{e, f\} \cap \{a, f\} = \{f\}$$
$$A^{(3,3)} = \{e, f\} \cap \{d, e\} = \{e\}$$
$$A^{(4,4)} = \{g, h, i\} \cap \{g, h\} = \{g, h\}$$
$$A^{(4,5)} = \{g, h, i\} \cap \{i\} = \{i\}$$

Hence,

$$\pi_1 \wedge \pi_2 = \{\{a\}, \{b\}, \{c\}, \{d\}, \{e\}, \{f\}, \{g, h\}, \{i\}\}$$

(Roughly speaking, $\pi_1 \wedge \pi_2$ is the partition obtained by retaining those "boundary lines" belonging to either π_1 or π_2.) □

Proof of Theorem 6-9 We first note that, by virtue of the definition of "connectedness," $\pi_1 \vee \pi_2 = \pi_2 \vee \pi_1$ and $\pi_1 \wedge \pi_2 = \pi_2 \wedge \pi_1$. Since, for $i = 1, 2, \ldots, r_1$, $S_i \subset A^{(i)}$, we have $\pi_1 \leq \pi_1 \vee \pi_2$. Similarly, $\pi_2 \leq \pi_2 \vee \pi_1 = \pi_1 \vee \pi_2$. Now, suppose $\pi_1 \leq \pi_3$ and $\pi_2 \leq \pi_3$, where $\pi_3 = \{U_1, U_2, \ldots, U_{r_3}\}$. Then every S_i and T_j is a subset of some U_k; hence, every $A_{j_\nu} \in \pi_1 \cup \pi_2$ is a subset of some U_k. Thus, no S_i can be connected to any $A_{j_\nu} \in \pi_i \cup \pi_2$ unless both are in the same U_k. This implies that $A^{(i)}$ is a subset of some U_k; hence, $\pi_1 \vee \pi_2 \leq \pi_3$. In conclusion:

$$\pi_1 \vee \pi_2 \geq \pi_1, \pi_1 \vee \pi_2 \geq \pi_2$$
$$(\pi_3 \geq \pi_1, \pi_3 \geq \pi_2) \Longrightarrow (\pi_3 \geq \pi_1 \vee \pi_2)$$

Hence, $\pi_1 \vee \pi_2$ is the lub of π_1 and π_2.

Since, for $i = 1, 2, \ldots, r_1$, $S_i \cap T_j \subset S_i$, we have $\pi_1 \wedge \pi_2 \leq \pi_1$. Similarly, $\pi_2 \wedge \pi_1 \leq \pi_2$ and, hence, $\pi_1 \wedge \pi_2 \leq \pi_2$. Now, suppose $\pi_3 \leq \pi_1$ and $\pi_3 \leq \pi_2$, where $\pi_3 = \{U_1, U_2, \ldots, U_{r_3}\}$. Then every U_k is a subset of exactly one S_i and one T_j, hence, must be included in some $S_i \cap T_j$. Thus, $\pi_3 \leq \pi_1 \wedge \pi_2$. In conclusion:

$$\pi_1 \wedge \pi_2 \leq \pi_1, \pi_1 \wedge \pi_2 \leq \pi_2$$
$$(\pi_3 \leq \pi_1, \pi_3 \leq \pi_2) \Longrightarrow (\pi_3 \leq \pi_1 \wedge \pi_2)$$

Hence, $\pi_1 \wedge \pi_2$ is the glb of π_1 and π_2. □

By Theorem 6-9, then, the algebraic system $\langle P(A); \vee, \wedge \rangle$ is a lattice, and all lattice properties (for example, those described in Theorems 6-4 through 6-7) are immediately attributable to it.

PROBLEMS

1. Draw the ordering diagram for the poset $P(\{a, b, c\})$ [whose partial ordering is defined in (6-9)].

2. Construct the operation tables for the lattice $\langle P(A); \vee, \wedge \rangle$, where $A = \{a, b, c\}$.

3. Consider the lattice $\langle P(A); \vee, \wedge \rangle$, where $A = \{a, b, c, d, e, f, g, h, i\}$. Compute $\pi_1 \vee \pi_2$ and $\pi_1 \wedge \pi_2$, where

$$\pi_1 = \{\{a, d\}, \{b, c\}, \{e, g, i\}, \{f\}, \{h\}\}$$
$$\pi_2 = \{\{a, f\}, \{b, c\}, \{d\}, \{e, h\}, \{g, i\}\}$$

4. Let $E(A)$ denote the set of all equivalence relations on the set A. For any $e_1, e_2 \in E(A)$, define $e_1 \leq e_2$ appropriately, and show that $\langle E(A); \vee, \wedge \rangle$ is a lattice (under the defined $\leq$).

6-5. DISTRIBUTIVE LATTICES

A lattice $\langle L; \vee, \wedge\rangle$ is said to be *distributive* if it obeys these *distributive laws*: For all $l_1, l_2, l_3 \in L$,

$$l_1 \wedge (l_2 \vee l_3) = (l_1 \wedge l_2) \vee (l_1 \wedge l_3)$$
$$l_1 \vee (l_2 \wedge l_3) = (l_1 \vee l_2) \wedge (l_1 \vee l_3)$$

Note that not every lattice is distributive. For example, the lattice whose partial ordering diagram is shown in Fig. 6-2 is not distributive. To verify this assertion, observe that

$$\begin{aligned}&\{\{a\}, \{b, c, d\}\} \vee (\{\{a, b, c\}, \{d\}\} \wedge \{\{a, d\}, \{b\}, \{c\}\})\\ &\quad = \{\{a\}, \{b, c, d\}\} \vee \{\{a\}, \{b\}, \{c\}, \{d\}\}\\ &\quad = \{\{a\}, \{b, c, d\}\}\end{aligned}$$

while

$$\begin{aligned}&(\{\{a\}, \{b, c, d\}\} \vee \{\{a, b, c\}, \{d\}\})\\ &\qquad \wedge (\{\{a\}, \{b, c, d\}\} \vee \{\{a, d\}, \{b\}, \{c\}\})\\ &\quad = \{\{a, b, c, d\}\} \wedge \{\{a, b, c, d\}\} = \{\{a, b, c, d\}\}\end{aligned}$$

If $\langle L; \vee, \wedge\rangle$ is a distributive lattice, then we can readily prove by induction that, for any $l, m_1, m_2, \ldots, m_k \in L$,

$$l \vee (\textstyle\bigwedge_{i=1}^{k} m_i) = \bigwedge_{i=1}^{k} (l \vee m_i)$$
$$l \wedge (\textstyle\bigvee_{i=1}^{k} m_i) = \bigvee_{i=1}^{k} (l \wedge m_i)$$

More generally, for any $l_1, l_2, \ldots, l_h, m_1, m_2, \ldots, m_k \in L$,

$$(\textstyle\bigwedge_{i=1}^{h} l_i) \vee (\bigwedge_{j=1}^{k} m_j) = \bigwedge_{i=1}^{h} (\bigwedge_{j=1}^{k} (l_i \vee m_j)) \qquad (6\text{-}10)$$
$$(\textstyle\bigvee_{i=1}^{h} l_i) \wedge (\bigvee_{j=1}^{k} m_j) = \bigvee_{i=1}^{h} (\bigvee_{j=1}^{k} (l_i \wedge m_j)) \qquad (6\text{-}11)$$

THEOREM 6-10

If l_1, l_2, and l_3 are elements of a distributive lattice $\langle L; \vee, \wedge\rangle$, then

$$(l_1 \vee l_2 = l_1 \vee l_3, l_1 \wedge l_2 = l_1 \wedge l_3) \Longleftrightarrow (l_2 = l_3)$$

Proof The right-to-left implication is trivial. To prove the left-to-right implication, we make use of the commutative, absorption, and distributive laws:

$$\begin{aligned}l_2 &= l_2 \vee (l_2 \wedge l_1) = l_2 \vee (l_3 \wedge l_1)\\ &= (l_2 \vee l_3) \wedge (l_2 \vee l_1)\\ &= (l_3 \vee l_2) \wedge (l_3 \vee l_1) = l_3 \vee (l_2 \wedge l_1)\\ &= l_3 \vee (l_3 \wedge l_1) = l_3\end{aligned}$$

□

In a distributive lattice $\mathfrak{L} = \langle L; \vee, \wedge \rangle$, a *polynomial over* $\mathfrak{L}$ is defined recursively as:

(i) (*Basis*). Any element of L is a polynomial over $\mathfrak{L}$.

(ii) (*Induction step*). If p_1 and p_2 are polynomials over $\mathfrak{L}$, so are (p_1), $(p_1) \vee (p_2)$, and $(p_1) \wedge (p_2)$. (Parentheses can be omitted under the convention that $\wedge$ has precedence over $\vee$.)

Using rules (i) and (ii), every polynomial over $\mathfrak{L}$ can be written either as a "join of meets" or as a "meet of joins." As an example, the following shows how a polynomial can be converted into a join of meets:

$$\begin{aligned}
&((((l_1 \wedge l_3) \vee (l_2 \wedge l_3) \vee l_2) \wedge l_1) \vee (l_2 \wedge l_3)) \wedge (l_1 \vee l_3) \\
&= ((l_1 \wedge l_3) \vee (l_1 \wedge l_2 \wedge l_3) \vee (l_1 \wedge l_2) \vee (l_2 \wedge l_3)) \wedge (l_1 \vee l_3) \\
&= (l_1 \wedge l_3) \vee (l_1 \wedge l_2 \wedge l_3) \vee (l_1 \wedge l_2) \vee (l_1 \wedge l_2 \wedge l_3) \\
&\quad \vee (l_1 \wedge l_3) \vee (l_1 \wedge l_2 \wedge l_3) \vee (l_1 \wedge l_2 \wedge l_3) \vee (l_2 \wedge l_3) \\
&= (l_1 \wedge l_3) \vee (l_1 \wedge l_2) \vee (l_2 \wedge l_3) \vee (l_1 \wedge l_2 \wedge l_3) \\
&= (l_1 \wedge l_3) \vee (l_1 \wedge l_2) \vee (l_2 \wedge l_3)
\end{aligned}$$

(where the last step is justified by the absorption law).

Let $\mathfrak{L}_n = \langle L; \vee, \wedge \rangle$ be a distributive lattice, where L is finite and of cardinality n. When a polynomial over $\mathfrak{L}_n$ is converted into a join of meets, we can invoke basic lattice laws (in particular, the idempotent and absorption laws) to insure that each meet will not involve more than n elements, and that the number of meets will not exceed

$$\binom{n}{1} + \binom{n}{2} + \dots + \binom{n}{n} = 2^n - 1$$

Thus, although the number of polynomials over $\mathfrak{L}_n$ is infinite, each one of them equals to one of a finite number of such polynomials. For example, every polynomial over $\mathfrak{L}_3 = \langle L; \vee, \wedge \rangle$, where $L = \{l_1, l_2, l_3\}$, is equal to exactly one of these 18 polynomials:

$$\begin{aligned}
&l_1, l_2, l_3, l_1 \vee l_2, l_1 \vee l_3, l_2 \vee l_3, l_1 \vee l_2 \vee l_3, l_1 \wedge l_2 \wedge l_3, \\
&l_1 \wedge l_2, l_1 \wedge l_3, l_2 \wedge l_3, (l_1 \wedge l_2) \vee l_3, (l_1 \wedge l_3) \vee l_2, \\
&(l_2 \wedge l_3) \vee l_1, (l_1 \wedge l_2) \vee (l_1 \wedge l_3), (l_1 \wedge l_2) \vee (l_2 \wedge l_3), \\
&(l_1 \wedge l_3) \vee (l_2 \wedge l_3), (l_1 \wedge l_3) \vee (l_1 \wedge l_2) \vee (l_2 \wedge l_3)
\end{aligned}$$

The smallest set of polynomials over $\mathfrak{L}_n$ that "generates" all such polynomials is denoted by G_n. It has been found, for example, that $\#G_3 = 18$, $\#G_4 = 166$, $\#G_5 = 7579$, and $\#G_6 = 7{,}828{,}352$. The general formula for $\#G_n$ is unknown however.

PROBLEMS

1. Figure 6-C shows the ordering diagrams of three lattices. Which of these lattices is distributive?

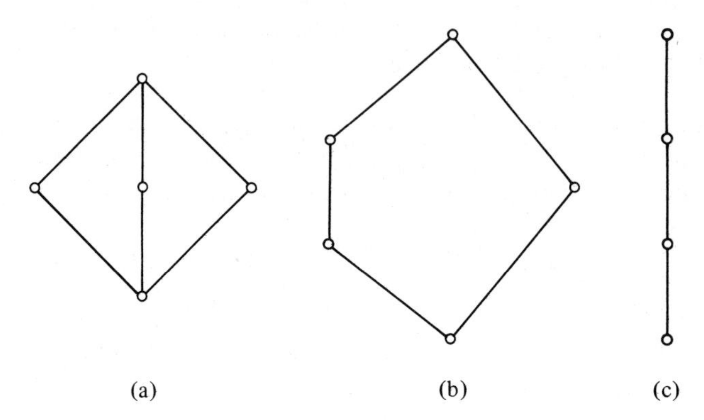

Fig. 6-C

2. Consider the lattice $\langle \mathbb{I}; \min, \max \rangle$, under the partial ordering $\leq$, where $\mathbb{I}$ is the set of integers, $\min(i, j)$ is the least of i and j, $\max(i, j)$ is the greatest of i and j, and $\leq$ is the ordinary "less than or equal to" relation. Prove that this lattice is distributive.

3. Are the lattices $\langle \mathbf{2}^U; \cup, \cap \rangle$ and $\langle \mathbb{N}; \text{gcd}, \text{lcm} \rangle$ of Examples 6-2 and 6-3 distributive? Prove your answer.

4. Show that every distributive lattice is modular (see Problem 2, Sec. 6-3).

5. A *chain* L is a poset where, for every $l_1, l_2 \in L$, either $l_1 \leq l_2$ or $l_2 \leq l_1$. Show that every chain forms a distributive lattice (under $\leq$).

6. Determine G_2—the smallest set of polynomials over $\mathfrak{L}_2 = \langle \{l_1, l_2\}; \vee, \wedge \rangle$ which generates all such polynomials.

7. Which of the elements of G_3, listed in Sec. 6-5, equals this polynomial over $\mathfrak{L}_3$:

$$(((l_1 \vee l_2) \wedge l_3) \vee (l_2 \wedge l_3)) \vee l_1$$

6-6. COMPLEMENTED LATTICES

Let $\mathfrak{L} = \langle L; \vee, \wedge \rangle$ be a lattice that contains the elements 1 and 0. By definition, for all $l \in L$,

$$l \leq 1, \qquad l \geq 0$$

Hence, using Theorem 6-3, for all $l \in L$,

$$l \vee 1 = 1, \qquad l \wedge 1 = l \tag{6-12}$$

$$l \wedge 0 = 0, \qquad l \vee 0 = l \tag{6-13}$$

Thus, 1 is the identity of $\mathcal{L}$ with respect to $\wedge$, and 0 is the identity of $\mathcal{L}$ with respect to $\vee$.

From the uniqueness of 1 and 0 (see Theorem 6-2) it follows that the ordering diagram of a lattice containing such elements exhibits a single vertex (labeled "1") at its uppermost level, and a single vertex (labeled "0") at its lowermost level. The 1 vertex can be reached from any other vertex via an ascending path, while the 0 vertex can be reached from any other vertex via a descending path.

A lattice $\langle L; \vee, \wedge \rangle$ with 1 and 0 is said to be *complemented* if, for every $l \in L$, there is an element, denoted by $\bar{l}$, such that

$$l \vee \bar{l} = 1, \qquad l \wedge \bar{l} = 0$$

$\bar{l}$ is called the *complement* of l. From (6-12) and (6-13) it follows that

$$\bar{0} = 1, \qquad \bar{1} = 0$$

THEOREM 6-11

In a complemented, distributive lattice $\langle L; \vee, \wedge \rangle$, the complement $\bar{l}$ of any element $l \in L$ is unique.

Proof Suppose there are two elements l_1 and l_2, such that

$$l \vee l_1 = 1, \qquad l \wedge l_1 = 0$$
$$l \vee l_2 = 1, \qquad l \wedge l_2 = 0$$

Hence,
$$\begin{aligned} l_1 &= l_1 \wedge 1 = l_1 \wedge (l \vee l_2) = (l_1 \wedge l) \vee (l_1 \wedge l_2) \\ &= 0 \vee (l_1 \wedge l_2) = l_1 \wedge l_2 = 0 \vee (l_2 \wedge l_1) \\ &= (l_2 \wedge l) \vee (l_2 \wedge l_1) = l_2 \wedge (l \vee l_1) \\ &= l_2 \wedge 1 = l_2 \end{aligned}$$

Thus, the complement of l is unique. □

THEOREM 6-12 (*Involution law*)

In a complemented, distributive lattice $\langle L; \vee, \wedge \rangle$, for every $l \in L$,

$$\bar{\bar{l}} = l$$

Proof Follows immediately from the definition and uniqueness of $\bar{l}$. □

THEOREM 6-13 (*De Morgan's laws*)

In a complemented, distributive lattice $\langle L; \vee, \wedge \rangle$, for all $l_1, l_2 \in L$,

$$\text{(a) } \overline{l_1 \vee l_2} = \bar{l}_1 \wedge \bar{l}_2 \qquad \text{(b) } \overline{l_1 \wedge l_2} = \bar{l}_1 \vee \bar{l}_2$$

Proof (a) Using the distributive law,

$$(l_1 \vee l_2) \wedge (\bar{l}_1 \wedge \bar{l}_2) = (l_1 \wedge \bar{l}_1 \wedge \bar{l}_2) \vee (l_2 \wedge \bar{l}_1 \wedge \bar{l}_2) = 0 \vee 0 = 0$$
$$(l_1 \vee l_2) \vee (\bar{l}_1 \wedge \bar{l}_2) = (l_1 \vee l_2 \vee \bar{l}_1) \wedge (l_1 \vee l_2 \vee \bar{l}_2) = 1 \wedge 1 = 1$$

which, together with Theorem 6-11, proves identity (a). (b) Follows from the duality principle. □

THEOREM 6-14

In a complemented, distributive lattice $\langle L; \vee, \wedge \rangle$, for all $l_1, l_2 \in L$,

$$(l_1 \leq l_2) \Longleftrightarrow (l_1 \wedge \bar{l}_2 = 0) \Longleftrightarrow (\bar{l}_1 \vee l_2 = 1)$$

Proof By Theorem 6-3,

$$(l_1 \leq l_2) \Longleftrightarrow (l_1 \wedge l_2 = l_1) \Longleftrightarrow (l_1 \vee l_2 = l_2)$$

Using De Morgan's laws,

$$(l_1 \leq l_2) \Longleftrightarrow (\bar{l}_1 \wedge \bar{l}_2 = \bar{l}_2) \Longleftrightarrow (\bar{l}_1 \vee \bar{l}_2 = \bar{l}_1)$$

Hence,

$$(l_1 \leq l_2) \Longrightarrow (l_1 \wedge \bar{l}_2 = 0)$$
$$(l_1 \leq l_2) \Longrightarrow (\bar{l}_1 \vee l_2 = 1)$$

On the other hand,

$$\begin{aligned}(l_1 \wedge \bar{l}_2 = 0) &\Longrightarrow (l_2 \vee (l_1 \wedge \bar{l}_2) = l_2)\\ &\Longrightarrow ((l_2 \vee l_1) \wedge (l_2 \wedge \bar{l}_2) = l_2)\\ &\Longrightarrow (l_2 \vee l_1 = l_2) \Longrightarrow (l_1 \leq l_2)\\ (\bar{l}_1 \vee l_2 = 1) &\Longrightarrow (l_1 \wedge (\bar{l}_1 \vee l_2) = l_1)\\ &\Longrightarrow ((l_1 \wedge \bar{l}_1) \vee (l_1 \wedge l_2) = l_1)\\ &\Longrightarrow (l_1 \wedge l_2 = l_1) \Longrightarrow (l_1 \leq l_2)\end{aligned}$$
□

PROBLEMS

1. Let $\mathfrak{L} = \langle L; \vee, \wedge \rangle$ be a lattice where $\#L > 1$. Prove that, if $\mathfrak{L}$ has 1 and 0 elements, then these elements must be distinct.
2. Which of the lattices whose ordering diagrams are shown in Fig. 6-C are complemented lattices?
3. Figure 6-D shows the ordering diagram of a complemented lattice $\langle L; \vee, \wedge \rangle$. Determine the complement of every element in L.

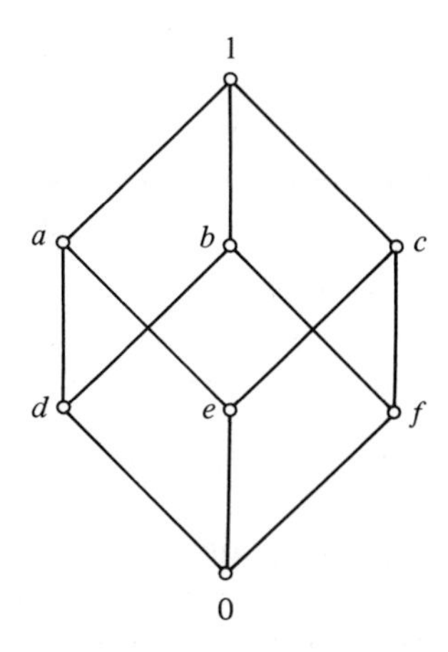

Fig. 6-D

4. Prove, by example, that not every complemented lattice is distributive, and that not every distributive lattice is complemented.

6-7. BOOLEAN ALGEBRAS

A lattice that contains the elements 1 and 0, and which is both distributive and complemented, is called a *Boolean algebra* (after the English mathematician George Boole, 1815–1864). Since in a complemented, distributive lattice the complement of every element is unique (see Theorem 6-11), complementation can be regarded as a bona fide operation over the domain of such a lattice. Accordingly, a Boolean algebra with the domain B is denoted by $\langle B; ^-, \vee, \wedge\rangle$ (where the two basic lattice operations $\vee$ and $\wedge$ are now supplemented by the complementation operation $^-$).

Tables 6-1 and 6-2 summarize the basic properties of $\langle B; ^-, \vee, \wedge\rangle$ (where x, y, and z are arbitrary elements of B). The laws listed in Table 6-1, together with the stipulation that 1 and 0 be distinct, can serve as the

Table 6-1 BOOLEAN ALGEBRA LAWS

Commutative Laws	1. $x \vee y = y \vee x$	1′. $x \wedge y = y \wedge x$
Associative Laws	2. $x \vee (y \vee z) = (x \vee y) \vee z$	2′. $x \wedge (y \wedge z) = (x \wedge y) \wedge z$
Distributive Laws	3. $x \wedge (y \vee z) =$ $(x \wedge y) \vee (x \wedge z)$	3′. $x \vee (y \wedge z) =$ $(x \vee y) \wedge (x \vee z)$
Identity Laws	4. $x \vee 0 = 0 \vee x = x$	4′. $x \wedge 1 = 1 \wedge x = x$
Complement Laws	5. $x \vee \bar{x} = 1$	5′. $x \wedge \bar{x} = 0$

Table 6-2 ADDITIONAL BOOLEAN ALGEBRA LAWS

Involution Law	1. (and 1′.) $\bar{\bar{x}} = x$	
Idempotent Laws	2. $x \vee x = x$	2′. $x \wedge x = x$
Null Laws	3. $x \vee 1 = 1$	3′. $x \wedge 0 = 0$
Absorption Laws	4. $x \vee (x \wedge y) = x$	4′. $x \wedge (x \vee y) = x$
De Morgan's Laws	5. $\overline{x \vee y} = \bar{x} \wedge \bar{y}$	5′. $\overline{x \wedge y} = \bar{x} \vee \bar{y}$

set of postulates for $\langle B; {}^-, \vee, \wedge \rangle$, from which all other properties can be derived.

The importance of the Boolean algebra $\langle B; {}^-, \vee, \wedge \rangle$ stems from the fact that many algebraic systems of interest in pure and applied mathematics are isomorphic to it. Comparing Tables 6-1 and 6-2 with Tables 1-3 and 1-4 reveals that the power set $\mathbf{2}^U$, with the symbols $'$, $\cup$, $\cap$, $\varnothing$, and U replacing ${}^-$, $\vee$, $\wedge$, 0, and 1, respectively, satisfies all the Boolean algebra postulates, and, hence, that $\langle \mathbf{2}^U; ', \cup, \cap \rangle$ is isomorphic to $\langle B; {}^-, \vee, \wedge \rangle$. Similarly, comparing Tables 6-1 and 6-2 with Tables 5-6 and 5-7 reveals that the set of propositions P_U, with the symbols $\Leftrightarrow$ and $\sim$ replacing $=$ and ${}^-$, respectively, satisfies all the Boolean algebra postulates, hence, that $\langle P_U; \sim, \vee, \wedge \rangle$ is isomorphic to $\langle B; {}^-, \vee, \wedge \rangle$. Thus, many of the results obtained in Chapters 1 and 4 for sets and in Chapter 5 for propositions (for example, the duality principle, the normal forms, etc.) could alternatively be obtained by deriving them for the general Boolean algebra $\langle B; {}^-, \vee, \wedge \rangle$ and then applying them to $\langle \mathbf{2}^U; ', \cup, \cap \rangle$ and $\langle P_U; \sim, \vee, \wedge \rangle$ simply by changing notation. (In this book we did not follow this route, preferring, for pedagogical reasons, to treat each system independently).

As already pointed out, the five laws and their duals listed in Table 6-1 can be taken as the postulates of Boolean algebra. The additional laws listed in Table 6-2 can be proved to be consequences of those in Table 6-1. Except for notation, the proofs are identical to those presented in Secs. 4-5 and 4-6, where the involution, idempotent, null, absorption, and De Morgan's laws were derived from the set postulates listed in Table 1-3.

The following two theorems state the interesting fact that subsystems of Boolean algebras and epimorphic images of Boolean algebras are themselves Boolean algebras.

THEOREM 6-15

Every subsystem of a Boolean algebra is a Boolean algebra.

Proof Let $\langle \tilde{B}; {}^-, \vee, \wedge \rangle$ be a subsystem of $\langle B; {}^-, \vee, \wedge \rangle$. From the definition of subsystem (see Sec. 4-2), it follows that the commutative, associative, and distributive laws are preserved in $\langle \tilde{B}; {}^-, \vee, \wedge \rangle$. If $x \in \tilde{B}$, then $\bar{x} \in \tilde{B}$; hence, $x \vee \bar{x}, x \wedge \bar{x} \in B$. Thus, the elements 0 and 1 are included in $\tilde{B}$; hence, $\langle \tilde{B}; {}^{-}, \vee, \wedge \rangle$ satisfies the identity and complement laws. □

THEOREM 6-16

Every epimorphic image of a Boolean algebra, where the image has at least two elements, is a Boolean algebra.

Proof Let $\langle B_0; {}^{-}, \vee, \wedge \rangle$ be the epimorphic image of the Boolean algebra $\langle B; {}^{-}, \vee, \wedge \rangle$ under the epimorphism h. By Theorem 4-11, the commutative, associative, and distributive laws are preserved in $\langle B_0; {}^{-}, \vee, \wedge \rangle$, and $h(1)$ and $h(0)$ are the one and zero, respectively, of $\langle B_0; {}^{-}, \vee, \wedge \rangle$. Now, suppose $h(0) = h(1)$. Since h is a surjection from B onto B_0, for every $x_0 \in B_0$ we have $x \in B$ such that $h(x) = x_0$, and we can write

$$\begin{aligned} x_0 &= h(x) = h(0 \vee x) = h(0) \vee h(x) \\ &= h(1) \vee h(x) = h(1 \vee x) = h(1) \end{aligned}$$

This implies that B_0 has only one element [namely, $h(1)$]—a contradiction. Thus, $h(1) \neq h(0)$ and $\langle B_0; {}^{-}, \vee, \wedge \rangle$ has distinct one and zero elements. To show that $\langle B_0; {}^{-}, \vee, \wedge \rangle$ satisfies the complement laws, again let $h(x) = x_0$, and write

$$\begin{aligned} h(0) = h(x \wedge \bar{x}) = h(x) \wedge h(\bar{x}) &= h(x) \wedge \overline{h(x)} \\ &= x_0 \wedge \bar{x}_0 \end{aligned}$$

Dually, $h(1) = x_0 \vee \bar{x}_0$. □

Example 6-5

Let $U = \{u_1, u_2, u_3\}$. The Boolean algebra $\langle 2^U; ', \cup, \cap \rangle$ has the subsystems

$$\begin{gathered} \langle \{\varnothing, U\}; ', \cup, \cap \rangle \\ \langle \{\varnothing, \{u_1\}, \{u_2, u_3\}, U\}; ', \cup, \cap \rangle \\ \langle \{\varnothing, \{u_2\}, \{u_1, u_3\}, U\}; ', \cup, \cap \rangle \\ \langle \{\varnothing, \{u_3\}, \{u_1, u_2\}, U\}; ', \cup, \cap \rangle \end{gathered}$$

Table 6-3 OPERATION TABLES FOR $\langle \{\varnothing, U\}; ', \cup, \cap \rangle$

X	$\bar{X}$
$\varnothing$	U
U	$\varnothing$

$\cup$	$\varnothing$	U
$\varnothing$	$\varnothing$	U
U	U	U

$\cap$	$\varnothing$	U
$\varnothing$	$\varnothing$	$\varnothing$
U	$\varnothing$	U

Table 6-4 OPERATION TABLES FOR $\langle\{\varnothing, S, T, U\}; ', \cup, \cap\rangle$

X	$\bar{X}$
$\varnothing$	U
S	T
T	S
U	$\varnothing$

$\cup$	$\varnothing$	S	T	U
$\varnothing$	$\varnothing$	S	T	U
S	S	S	U	U
T	T	U	T	U
U	U	U	U	U

$\cap$	$\varnothing$	S	T	U
$\varnothing$	$\varnothing$	$\varnothing$	$\varnothing$	$\varnothing$
S	$\varnothing$	S	$\varnothing$	S
T	$\varnothing$	$\varnothing$	T	T
U	$\varnothing$	S	T	U

all of which are Boolean algebras. Tables 6-3 and 6-4 define the operations of these systems (with the domains of the last three being denoted, for convenience, by $\{\varnothing, S, T, U\}$). □

Example 6-6

Let $U = \{u_1, u_2, u_3\}$, and define the following relation ρ on $\mathbf{2}^U$: $X\rho Y$ if and only if $\{u_1\} \cap X = \{u_1\} \cap Y$. Clearly, if $\{u_1\} \cap X = \{u_1\} \cap Y$, $\{u_1\} \cap X' = \{u_1\} \cap Y'$. Hence, $X\rho Y$ implies $X'\rho Y'$. Also, if $\{u_1\} \cap X_1 = \{u_1\} \cap Y_1$ and $\{u_1\} \cap X_2 = \{u_1\} \cap Y_2$, then $\{u_1\} \cap (X_1 \cup X_2) = \{u_1\} \cap (Y_1 \cup Y_2)$ and $\{u_1\} \cap (X_1 \cap X_2) = \{u_1\} \cap (Y_1 \cap Y_2)$. Hence, $X_1\rho Y_1$ and $X_2\rho Y_2$ imply $(X_1 \cup X_2)\rho(Y_1 \cup Y_2)$ and $(X_1 \cap X_2)\rho(Y_1 \cap Y_2)$. Thus, ρ is a congruence relation on $\langle\mathbf{2}^U: ', \cup, \cap\rangle$ (see Sec. 4-8). This congruence relation induces on $\mathbf{2}^U$ the equivalence partition

$$\begin{aligned}\pi_\rho^{\mathbf{2}^U} &= \{\{\varnothing, \{u_2\}, \{u_3\}, \{u_2, u_3\}\}, \{\{u_1\}, \{u_1, u_2\}, \{u_1, u_3\}, U\}\} \\ &= \{\mathcal{E}_\rho(\varnothing), \mathcal{E}_\rho(U)\}\end{aligned}$$

Correspondingly, there is an epimorphism from $\langle\mathbf{2}^U; ', \cup, \cap\rangle$ onto $\langle\{\mathcal{E}_\rho(\varnothing), \mathcal{E}_\rho(U)\}; ', \cap, \cup\rangle$ given by

$$h: \mathbf{2}^U \longrightarrow \{\mathcal{E}_\rho(\varnothing), \mathcal{E}_\rho(U)\}$$

where

$$h(X) = \mathcal{E}_\rho(X)$$

and where the operations for $\langle\mathcal{E}_\rho(\varnothing), \mathcal{E}_\rho(U)\}; ', \cup, \cap\rangle$ are as specified in Table 6-5. For example,

$$\begin{aligned}\mathcal{E}_\rho(\varnothing) \cup \mathcal{E}_\rho(U) &= h(\{u_2\}) \cup h(\{u_1, u_2\}) \\ &= h(\{u_2\} \cup \{u_1, u_2\}) = h(\{u_1, u_2\}) = \mathcal{E}_\rho(U)\end{aligned}$$

Comparing Tables 6-3 and 6-5, it is apparent that $\langle\{\mathcal{E}_\rho(\varnothing), \mathcal{E}_\rho(U)\}; ', \cup, \cap\rangle$ is isomorphic to $\langle\{\varnothing, U\}; ', \cup, \cap\rangle$ and, hence, is a Boolean algebra. Since $\langle\mathbf{2}^U; ', \cup, \cap\rangle$ is a Boolean algebra, this fact can also be deduced immediately from Theorem 6-16. □

Table 6-5 OPERATION TABLES FOR $\langle\{\mathcal{E}_\rho(\varnothing), \mathcal{E}_\rho(U)\}; ', \cup, \cap\rangle$

X	$\bar{X}$
$\mathcal{E}_\rho(\varnothing)$	$\mathcal{E}_\rho(U)$
$\mathcal{E}_\rho(U)$	$\mathcal{E}_\rho(\varnothing)$

$\cup$	$\mathcal{E}_\rho(\varnothing)$	$\mathcal{E}_\rho(U)$
$\mathcal{E}_\rho(\varnothing)$	$\mathcal{E}_\rho(\varnothing)$	$\mathcal{E}_\rho(U)$
$\mathcal{E}_\rho(U)$	$\mathcal{E}_\rho(U)$	$\mathcal{E}_\rho(U)$

$\cap$	$\mathcal{E}_\rho(\varnothing)$	$\mathcal{E}_\rho(U)$
$\mathcal{E}_\rho(\varnothing)$	$\mathcal{E}_\rho(\varnothing)$	$\mathcal{E}_\rho(\varnothing)$
$\mathcal{E}_\rho(U)$	$\mathcal{E}_\rho(\varnothing)$	$\mathcal{E}_\rho(U)$

PROBLEMS

1. Consider the algebraic system $\langle F; ^-, \vee, \wedge\rangle$, where $F = \{f \mid f: \mathbb{N} \longrightarrow \{0, 1\}\}$ and, for every $f_1, f_2 \in F$,

$$\bar{f}_1(n) = 1 \text{ if and only if } f_1(n) = 0$$

$$(f_1 \vee f_2)(n) = 1 \text{ if and only if } f_1(n) = 1 \text{ or } f_2(n) = 1$$

$$(f_1 \wedge f_2)(n) = 1 \text{ if and only if } f_1(n) = 1 \text{ and } f_2(n) = 1$$

Verify that $\langle F; ^-, \vee, \wedge\rangle$ is a Boolean algebra.

2. Let $U = \{u_1, u_2, u_3, u_4\}$. Show that the algebraic system

$$\langle\{\varnothing, \{u_1, u_2\}, \{u_3, u_4\}, U\}; ', \cup, \cap\rangle$$

is a Boolean algebra. List all other subsystems of $\langle \mathbf{2}^U; ', \cup, \cap\rangle$.

3. Let $U = \{u_1, u_2, \ldots, u_n\}$. For some fixed $S \subset U$, define the relation ρ on $\mathbf{2}^U$ as $X \rho Y$ if and only if $S \cap X = S \cap Y$. Prove that ρ is a congruence relation. On the basis of this relation, define a Boolean algebra epimorphic to $\langle \mathbf{2}^U; ', \cup, \cap\rangle$. (This is a generalization of Example 6-6.)

4. Let $X_1, X_2, \ldots, X_r$ be any subsets of some universal set U, and S the set of all sets generated by $X_1, X_2, \ldots, X_r$. Show that $\langle S; ', \cup, \cap\rangle$ is a Boolean algebra.

6-8. ATOMIC REPRESENTATION OF BOOLEAN ALGEBRAS

Let $\langle B; ^-, \vee, \wedge\rangle$ be a Boolean algebra. An element $a \in B$ is called an *atom* if $a \neq 0$ and if, for every $x \in B$,

$$x \wedge a = a \qquad \text{or} \qquad x \wedge a = 0$$

By Theorem 6-3, then, a is an atom if $a \neq 0$ and if, for every $x \in B$,

$$a \leq x \qquad \text{or} \qquad x \wedge a = 0$$

THEOREM 6-17

Let $\langle B; ^-, \vee, \wedge\rangle$ be a *finite* Boolean algebra. Then, for every *nonzero* $x \in B$, there is an atom a such that $x \wedge a = a$ (or $a \leq x$).

Proof Consider the (finite) ordering diagram of B, and the set of vertices $a_1, a_2, \ldots, a_k$ reachable from the 0 vertex via a single edge (see Fig. 6-5). Since the 0 vertex is reachable from every other vertex via a descending path, every vertex x, where $x \neq 0$, is either an a_i or reaches an a_i via a descending path. Thus, either $x = 0$ or $x \wedge a_i = a_i$. Also, if vertex x does not reach a given vertex a_j, then $x \wedge a_j = 0$. By definition, then, the a are atoms of $\langle B; {}^-, \wedge, \vee \rangle$. □

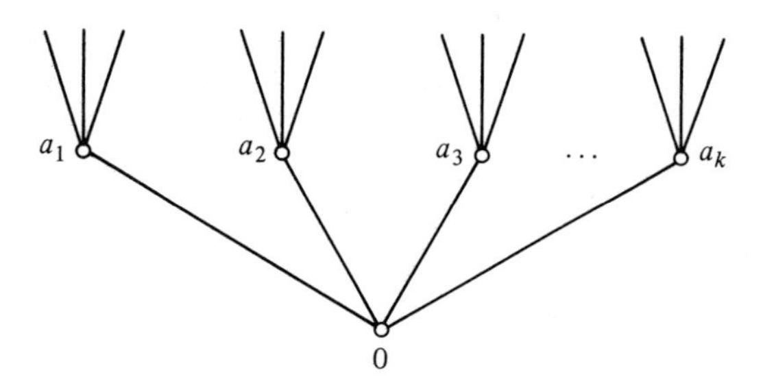

Fig. 6-5 For proof of Theorem 6-17

THEOREM 6-18

If a_1 and a_2 are atoms in the Boolean algebra $\langle B; {}^-, \vee, \wedge \rangle$, and if $a_1 \wedge a_2 \neq 0$, then $a_1 = a_2$.

Proof By definition, $a_1 \wedge a_2 = a_2$ and $a_2 \wedge a_1 = a_1$; hence, $a_1 = a_2$. □

THEOREM 6-19

Let $\langle B; {}^-, \vee, \wedge \rangle$ be a finite Boolean algebra. Let x be any nonzero element in B, and $a_1, a_2, \ldots, a_h$ be all those atoms a_i of $\langle B; {}^-, \vee, \wedge \rangle$ such that $a_i \leq x$. Then $x = a_1 \vee a_2 \vee \ldots \vee a_h$.

Proof Let $y = a_1 \vee a_2 \vee \ldots \vee a_h$. By (6-5), $y \leq x$, and all that remains to be shown is that $x \leq y$, or—by Theorem 6-14—that $x \wedge \bar{y} = 0$. Now, hypothesize that $x \wedge \bar{y} \neq 0$; then, by Theorem 6-17, there exists an atom a such that $a \leq x \wedge \bar{y}$. By (6-4′), we have $x \wedge \bar{y} \leq x$ and $x \wedge \bar{y} \leq y$ which, by transitivity, implies $a \leq x$ and $a \leq \bar{y}$. Since $a \leq x$, there is some a_i such that $a = a_i$ and, hence, by (6-4), $a \leq a_1 \vee a_2 \vee \ldots \vee a_h = y$. We thus have $a \leq y$ and $a \leq \bar{y}$, which, by (6-5′), implies $a \leq y \wedge \bar{y} = 0$. In conclusion, $a = 0$, which is a contradiction. The hypothesis $x \wedge \bar{y} \neq 0$, therefore, must be false. □

THEOREM 6-20

Let $a, a_1, a_2, \ldots, a_h$ be any atoms in the finite Boolean algebra $\langle B; {}^-, \vee, \wedge \rangle$. If $x = a_1 \vee a_2 \vee \ldots \vee a_h$ and $a \leq x$, then there is some a_i such that $a = a_i$.

Proof Since a is an atom, $x \wedge a = a$ or $x \wedge a = 0$. By Theorem 6-3, $a \leq x$ implies $x \wedge a = a$. Hence, if $x \wedge a = 0$, we must have $a = 0$ — a contradiction. Thus $x \wedge a = a$, and we can write

$$\begin{aligned} & a \wedge (a_1 \vee a_2 \vee \ldots \vee a_h) \\ &= (a \wedge a_1) \vee (a \wedge a_2) \vee \ldots \vee (a \wedge a_h) = a \end{aligned}$$

which implies that, for some a_i, $a \wedge a_i \neq 0$. By Theorem 6-18, then, $a = a_i$. □

Theorems 6-19 and 6-20 imply that every nonzero $x \in B$ can be expressed *uniquely* in the form

$$x = a_1 \vee a_2 \vee \ldots \vee a_h$$

where $\{a_1, a_2, \ldots, a_h\}$ is the set of all atoms of $\langle B; ^-, \vee, \wedge \rangle$ such that $a_i \leq x$. (By Theorem 6-17, this set is necessarily nonempty.)

THEOREM 6-21

Let $\langle B; ^-, \vee, \wedge \rangle$ be a finite Boolean algebra, and let M denote the set of all atoms in this algebra. Then $\langle B; ^-, \vee, \wedge \rangle$ is isomorphic to $\langle \mathbf{2}^M; ', \cup, \cap \rangle$.

Proof Consider the function

$$h: B \longrightarrow \mathbf{2}^M$$

where
$$h(x) = \begin{cases} \varnothing & (x = 0) \\ \{a \in M \mid a \leq x\} & (x \neq 0) \end{cases}$$

From Theorems 6-19 and 6-20 it follows that h is a bijection. Now, consider any nonzero elements $x_1, x_2 \in B$, and let

$$\begin{aligned} h(x_1) &= M_1 = \{a_{11}, a_{12}, \ldots, a_{1k_1}\} \\ h(x_2) &= M_2 = \{a_{21}, a_{22}, \ldots, a_{2k_2}\} \end{aligned}$$

Hence,
$$\begin{aligned} x_1 &= a_{11} \vee a_{12} \vee \ldots \vee a_{1k_1} \\ x_2 &= a_{21} \vee a_{22} \vee \ldots \vee a_{2k_2} \\ x_1 \vee x_2 &= a_{11} \vee a_{12} \vee \ldots \vee a_{1k_1} \vee a_{21} \vee a_{22} \vee \ldots \vee a_{2k_2} \end{aligned}$$

Thus,
$$h(x_1 \vee x_2) = M_1 \cup M_2 \tag{6-14}$$

Next, using the distributivity property, we can write

$$\begin{aligned} x_1 \wedge x_2 &= (a_{11} \vee a_{12} \vee \ldots \vee a_{1k_1}) \wedge (a_{21} \vee a_{22} \vee \ldots \vee a_{2k_2}) \\ &= \textstyle\bigvee_{i=1}^{k_1} (\bigvee_{j=1}^{k_2} (a_{1i} \wedge a_{2j})) \end{aligned}$$

By Theorem 6-18,

$$a_{1i} \wedge a_{2j} = \begin{cases} a_{1i} = a_{2j} & \text{if } a_{1i} = a_{2j} \\ 0 & \text{otherwise} \end{cases}$$

Hence, $x_1 \wedge x_2$ equals the join of all a_{1i} (or a_{2j}) such that $a_{1i} = a_{2j}$, and consequently,

$$h(x_1 \wedge x_2) = M_1 \cap M_2 \tag{6-15}$$

Finally, suppose $x_2 = \bar{x}_1$. Then $x_1 \vee x_2 = 1$; hence,

$$h(x_1 \vee x_2) = M_1 \cup M_2 = M$$

Also, $x_1 \wedge x_2 = 0$; hence,

$$h(x_1 \wedge x_2) = M_1 \cap M_2 = \varnothing$$

Consequently, $M_2 = M'_1$, or

$$h(\bar{x}_1) = M'_1 \tag{6-16}$$

When the assumption that x_1 and x_2 are nonzero is removed, and either $x_1 = 0$ or $x_2 = 0$, then $M_1 = \varnothing$ or $M_2 = \varnothing$, and (6-14), (6-15), and (6-16) are immediate. From (6-14), (6-15), and (6-16), then, it follows that h is an isomorphism from $\langle B; ^-, \vee, \wedge \rangle$ onto $\langle \mathbf{2}^M; ', \cup, \cap \rangle$; hence, the two algebraic systems are isomorphic. □

Theorem 6-21 is of great significance, inasmuch as it permits us to represent every finite Boolean algebra $\langle B; ^-, \vee, \wedge \rangle$ by some algebra of sets $\langle \mathbf{2}^M; ', \cup, \cap \rangle$. An immediate corollary of this result is that

$$\#B = 2^{\#M}$$

This corollary, in turn, leads to the following conclusion: lf the domains of two finite Boolean algebras $\mathcal{B}_1$ and $\mathcal{B}_2$ have the same cardinality, then their sets of atoms must have the same cardinality; thus, the algebras of sets representing $\mathcal{B}_1$ and $\mathcal{B}_2$ must be isomorphic. Summarizing, we have:

THEOREM 6-22

The cardinality of the domain of every finite Boolean algebra is a power of 2. Boolean algebras whose domains have the same cardinality must be isomorphic.

Example 6-7

Let $X_1, X_2, \ldots, X_r$ be subsets of some universal set U. If S denotes the set of all sets generated by $X_1, X_2, \ldots, X_r$, then $\langle S; ', \cup, \cap \rangle$ is a Boolean algebra (see Problem 4 of Sec. 6-7). Since a minset generated by $X_1, X_2, \ldots, X_r$ is either included in an element of S or forms an empty intersection with that element, the minsets generated by $X_1, X_2, \ldots, X_r$ are precisely the atoms of $\langle S; ', \cup, \cap \rangle$. By Theorem 6-21, then, $\langle S; ', \cup, \cap \rangle$ is isomorphic to the Boolean algebra $\langle \mathbf{2}^M; ', \cup, \cap \rangle$, where M is the set of all minsets generated by $X_1, X_2, \ldots, X_r$. For example, if S is the set of all sets generated by X, Y (see Fig. 6-6), then $\langle S; ', \cup, \cap \rangle$ is isomorphic to the Boolean

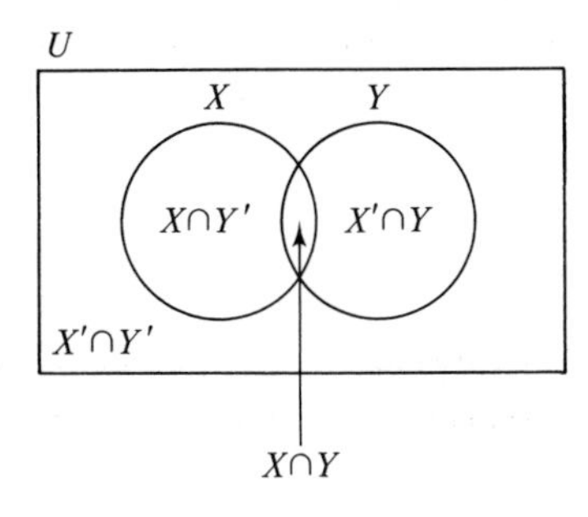

Fig. 6-6 Atoms of $\langle S; ', \cup, \cap \rangle$ of Example 6-7

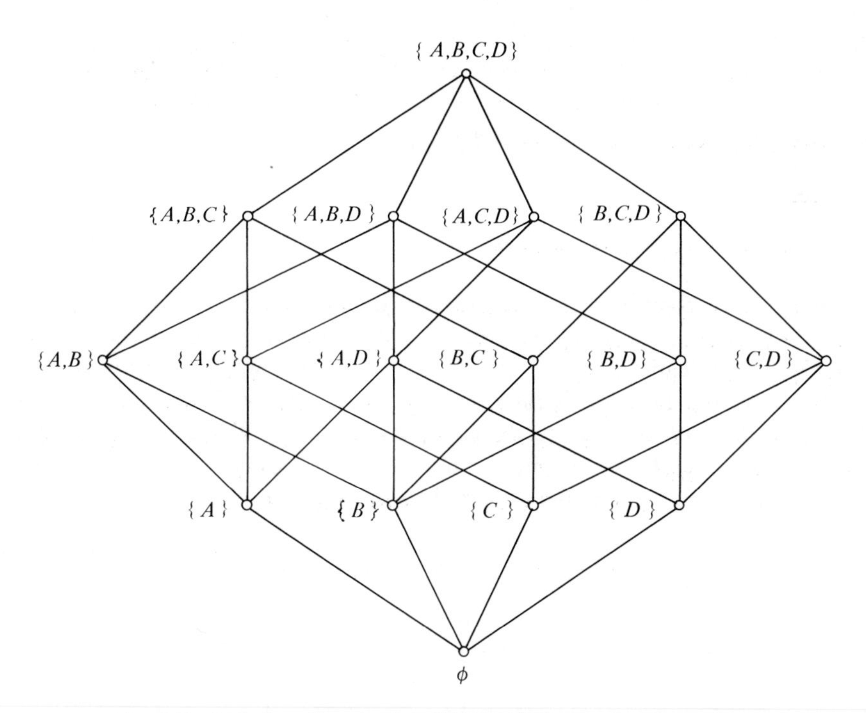

Fig. 6-7 Ordering diagram for $\langle \mathbf{2}^{\{A,B,C,D\}}; ', \cup, \cap \rangle$, where $A = X \cap Y$, $B = X \cap Y'$, $C = X' \cap Y$, and $D = X' \cap Y'$

algebra $\langle \mathbf{2}^M; ', \cup, \cap \rangle$, where

$$M = \{X \cap Y, X \cap Y', X' \cap Y, X' \cap Y')$$

The ordering diagram of $\langle \mathbf{2}^M; ', \cup, \cap \rangle$ is shown in Fig. 6-7. □

It should be mentioned that Theorem 6-21 holds also for *infinite* Boolean algebra. (see Problem 1, Sec. 6-7 for an example of such an algebra). However, the proof of the more general version of this theorem is beyond the scope of this book.

PROBLEMS

1. Let $\langle B; ^-, \vee, \wedge \rangle$ be a Boolean algebra. An element $a \in B$ is said to be *minimal* if $a \neq 0$ and if, for every $x \in B$, $x \leq a$ implies $x = a$ or $x = 0$. Show that a is an atom if and only if it is minimal.
2. Without making use of Theorem 6-22, show that there cannot exist a Boolean algebra whose domain has cardinality 3.
3. Let $\mathcal{B} = \langle \{\varnothing, \{u_1, u_2\}, \{u_3, u_4\}, U\}; ', \cup, \cap \rangle$ be the Boolean algebra of Problem 2, Sec. 6-7. What is the set M of atoms of $\mathcal{B}$? Draw the ordering diagram for $\mathcal{B}$ and for the Boolean algebra $\langle \mathbf{2}^M; ', \cup, \cap \rangle$ which is isomorphic to $\mathcal{B}$.
4. Specialize Example 6-7 to the case $r = 1$.

6-9. THE BOOLEAN ALGEBRAS $\mathcal{B}_2^r$

A Boolean algebra with n elements is denoted by $\mathcal{B}_n = \langle B_n; ^-, \vee, \wedge \rangle$. From Theorem 6-22 it follows that n must be a power of 2; hence, the "smallest" Boolean algebra is $\mathcal{B}_2 = \langle B_2; ^-, \vee, \wedge \rangle$, whose domain is $B = \{0, 1\}$. Using the identity, idempotent, and null laws of Tables 6-1 and 6-2, the operations of $\mathcal{B}_2$ can be displayed as shown in Table 6-6. Table 6-7 defines the operations of $\mathcal{B}_4 = \langle B_4; ^-, \vee, \wedge \rangle$, where $B_4 = \{0, \alpha, \beta, 1\}$. (Compare Tables 6-6 and 6-7 with Tables 6-3 and 6-4!)

Table 6-6 OPERATION TABLES FOR $\mathcal{B}_2$

x	$\bar{x}$
0	1
1	0

$\vee$	0	1
0	0	1
1	1	1

$\wedge$	0	1
0	0	0
1	0	1

Table 6-7 OPERATION TABLES FOR $\mathcal{B}_4$

x	$\bar{x}$
0	1
α	β
β	α
1	0

$\vee$	0	α	β	1
0	0	α	β	1
α	α	α	1	1
β	β	1	β	1
1	1	1	1	1

$\wedge$	0	α	β	1
0	0	0	0	0
α	0	α	0	α
β	0	0	β	β
1	0	α	β	1

An algebraic system of special interest is $\mathcal{B}_2 \times \mathcal{B}_2 \times \ldots \times \mathcal{B}_2$ (r times); that is, the direct product of r identical Boolean algebras $\mathcal{B}_2$ (see Sec. 4-9). This system is denoted by $\mathcal{B}_2^r$, and the symbols used for its operations are the same as the corresponding ones in $\mathcal{B}_2$. That is, $\mathcal{B}_2^r = \langle B_2^r; {}^-, \vee, \wedge \rangle$ where, for every $(\beta_1, \beta_2, \ldots, \beta_r), (\beta'_1, \beta'_2, \ldots, \beta'_r) \in B_2^r$,

$$\overline{(\beta_1, \beta_2, \ldots, \beta_r)} = (\bar{\beta}_1, \bar{\beta}_2, \ldots, \bar{\beta}_r)$$

$$(\beta_1, \beta_2, \ldots, \beta_r) \vee (\beta'_1, \beta'_2, \ldots, \beta'_r) = (\beta_1 \vee \beta'_1, \beta_2 \vee \beta'_2, \ldots, \beta_r \vee \beta'_r)$$

$$(\beta_1, \beta_2, \ldots, \beta_r) \wedge (\beta'_1, \beta'_2, \ldots, \beta'_r) = (\beta_1 \wedge \beta'_1, \beta_2 \wedge \beta'_2, \ldots, \beta_r \wedge \beta'_r)$$

From Theorem 4-17 we know that the commutative, associative, distributive, and identity laws of $\mathcal{B}_2$ are preserved in $\mathcal{B}_2^r$ [the identity and zero of $\mathcal{B}_2^r$ being $(1, 1, \ldots, 1)$ and $(0, 0, \ldots, 0)$, respectively]. It can also be shown readily that the complement laws hold in $\mathcal{B}_2^r$. We can thus conclude that $\mathcal{B}_2^r$ is a Boolean algebra. By Theorem 6-22, then, we have:

THEOREM 6-23

The Boolean algebras $\mathcal{B}_2^r$ and $\mathcal{B}_{2^r}$ are isomorphic, and every Boolean algebra is isomorphic to some Boolean algebra $\mathcal{B}_2^r$.

Example 6-8

Table 6-8 defines the operations of $\mathcal{B}_2^2 = \langle B_2^2; {}^-, \vee, \wedge \rangle$. Comparison with Table 6-7 verifies that $\mathcal{B}_2^2$ and $\mathcal{B}_4$ are isomorphic. □

Table 6-8 OPERATION TABLES FOR $\mathcal{B}_2^2$

x	$\bar{x}$
(0, 0)	(1, 1)
(0, 1)	(1, 0)
(1, 0)	(0, 1)
(1, 1)	(0, 0)

Table 6-8 (cont.)

$\vee$	(0, 0)	(0, 1)	(1, 0)	(1, 1)
(0, 0)	(0, 0)	(0, 1)	(1, 0)	(1, 1)
(0, 1)	(0, 1)	(0, 1)	(1, 1)	(1, 1)
(1, 0)	(1, 0)	(1, 1)	(1, 0)	(1, 1)
(1, 1)	(1, 1)	(1, 1)	(1, 1)	(1, 1)

$\wedge$	(0, 0)	(0, 1)	(1, 0)	(1, 1)
(0, 0)	(0, 0)	(0, 0)	(0, 0)	(0, 0)
(0, 1)	(0, 0)	(0, 1)	(0, 0)	(0, 1)
(1, 0)	(0, 0)	(0, 0)	(1, 0)	(1, 0)
(1, 1)	(0, 0)	(0, 1)	(1, 0)	(1, 1)

It is interesting to note that there is a natural isomorphism between the algebraic system $\langle \mathbf{2}^U; ', \cup, \cap \rangle$, where $U = \{u_1, u_2, \ldots, u_r\}$, and the Boolean algebra $\mathcal{B}_2^r = \langle B_2^r; ^-, \vee, \wedge \rangle$. This isomorphism is given by

$$h: \mathbf{2}^U \longrightarrow B_2^r$$

where

$$h(S) = (\beta_1, \beta_2, \ldots, \beta_r)$$

with

$$\beta_i = \begin{cases} 1 & \text{if } u_i \in S \\ 0 & \text{otherwise} \end{cases} \qquad (i = 1, 2, \ldots, r)$$

[For example, when $r = 4$, $h(\{u_1, u_3, u_4\}) = (1, 0, 1, 1)$.] The proof that h is an isomorphism is left to the reader. The existence of this isomorphism confirms what we have already established in Sec. 6-8, namely, that every Boolean algebra is isomorphic to some algebra of sets.

PROBLEMS

1. Show that $\mathcal{B}_2^r$ satisfies the complement laws.
2. Verify the correctness of Table 6-7.
3. The operations of the algebraic system $\langle A; ', +, \cdot \rangle$, where $A = \{\alpha, \beta, \gamma, \delta\}$ are defined as:

a	a'
α	δ
β	γ
γ	β
δ	α

$+$	α	β	γ	δ
α	α	α	γ	γ
β	α	β	γ	δ
γ	γ	γ	γ	γ
δ	γ	δ	γ	δ

$\cdot$	α	β	γ	δ
α	α	β	α	β
β	β	β	β	β
γ	α	β	γ	δ
δ	β	β	δ	δ

Is this system a Boolean algebra? Prove your answer.

4. Prove that the function h specified at the end of Sec. 6-9 is an isomorphism from $\langle 2^U; ', \cup, \cap \rangle$ onto $\langle B_2^r; ^-, \vee, \wedge \rangle$. Construct h for the case $U = \{u_1, u_2, u_3\}$.

6-10. BOOLEAN EXPRESSIONS

A *Boolean expression generated by* $x_1, x_2, \ldots, x_n$ *over the Boolean algebra* $\langle B; ^-, \vee, \wedge \rangle$ is defined recursively as:

(i) (*Basis*). Any element of B and any of the symbols $x_1, x_2, \ldots, x_n$ (which should not coincide with names of elements of B) are Boolean expressions generated by $x_1, x_2, \ldots, x_n$ over $\langle B; ^-, \vee, \wedge \rangle$.

(ii) (*Induction step*). If e_1 and e_2 are expressions generated by $x_1, x_2, \ldots, x_n$ over $\langle B; ^-, \vee, \wedge \rangle$, so are (e_1), $\bar{e}_1$, $(e_1) \vee (e_2)$ and $(e_1) \wedge (e_2)$. (Parentheses can be omitted under the convention that $\wedge$ has precedence over $\vee$.)

For example, $0 \wedge \bar{1}$, $1 \vee (\alpha \wedge x_1) \vee (\bar{x}_2 \wedge x_3)$ and $(\overline{\bar{\beta} \vee x_1 \vee x_3}) \wedge 0$ are Boolean expressions generated by $x_1, x_2, \ldots, x_n$ over the Boolean algebra $\langle \{0, \alpha, \beta, 1\}; ^-, \vee, \wedge \rangle$.

If $x_1, x_2, \ldots, x_n$ are interpreted as variables that can assume only values from B, then Boolean expressions generated by $x_1, x_2, \ldots, x_n$ over $\langle B; ^-, \vee, \wedge \rangle$ are seen to represent elements in B. Thus, such expressions can be interpreted as functions of the type

$$f : B^n \longrightarrow B$$

where $f(x_1, x_2, \ldots, x_n)$, for any particular argument $(x_1, x_2, \ldots, x_n)$, can be determined from the operation tables of $^-$, $\vee$, and $\wedge$ for $\langle B; ^-, \vee, \wedge \rangle$. Boolean expressions generated by $x_1, x_2, \ldots, x_n$ over $\langle B; ^-, \vee, \wedge \rangle$ are sometimes referred to as *Boolean functions of n variables over* $\langle B; ^-, \vee, \wedge \rangle$.

Example 6-9

The following is a Boolean expression generated by x, y (or a Boolean function of 2 variables) over the Boolean algebra $\langle B; ^-, \vee, \wedge \rangle$, where $B = \{0, \alpha, \beta, 1\}$:

$$f(x, y) = (\beta \wedge \bar{x} \wedge y) \vee (\beta \wedge x \wedge (\overline{x \vee \bar{y}})) \vee (\alpha \wedge (x \vee (\bar{x} \wedge y)))$$

Using Table 6-7, we have, for example:

$$\begin{aligned} f(\alpha, 0) &= (\beta \wedge \beta \wedge 0) \vee (\beta \wedge \alpha \wedge (\overline{\alpha \vee 1})) \vee (\alpha \wedge (\alpha \vee (\beta \wedge 0))) \\ &= (\beta \wedge \alpha \wedge (\beta \wedge 0)) \vee (\alpha \wedge \alpha) = \alpha \end{aligned}$$

Table 6-9 lists values of $f(x, y)$ for all arguments $(x, y) \in B^2$. □

Table 6-9 BOOLEAN FUNCTION $f(x, y)$ OF EXAMPLE 6-9

x	y	$f(x, y)$
0	0	0
0	α	α
0	β	β
0	1	1
α	0	α
α	α	α
α	β	1
α	1	1
β	0	0
β	α	α
β	β	0
β	1	α
1	0	α
1	α	α
1	β	α
1	1	α

In what follows, a Boolean expression generated by $x_1, x_2, \ldots, x_n$ over $\langle B; ^-, \vee, \wedge \rangle$, which has the form $\hat{x}_1 \wedge \hat{x}_2 \wedge \ldots \wedge \hat{x}_n$, where $\hat{x}_i$ is either x_i or $\bar{x}_i$, will be called a *minterm generated by* $x_1, x_2, \ldots, x_n$. It is denoted by $m_{\delta_1 \delta_2 \ldots \delta_n}$, where

$$\delta_i = \begin{cases} 1 & \text{if } \hat{x}_i = x_i \\ 0 & \text{if } \hat{x}_i = \bar{x}_i \end{cases}$$

A Boolean expression of the form $\hat{x}_1 \vee \hat{x}_2 \vee \ldots \vee \hat{x}_n$ is called a *maxterm generated by* $x_1, x_2, \ldots, x_n$. It is denoted by $\tilde{m}_{\delta_1 \delta_2 \ldots \delta_n}$, where

$$\delta_i = \begin{cases} 0 & \text{if } \hat{x}_i = x_i \\ 1 & \text{if } \hat{x}_i = \bar{x}_i \end{cases}$$

THEOREM 6-24

Every Boolean expression $f(x_1, x_2, \ldots, x_n)$ over $\langle B; ^-, \vee, \wedge \rangle$ can be written in the forms

$$f(x_1, x_2, \ldots, x_n) = \bigvee_{k=00\ldots0}^{11\ldots1} (c_k \wedge m_k) \tag{6-17}$$

$$f(x_1, x_2, \ldots, x_n) = \bigwedge_{k=00\ldots0}^{11\ldots1} (\tilde{c}_k \vee \tilde{m}_k) \tag{6-18}$$

where k assumes all 2^n possible configurations $\delta_1 \delta_2 \ldots \delta_n$ ($\delta_i \in \{0, 1\}$), and where

$$c_{\delta_1 \delta_2 \ldots \delta_n} = \tilde{c}_{\delta_1 \delta_2 \ldots \delta_n} = f(\delta_1, \delta_2, \ldots, \delta_n) \tag{6-19}$$

Proof From the definition of $m_{\delta_1\delta_2\ldots\delta_n}$ and $\tilde{m}_{\delta_1\delta_2\ldots\delta_n}$ it follows that $m_{\delta_1\delta_2\ldots\delta_n} = 1$ and $\tilde{m}_{\delta_1\delta_2\ldots\delta_n} = 0$ if and only if $x_1 = \delta_1, x_2 = \delta_2, \ldots, x_n = \delta_n$. Hence,

$$f(x_1, x_2, \ldots, x_n) = \begin{cases} c_{\delta_1\delta_2\ldots\delta_n} \wedge m_{\delta_1\delta_2\ldots\delta_n} & \text{if } x_1x_2\ldots x_n = \delta_1\delta_2\ldots\delta_n \\ 0 & \text{otherwise} \end{cases}$$

$$f(x_1, x_2, \ldots, x_n) = \begin{cases} \tilde{c}_{\delta_1\delta_2\ldots\delta_n} \vee \tilde{m}_{\delta_1\delta_2\ldots\delta_n} & \text{if } x_1x_2\ldots x_n = \delta_1\delta_2\ldots\delta_n \\ 1 & \text{otherwise} \end{cases}$$

which implies (6-17) and (6-18). □

Thus, every Boolean expression over $\langle B; ^-, \vee, \wedge \rangle$ can be expressed as a "weighted" join of all possible minterms or a "weighted" meet of all possible maxterms, where the "weights" (that is, $c_{\delta_1\delta_2\ldots\delta_n}$ and $\tilde{c}_{\delta_1\delta_2\ldots\delta_n}$) are elements of B. These forms, called the *minterm normal form* and *maxterm normal form*, respectively, are unique, since the weights are unique. They can be computed directly from the Boolean expression by using equation (6-19).

Example 6-10

Consider the expression

$$f(x, y) = (\beta \wedge \bar{x} \wedge y) \vee (\beta \wedge x \wedge \overline{(x \vee \bar{y})}) \vee (\alpha \wedge (x \vee (\bar{x} \wedge y)))$$

of Example 6-9. Using Table 6-9, we have:

$$\begin{aligned} f(x, y) &= (c_{00} \wedge m_{00}) \vee (c_{01} \wedge m_{01}) \vee (c_{10} \wedge m_{10}) \vee (c_{11} \wedge m_{11}) \\ &= (f(0, 0) \wedge \bar{x} \wedge \bar{y}) \vee (f(0, 1) \wedge \bar{x} \wedge y) \\ &\quad \vee (f(1, 0) \wedge x \wedge \bar{y}) \vee (f(1, 1) \wedge x \wedge y) \\ &= (0 \wedge \bar{x} \wedge \bar{y}) \vee (1 \wedge \bar{x} \wedge y) \vee (\alpha \wedge x \wedge \bar{y}) \vee (\alpha \wedge x \wedge y) \\ f(x, y) &= (\tilde{c}_{00} \vee \tilde{m}_{00}) \wedge (\tilde{c}_{01} \vee \tilde{m}_{01}) \wedge (\tilde{c}_{10} \vee \tilde{m}_{10}) \wedge (\tilde{c}_{11} \vee \tilde{m}_{11}) \\ &= (f(0, 0) \vee x \vee y) \wedge (f(0, 1) \vee x \vee \bar{y}) \\ &\quad \wedge (f(1, 0) \vee \bar{x} \vee y) \wedge (f(1, 1) \vee \bar{x} \vee \bar{y}) \\ &= (0 \vee x \vee y) \wedge (1 \vee x \vee \bar{y}) \wedge (\alpha \vee \bar{x} \vee y) \wedge (\alpha \vee \bar{x} \vee \bar{y}) \end{aligned}$$

□

Note that sets generated by $A_1, A_2, \ldots, A_r$ (see Sec. 1-3) are simply Boolean expressions generated by $A_1, A_2, \ldots, A_r$ over the Boolean algebra $\langle \{\varnothing, U\}; ', \cup, \cap \rangle$. Thus, the minset and maxset normal forms derived in Secs. 1-8 and 1-9 are merely special cases of Theorem 6-24 (with the "weights" being restricted to either $\varnothing$ or U). Similarly, propositions generated by $p_1, p_2, \ldots, p_r$ (see Sec. 5-5) are simply Boolean expressions generated by

$p_1, p_2, \ldots, p_r$ over the Boolean algebra $\langle\{0, 1\}; \sim, \vee, \wedge\rangle$. Thus, the disjunctive and conjunctive normal forms presented in Sec. 5-5 are, again, special cases of Theorem 6-24 (with the "weights" being restricted to either 0 or 1).

The normal forms of Boolean expressions can also be obtained by a method analogous to that outlined for sets in Algorithms 1-3 and 1-4. The generalized procedure is:

ALGORITHM 6-1 (or 2)

To find the minterm (or maxterm) normal form of a given Boolean expression $f(x_1, x_2, \ldots, x_n)$ over $\langle B; {}^-, \vee, \wedge\rangle$:

(i) Use the laws in Tables 6-1 and 6-2 to express $f(x_1, x_2, \ldots, x_n)$ as a join (or meet) of distinct meets (or joins) of the form $c \wedge \hat{x}_{i_1} \wedge \hat{x}_{i_2} \wedge \ldots \wedge \hat{x}_{i_k}$ (or $c \vee \hat{x}_{i_1} \vee \hat{x}_{i_2} \vee \ldots \vee \hat{x}_{i_k}$) where $c \in B$ and $i_1 < i_2 < \ldots . i_k$.

(ii) If every meet (or join) is of the form $c \wedge m$ (or $c \vee m$), where $c \in B$ and m is a minterm (or maxterm), augment the expression with $(0 \wedge m_1) \vee (0 \wedge m_2) \wedge \ldots$ [or $(1 \vee m_1) \wedge (1 \vee m_2) \wedge \ldots$], where $m_1, m_2, \ldots$ are the minterms (or maxterms) originally absent from the expression. The augmented expression is in the desired form. Otherwise:

(iii) From the expression last obtained, select a meet (or a join) of the form $c \wedge \hat{x}_{i_1} \wedge \hat{x}_{i_2} \wedge \ldots \wedge \hat{x}_{i_k}$ (or $c \vee \hat{x}_{i_1} \vee \hat{x}_{i_2} \vee \ldots \vee \hat{x}_{i_k}$) where $k < n$ and where, for some h, $\hat{x}_{i_h}$ does not appear, and replace it with

$$(c \wedge \hat{x}_{i_1} \wedge \hat{x}_{i_2} \wedge \ldots \wedge \hat{x}_{i_k} \wedge x_{i_h}) \vee (c \wedge \hat{x}_{i_1} \wedge \hat{x}_{i_2} \wedge \ldots \wedge \hat{x}_{i_k} \wedge \bar{x}_{i_h})$$
$$[\text{or } (c \vee \hat{x}_{i_1} \vee \hat{x}_{i_2} \vee \ldots \vee \hat{x}_{i_k} \vee x_{i_h}) \wedge (c \vee \hat{x}_{i_1} \vee \hat{x}_{i_2} \vee \ldots \vee \hat{x}_{i_k} \vee \bar{x}_{i_h})]$$

which is permitted by virtue of the commutative, distributive, and complement laws in Table 6-1. Rearrange the $\hat{x}_{i_j}$ in the new meets (or joins) in order of ascending subscript, and replace groups of meets (or joins) such as $c_1 \wedge m$, $c_2 \wedge m, \ldots$ (or $c_1 \vee m, c_2 \vee m, \ldots$) with $(c_1 \vee c_2 \vee \ldots) \wedge m$ [or $(c_1 \wedge c_2 \wedge \ldots) \vee m$]. Return to step (ii). □

Example 6-11

The following illustrates the use of Algorithm 6-1 for constructing the minterm normal form of $f(x, y)$ of Examples 6-9 and 6-10:

$$\begin{aligned}
f(x, y) &= (\beta \wedge \bar{x} \wedge y) \vee (\beta \wedge x \wedge (\overline{x \vee \bar{y}})) \vee (\alpha \wedge (x \vee (\bar{x} \wedge y))) \\
&= (\beta \wedge \bar{x} \wedge y) \vee (\beta \wedge x \wedge \bar{x} \wedge y) \vee (\alpha \wedge x) \vee (\alpha \wedge \bar{x} \wedge y) \\
&= (\bar{x} \wedge y) \vee (\alpha \wedge x) \\
&= (\bar{x} \wedge y) \vee (\alpha \wedge x \wedge y) \vee (\alpha \wedge x \wedge \bar{y}) \\
&= (0 \wedge \bar{x} \wedge \bar{y}) \vee (1 \wedge \bar{x} \wedge y) \vee (\alpha \wedge x \wedge \bar{y}) \vee (\alpha \wedge x \wedge y)
\end{aligned}$$

Using Algorithm (6-2) for constructing the maxterm normal form of the same expression, we have:

$$\begin{aligned} f(x, y) &= (\beta \wedge \bar{x} \wedge y) \vee (\beta \wedge x \wedge \overline{(x \vee \bar{y})}) \vee (\alpha \wedge (x \vee (\bar{x} \wedge y))) \\ &= (\bar{x} \wedge y) \vee (\alpha \wedge x) \\ &= (\alpha \vee \bar{x}) \wedge (\alpha \vee y) \wedge (x \vee y) \\ &= (\alpha \vee \bar{x} \vee y) \wedge (\alpha \vee \bar{x} \vee \bar{y}) \wedge (\alpha \vee x \vee y) \wedge (x \vee y) \\ &= (0 \vee x \vee y) \wedge (\alpha \vee \bar{x} \vee y) \wedge (\alpha \vee \bar{x} \vee \bar{y}) \\ &= (0 \vee x \vee y) \wedge (1 \vee x \vee \bar{y}) \wedge (\alpha \vee \bar{x} \vee y) \wedge (\alpha \vee \bar{x} \vee \bar{y}) \end{aligned}$$

□

Since a Boolean expression $f(x_1, x_2, \ldots, x_n)$ over $\langle B; ^-, \vee, \wedge \rangle$ is uniquely determined by its 2^n "weights," and since each weight is an element of B, it follows that there are $(\#B)^{2^n}$ distinct Boolean expressions $f(x_1, x_2, \ldots, x_n)$ over $\langle B; ^-, \vee, \wedge \rangle$. On the other hand, the number of distinct functions of the type f: $B^n \longrightarrow B$ equals the number of ways in which the $(\#B)^n$ values of $(f(x_1, x_2, \ldots, x_n)$ can be assigned, namely, $(\#B)^{(\#B)^n}$. Thus, when $\#B > 2$, functions of type $B^n \longrightarrow B$ must exist, which are not Boolean functions. For example, the function f: $B^2 \longrightarrow B$, where $B = \{0, \alpha, \beta, 1\}$ and where

$$f(0, 0) = 0,\ f(0, 1) = 1,\ f(1, 0) = f(1, 1) = \alpha,\ f(0, \alpha) = \beta$$

is not a Boolean function. (Why?)

PROBLEMS

1. The following is a Boolean expression generated by x, y over the Boolean algebra $\langle B; ^-, \vee, \wedge \rangle$, where $B = \{0, \alpha, \beta, 1\}$:

$$f(x, y) = (x \wedge (\alpha \vee y)) \vee (\bar{x} \wedge \bar{y})$$

Tabulate the values of $f(x, y)$ for all arguments $(x, y) \in B^2$.

2. Using the table constructed in Problem 1, write directly the minterm and maxterm normal forms of $f(x, y)$ specified in that problem.

3. The following is a Boolean expression generated by x, y, z over the Boolean algebra $\langle B; ^-, \vee, \wedge \rangle$, where $B = \{0, \alpha, \beta, 1\}$:

$$f(x, y, z) = \overline{((\beta \wedge \bar{x}) \vee (y \wedge (\alpha \vee \bar{z}))} \vee (\bar{y} \wedge \bar{z})$$

Using Theorem 6-24, find the minterm and maxterm normal forms of this expression.

4. Express $f(x, y, z)$ of Problem 3 in the minterm and maxterm normal forms, using Algorithms 6-1 and 6-2.

5. The minterm normal form of a Boolean expression generated by x, y over the Boolean algebra $\langle B; {}^-, \vee, \wedge \rangle$, where $B = \{0, \alpha, \beta, 1\}$, is given by

$$f(x, y) = (\alpha \wedge \bar{x} \wedge \bar{y}) \vee (\beta \wedge \bar{x} \wedge y) \vee (0 \wedge x \wedge \bar{y}) \vee (1 \wedge x \wedge y)$$

Write, by inspection, the maxterm normal form of $f(x, y)$.

6. Show that the function $f\colon B^2 \longrightarrow B$, where $B = \{0, \alpha, \beta, 1\}$ and

$$f(0, 0) = 0, f(0, 1) = 1$$
$$f(1, 0) = f(0, 1) = \alpha, f(0, \alpha) = \beta$$

is not a Boolean function. Enumerate three other functions $f\colon B^2 \longrightarrow B$ that are not Boolean.

7. Enumerate all 16 Boolean functions of one variable over $\langle B; {}^-, \vee, \wedge \rangle$, where $B = \{0, \alpha, \beta, 1\}$. How many functions of the type $B \longrightarrow B$ are not Boolean? Produce one.

8. Show that, for any Boolean expression $f(x, y)$,

$$f(x, y) = (x \wedge f(1, y)) \vee (\bar{x} \wedge f(0, y))$$
$$= (x \vee f(0, y)) \wedge (\bar{x} \vee f(1, y))$$

REFERENCES

ABBOTT, J. C., *Sets, Lattices and Boolean Algebras*. Boston: Allyn and Bacon, 1969.

BIRKHOFF, G., *Lattice Theory*. Providence, RI: American Mathematical Society, 1967.

GRATZER, G., *Lattice Theory*. San Francisco: Freeman, 1971.

HALMOS, P. R., *Lectures on Boolean Algebras*. Princeton, NJ: Van Nostrand, 1967.

MACLANE, S., and G. BIRKHOFF, *Algebra*. New York: Macmillan, 1967.

RUTHERFORD, D. E., *Introduction to Lattice Theory*. New York: Hafner, 1965.

7 COMBINATIONAL AND SEQUENTIAL NETWORKS

In this chapter we move for the first time into the area of applications. We start by defining *combinational networks* as realizations of functions f: $B_2^n \longrightarrow B_2^m$, with *gates* (combinational networks realizing functions f: $B_2^n \longrightarrow B_2$) and *standard combinational networks* (combinational networks consisting of NOT, OR, and AND gates only) being special cases. A *switching algebra* is developed for the analysis and synthesis of such networks, and it is demonstrated that this algebra is isomorphic to a Boolean algebra with a two-element domain. Methods are developed for synthesizing functions $B_2^n \longrightarrow B_2^m$ by means of standard and nonstandard combinational networks, and the concept of *functional completeness* is studied in this connection. Subsequently, *delay units* are added to combinational networks to form *sequential networks* that are endowed with the capacity to "remember" past events. This leads to the concept of a *finite-state machine*—a mathematical model useful in the representation of a wide variety of systems and processes. It is shown how every finite-state machine can be "simulated" by a sequential network, and conversely.

7-1. COMBINATIONAL NETWORKS

High-speed digital computers are built predominantly out of "binary" components—that is, components that can assume only two possible distinct conditions. Examples of binary components are an electric switch or a light bulb, which can be either on or off; a transistor whose current can be either present or absent; a magnetic core which can be magnetized either clockwise or counterclockwise; a spot on a card, which can be either punched or unpunched. The reason for employing binary components is practicability:

they are faster to operate, cheaper to manufacture, and more reliable than nonbinary components.

Thus, various functional units in a digital computer can be viewed as "boxes," called *combinational networks* (also *switching networks*, or *logic networks*), which accept a collection of "inputs" and generate a collection of "outputs." Each input and output is "binary"—that is, capable of assuming only two distinct values which, for convenience, are designated by 0 and 1. The outputs generated at any time depend only on the inputs applied at that time, according to some fixed dependence rule. More formally, a *combinational network realizing the function* $f: B_2^n \longrightarrow B_2^m$ (where $B_2 = \{0, 1\}$) is a box with n binary inputs, denoted by $x_1, x_2, \ldots, x_n$, and m binary outputs, denoted by $z_1, z_2, \ldots, z_m$, such that

$$(z_1, z_2, \ldots, z_m) = f(x_1, x_2, \ldots, x_n)$$

(see Fig. 7-1). Usually, f (and, hence, the combinational network realizing it) is specified by means of a *truth table*, which lists the values of $z_1, z_2, \ldots, z_m$ for each of the 2^n possible arguments $(x_1, x_2, \ldots, x_n)$ of f. Two combinational networks are said to be *equivalent* if they realize the same function.

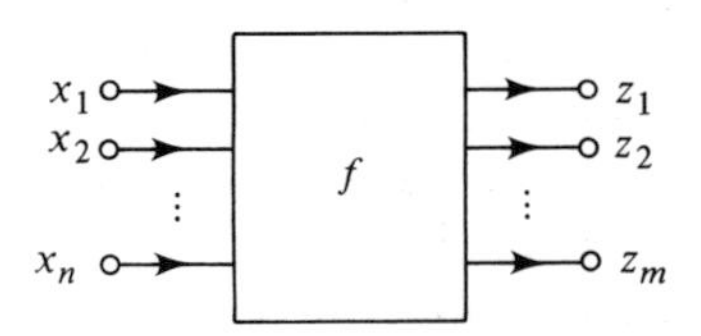

Fig. 7-1 Combinational network realizing $f: B_2^n \longrightarrow B_2^m$

Example 7-1

Figure 7-2(a) shows the schematic representation and truth table of a combinational network called an OR *gate*, whose output z is 0 when both inputs x and y are 0, and 1 otherwise.

Figure 7-2(b) shows the schematic representation and truth table of a combinational network called a *half adder*, whose purpose is to add two one-digit binary integers. The two integers are applied as inputs x and y, their sum is generated as output s, and the carry digit is generated as output c. According to the rules of binary addition, $0 + 0 = 0$ (carry 0), $0 + 1 = 1 + 0 = 1$ (carry 0), and $1 + 1 = 0$ (carry 1).

Figure 7-2(c) shows the schematic representation and truth table of a combinational network called a *full adder*, whose purpose is to add two binary integers of any magnitude. Starting with the least significant digits, corresponding digits of the two integers are applied as inputs x and y at successive instants of time, say, $t = 0, 1, 2, \ldots$; at $t = 0$ input c is 0, while at $t = 1, 2, \ldots,$ input c equals the carry digit generated by the addition

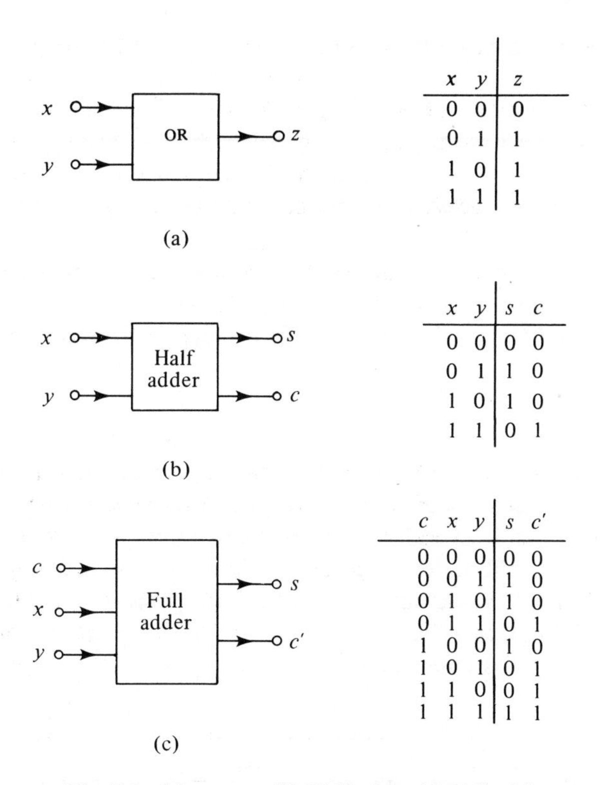

x	y	z
0	0	0
0	1	1
1	0	1
1	1	1

x	y	s	c
0	0	0	0
0	1	1	0
1	0	1	0
1	1	0	1

c	x	y	s	c′
0	0	0	0	0
0	0	1	1	0
0	1	0	1	0
0	1	1	0	1
1	0	0	1	0
1	0	1	0	1
1	1	0	0	1
1	1	1	1	1

Fig. 7-2 (a) OR gate (b) Half adder (c) Full adder

operation completed at $t - 1$. Output s is the sum (binary addition) of x, y, and c; output c' is the carry digit corresponding to this sum (thus, c' at time t equals c at time $t + 1$). For example, the following are possible sequences of x, y, c, s, and c' digits (ordered right-to-left) in a full adder:

$$\begin{array}{rl} c: & 1101100 \\ x: & 0110011 \\ y: & \underline{0110110} \\ s: & 1101001 \\ c': & 0110110 \end{array}$$

□

Combinational networks can serve as building blocks in the construction of more elaborate such networks. More specifically, given the combinational networks realizing the functions $f_1, f_2, \ldots, f_r$, a new combinational network that realizes a function f can be constructed by these rules:

(a) An output of a network f_i can serve as input to any number of networks f_j.

(b) An input of a network f_j can be obtained from the output of at most one network f_i.

(c) In applying rules (a) and (b), it should be ascertained that there is no sequence of networks $f_{i_1}, f_{i_2}, \ldots, f_{i_k}$ such that f_{i_ν} provides an input to $f_{i_{\nu+1}}$ ($\nu = 1, 2, \ldots, k-1$), and f_{i_k} provides an input to f_{i_1}. This so-called "no-feedback rule" is necessary to prevent indeterminacy in the operation of the f_{i_ν} networks.

(d) The set of network inputs that are not obtained from any network outputs are the over-all inputs of the new network f.

(e) Outputs of the new network f can be tapped from any point in this network.

Example 7-2

Figure 7-3 shows a combinational network constructed from an interconnection of two half adders and an OR gate. From the truth tables of Fig. 7-2(a) and (b) it can be verified that the truth table of the new network is precisely that of a full adder, as given in Fig. 7-2(c). For example, when $c = 0$, $x = 1$, and $y = 1$, we have $s^* = 1$ and $c^* = 0$ [Fig. 7-2(b)], $s = 0$ and $c^{**} = 1$ [Fig. 7-2(b)], and, finally, $c' = 1$ [Fig. 7-2(a)]. Thus, the networks of Figs. 7-3 and Fig. 7-2(c) are equivalent. □

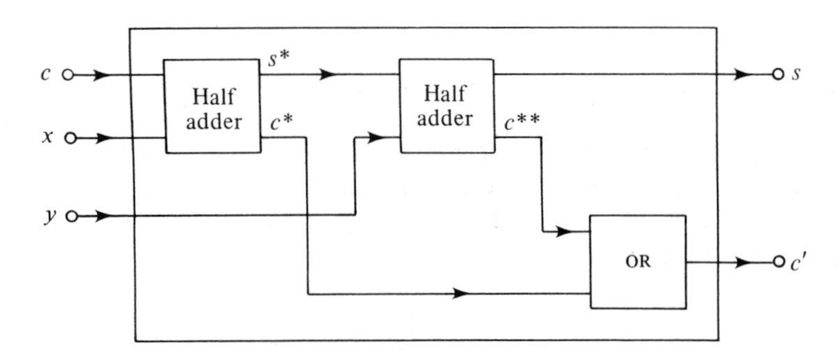

Fig. 7-3 Full adder from half adder and OR gate

PROBLEMS

1. A combinational network has 3 inputs and 4 outputs. When the inputs represent the binary number n ($0 \leq n \leq 7$), the outputs represent the number $2n$. Construct the truth table of this network.

2. Figure 7-A depicts a combinational network employed to compute $n_1 - n_2$, where $0 \leq n_1 \leq 3$ and $0 \leq n_2 \leq 3$. Inputs x_1 and x_2 represent n_1, and inputs

x_3 and x_4 represent n_2. Outputs z_1 and z_2 represent $|n_1 - n_2|$, and output z_3 represents the sign of $n_1 - n_2$ (say, $z_3 = 1$ if and only if the sign is negative). Construct the truth table of this network.

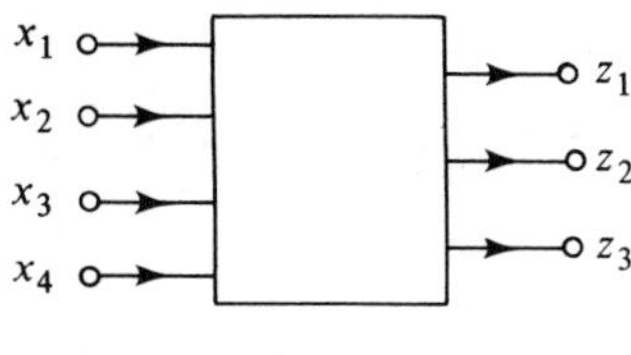

Fig. 7-A

3. What function is realized by the combinational network shown in Fig. 7-B?

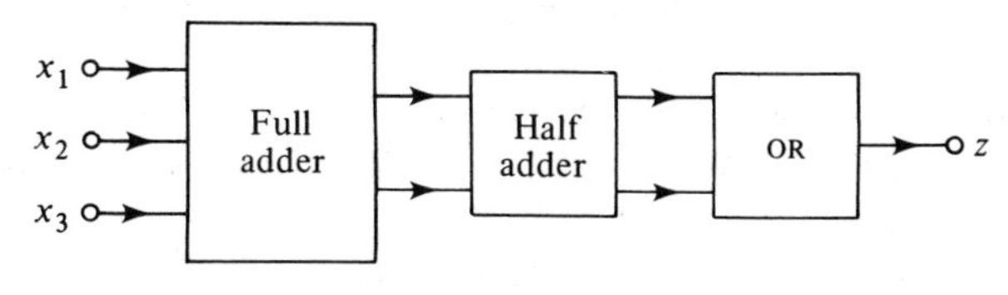

Fig. 7-B

4. Use two half adders to realize a function $B_2^2 \longrightarrow B_2$ whose truth table is given by

x	y	z
0	0	0
0	1	0
1	0	1
1	1	1

5. Show that functions realized by the networks shown in Fig. 7-1 are unchanged by any permutation of their variables. (Such functions are called *symmetric functions.*)

6. Verify that the combinational networks shown in Figs. 7-2(c) and 7-3 are equivalent.

7-2. STANDARD COMBINATIONAL NETWORKS

A *gate* is a combinational network with a single output. More precisely, a *gate realizing the function* f: $B_2^n \longrightarrow B_2$ (where $B_2 = \{0, 1\}$) is a box with n binary inputs, denoted by $x_1, x_2, \ldots, x_n$, and a binary output z

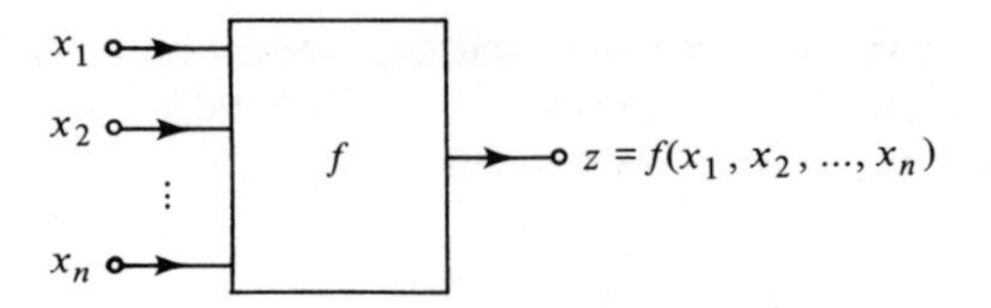

Fig. 7-4 Gate realizing f: $B_2^n \longrightarrow B_2$

such that

$$z = f(x_1, x_2, \ldots, x_n)$$

(See Fig. 7-4.)

Three of the most commonly used gates are those known as "NOT gate," "OR gate" (already introduced in Sec. 7-1), and "AND gate":

(a) A NOT gate has one input; its output is 1 if and only if the input is 0. If the input is x, the output is denoted by $\bar{x}$.
(b) An OR gate has two inputs; its output is 1 if and only if the first input is 1 *or* the second input is 1. If the inputs are x and y, the output is denoted by $x + y$.

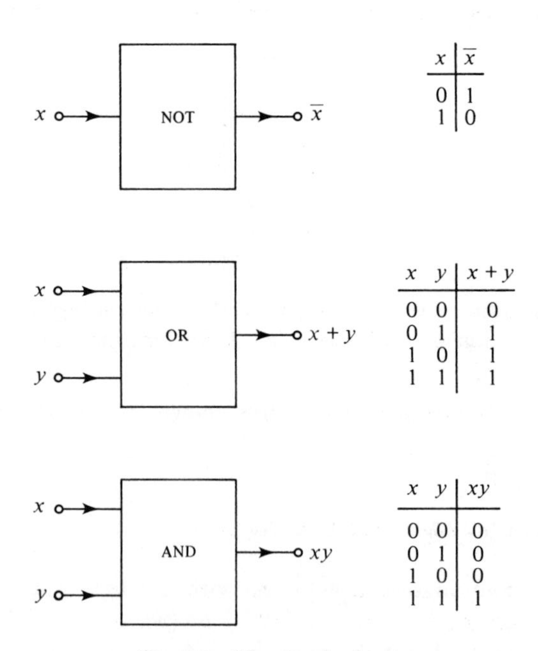

x	$\bar{x}$
0	1
1	0

x	y	$x + y$
0	0	0
0	1	1
1	0	1
1	1	1

x	y	xy
0	0	0
0	1	0
1	0	0
1	1	1

Fig. 7-5 The standard gates

(c) An AND gate has two inputs; its output is 1 if and only if the first input is 1 *and* the second input is 1. If the inputs are x and y, the output is denoted by $x \cdot y$, or simply xy.

These three gates are referred to as the *standard gates*. Their schematic representations and truth tables are shown in Fig. 7-5.

Let us consider now a combinational network with n inputs and m outputs (see Fig. 7-1), constructed from standard gates only (according to the rules spelled out in Sec. 7-1). A combinational network of this type is called a *standard combinational network*. In such a network, any output z_i can be expressed by a formula that involves the inputs $x_1, x_2, \ldots, x_n$ and the symbols $-$, $+$, and $\cdot$ (and parentheses, where needed to avoid ambiguity). The formula can be derived directly from the diagram of the network, by tracing the paths that lead from inputs $x_1, x_2, \ldots, x_n$ to output z_i. In the tracing process, a variable x is replaced by $\bar{x}$ whenever x traverses a NOT gate; x and y are replaced by $x + y$ when they traverse an OR gate and by xy when they traverse an AND gate. The resulting expression is called a *transmission function for z_i, generated by $x_1, x_2, \ldots, x_n$*. This transmission function, together with the truth tables of Fig. 7-5, can be used to evaluate z_i for any prescribed values of $x_1, x_2, \ldots, x_n$.

Example 7-3

Figure 7-6 shows a standard combinational network with three inputs, denoted x_1, x_2, x_3, and two outputs, denoted z_1, z_2. As indicated in the figure, the transmission functions for z_1 and z_2 can be derived to be $\overline{(x_1 + \bar{x}_2)x_3}$ and $x_3 + \overline{(x_1 + \bar{x}_2)}$, respectively. When $x_1 = 1$, $x_2 = 0$, and $x_3 = 0$, then z_1

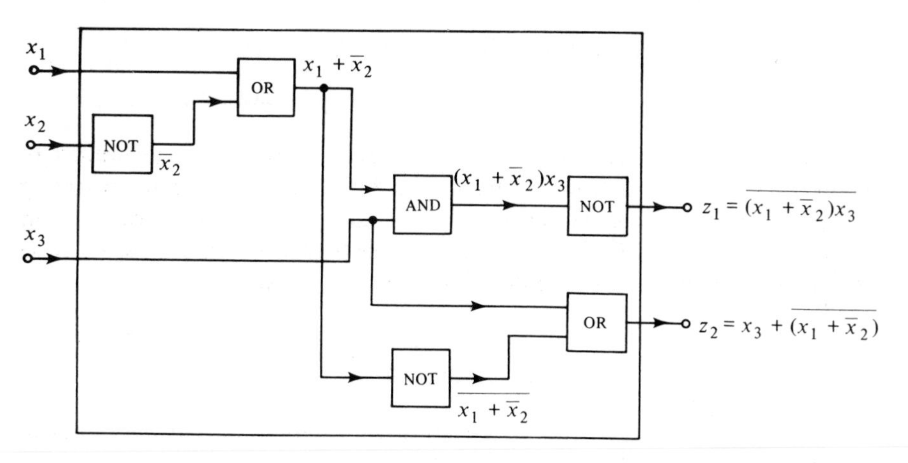

Fig. 7-6 A standard combinational network

Table 7-1 TRUTH TABLE OF COMBINATIONAL NETWORK OF FIG. 7-6

x_1	x_2	x_3	$\bar{x}_2$	$x_1 + \bar{x}_2$	$(x_1 + \bar{x}_2)x_3$	$\overline{x_1 + \bar{x}_2}$	$z_1 = \overline{(x_1 + \bar{x}_2)x_3}$	$z_2 = x_3 + \overline{(x_1 + \bar{x}_2)}$
0	0	0	1	1	0	0	1	0
0	0	1	1	1	1	0	0	1
0	1	0	0	0	0	1	1	1
0	1	1	0	0	0	1	1	1
1	0	0	1	1	0	0	1	0
1	0	1	1	1	1	0	0	1
1	1	0	0	1	0	0	1	0
1	1	1	0	1	1	0	0	1

and z_2 can be evaluated from these transmission functions and the truth tables of Fig. 7-5 as:

$$z_1 = \overline{(x_1 + \bar{x}_2)x_3} = \overline{(1 + \bar{0})0} = \overline{(1 + 1)0} = \overline{10} = \bar{0} = 1$$

$$z_2 = x_3 + \overline{(x_1 + \bar{x}_2)} = 0 + \overline{(1 + \bar{0})} = 0 + \overline{(1 + 1)} = 0 + \bar{1} = 0 + 0 = 0$$

Table 7-1 exhibits the evaluation of z_1 and z_2 for all eight values of (x_1, x_2, x_3). The first three and last two columns of Table 7-1 constitute the truth table for the given combinational network. □

The foregoing can be summarized: Given a standard combinational network with n inputs and m outputs, the function $f: B_2^n \longrightarrow B_2^m$ realized by this network can be expressed by means of m transmission functions. Specifically, we can write

$$f(x_1, x_2, \ldots, x_n) = (f_1(x_1, x_2, \ldots, x_n), \\ f_2(x_1, x_2, \ldots, x_n), \ldots, f_m(x_1, x_2, \ldots, x_m))$$

where $f_i: B_2^n \longrightarrow B_2$ is the transmission function for output z_i generated by inputs $x_1, x_2, \ldots, x_n$. Each transmission function is a formula involving $x_1, x_2, \ldots, x_n$, $-$, $+$, $\cdot$, and parentheses, which can be derived directly from the network diagram. The truth table for f can be derived directly from these transmission functions and the definitions of $-$, $+$, and $\cdot$.

PROBLEMS

1. Find the transmission function and construct the truth table of the combinational network shown in Fig. 7-C.

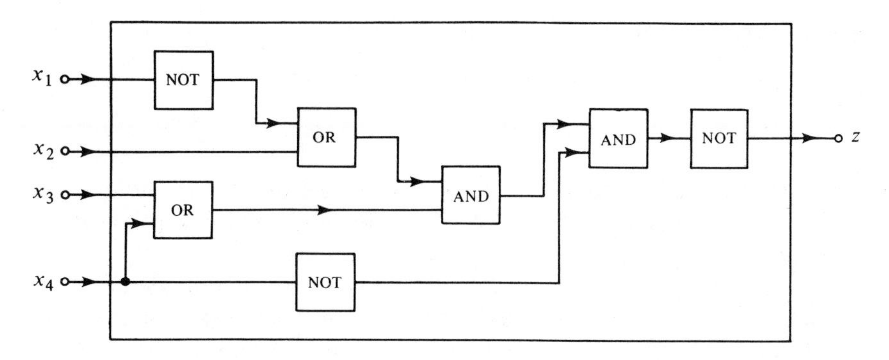

Fig. 7-C

2. Find the transmission function of the combinational network shown in Fig. 7-D. Verify that this network can serve as a half adder.

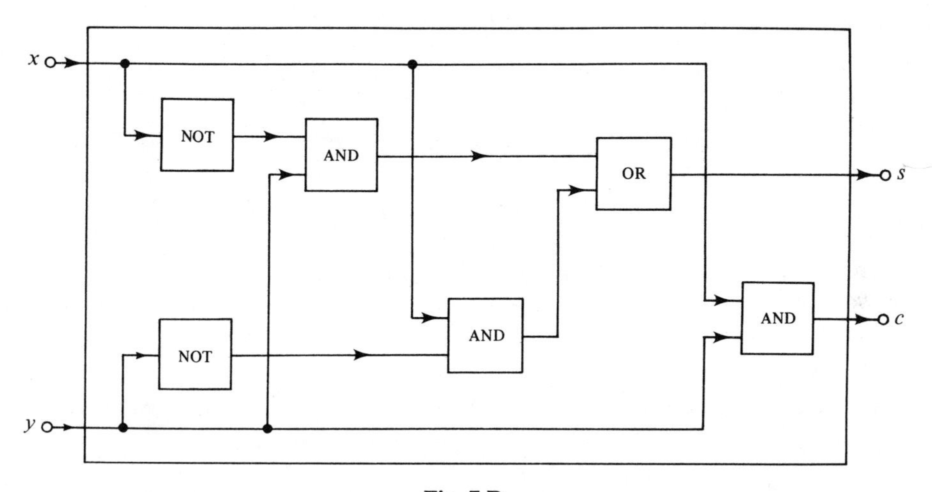

Fig. 7-D

3. How many distinct transmission functions are generated by $x_1, x_2, \ldots, x_n$?
4. How many distinct symmetric transmission functions are generated by $x_1, x_2, \ldots, x_n$? (See Problem 5 of Sec. 7-1.)

7-3. SWITCHING ALGEBRA

Two transmission functions generated by $x_1, x_2, \ldots, x_n$ are said to be *equal* if they represent the same function $f: B_2^n \longrightarrow B_2$. Thus, transmission functions are equal if and only if their truth tables are identical. For example,

Table 7-2 verifies that the transmission functions

$$f_1(x, y, z) = xy + \bar{x}z$$

and

$$f_2(x, y, z) = xy + \bar{x}z + yz$$

are equal.[1] Tables 7-3 and 7-4 list a number of identities which hold among transmission functions generated by x, y, z. Each identity can be verified in a straightforward manner by means of a truth table.

It is helpful to note that the commutative, associative, and identity laws, as well as distributive law 3, are the same as those prevailing among integers or real numbers and, hence, quite easy to memorize and apply. Also, the

Table 7-2 TRUTH TABLE FOR $xy + \bar{x}z$ AND $xy + \bar{x}z + yz$

x	y	z	$\bar{x}$	xy	$\bar{x}z$	yz	$xy + \bar{x}z$	$xy + \bar{x}z + yz$
0	0	0	1	0	0	0	0	0
0	0	1	1	0	1	0	1	1
0	1	0	1	0	0	0	0	0
0	1	1	1	0	1	1	1	1
1	0	0	0	0	0	0	0	0
1	0	1	0	0	0	0	0	0
1	1	0	0	1	0	0	1	1
1	1	1	0	1	0	1	1	1

Table 7-3 SWITCHING ALGEBRA LAWS

Commutative Laws	1. $x + y = y + x$	1′. $xy = yx$
Associative Laws	2. $x + (y + z) = (x + y) + z$	2′. $x(yz) = (xy)z$
Distributive Laws	3. $x(y + z) = xy + xz$	3′. $x + yz = (x + y)(x + z)$
Identity Laws	4. $x + 0 = 0 + x = x$	4′. $x1 = 1x = x$
Complement Laws	5. $x + \bar{x} = 1$	5′. $x\bar{x} = 0$

Table 7-4 ADDITIONAL SWITCHING ALGEBRA LAWS

Involution Law	1. (and 1′.) $\bar{\bar{x}} = x$	
Idempotent Laws	2. $x + x = x$	2′. $xx = x$
Null Laws	3. $x + 1 = 1$	3′. $x0 = 0$
Absorption Laws	4. $x + xy = x$	4′. $x(x + y) = x$
De Morgan's Laws	5. $\overline{x + y} = \bar{x}\bar{y}$	5′. $\overline{xy} = \bar{x} + \bar{y}$

[1]To avoid proliferation of parentheses, we assume that, in every transmission function, $\cdot$ has precedence over $+$.

associative laws imply that we can unambiguously interpret the parentheses-free expressions

$$x_1 + x_2 + \ldots + x_n = \sum_{i=1}^{n} x_i$$
$$x_1 x_2 \ldots x_n = \prod_{i=1}^{n} x_i$$

Correspondingly, combinational networks that consist of $n - 1$ OR gates or $n - 1$ AND gates, connected as shown in Fig. 7-7, can be represented, for convenience, by a single (generalized) OR gate or AND gate, respectively, which is capable of accepting any number $n \geq 2$ of inputs.

Comparing Tables 7-3 and 7-4 with Tables 6-1 and 6-2, it is apparent that transmission functions generated by $x_1, x_2, \ldots, x_n$ are simply Boolean expressions generated by $x_1, x_2, \ldots, x_n$ over the Boolean algebra $\langle B; {}^-, \vee, \wedge \rangle$, where $B = B_2 = \{0, 1\}$, and where the operational symbols $\vee$ and $\wedge$ are replaced by $+$ and $\cdot$, respectively. This algebra, denoted by $\langle B_2; {}^-, +, \cdot \rangle$, is referred to as the *switching algebra*. As pointed out in Sec. 6-10, the switching algebra is isomorphic to the algebra $\langle \{\varnothing, U\}; ', \cup, \cap \rangle$, hence, results obtained in the algebra of sets can be directly applied (after appropriate notational changes) to switching algebra.

The laws listed in Tables 7-3 and 7-4 are often useful in the conversion of transmission functions from one form to another. A common objective is to express a given transmission function with the least possible number of $+$ and $\cdot$ operations—an objective often achievable by judicious use of the switching algebra laws.

Example 7-4

The transmission function

$$f(x, y, z) = xyz + xy\bar{z} + \bar{x}y\bar{z}$$

can be expressed with a minimum number of $+$ and $\cdot$ operations by this series of conversions:

$$\begin{aligned}
& xyz + xy\bar{z} + \bar{x}y\bar{z} & \\
&= xyz + xy\bar{z} + xy\bar{z} + \bar{x}y\bar{z} & \text{(Idempotent)} \\
&= xyz + xy\bar{z} + y\bar{z}x + y\bar{z}\bar{x} & \text{(Commutative)} \\
&= xy(z + \bar{z}) + y\bar{z}(x + \bar{x}) & \text{(Distributive)} \\
&= xy1 + y\bar{z}1 & \text{(Complement)} \\
&= xy + y\bar{z} & \text{(Identity)} \\
&= yx + y\bar{z} & \text{(Commutative)} \\
&= y(x + \bar{z}) & \text{(Distributive)}
\end{aligned}$$

□

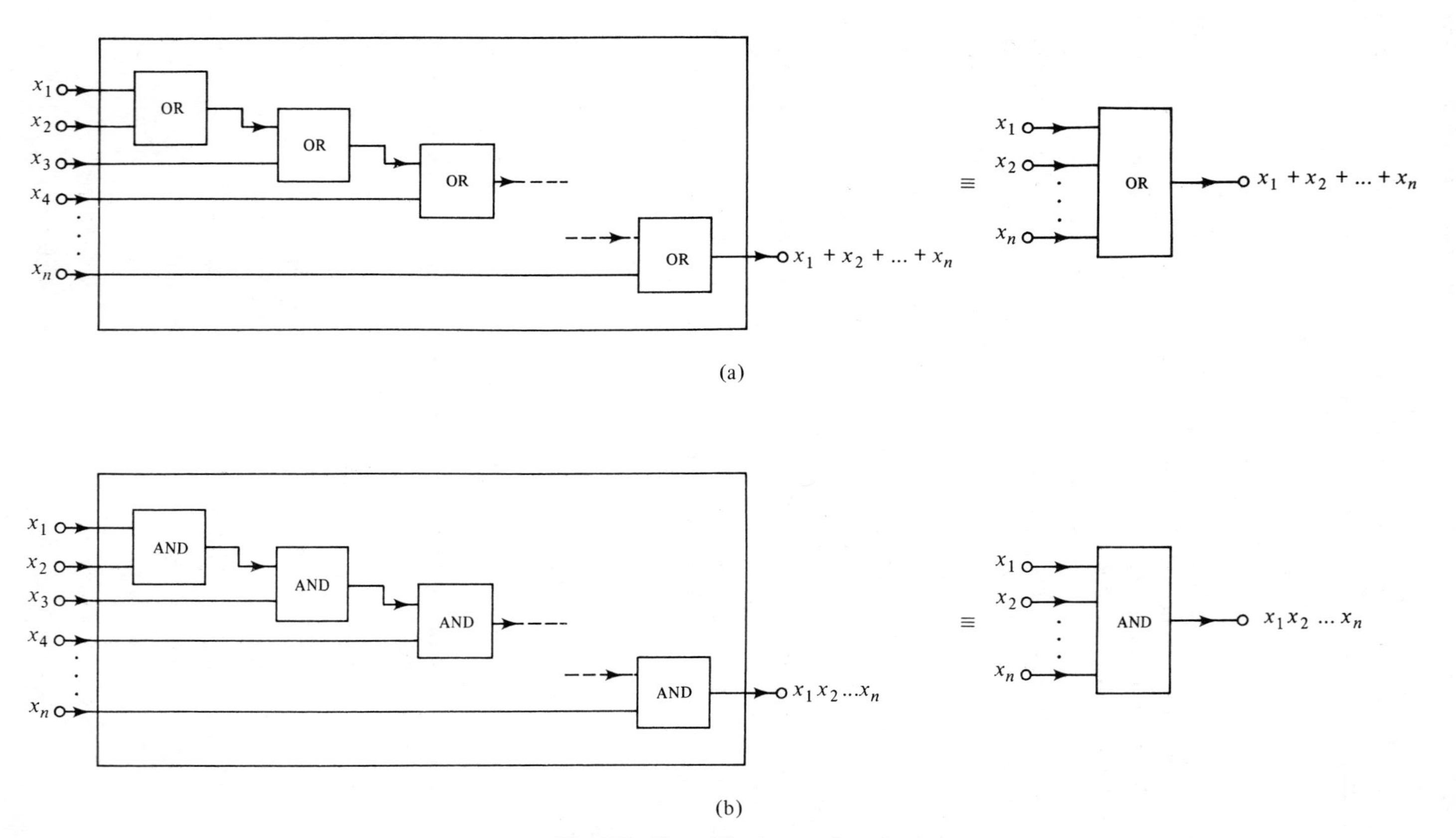

Fig. 7-7 Generalized OR and AND gates

The process exemplified here is referred to as the "minimization" of transmission functions. As it is discussed in detail in every textbook on logical design, it will not be included in this book.

PROBLEMS

1. Verify the identities of Tables 7-3 and 7-4.
2. Using truth tables, verify these identities:
   (a) $xy + xz + yz = \bar{x}yz + x\bar{y}z + xy$
   (b) $\bar{x}\bar{z} + \bar{x}y + x\bar{y} = (\bar{x} + \bar{y})(x + y + \bar{z})$
   (c) $\bar{a}\bar{b}\bar{c} + \bar{a}bd + abc + a\bar{b}\bar{d} = \bar{b}\bar{c}\bar{d} + \bar{a}\bar{c}d + bcd + ac\bar{d}$
3. Using the switching algebra laws of Tables 7-3 and 7-4, show that:
   (a) $\overline{(x + \bar{y}) + z(x + \bar{y})} = \bar{x}y$
   (b) $\overline{(x + \bar{z})(y + \bar{x}z)} = \bar{x}z$
   (c) $xy + \bar{x}y + x\bar{y} = x + y$
4. Using the switching algebra laws of Tables 7-3 and 7-4, express these transmission functions with as few $+$ and $\cdot$ operations as you can:
   (a) $f(x, y, z) = xy\bar{z} + y\bar{z} + xz$
   (b) $f(x, y) = (x + y)(x + \bar{y}) + y$
   (c) $f(x, y, z) = xz + xy + \bar{x}\bar{y}$

7-4. NORMAL FORMS OF TRANSMISSION FUNCTIONS

One important result carried over from the algebra of sets to the switching algebra is that concerning minset and maxset normal forms. This result enables us to express any transmission function generated by $x_1, x_2, \ldots, x_n$ as a sum of products of the form $\hat{x}_1\hat{x}_2 \ldots \hat{x}_n$, or as a product of sums of the form $\hat{x}_1 + \hat{x}_2 + \ldots + \hat{x}_n$, where $\hat{x}_i$ is either x_i or $\bar{x}_i$. These forms are called, respectively, the *minterm normal form* (where each product is called a *minterm generated by* $x_1, x_2, \ldots, x_n$) and the *maxterm normal form* (where each sum is called a *maxterm generated by* $x_1, x_2, \ldots, x_n$). The next algorithm for constructing these forms follows immediately from Algorithms 1-1 and 1-2. In this algorithm, row $\delta_1\delta_2 \ldots \delta_n$ (where δ_i is either 0 or 1) denotes a truth-table row that contains the digits $\delta_1, \delta_2, \ldots, \delta_n$ in the columns labeled $x_1, x_2, \ldots, x_n$, respectively.

ALGORITHM 7-1

To find the minterm and maxterm normal forms of a given transmission function $f: B_2^n \longrightarrow B_2$ generated by $x_1, x_2, \ldots, x_n$:

(i) Construct the truth table of f. If the column labeled $f(x_1, x_2, \ldots, x_n)$ contains no 1s, then the minterm normal form is $f(x_1, x_2, \ldots, x_n) = 0$; if this column contains no 0s, then the maxterm normal form is $f(x_1, x_2, \ldots, x_n) = 1$. Otherwise:

(ii) If column $f(x_1, x_2, \ldots, x_n)$ contains 1s in rows $\delta_{11}\delta_{12} \ldots \delta_{1n}$, $\delta_{21}\delta_{22} \ldots \delta_{2n}, \ldots, \delta_{l1}\delta_{l2} \ldots \delta_{ln}$, then the minterm normal form of f is given by

$$f(x_1, x_2, \ldots, x_n) = \textstyle\sum_{i=1}^{l} (\prod_{j=1}^{n} x_{ij})$$

where

$$x_{ij} = \begin{cases} \bar{x}_i & \text{if} \quad \delta_{ij} = 0 \\ x_i & \text{if} \quad \delta_{ij} = 1 \end{cases}$$

If column $f(x_1, x_2, \ldots, x_n)$ contains 0s in these rows, then the maxterm normal form of f is given by

$$f(x_1, x_2, \ldots, x_n) = \textstyle\prod_{i=1}^{l} (\sum_{j=1}^{n} x_{ij})$$

where

$$x_{ij} = \begin{cases} x_i & \text{if} \quad \delta_{ij} = 0 \\ \bar{x}_i & \text{if} \quad \delta_{ij} = 1 \end{cases}$$ □

Alternatively, we can use the switching-algebra version of Algorithm 1-3 (or 1-4):

ALGORITHM 7-2 (or 7-3)

To find the minterm (or maxterm) normal form of a given transmission function $f: B_2^n \longrightarrow B_2$ generated by $x_1, x_2, \ldots, x_n$:

(i) Use the laws in Tables 7-3 and 7-4 to express f as 0 (or 1), or as a sum (or product) of distinct products (or sums) of the form $\hat{x}_{i_1}\hat{x}_{i_2} \ldots \hat{x}_{i_k}$ (or $\hat{x}_{i_1} + \hat{x}_{i_2} + \ldots + \hat{x}_{i_k}$), where $i_1 < i_2 < \ldots < i_k$. If $f(x_1, x_2, \ldots, x_n)$ is 0 (or 1), it is in the desired form. Otherwise:

(ii) If every product (or sum) is a minterm (or maxterm), then f is in the desired form. Otherwise:

(iii) From the expression last obtained, select a product (or sum) of the form $\hat{x}_{i_1}\hat{x}_{i_2} \ldots \hat{x}_{i_k}$ (or $\hat{x}_{i_1} + \hat{x}_{i_2} + \ldots + \hat{x}_{i_k}$), where $k < n$ and where, for some h, $\hat{x}_{i_h}$ does not appear, and replace it with

$$\hat{x}_{i_1}\hat{x}_{i_2} \ldots \hat{x}_{i_k}\hat{x}_{i_h} + \hat{x}_{i_1}\hat{x}_{i_2} \ldots \hat{x}_{i_k}\bar{x}_{i_h}$$

$$[\text{or } (\hat{x}_{i_1} + \hat{x}_{i_2} + \ldots + \hat{x}_{i_k} + x_{i_h})(\hat{x}_{i_1} + \hat{x}_{i_2} + \ldots \hat{x}_{i_k} + \bar{x}_{i_h})]$$

which is permitted by virtue of the commutative, distributive, and complement laws in Table 7-3. Rearrange the x_{i_j} in the new products (or sums) in order of ascending subscript and delete all duplicate products (or sums). Return to step (ii). □

Example 7-5

Consider the transmission function

$$f(x_1, x_2, x_3) = \overline{(x_1 + \bar{x}_2)x_3}$$

whose truth table is given in Table 7-1. Using Algorithm 7-1, the minterm and maxterm normal forms of f can be derived directly from the truth table:

$$\begin{aligned} f(x_1, x_2, x_3) &= \bar{x}_1\bar{x}_2\bar{x}_3 + \bar{x}_1x_2\bar{x}_3 + \bar{x}_1x_2x_3 + x_1\bar{x}_2\bar{x}_3 + x_1x_2\bar{x}_3 \\ &= (x_1 + x_2 + \bar{x}_3)(\bar{x}_1 + x_2 + \bar{x}_3)(\bar{x}_1 + \bar{x}_2 + \bar{x}_3) \end{aligned}$$

When the truth table is not given, normal forms can be derived with the aid of Algorithms 7-2 and 7-3:

$$\begin{aligned} f(x_1, x_2, x_3) &= \overline{(x_1 + \bar{x}_2)x_3} \\ &= \overline{(x_1 + \bar{x}_2)} + \bar{x}_3 = \bar{x}_1x_2 + \bar{x}_3 \\ &= \bar{x}_1x_2x_3 + \bar{x}_1x_2\bar{x}_3 + x_1x_2\bar{x}_3 + x_1\bar{x}_2\bar{x}_3 + \bar{x}_1x_2\bar{x}_3 + \bar{x}_1\bar{x}_2\bar{x}_3 \\ &= \bar{x}_1\bar{x}_2\bar{x}_3 + \bar{x}_1x_2\bar{x}_3 + \bar{x}_1x_2x_3 + x_1\bar{x}_2\bar{x}_3 + x_1x_2\bar{x}_3 \end{aligned}$$

$$\begin{aligned} f(x_1, x_2, x_3) &= \overline{(x_1 + \bar{x}_2)x_3} = \overline{x_1x_3 + \bar{x}_2x_3} \\ &= \overline{(x_1x_3)}\,\overline{(\bar{x}_2x_3)} = (\bar{x}_1 + \bar{x}_3)(x_2 + \bar{x}_3) \\ &= (\bar{x}_1 + x_2 + \bar{x}_3)(\bar{x}_1 + \bar{x}_2 + \bar{x}_3)(x_1 + x_2 \\ &\quad + \bar{x}_3)(\bar{x}_1 + \bar{x}_2 + \bar{x}_3) \\ &= (x_1 + x_2 + \bar{x}_3)(\bar{x}_1 + x_2 + \bar{x}_3)(\bar{x}_1 + \bar{x}_2 + \bar{x}_3) \end{aligned}$$

Clearly, once either of the normal forms of f is available, the truth table for f can be constructed by inspection. □

Using Algorithm 7-1, we can always start from a truth table of an arbitrary function f: $B_2^n \longrightarrow B_2$ and construct a transmission function (in the minterm or maxterm normal form) that equals f. Thus, *every function* f: $B_2^n \longrightarrow B_2$ *can be expressed as a transmission function.* More generally, we have:

THEOREM 7-1

Given a function f: $B_2^n \longrightarrow B_2^m$, where

$$f(x_1, x_2, \ldots, x_n) = (z_1, z_2 \ldots, z_m)$$

there always exist m transmission functions $f_1(x_1, x_2, \ldots, x_n), f_2(x_1, x_2, \ldots, x_n), \ldots, f_m(x_1, x_2, \ldots, x_n)$ such that

$$z_i = f_i(x_1, x_2, \ldots, x_n) \qquad (i = 1, 2, \ldots, m)$$

Example 7-6

Consider the full adder of Fig. 7-2(c). Here the given function is $f: B_2^3 \longrightarrow B_2^2$, where

$$f(c, x, y) = (s, c')$$

From the truth table of f, we have

$$s = \bar{c}\bar{x}y + \bar{c}x\bar{y} + c\bar{x}\bar{y} + cxy$$
$$c' = \bar{c}xy + c\bar{x}y + cx\bar{y} + cxy$$

□

PROBLEMS

1. Using Algorithm 7-1, find the minterm and maxterm normal forms of these transmission functions, generated by x, y:
   (a) 1 (b) 0 (c) x (d) $\bar{x}$ (e) $x + y$ (f) xy
2. Using Algorithm 7-1, find the minterm and maxterm normal forms of:
   (a) $f(x, y, z) = x + \bar{y} + y(x + \bar{z})$
   (b) $f(a, b, c, d) = (a + \bar{d})(\overline{\bar{b} + c}) + abd$
3. Using Algorithms 7-2 and 7-3, find the minterm and maxterm normal forms of these transmission functions, generated by x, y, z:
   (a) 1 (b) x (c) $y\bar{z}$ (d) $x + \bar{y}z$
   (e) $x\bar{y} + y(x + \bar{z})$
4. Find the minterm and maxterm normal forms of the function realized by the combinational network shown in Fig. 7-C.
5. (a) Prove the following *expansion theorem:* If f is a transmission function generated by $x_1, x_2, \ldots, x_n$, then

$$f(x_1, x_2, \ldots, x_n) = x_1 f(1, x_2, \ldots, x_n) + \bar{x}_1 f(0, x_2, \ldots, x_n)$$
$$= [\bar{x}_1 + f(1, x_2, \ldots, x_n)][x_1 + f(0, x_2, \ldots, x_n)]$$

   (b) Use the preceding theorem to show that

$$f(x_1, x_2, x_3) = f(0, 0, 0)\bar{x}_1\bar{x}_2\bar{x}_3 + f(0, 0, 1)\bar{x}_1\bar{x}_2x_3 + \ldots + f(1, 1, 1)x_1x_2x_3$$

   (This is a special case of Theorem 6-24)

6. It is common to denote the minterm corresponding to row $\delta_1\delta_2 \ldots \delta_n$ of the truth table of $f: B_2^n \longrightarrow B_2$ by $m_{|\delta_1\delta_2\ldots\delta_n|}$, where $|\delta_1\delta_2 \ldots \delta_n|$ is the decimal value of the binary integer $\delta_1\delta_2 \ldots \delta_n$. Thus, the minterm normal form of f can be written as

$$f(x_1, x_2, \ldots, x_n) = \textstyle\sum_{i \in I_f} m_i \qquad (I_f \subset \{0, 1, \ldots, 2^n - 1\})$$

Similarly, the minterm normal form of $g\colon B_2^n \longrightarrow B_2$ can be written as

$$g(x_1, x_2, \ldots, x_n) = \sum_{i \in I_g} m_i \qquad (I_g \subset \{0, 1, \ldots, 2^n - 1\})$$

Show that:
(a) $f(x_1, x_2, \ldots, x_n) + g(x_1, x_2, \ldots, x_n) = \sum_{i \in I_f \cup I_g} m_i$
(b) $f(x_1, x_2, \ldots, x_n)g(x_1, x_2, \ldots, x_n) = \sum_{i \in I_f \cap I_g} m_i$
(c) $\bar{f}(x_1, x_2, \ldots, x_n) = \sum_{i \in I'_f} m_i$
[where $\bar{f}(x_1, x_2, \ldots, x_n) = 1$ if and only if $f(x_1, x_2, \ldots, x_n) = 0$, and $I'_f = \{0, 1, \ldots, 2^n - 1\} - I_f$].

7-5. SYNTHESIS OF COMBINATIONAL NETWORKS

Given a function $f\colon B_2^n \longrightarrow B_2^m$ and a set of gates $G = \{g_1, g_2, \ldots, g_h\}$, we say that *f is realizable with G* if there exists a combinational network realizing f and consisting of elements of G only. Such a network is referred to as a *realization of f with G*, and the process of constructing it as a *synthesis of f with G*.

When f is a transmission function $f\colon B_2^n \longrightarrow B_2$, and $G = \{$NOT, OR, AND$\}$ (that is, the set of standard gates), f is always realizable with G. In this case, the synthesis of f with G is straightforward: In the expression representing f, each subexpression of the form $\bar{x}$ corresponds to a NOT gate with input x, each subexpression of the form $x + y$ corresponds to an OR gate with inputs x and y, and each subexpression of the form xy corresponds to an AND gate with inputs x and y. As an example, Fig. 7-8 shows the realization of the transmission function

$$f(x_1, x_2, x_3) = x_2(x_1 + \bar{x}_3)$$

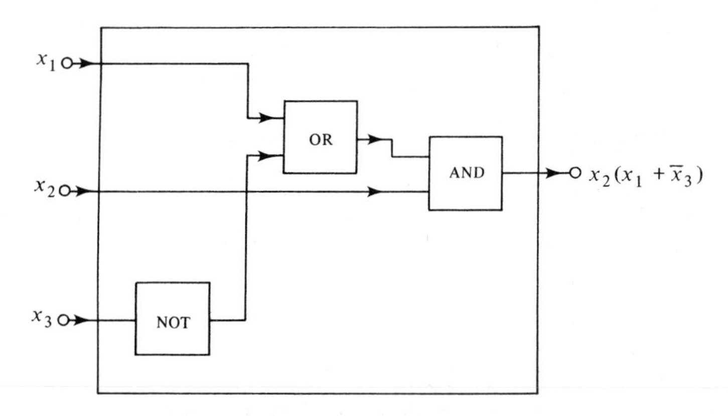

Fig. 7-8 Realization of $x_2\,(x_1 + \bar{x}_3)$

When $f\colon B_2^n \longrightarrow B_2$ is a single minterm $\hat{x}_1\hat{x}_2 \ldots \hat{x}_n$ or a single maxterm $\hat{x}_1 + \hat{x}_2 + \ldots + \hat{x}_n$ generated by $x_1, x_2, \ldots, x_n$, the realization of f consists of a single (generalized) AND or OR gate, respectively, and n or fewer NOT gates (see Fig. 7-9). When $f\colon B_2^n \longrightarrow B_2$ is a transmission function in the minterm normal form $\sum_{i=1}^{l} m_i$ (where the m_i are minterms) or the maxterm normal form $\prod_{i=1}^{l} \tilde{m}_i$ (where the $\tilde{m}_i$ are maxterms), the realization of f consists of l minterm realizations with their outputs "added" by a (generalized) OR gate, or l maxterm realizations with their outputs "multiplied" by a (generalized) AND gate. The resulting realizations, shown in Fig. 7-10, are called a *minterm normal combinational network* and a *maxterm normal combinational network*,

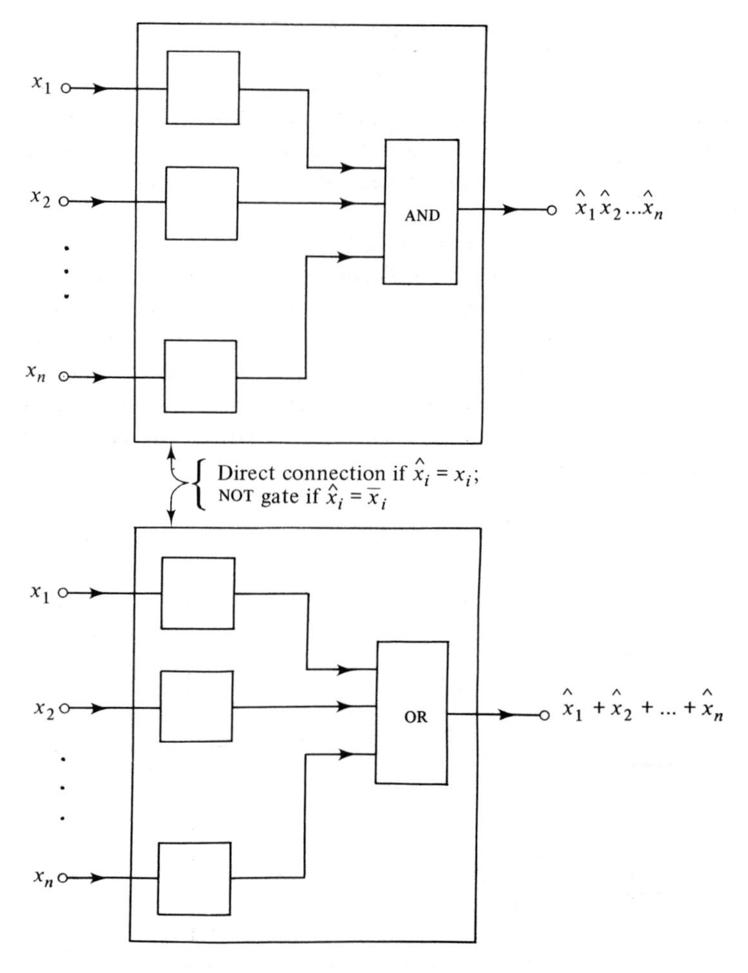

Fig. 7-9 Realization of a minterm and a maxterm

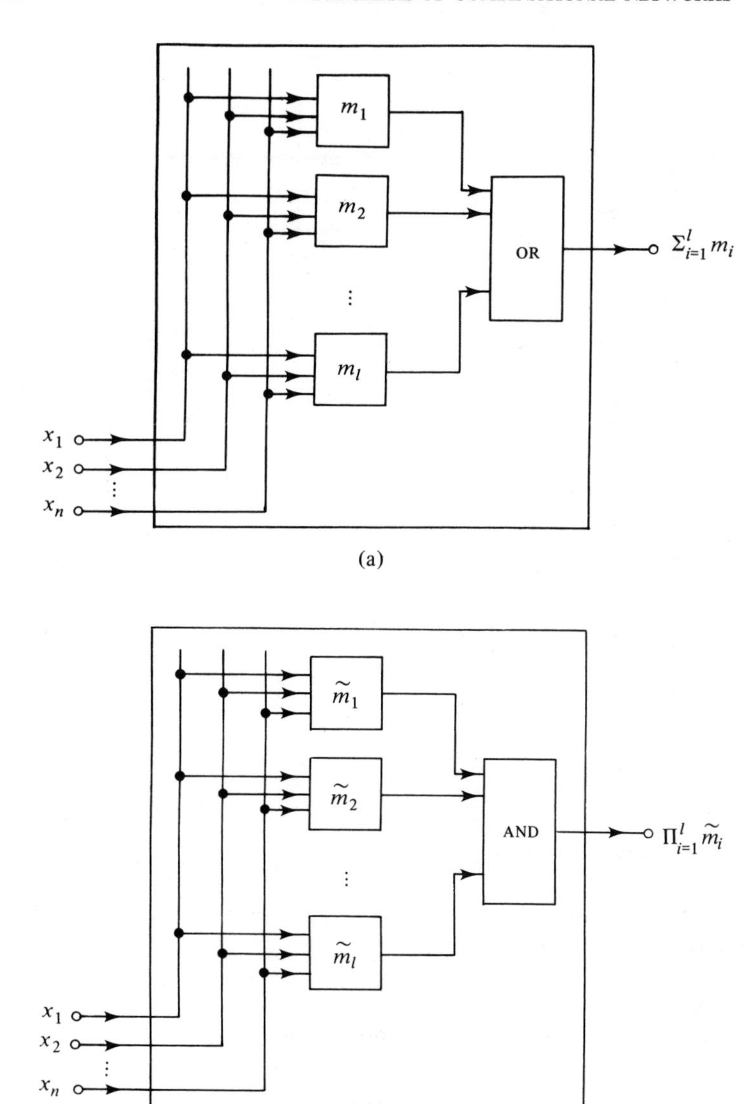

Fig. 7-10 (a) Minterm normal combinational network (b) Maxterm normal combinational network

respectively. For example, Fig. 7-11 shows the minterm normal combinational network realizing the transmission function

$$f(x_1, x_2, x_3) = x_1 x_2 x_3 + x_1 x_2 \bar{x}_3 + \bar{x}_1 x_2 \bar{x}_3$$

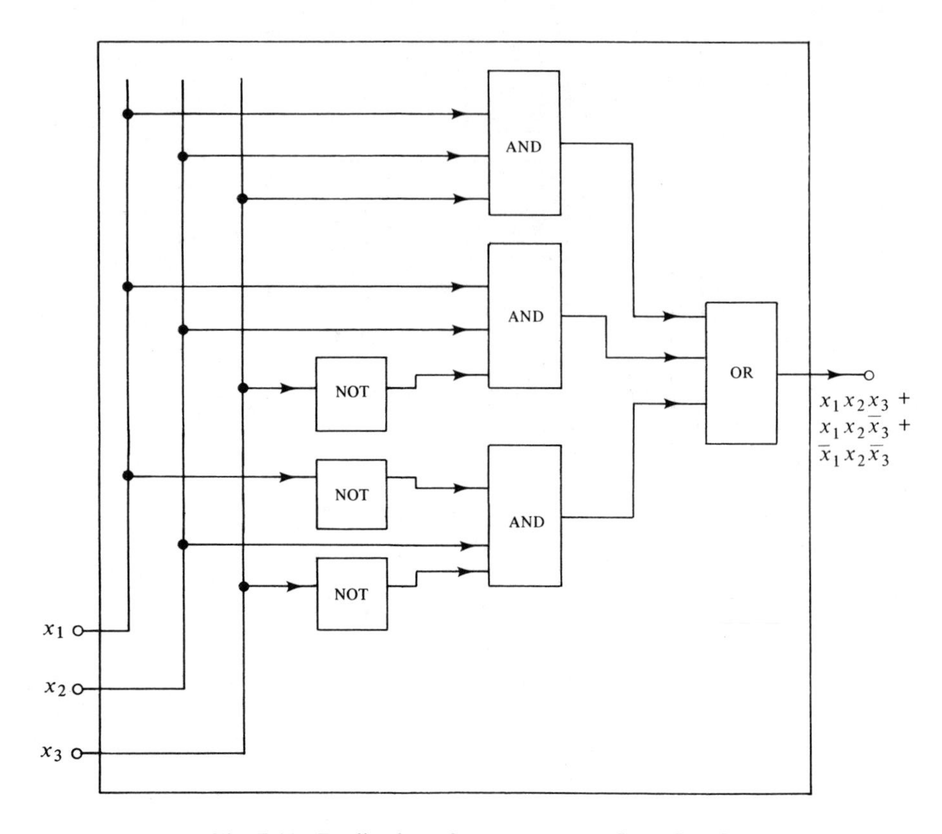

Fig. 7-11 Realization of $x_1x_2x_3 + x_1x_2\bar{x}_3 + \bar{x}_1x_2\bar{x}_3$

We are now in a position to formulate a synthesis procedure for an arbitrary function $f: B_2^n \longrightarrow B_2^m$:

ALGORITHM 7-4

To synthesize a function $f: B_2^n \longrightarrow B_2^m$, where

$$f(x_1, x_2, \ldots, x_n) = (z_1, z_2, \ldots, z_m)$$

(specified by a truth table):

(i) Using Algorithm 7-1, 7-2, or 7-3, construct m transmission functions $f_i(x_1, x_2, \ldots, x_n)$ $(i = 1, 2, \ldots, m)$ such that

$$z_i = f_i(x_1, x_2, \ldots, x_n)$$

(ii) Realize each transmission function f_i with {NOT, OR, AND}. The union of these realizations is the desired realization of f. (See Fig. 7-12.) □

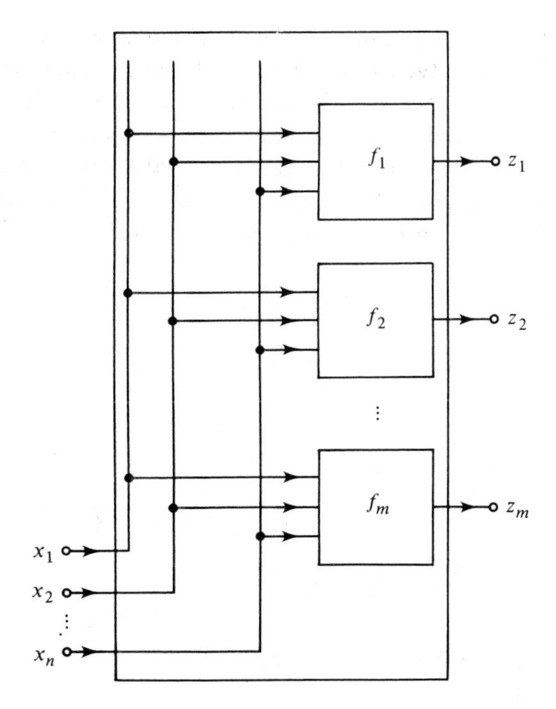

Fig. 7-12 Realization of f: $B_2^n \longrightarrow B_2^m$

We can thus draw the following conclusion:

THEOREM 7-2

Every function f: $B_2^n \longrightarrow B_2^m$ is realizable with {NOT, OR, AND}. (Hence, every combinational network is equivalent to a standard combinational network.)

The realization of f obtained by Algorithm 7-4 is a collection of minterm or maxterm combinational networks. Each one of these networks can, of course, be replaced by any network equivalent to it. In particular, it is often desirable—for reasons of economy—to construct networks that require as few OR and AND gates as possible (NOT gates are relatively inexpensive). This can be achieved by "minimizing" each transmission function f_i—that is, expressing each f_i with as few $+$ and $\cdot$ operations as possible. For example, given

$$f_i(x_1, x_2, x_3) = x_1x_2x_3 + x_1x_2\bar{x}_3 + \bar{x}_1x_2\bar{x}_3$$

we can write (see Example 7-4)

$$f_i(x_1, x_2, x_3) = x_2(x_1 + \bar{x}_3)$$

which demonstrates that f_i can be realized with a single OR and a single AND gate, rather than one (generalized) OR and three (generalized) AND gates required by the minterm normal realization. It should also be noted that even when normal realizations are insisted upon, the minterm realization may be more economical than the maxterm realization (for example, when the "f_i" column in the truth table has fewer 1s than 0s), or conversely. Note also that intermediate outputs produced by the realization of any f_i can be

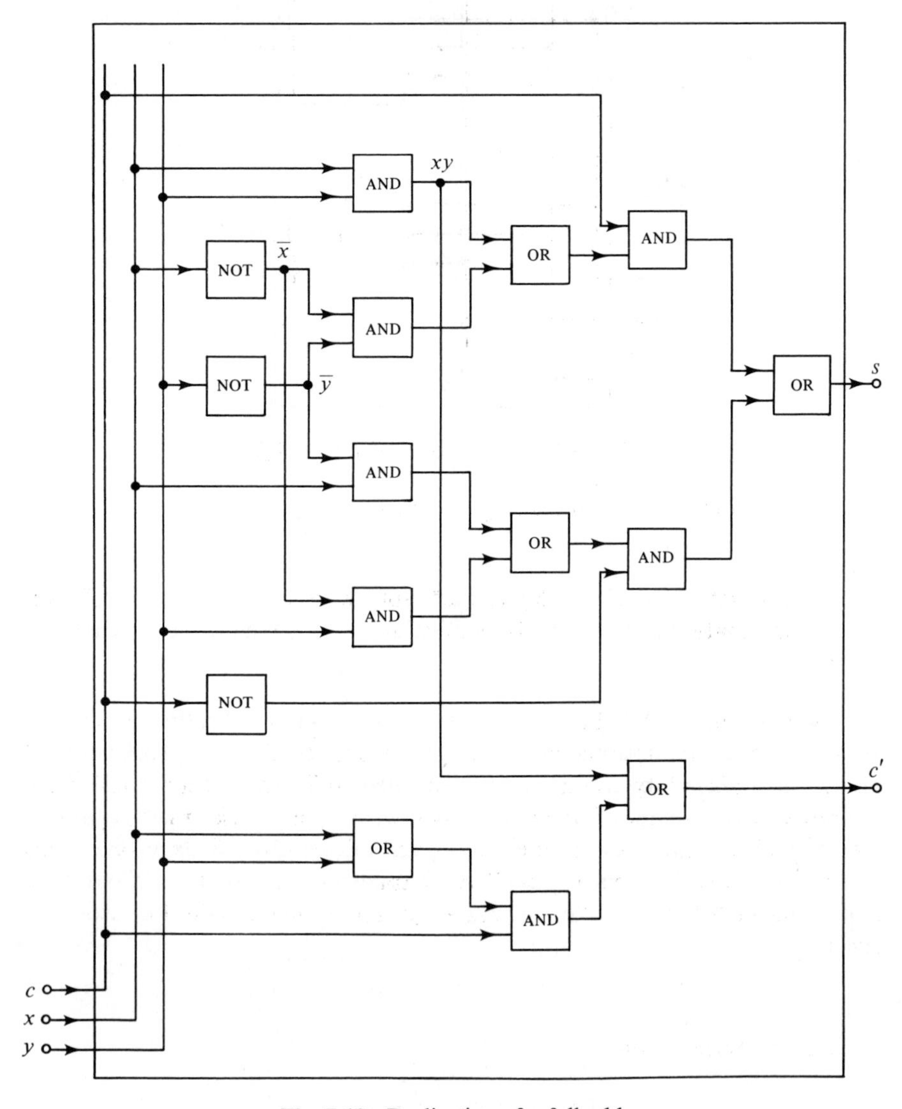

Fig. 7-13 Realization of a full adder

used freely in the realization of any f_j ($j \neq i$), with an attendant saving of gates in the latter.

Example 7-7

For the full adder of Fig. 7-2(c), we have

$$\begin{aligned} s &= \bar{c}\bar{x}y + \bar{c}x\bar{y} + c\bar{x}\bar{y} + cxy \\ &= c(xy + \bar{x}\bar{y}) + \bar{c}(x\bar{y} + \bar{x}y) \\ c' &= \bar{c}xy + c\bar{x}y + cx\bar{y} + cxy \\ &= (\bar{c}xy + cxy) + (cxy + c\bar{x}y) + (cxy + cx\bar{y}) \\ &= xy + cy + cx = xy + c(x + y) \end{aligned}$$

The standard combinational network realizing a full adder is shown in Fig. 7-13. □

PROBLEMS

1. Construct standard combinational networks realizing:
   (a) $f(x_1, x_2) = x_1 + \bar{x}_1 x_2$
   (b) $f(x_1, x_2, x_3) = (x_1 + \bar{x}_3)(x_2 + x_1 x_3)$
   (c) $f(x_1, x_2, x_3, x_4) = x_1(\bar{x}_2 + x_3 x_4) + \bar{x}_1 x_2(x_3 + x_4)$
2. Using as few OR and AND gates as you can, construct the combinational networks described in Problems 2 and 3 in Sec 7-1.
3. Construct a standard combinational network with three inputs and one output, whose output is 1 if and only if the majority of the inputs are 1.
4. Realize a half adder with NOT and AND gates only.

7-6. COMPLEMENTARY AND DUAL TRANSMISSION FUNCTIONS

Given a transmission function $f(x_1, x_2, \ldots, x_n)$, the *complement* of f, denoted by $\bar{f}$, is a transmission function defined by: $\bar{f}(x_1, x_2, \ldots, x_n) = 0$ if and only if $f(x_1, x_2, \ldots, x_n) = 1$. Thus, for all $(x_1, x_2, \ldots, x_n) \in B_2^n$,

$$\bar{f}(x_1, x_2, \ldots, x_n) = \overline{f(x_1, x_2, \ldots, x_n)}$$

For example, if $f(x_1, x_2, x_3) = x_1(x_2 + \bar{x}_3)$, then $\bar{f}(x_1, x_2, x_3) = \overline{x_1(x_2 + \bar{x}_3)}$. The truth table of $\bar{f}$ is obtainable from that of f simply by replacing, in the column labeled "$f(x_1, x_2, \ldots, x_n)$," every 0 with 1 and every 1 with 0.

Clearly, if $\bar{f}$ is the complement of f, then f is the complement of $\bar{f}$. Thus, f and $\bar{f}$ (and the combinational networks realizing them) are said to be *complementary* (to each other).

Given a transmission function $f(x_1, x_2, \ldots, x_n)$, the *dual* of f, denoted by $\tilde{f}$, is a transmission function constructible from f according to the following recursive rule:

(i) (*Basis*). $\tilde{0} = 1$, $\tilde{1} = 0$, $\tilde{x}_i = x_i$ $(i = 1, 2, \ldots, n)$.
(ii) (*Induction step*). If f_1 and f_2 are transmission functions, then

$$\tilde{\bar{f}}_1 = \bar{\tilde{f}}_1$$
$$\widetilde{f_1 + f_2} = \tilde{f}_1\tilde{f}_2$$
$$\widetilde{f_1 f_2} = \tilde{f}_1 + \tilde{f}_2$$

For example, if $f(x_1, x_2, x_3) = x_1(x_2 + \bar{x}_3)$, then

$$\begin{aligned}\tilde{f}(x_1, x_2, x_3) &= \widetilde{x_1(x_2 + \bar{x}_3)} = \tilde{x}_1 + \widetilde{(x_2 + \bar{x}_3)} \\ &= x_1 + \tilde{x}_2\tilde{\bar{x}}_3 = x_1 + x_2\bar{\tilde{x}}_3 = x_1 + x_2\bar{x}_3\end{aligned}$$

It can be observed that $\tilde{f}$ is obtainable from f simply by replacing every 0 with 1, and conversely; and every sum with a product, and conversely. Duals of complicated expressions can thus be constructed by inspection. For example, if

$$f(x_1, x_2, x_3) = (x_1 + \bar{x}_2 + x_3)\overline{(x_1 + x_2\bar{x}_3)} + x_2(x_3 + \bar{x}_1(x_2 + \bar{x}_3))$$

then

$$\tilde{f}(x_1, x_2, x_3) = (x_1\bar{x}_2x_3 + \overline{x_1(x_2 + \bar{x}_3)})(x_2 + x_3(\bar{x}_1 + x_2\bar{x}_3))$$

When f is given in its minterm (or maxterm) normal form, $\tilde{f}$ can be immediately constructed in its maxterm (or minterm) normal form. For example, the dual of

$$f(x_1, x_2, x_3) = \bar{x}_1x_2x_3 + \bar{x}_1\bar{x}_2x_3 + \bar{x}_1x_2\bar{x}_3$$

is given by

$$\tilde{f}(x_1, x_2, x_3) = (\bar{x}_1 + x_2 + x_3)(\bar{x}_1 + \bar{x}_2 + x_3)(\bar{x}_1 + x_2 + \bar{x}_3)$$

THEOREM 7-3
Let $f(x_1, x_2, \ldots, x_n)$ be a transmission function. Then

$$\bar{\tilde{f}}(x_1, x_2, \ldots, x_n) = \tilde{f}(\bar{x}_1, \bar{x}_2, \ldots, \bar{x}_n) \tag{7-1}$$
$$\tilde{f}(\bar{x}_1, \bar{x}_2, \ldots \bar{x}_n) = \bar{\tilde{f}}(x_1, x_2, \ldots, x_n) \tag{7-2}$$

Proof Identity (7-2) can be derived from (7-1) simply by complementing each x_i in the argument. We shall now prove (7-1) by induction on the shape of f. (*Basis*). When $f(x_1, \ldots, x_n) = 0$,

$$\tilde{f}(x_1, \ldots, x_n) = \tilde{0} = 1 = \bar{0} = \bar{f}(\bar{x}_1, \ldots, \bar{x}_n)$$

When $f(x_1, \ldots, x_n) = 1$,

$$\tilde{f}(x_1, \ldots, x_n) = \tilde{1} = 0 = \bar{1} = \bar{f}(\bar{x}_1, \ldots, \bar{x}_n)$$

When $f(x_1, \ldots, x_n) = x_i \ (1 \leq i \leq n)$,

$$\tilde{f}(x_1, \ldots, x_n) = \tilde{x}_i = x_i = \bar{\bar{x}}_i = \bar{f}(\bar{x}_1, \ldots, \bar{x}_n)$$

(*Induction step*). Use the following induction hypothesis for $g_1(x_1, \ldots, x_n)$ and $g_2(x_1, \ldots, x_n)$:

$$\tilde{g}_1(x_1, \ldots, x_n) = \bar{g}_1(\bar{x}_1, \ldots, \bar{x}_n)$$
$$\tilde{g}_2(x_1, \ldots, x_n) = \bar{g}_2(\bar{x}_1, \ldots, \bar{x}_n)$$

If
$$f(x_1, \ldots, x_n) = \bar{g}_1(x_1, \ldots, x_n)$$

then
$$\begin{aligned} &\tilde{f}(x_1, \ldots, x_n) \\ &= \tilde{\bar{g}}_1(x_1, \ldots, x_n) && \text{(Definition of } f) \\ &= \bar{\tilde{g}}_1(x_1, \ldots, x_n) && (\tilde{\bar{g}}_1 = \bar{\tilde{g}}_1) \\ &= \bar{\bar{g}}_1(\bar{x}_1, \ldots, \bar{x}_n) && \text{(Induction hypoth.)} \\ &= \bar{f}(\bar{x}_1, \ldots, \bar{x}_n) && \text{(Definition of } f) \end{aligned}$$

If
$$f(x_1, \ldots, x_n) = g_1(x_1, \ldots, x_n) + g_2(x_1, \ldots, x_n)$$

then
$$\begin{aligned} &\tilde{f}(x_1, \ldots, x_n) \\ &= \widetilde{g_1(x_1, \ldots, x_n) + g_2(x_1, \ldots, x_n)} && \text{(Definition of } f) \\ &= \tilde{g}_1(x_1, \ldots, x_n)\tilde{g}_2(x_1, \ldots, x_n) && (\widetilde{g_1 + g_2} = \tilde{g}_1\tilde{g}_2) \\ &= \bar{g}_1(\bar{x}_1, \ldots, \bar{x}_n)\bar{g}_2(\bar{x}_1, \ldots, \bar{x}_n) && \text{(Induction hypoth.)} \\ &= \overline{g_1(\bar{x}_1, \ldots, \bar{x}_n) + g_2(\bar{x}_1, \ldots, \bar{x}_n)} && \text{(De Morgan)} \\ &= \bar{f}(\bar{x}_1, \ldots, \bar{x}_n) && \text{(Definition of } f) \end{aligned}$$

The remainder of the proof, where

$$f(x_1, \ldots, x_n) = g_1(x_1, \ldots, x_n)g_2(x_1, \ldots, x_n)$$

is analogous to the preceding, and is left to the reader. □

Equation (7-2) suggests this procedure for constructing the complement of a given transmission function f: Replace every x_i with $\bar{x}_i$ and construct the dual of the resulting function. For example, if $f(x_1, x_2, x_3) = x_1(x_2 + \bar{x}_3)$, then

$$\bar{f}(x_1, x_2, x_3) = \tilde{f}(\bar{x}_1, \bar{x}_2, \bar{x}_3) = \bar{x}_1 + \bar{x}_2 x_3$$

When f is given in its minterm (or maxterm) normal form, $\bar{f}$ can be immediately constructed in its maxterm (or minterm) normal form. For example, the complement of

$$f(x_1, x_2, x_3) = \bar{x}_1 x_2 x_3 + \bar{x}_1 \bar{x}_2 x_3 + \bar{x}_1 x_2 \bar{x}_3$$

is given by

$$\bar{f}(x_1, x_2, x_3) = (x_1 + \bar{x}_2 + \bar{x}_3)(x_1 + x_2 + \bar{x}_3)(x_1 + \bar{x}_2 + x_3)$$

Theorem 7-3 also implies that two combinational networks are complementary when one can be obtained from the other by replacing every input x_i with its complement $\bar{x}_i$, every OR gate with an AND gate, and every AND gate with an OR gate.

PROBLEMS

1. Complete the proof of Theorem 7-3.
2. Construct the duals of the following transmission functions:
   (a) $f(x, y, z) = (x + y + \bar{z})(\bar{y}z + x)$
   (b) $f(x, y, z) = xy + \bar{x}\bar{y}z(x + \bar{x}yz)$
   (c) $f(x, y, z) = (x + y(\bar{x} + x\bar{z}))(xy + \bar{y}z + \bar{z})$
   In each case, verify that $\tilde{f}(x, y, z) = \bar{f}(\bar{x}, \bar{y}, \bar{z})$.
3. The transmission function $f(x, y)$ is specified by the truth table shown below. Write, by inspection, the minterm and maxterm normal forms of f, $\bar{f}$, and $\tilde{f}$.

x	y	$f(x, y)$
0	0	1
0	1	0
1	0	0
1	1	0

4. The network shown in Fig. 7-E realizes the transmission function $f(x_1, x_2, x_3)$.
   (a) Determine $f(x_1, x_2, x_3)$ and $\tilde{f}(x_1, x_2, x_3)$.
   (b) Construct a network complementary to that shown, without using NOT gates.

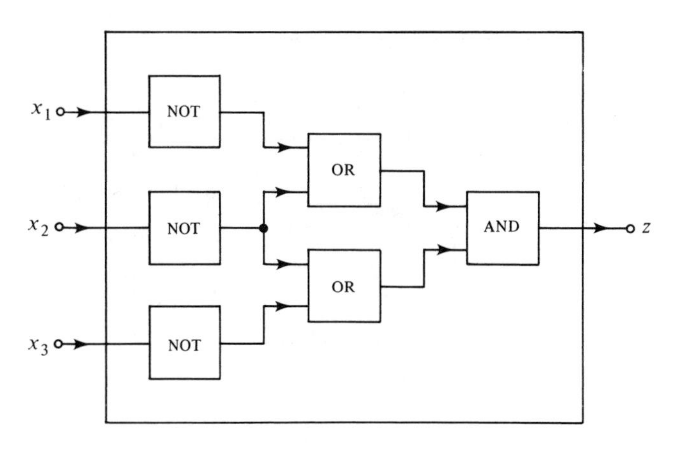

Fig. 7-E

5. A transmission function $f(x_1, x_2, \ldots, x_n)$ is said to be *self-dual* if

$$f(x_1, x_2, \ldots, x_n) = \tilde{f}(x_1, x_2, \ldots, x_n)$$

(a) Construct all self-dual transmission functions generated by x and y.
(b) How many distinct self-dual transmission functions are generated by $x_1, x_2, \ldots, x_n$?

6. Let $f_1(x_1, x_2, \ldots, x_n)$ and $f_2(x_1, x_2, \ldots, x_n)$ be self-dual transmission functions (see Problem 5). Are the following transmission functions self-dual?
(a) $f_1(x_1, x_2, \ldots, x_n) + f_2(x_1, x_2, \ldots, x_n)$
(b) $f_1(x_1, x_2, \ldots, x_n) f_2(x_1, x_2, \ldots, x_n)$
(c) $\bar{f}_1(x_1, x_2, \ldots, x_n)$
Prove your answers.

7. Given $g(x_1, x_2, \ldots, x_n)$, a transmission function generated by $x_1, x_2, \ldots, x_n$, define:

$$g_0(x_1, x_2, \ldots, x_n) = \bar{x}_n g(x_1, x_2, \ldots, x_{n-1}) + x_n \bar{g}(\bar{x}_1, \bar{x}_2, \ldots, \bar{x}_{n-1})$$

(a) Show that $g_0(x_1, x_2, \ldots, x_n)$ is self-dual (see Problem 5).
(b) Show that the set of all functions g_0, as defined above, is precisely the set of all self-dual transmission functions generated by $x_1, x_2, \ldots, x_n$.

7-7. FUNCTIONAL COMPLETENESS

A set of gates $G = \{g_1, g_2, \ldots, g_h\}$ is said to be *functionally complete*, if every possible transmission function is realizable with G. We have already established that the set of gates {NOT, OR, AND} is functionally complete. In fact, the subsets {NOT, OR} and {NOT, AND} of this set are also functionally complete, since an AND gate is always realizable with {NOT, OR} and an OR

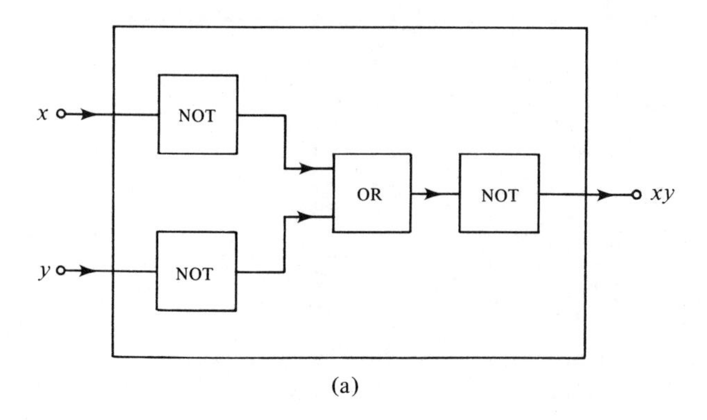

(a)

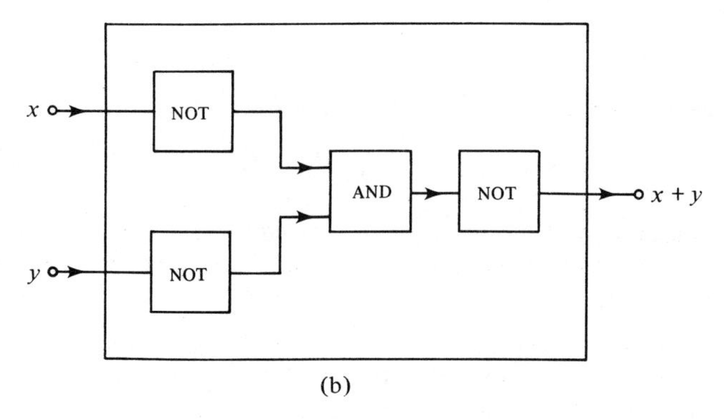

(b)

Fig. 7-14 Realization of: (a) AND gate with {NOT, OR} (b) OR gate with {NOT, AND}

gate with {NOT, AND} by applying De Morgan's laws:

$$xy = \overline{\bar{x} + \bar{y}}$$

$$x + y = \overline{\bar{x}\bar{y}}$$

(See Fig. 7-14.)

The question now arises whether there is a *single* gate with which every transmission function is realizable. The answer is in the affirmative; in fact, there are a number of possible gates—called *universal gates*—which satisfy this requirement. The most commonly used is the NAND (NOT AND) gate which has two inputs and whose truth table is shown in Fig. 7-15(a). If the inputs of a NAND gate are x and y, then the output, denoted by $x \uparrow y$, is given by

$$x \uparrow y = \overline{xy}$$

(The operation $\uparrow$ is known as the *Sheffer stroke* operation.) To prove that

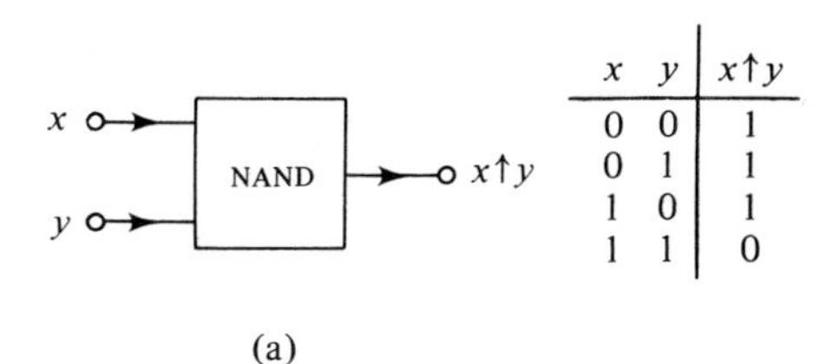

x	y	$x \uparrow y$
0	0	1
0	1	1
1	0	1
1	1	0

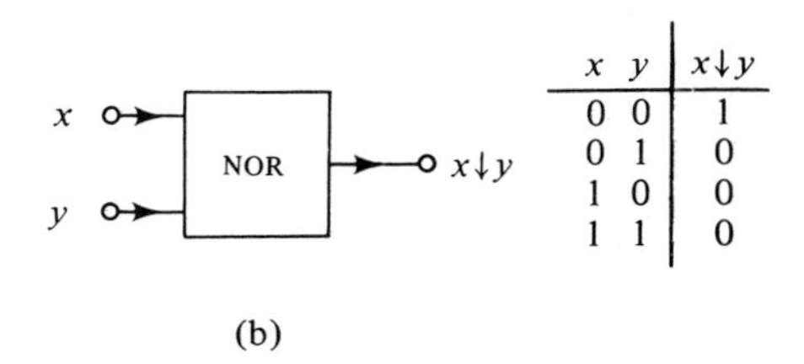

x	y	$x \downarrow y$
0	0	1
0	1	0
1	0	0
1	1	0

Fig. 7-15 (a) NAND gate (b) NOR gate

the NAND gate is a universal gate, it is sufficient to show that the NOT and OR gates (or the NOT and AND gates) are realizable with NAND gates only. This fact becomes apparent from these identities:

$$\bar{x} = \overline{xx} = x \uparrow x$$
$$x + y = \overline{\bar{x}\bar{y}} = \bar{x} \uparrow \bar{y} = (x \uparrow x) \uparrow (y \uparrow y)$$
$$xy = \overline{\overline{xy}} = \overline{x \uparrow y} = (x \uparrow y) \uparrow (x \uparrow y)$$

(See Fig. 7-16.)

A gate "dual" to NAND is the NOR (NOT OR) *gate* whose truth table is shown in Fig. 7-15(b). If the inputs of a NOR gate are x and y, then the output, denoted by $x \downarrow y$, is given by

$$x \downarrow y = \overline{x + y}$$

(The operation $\downarrow$ is sometimes called the *dagger* operation.) The proof that the NOR gate is a universal gate is left to the reader.

A *threshold gate* with *coefficients* $\alpha_1, \alpha_2, \ldots, \alpha_n$ and *threshold* θ (the α_i and θ being real numbers) is a gate with n inputs, realizing the function $f: B_2^n \longrightarrow B$, where

$$(f(x_1, x_2, \ldots, x_n) = 1) \Longleftrightarrow (\textstyle\sum_{i=1}^{n} \alpha_i x_i \geq \theta)^1$$

(Such a function is said to be *linearly separable.*) The schematic representation of a threshold gate is shown in Fig. 7-17. When $n = 1$, $\alpha_1 = -1$, and $\theta = 0$, we have

$$(f(x_1) = 1) \Longleftrightarrow (-x_1 \geq 0) \Longleftrightarrow (x_1 = 0)$$

[1]Here all additions and multiplications are the ordinary operations over real numbers.

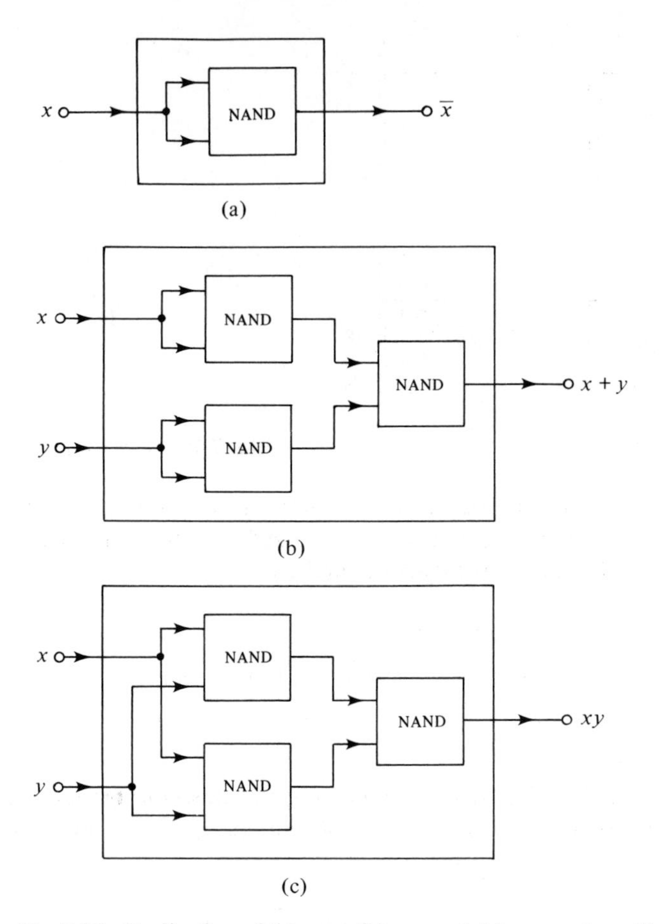

Fig. 7-16 Realization of (a) NOT, (b) OR, and (c) AND gates with NAND gates

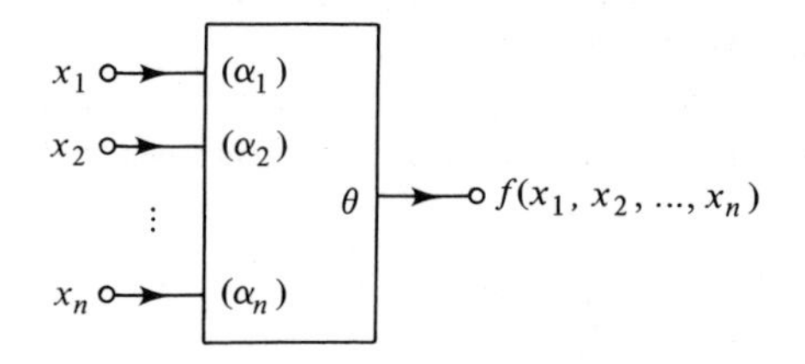

Fig. 7-17 A threshold gate

When $n = 2$, $\alpha_1 = \alpha_2 = 1$, and $\theta = 1$, we have

$$(f(x_1, x_2) = 1) \Longleftrightarrow (x_1 + x_2 \geq 1) \Longleftrightarrow (x_1 = 1 \quad \text{or} \quad x_2 = 1)$$

When $n = 2$, $\alpha_1 = \alpha_2 = 1$, and $\theta = 2$, we have

$$(f(x_1, x_2) = 1) \Longleftrightarrow (x_1 + x_2 \geq 2) \Longleftrightarrow (x_1 = 1 \quad \text{and} \quad x_2 = 1)$$

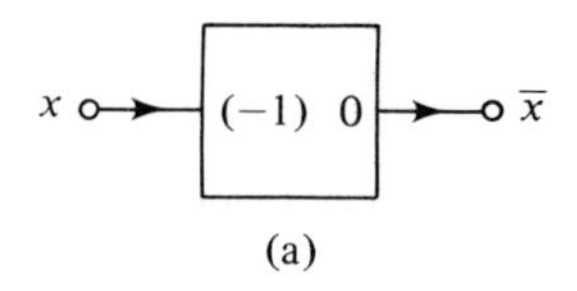

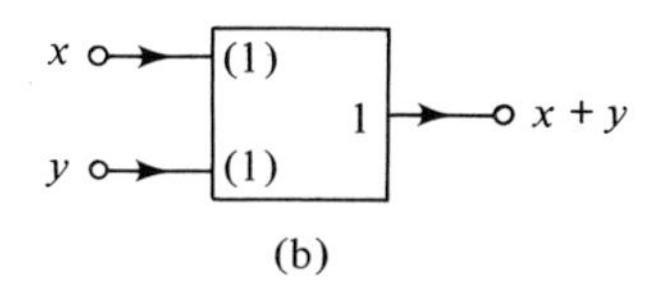

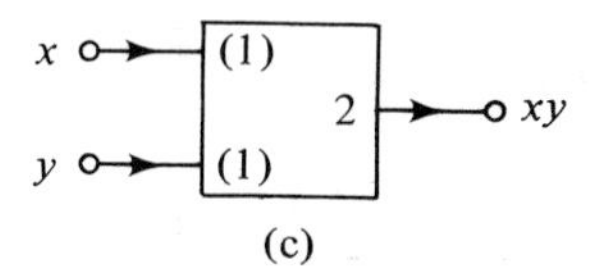

Fig. 7-18 Realization of (a) NOT, (b) OR, and (c) AND gates with threshold gates

This demonstrates that NOT, OR, and AND gates can be realized with the three threshold gates shown in Fig. 7-18, hence, that this set of threshold gates is functionally complete.

Sometimes we can achieve a reduction in the number of gate types needed for realization by providing the combinational networks with sources of constant 0 and 1. We say that the set of gates $G = \{g_1, g_2, \ldots, g_h\}$ is *functionally quasi-complete* if every transmission function is realizable with G and with sources of constant 0 and 1. Correspondingly, a gate g is called a *quasi-universal gate* if $\{g\}$ is functionally quasi-complete.

An example of a quasi-universal gate is the *inhibitor gate* whose truth table is shown in Fig. 7-19. If the inputs of an inhibitor gate are x and y, then the output, denoted by $x * y$, is given by

$$x * y = x\bar{y}$$

The following demonstrates that the inhibitor gate is quasi-universal:

$$\bar{x} = 1\bar{x} = 1 * x$$

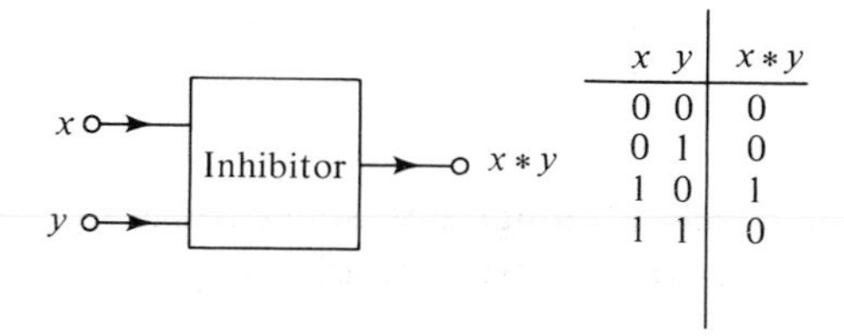

x	y	$x * y$
0	0	0
0	1	0
1	0	1
1	1	0

Fig. 7-19 Inhibitor gate

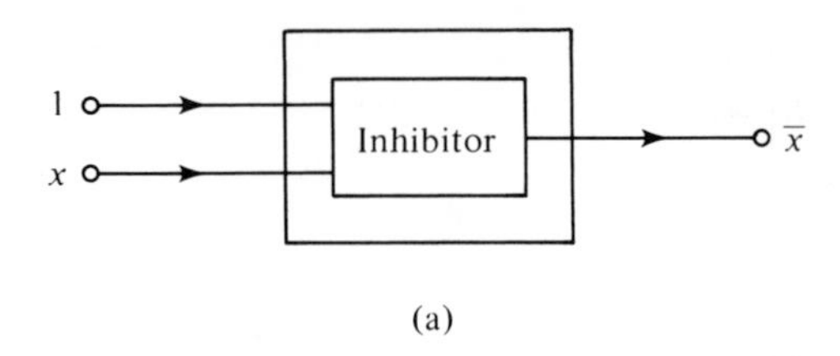

(a)

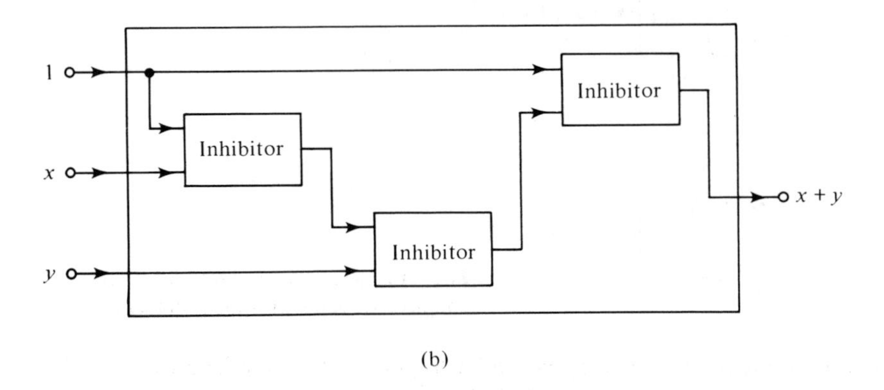

(b)

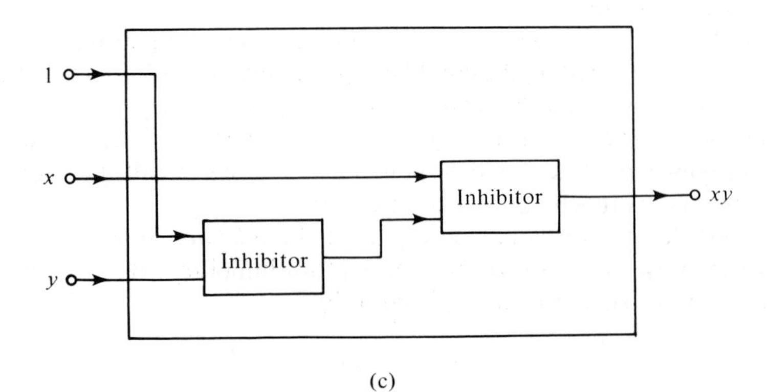

(c)

Fig. 7-20 Realization of (a) NOT, (b) OR, and (c) AND gates with inhibitor gates

$$x + y = \overline{\bar{x}\bar{y}} = 1 * (\bar{x}\bar{y}) = 1 * ((1 * x) * y)$$
$$xy = x\bar{\bar{y}} = x * \bar{y} = x * (1 * y)$$

(See Fig. 7-20.)

In closing, we shall mention a gate which, although not universal, is of great usefulness in many applications; it is the EXCOR (exclusive OR) gate.

If the inputs of an EXCOR gate are x and y, then the output, denoted by $x + y$, is given by

$$x + y = \bar{x}y + x\bar{y}$$

PROBLEMS

1. Prove that the NOR gate is a universal gate, by realizing the NOT, OR, and AND gates with {NOR}.
2. Are the operations ↑ (NAND), ↓ (NOR) and * (inhibitor) commutative? Are they associative?
3. Realize the half adder of Fig. 7-2(b) with:
   (a) NOT and OR gates. (b) NOT and AND gates.
   (c) NAND gates. (d) NOR gates.
   (e) Threshold gates. (f) Inhibitor gates and 1 source.
4. Specify an n-input threshold gate (where n is odd), whose output is 1 if and only if:
   (a) The majority of the inputs are 1.
   (b) The minority of the inputs are 1.
5. Prove that the three-input threshold gate with coefficient $\alpha_1 = -1$, $\alpha_2 = \alpha_3 = 1$, and threshold $\theta = 0$ is a quasi-universal gate.
6. Prove that {NOT, inhibitor} is functionally complete.

7-8. SEQUENTIAL NETWORKS

We now introduce a component, called a *delay unit*, that, when incorporated with combinational networks, enables them to have *memory*. That is, combinational networks synthesized in conjunction with such components, are capable of generating outputs that depend not only on the current inputs, but also on past inputs. These enriched combinational networks, known as *sequential networks*, are assumed to be supplied with inputs and generate outputs at *discrete instants of time* only. In practice, these instants can assume any values, and need not appear at uniform intervals. In our discussions, however, we attach to these instants integral values, with $t = 0$ usually taken as the instant at which the network is first placed under observation.

More specifically, a *delay unit* (see Fig. 7-21) is a box with a single input and a single output, where the output at any instant of time equals the input

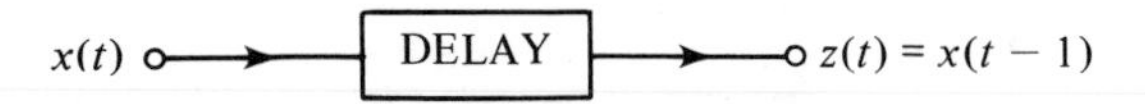

Fig. 7-21 Delay unit

at the preceding instant of time. Thus, if the input and output at time t are denoted by $x(t)$ and $z(t)$, respectively, then,

$$z(t) = x(t - 1)$$

It is seen that a delay unit is capable of "remembering" at any time t what occurred one unit of time earlier.

A *sequential network* is any interconnection of combinational networks and delay units, constructed by these rules:

(a) An output of a combinational network or a delay unit can serve as input to any number of combinational networks or delay units.

(b) An input of a combinational network or a delay unit can be obtained from the output of at most one combinational network or one delay unit.

(c) In applying rules (a) and (b), it should be ascertained that every closed loop (that is, a sequence of combinational networks and/or delay units which closes upon itself) contains at least one delay unit (in order to prevent violation of the "no-feedback rule" of Sec. 7-1).

(d) The set of inputs that are not obtained from any combinational network or delay unit outputs, are the overall inputs of the sequential network.

(e) Outputs of the new network can be tapped from any point in the network.

Example 7-8

Figure 7-22 shows an example of a sequential network, called a *serial adder*, which can be used to add two binary numbers fed "serially" (digit by digit) as inputs $x(t)$ and $y(t)$, with the least significant digit first and the most significant digit last. The sum appears as the output $s(t)$, with the least significant digit first and the most significant digit last. As shown, the serial adder is constructed from the full adder described in Sec. 7-1 and a single delay unit. The function of this unit is simply to enforce the condition that the carry digit generated by the full adder at any time t be applied as an input to the full adder at time $t + 1$. □

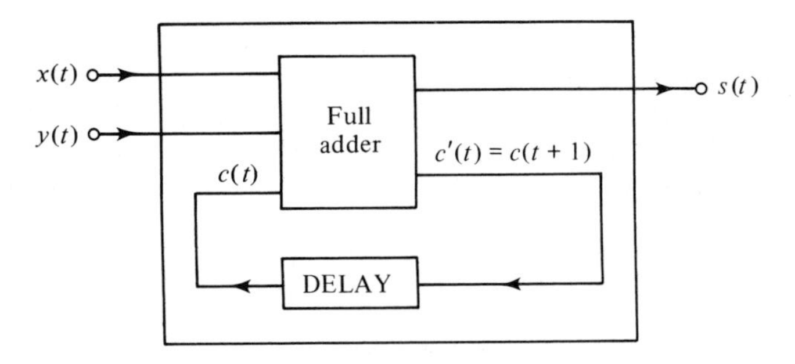

Fig. 7-22 Serial adder

Example 7-9

Figure 7-23 shows a sequential network consisting of a threshold gate with coefficients 1 and threshold 2, and three delay units. The outputs of the delay units at time t are denoted by $y_i(t)$ $(i = 1, 2, 3)$, and the output of the threshold gate by majority$(y_1(t), x(t), y_3(t))$ [which equals 1 if and only if $y_1(t) + x(t) + y_3(t) \geq 2$]. Thus, from the network diagram, we have:

$$z(t) = \text{majority}(y_1(t), x(t), y_3(t))$$
$$y_1(t+1) = x(t)$$
$$y_2(t+1) = \text{majority}(y_1(t), x(t), y_3(t))$$
$$y_3(t+1) = y_2(t)$$

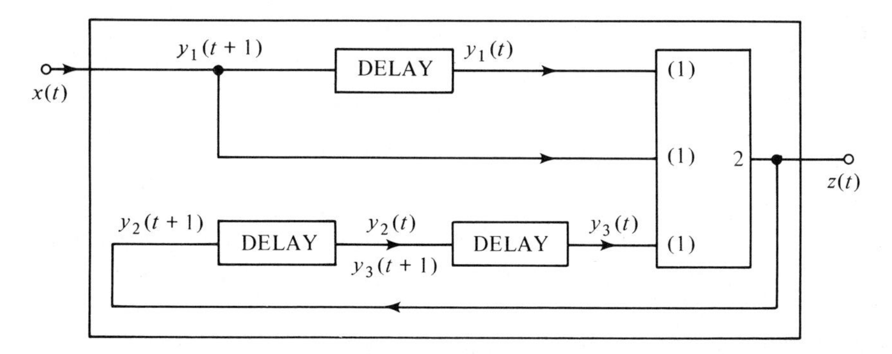

Fig. 7-23 A sequential network

Table 7-5 FOR SEQUENTIAL NETWORK OF FIG. 7-23

$x(t)$	$y_1(t)$	$y_2(t)$	$y_3(t)$	$z(t)$	$y_1(t+1)$	$y_2(t+1)$	$y_3(t+1)$
0	0	0	0	0	0	0	0
0	0	0	1	0	0	0	0
0	0	1	0	0	0	0	1
0	0	1	1	0	0	0	1
0	1	0	0	0	0	0	0
0	1	0	1	1	0	1	0
0	1	1	0	0	0	0	1
0	1	1	1	1	0	1	1
1	0	0	0	0	1	0	0
1	0	0	1	1	1	1	0
1	0	1	0	0	1	0	1
1	0	1	1	1	1	1	1
1	1	0	0	1	1	1	0
1	1	0	1	1	1	1	0
1	1	1	0	1	1	1	1
1	1	1	1	1	1	1	1

It is seen that, at any time t, $x(t)$, $y_1(t)$, $y_2(t)$, and $y_3(t)$ uniquely determine $z(t)$ and $y_i(t+1)$ $(i = 1, 2, 3)$. This dependence is tabulated in Table 7-5. □

PROBLEMS

1. For the sequential network shown in Fig. 7-F, tabulate $z(t)$, $y_1(t+1)$, $y_2(t+1)$ and $y_3(t+1)$ for all values of $x(t)$, $y_1(t)$, $y_2(t)$, and $y_3(t)$.

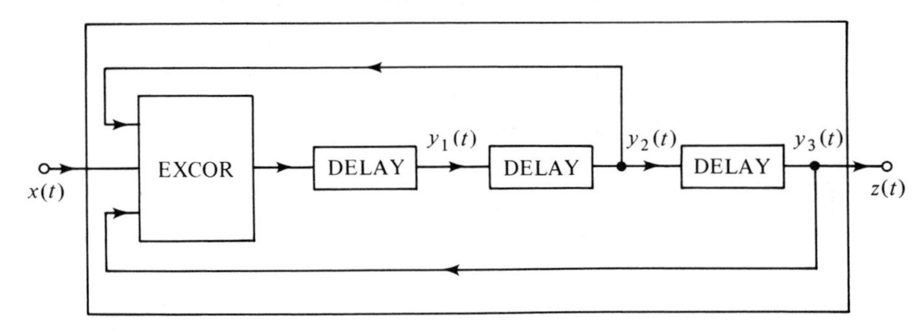

Fig. 7-F

2. In the network of Fig. 7-F, assume that $y_1(0) = y_2(0) = y_3(0) = 1$ and that $x(t) = 0$ for all $t \geq 0$. Show that the output $z(t)$ is periodic with period 7 [that is, $z(t+7) = z(t)$ for all $t \geq 0$].

3. Synthesize a sequential network with input $x(t)$ and output $z(t)$ such that

$$z(t) = z(t-1)(x(t) + x(t-1)).$$

(where all operations are over $\langle B_2; {}^-, +, \cdot \rangle$).

7-9. FINITE-STATE MACHINES

Sequential networks are examples of more general systems known as *finite-state machines*. Roughly, a finite-state machine is a box that is fed with inputs and generates outputs at discrete instants of time, and whose output at any time depends on the input and the internal condition (called the *state*) of the box at that time; in addition, the input and internal condition at any time determine the internal condition at the next instant of time.

Formally, a *finite-state machine* consists of:

(a) A finite nonempty set X called the *input alphabet* (and consisting of *input symbols*).

(b) A finite nonempty set Z called the *output alphabet* (and consisting of *output symbols*).

(c) A finite nonempty set S called the *state set* (and consisting of *states*).
(d) A function f: $S \times X \longrightarrow S$ called the *next-state function*.
(e) A function g: $S \times X \longrightarrow Z$ called the *output function*.

We shall denote a finite-state machine $\mathfrak{M}$ consisting of these entities by

$$\mathfrak{M} = \langle X, Z, S, f, g \rangle$$

The letters ξ, ζ, and σ (with possible subscripts or superscripts) will be used to denote elements of X, Z, and S, respectively.

"Physically," $\mathfrak{M}$ can be interpreted as a box (see Fig. 7-24) whose input, output, and state (a description of the box's internal condition) at time t are denoted by $x(t)$, $z(t)$, and $s(t)$, respectively, and are elements of X, Z, and S, respectively. The quantities $x(t)$, $z(t)$, and $s(t)$ are defined for integral values of t only, and their values are determined by f and g in the following manner:

$$s(t + 1) = f(s(t), x(t)) \tag{7-3}$$

$$z(t) = g(s(t), x(t)) \tag{7-4}$$

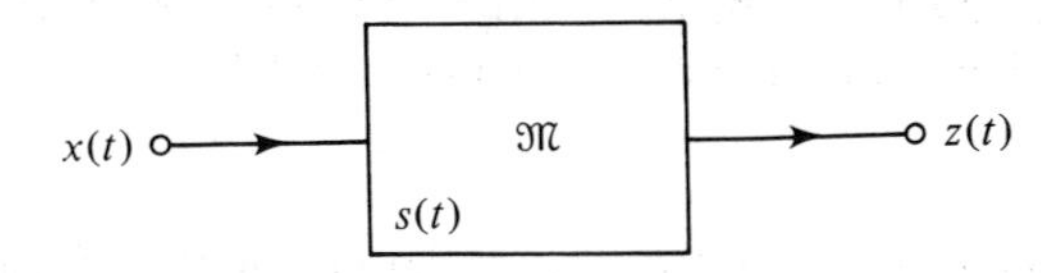

Fig. 7-24 Finite-state machine

We commonly assume that observation of $\mathfrak{M}$ commences at $t = 0$, and correspondingly refer to $s(0)$ as the *initial state* of $\mathfrak{M}$. When the initial state and the first k inputs of $\mathfrak{M}$ [that is, $x(0), x(1), \ldots, x(k - 1)$] are known, the first k outputs of $\mathfrak{M}$ [that is, $z(0), z(1), \ldots, z(k - 1)$] can be determined successively via (7-3) and (7-4):

$$
\begin{aligned}
z(0) &= g(s(0), x(0)), & s(1) &= f(s(0), x(0)) \\
z(1) &= g(s(1), x(1)), & s(2) &= f(s(1), x(1)) \\
&\;\;\vdots \\
z(k-2) &= g(s(k-2), x(k-2)), & s(k-1) &= f(s(k-2), x(k-2)) \\
z(k-1) &= g(s(k-1), x(k-1))
\end{aligned}
$$

A common method of specifying the f and g functions of $\mathfrak{M}$ is by means of a *transition table* whose general form is shown in Table 7-6. An alternative method of specification is provided by a *transition diagram*, which is a graph consisting of $\#S$ vertices, labeled as the $\#S$ states of $\mathfrak{M}$. The edge pointing

Table 7-6 TRANSITION TABLE

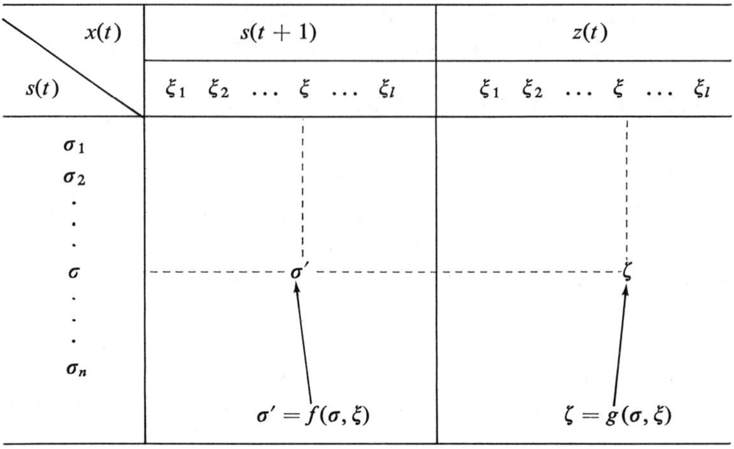

$x(t)$ / $s(t)$	$s(t+1)$						$z(t)$					
	ξ_1	ξ_2	$\dots$	ξ	$\dots$	ξ_l	ξ_1	ξ_2	$\dots$	ξ	$\dots$	ξ_l
σ_1												
σ_2												
$\cdot$												
σ				σ'						ζ		
$\cdot$												
σ_n												
				$\sigma' = f(\sigma, \xi)$						$\zeta = g(\sigma, \xi)$		

from vertex σ to vertex σ' is labeled with all expressions of the form ξ/ζ, where $f(\sigma, \xi) = \sigma'$ and $g(\sigma, \xi) = \zeta$. The following examples illustrate these two methods of specification, as well as demonstrate processes for which finite-state machines can serve as convenient models.

Example 7-10

An English text, composed of the 26 letters of the alphabet and blank spaces, is scanned with the purpose of counting the number of words starting with the letters UN and terminating with the letter D. We wish to define a finite-state machine $\mathfrak{M} = \langle X, Z, S, f, g \rangle$ capable of "simulating" this process. This can be done as follows: The input alphabet of $\mathfrak{M}$ is

$$X = \{ b\!\!\!/, \mathrm{D}, \mathrm{N}, \mathrm{U}, l \}$$

where $b\!\!\!/$ stands for blank space and l for all letters other than D, N, and U (all such letters l play an equivalent role in the process). The output alphabet is

$$Z = \{0, 1\}$$

where 0 stands for "no count" and 1 for "increment count by 1." The state set is

$$S = \{1, 2, 3, 4, 5\}$$

where the five states are interpreted as follows: $\mathfrak{M}$ enters state 1 after scanning a $b\!\!\!/$, state 2 after scanning $b\!\!\!/$U, state 3 after scanning $b\!\!\!/$UN, state 4 after scanning $b\!\!\!/$UN . . . D, and state 5 after scanning any letter that rules out the current word as one which can be possibly counted. The transition table based on

Table 7-7 TRANSITION TABLE FOR EXAMPLE 7-10

$s(t)$ \ $x(t)$	$s(t+1)$					$z(t)$				
	ƀ	D	N	U	*l*	ƀ	D	N	U	*l*
1	1	5	5	2	5	0	0	0	0	0
2	1	5	3	5	5	0	0	0	0	0
3	1	4	3	3	3	0	0	0	0	0
4	1	4	3	3	3	1	0	0	0	0
5	1	5	5	5	5	0	0	0	0	0

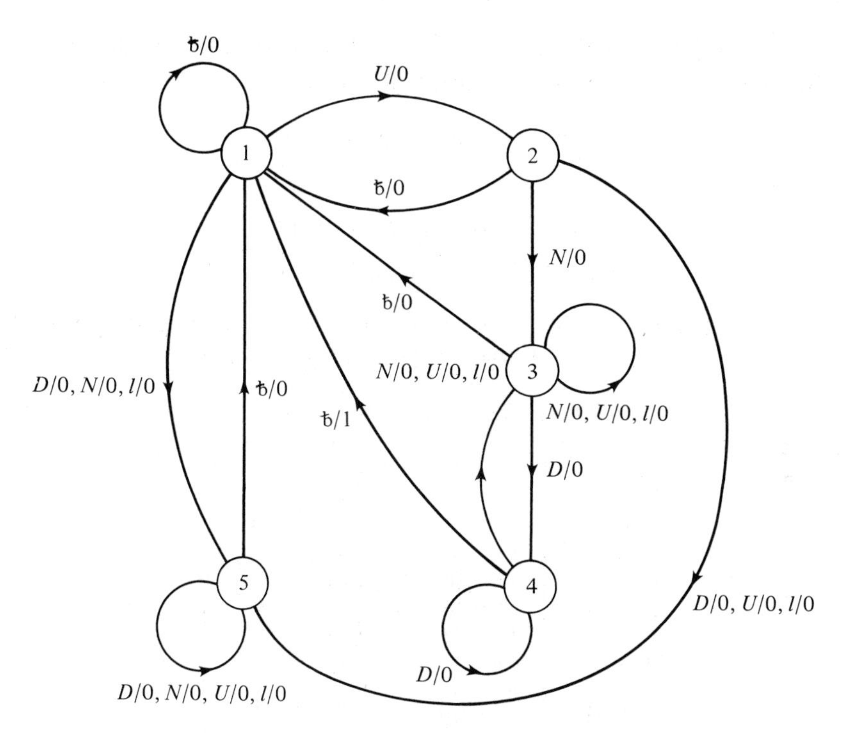

Fig. 7-25 Transition diagram for Example 7-10

this choice of state set is shown in Table 7-7; the corresponding transition diagram is shown in Fig. 7-25.

Given the initial state and any sequence of inputs, the corresponding outputs can be easily read off the diagram by tracing the path labeled with these inputs, starting at the vertex labeled with the initial state. For example, if $\mathcal{M}$ is in state 1 and the text is

THE UNITED STATES UNDERSTANDS THE UNDERLYING DIFFICULTIES

then the sequence of states is seen to be

155512333341555555123433333343155512343333333155555555555

and the sequence of outputs

00000000001000 . . . 0 □

Example 7-11

For the sequential network of Fig. 7-23 we can naturally choose

$$X = Z = \{0, 1\}$$

Table 7-5 reveals that $z(t)$ and $y_i(t + 1)$ $(i = 1, 2, 3)$ are uniquely determined by $x(t)$ and the $y_i(t)$. Hence, an adequate choice for $s(t)$ would be:

$$s(t) = (y_1(t), y_2(t), y_3(t))$$

Correspondingly,

$$S = \{(0, 0, 0), (0, 0, 1), (0, 1, 0), (0, 1, 1), (1, 0, 0), (1, 0, 1), (1, 1, 0), (1, 1, 1)\}$$

The transition table for the finite-state machine representing the network can be derived directly from Table 7-5 and is shown in Table 7-8. □

Table 7-8 TRANSITION TABLE FOR EXAMPLE 7-11

$s(t)$ \ $x(t)$	$s(t+1)$: 0	$s(t+1)$: 1	$z(t)$: 0	$z(t)$: 1
(0, 0, 0)	(0, 0, 0)	(1, 0, 0)	0	0
(0, 0, 1)	(0, 0, 0)	(1, 1, 0)	0	1
(0, 1, 0)	(0, 0, 1)	(1, 0, 1)	0	0
(0, 1, 1)	(0, 0, 1)	(1, 1, 1)	0	1
(1, 0, 0)	(0, 0, 0)	(1, 1, 0)	0	1
(1, 0, 1)	(0, 1, 0)	(1, 1, 0)	1	1
(1, 1, 0)	(0, 0, 1)	(1, 1, 1)	0	1
(1, 1, 1)	(0, 1, 1)	(1, 1, 1)	1	1

In what follows, X^* denotes the set of all sequences ("strings") of elements of X; λ denotes the "empty string" — the string with no symbols. (More on this subject will be said in Sec. 8-1.)

Given a finite-state machine $\mathfrak{M} = \langle X, Z, S, f, g\rangle$, the *extended next-state function* f^* of $\mathfrak{M}$ is the function

$$f^*: \; S \times X^* \longrightarrow S$$

where, for all $\sigma \in S$, $\xi \in X$ and $w_1, w_2 \in X^*$,

$$f^*(\sigma, \lambda) = \sigma$$
$$f^*(\sigma, \xi) = f(\sigma, \xi)$$
$$f^*(\sigma, w_1 w_2) = f^*(f^*(\sigma, w_1), w_2)$$

Thus, for any $\sigma \in S$ and $w \in X^*$, $f^*(\sigma, w)$ is the state into which $\mathfrak{M}$ enters when the input string w is applied to $\mathfrak{M}$. Since f and f^* are identical in the domain of f, both functions are usually denoted by the same symbol f.

PROBLEMS

1. Choosing appropriate input alphabets, output alphabets, and state sets, construct the transition tables and diagrams of the finite-state machines that simulate these sequential networks:
   (a) A delay unit (Fig. 7-21).
   (b) A network whose output at time t is 1 if and only if the cumulative number of input 1s up to time t is odd. (This network is called a *parity checker*.)
   (c) A network such that if the total number of 1s applied to it in any period of time is $2k$ or $2k + 1$ ($k = 0, 1, 2, \ldots$), then the total number of 1s it generates in that period is k. (This network is called a *binary scaler*.)
   (d) The network of Problem 1, Sec 7-8.
   (e) The network of Problem 3, Sec. 7-8.
   (f) The network of Fig. 7-G.

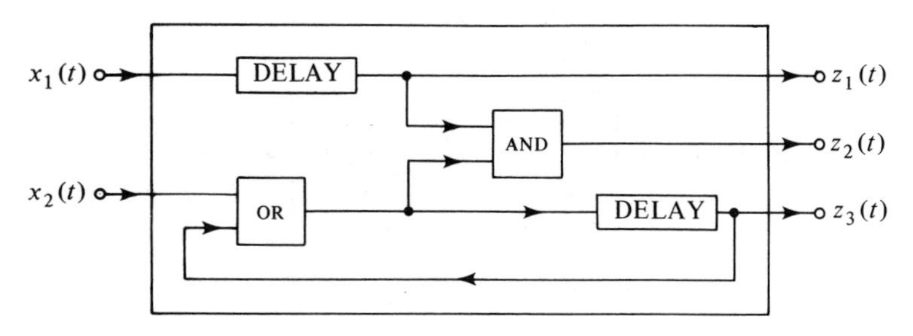

Fig. 7-G

2. Choosing appropriate state sets, construct the transition tables and diagrams of the finite-state machines that simulate these processes:
   (a) A coin is tossed repeatedly, and a point is scored for every other tail in a sequence of successive tails, and for every other head—whether in a sequence of successive heads or not. [$x(t)$ = coin face at time t; $z(t)$ = score at time t.]
   (b) Two chips, each of which has the number 1 marked on one side and 2 on the other, are thrown repeatedly and simultaneously. After each throw, the

sum modulo 2 is computed of the present faces, the faces obtained in the last throw, and the last computed sum. [$x(t)$ = combination of chip faces at time t; $z(t)$ = sum computed at time t.]

(c) A (rather primitive) freight elevator serving a three-story warehouse with a call button on each floor, operates according to these rules: Buttons are pushed only when the elevator is at rest. If a single button is pushed, the elevator moves to the floor on which the pushed button is located; if two or more buttons are pushed, the elevator moves to the lowest calling floor. [$x(t)$ = combination of calling floors at time t; $z(t)$ = direction of motion and number of floors to be traversed following the calls at time t.]

(d) An English text, composed of the 26 letters of the alphabet and blank spaces, is scanned with the purpose of counting the number of words that end with ART. [$x(t)$ = character scanned at time t; $z(t)$ = increment to count at time t.]

3. $\mathfrak{M}$ is a finite-state machine whose transition diagram is shown in Fig. 7-H(a). Draw transition diagrams of the finite-state machines $\mathfrak{M}_1$ and $\mathfrak{M}_2$ shown in Figs. 7-H(b) and (c).

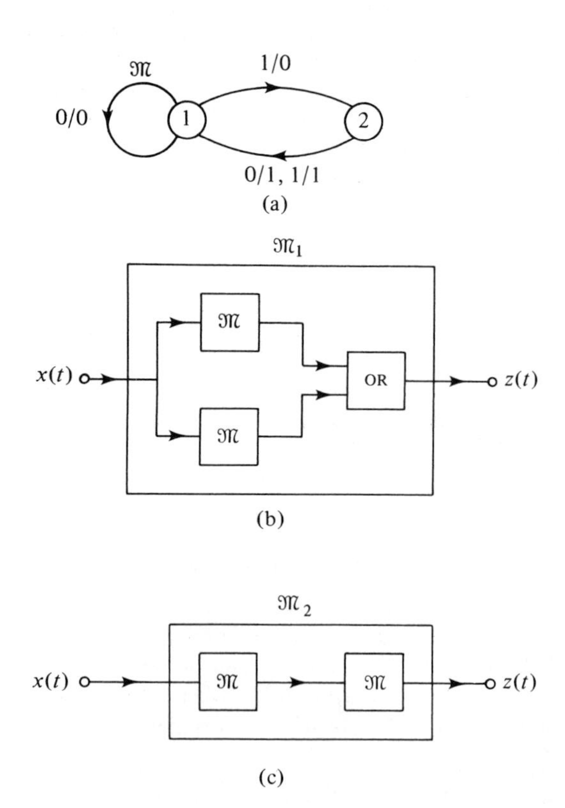

Fig. 7-H

7-10. SEQUENTIAL NETWORKS AND FINITE-STATE MACHINES

When a finite-state machine model is sought for any given system, the input and output alphabets can usually be deduced from the description of the system (input is identified with the variables imposed by external sources, and the output with the variables selected for observation). In general, however, there is no simple way of recognizing the state set, other than trial and error (which can be often facilitated by intuition and first-hand familiarity with the process at hand). Moreover, the choice of a state set is not unique, and different lines of reasoning may lead to different state sets, not necessarily of equal cardinality.

Fortunately, in the case of sequential networks, there *is* a systematic procedure for constructing the finite-state machine representation, and no trial-and-error search is necessary. Given a sequential network with inputs $x_1(t), x_2(t), \ldots, x_u(t)$ and outputs $z_1(t), z_2(t), \ldots, z_v(t)$ (see Fig. 7-26), the

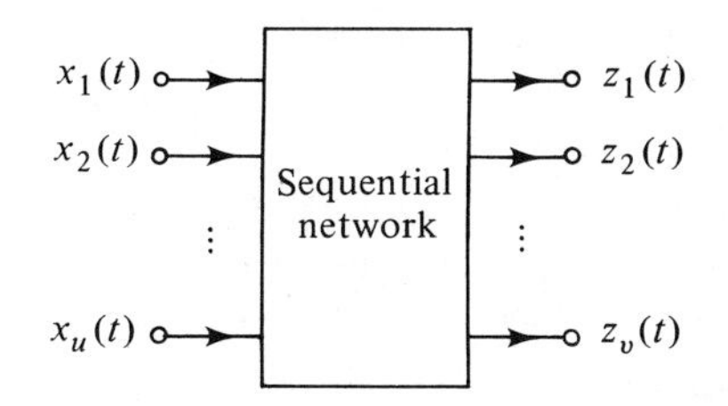

Fig. 7-26 A sequential network

finite-state machine $\mathfrak{M} = \langle X, Z, S, f, g \rangle$ that simulates the network can be constructed as follows:

(i) Let $X = \{0, 1\}^u$ and $x(t) = (x_1(t), x_2(t), \ldots, x_u(t))$.
(ii) Let $Z = \{0, 1\}^v$ and $z(t) = (z_1(t), z_2(t), \ldots, z_v(t))$.
(iii) If the network contains w delay units (Fig. 7-27 shows these units "pulled away" from the rest of the network), denote their outputs by $s_1(t), s_2(t), \ldots, s_w(t)$. Let $S = \{0, 1\}^w$ and $s(t) = (s_1(t), s_2(t), \ldots, s_w(t))$.

To verify that the choice in (iii) is adequate, note that $z_1(t), \ldots, z_v(t), s_1(t+1), \ldots, s_w(t+1)$ are outputs of a combinational network (that which remains after the delay units are pulled away from the original sequential network), whose inputs are $x_1(t), \ldots, x_u(t), s_1(t), \ldots, s_w(t)$. Hence, $z_1(t), \ldots, z_v(t), s_1(t+1), \ldots, s_w(t+1)$ are functions of $x_1(t), \ldots, x_u(t), s_1(t), \ldots, s_w(t)$. This implies the existence of functions f and g such that

$$s(t+1) = f(s(t), x(t))$$
$$z(t) = g(s(t), x(t))$$

where $x(t)$, $z(t)$, and $s(t)$ are as defined in (i), (ii), and (iii) above.

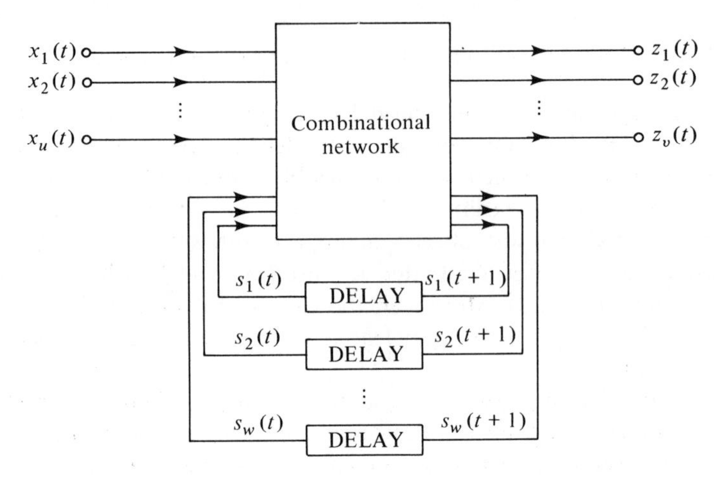

Fig. 7-27 A sequential network with delay units "pulled away"

Example 7-11 illustrates this construction for the sequential network of Fig. 7-23.

While every sequential network can be modeled by a finite-state machine, it is also true that every finite-state machine $\mathfrak{M} = \langle X, Z, S, f, g \rangle$ can be simulated by some sequential network, after an appropriate encoding of its input and output symbols. Assuming that $\#X = l > 1$, $\#Z = m > 1$, and $\#S = n > 1$, this can be done as follows:

(i) Represent the l element of X by l distinct elements from $\{0, 1\}^u$ (where $u = \lceil \log_2 l \rceil$).[1]

(ii) Represent the m elements of Z by m distinct elements from $\{0, 1\}^v$ (where $v = \lceil \log_2 m \rceil$).

(iii) Represent the n elements of S by n distinct elements from $\{0, 1\}^w$ (where $w = \lceil \log_2 n \rceil$).

(iv) Construct a table with $u + v + 2w$ columns, labeled (left to right) $x_1(t), x_2(t), \ldots, x_u(t), s_1(t), s_2(t), \ldots, s_w(t), z_1(t), z_2(t), \ldots, z_v(t), s_1(t + 1), s_2(t + 1), \ldots, s_w(t + 1)$. Fill the first $u + w$ columns with all $(u + w)$-tuples such that the first u elements of each tuple represent an element of X, and the last w elements represent an element of S. If, at any given row, $(x_1(t), \ldots, x_u(t))$ and $(s_1(t), \ldots, s_w(t))$ represent $\xi \in X$ and $\sigma \in S$, respectively, fill the remainder of this row with $(z_1(t), \ldots, z_v(t))$ and $(s_1(t + 1), \ldots, s_w(t + 1))$ representing $g(\sigma, \xi)$ and $f(\sigma, \xi)$, respectively.

(v) Use the table constructed in (iv) as the truth table of a combinational network with inputs $x_1(t), \ldots, x_u(t), s_1(t), \ldots, s_w(t)$ and outputs $z_1(t), \ldots, z_v(t), s_1(t + 1), \ldots, s_w(t + 1)$; realize the corresponding network.

[1] $\lceil a \rceil$ denotes the least integer that equals or exceeds a.

(vi) Incorporate the combinational network constructed in (v) with w delay units as shown in Fig. 7-27. The resulting sequential network simulates $\mathfrak{M}$, with $x(t)$, $z(t)$, and $s(t)$ being represented by $(x_1(t), \ldots, x_u(t))$, $(z_1(t), \ldots, z_v(t))$, and $(s_1(t), \ldots, s_w(t))$, respectively.

Example 7-12

Consider the following game: A coin is tossed repeatedly, and a point is scored for the first tail in every sequence of successive tails, and for each head—except for the first two—in every sequence of successive heads. For example, if the outcomes of the tosses are

$$HTTTHHHHTHTHTTHHH\ldots$$

(where H and T stand for "head" and "tail," respectively), then the corresponding scores (assuming that the initial H is the first in a sequence of successive heads) are

$$NPNNNNPPPNPNPNNNP\ldots$$

(where N and P stand for "no point" and "point," respectively). This game can be modeled by $\mathfrak{M} = \langle X, Z, S, f, g\rangle$, where $X = \{H, T\}$, $Z = \{N, P\}$, and $S = \{1, 2, 3\}$, with the following interpretation: $\mathfrak{M}$ enters state 1 if first tail has appeared in a sequence of tails; $\mathfrak{M}$ enters state 2 when first head appears in a sequence of heads; $\mathfrak{M}$ enters state 3 if two or more heads have appeared in a sequence of heads. The transition table of $\mathfrak{M}$ is shown in Table 7-9.

Table 7-9 TRANSITION TABLE FOR EXAMPLE 7-12

$s(t)$ \ $x(t)$	$s(t+1)$		$z(t)$	
	H	T	H	T
1	2	1	N	N
2	3	1	N	P
3	3	1	P	P

For purpose of simulation, we can encode X, Z, and S as follows:

$$H \longrightarrow 0,\ T \longrightarrow 1$$

$$N \longrightarrow 0,\ P \longrightarrow 1$$

$$1 \longrightarrow 00,\ 2 \longrightarrow 01,\ 3 \longrightarrow 10$$

The transition table of $\mathfrak{M}$, modified by these encodings, is shown in Table 7-10. Table 7-11 shows the truth table that relates $s(t+1)$ and $z(t)$ to $x(t)$

Table 7-10 ENCODED TRANSITION TABLE FOR EXAMPLE 7-12

$x(t)$ \ $s(t)$	$s(t+1)$		$z(t)$	
	0	1	0	1
(0, 0)	(0, 1)	(0, 0)	0	0
(0, 1)	(1, 0)	(0, 0)	0	1
(1, 0)	(1, 0)	(0, 0)	1	1

Table 7-11 TRUTH TABLE FOR EXAMPLE 7-12

	$s(t)$			$s(t+1)$	
$x(t)$	$s_1(t)$	$s_2(t)$	$z(t)$	$s_1(t+1)$	$s_2(t+1)$
0	0	0	0	0	1
0	0	1	0	1	0
0	1	0	1	1	0
1	0	0	0	0	0
1	0	1	1	0	0
1	1	0	1	0	0

and $s(t)$. From this truth table, we have:

$$\begin{aligned} z(t) &= \overline{x(t)}s_1(t)\overline{s_2(t)} + x(t)\overline{s_1(t)}s_2(t) + x(t)s_1(t)\overline{s_2(t)} \\ &= s_1(t)\overline{s_2(t)} + x(t)\overline{s_1(t)}s_2(t) \\ s_1(t+1) &= \overline{x(t)}\,\overline{s_1(t)}s_2(t) + \overline{x(t)}s_1(t)\overline{s_2(t)} \\ &= \overline{x(t)}(s_1(t) \oplus s_2(t)) \\ s_2(t+1) &= \overline{x(t)}\,\overline{s_1(t)}\,\overline{s_2(t)} \end{aligned}$$

The sequential network that simulates $\mathfrak{M}$ is shown in Fig. 7-28. □

Note that, if $\#X$ and $\#S$ are not powers of 2, some $(u + w)$-tuples may be missing from the first $u + w$ columns of the truth table [for example, (0, 1, 1) and (1, 1, 1) are missing from the first three columns of Table 7-11]. These so-called "don't care" combinations may be ignored in the synthesis procedure, since they correspond to input/state combinations that can never occur. Alternatively, the missing rows may be included and completed in such a way as to enhance the simplification of the transmission functions for the $z_i(t)$ and the $s_i(t + 1)$. It should also be pointed out that a judicious choice for the encoding of X, Z, and S may, likewise, lead to simple transmission functions.

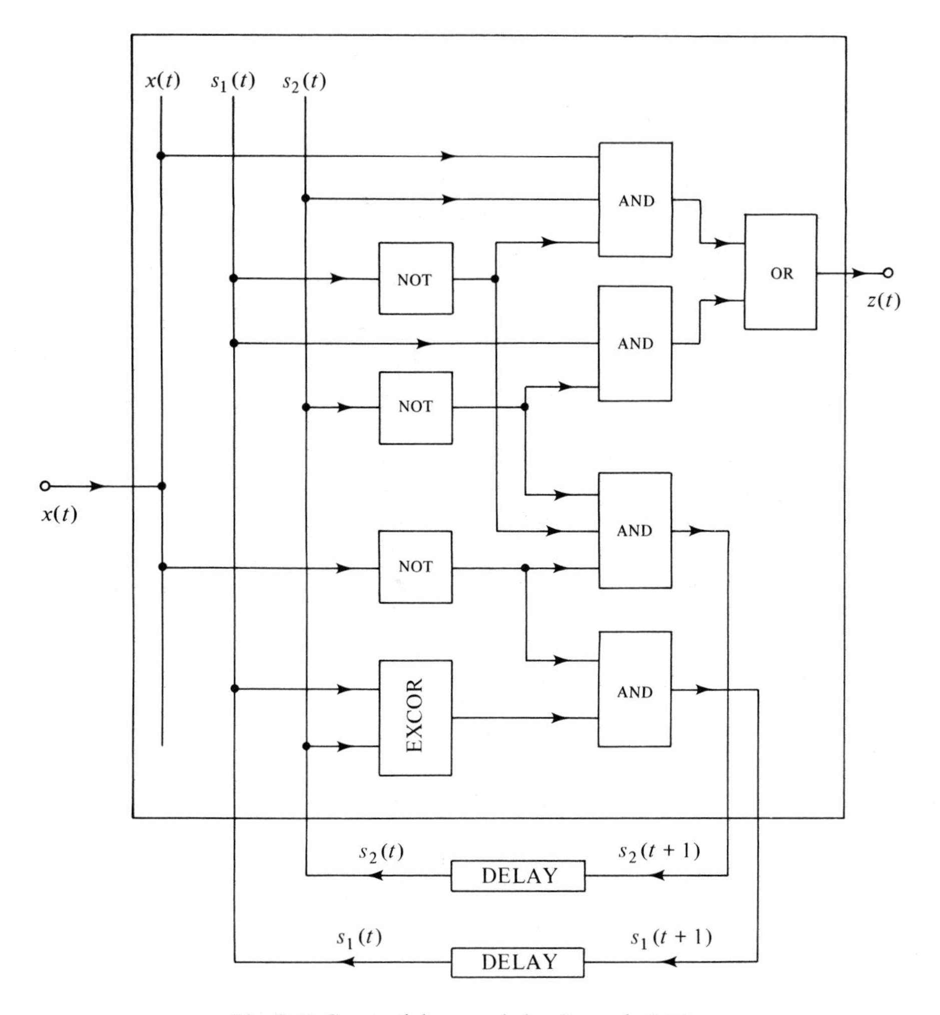

Fig. 7-28 Sequential network for Example 7-12

PROBLEMS

1. Construct sequential networks that simulate the finite-state machines obtained in Problem 2, Sec. 7-9.

2. The box shown in Fig. 7-I is a sequential network characterized as follows:

$$z_1(t) = \begin{cases} 1 & \text{if } x(t) = x(t-1) \\ 0 & \text{otherwise} \end{cases}$$

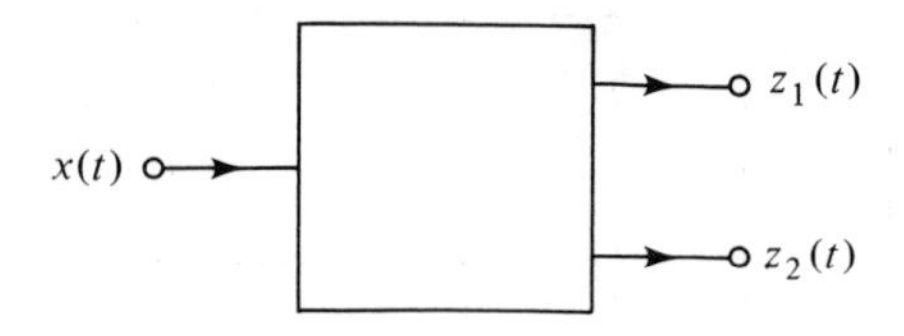

Fig. 7-I

$$z_2(t) = \begin{cases} 1 & \text{if } x(t) = z_1(t-1) \\ 0 & \text{otherwise} \end{cases}$$

(a) Construct a transition table and diagram for a finite-state machine that simulates this network.
(b) Construct the network with standard gates and delay units.
(c) Construct the network with comparator gates and delay units. (A *comparator gate* has two inputs and an output that equals 1 if and only if the inputs are equal.)

3. Construct a sequential network that simulates the finite-state machine of Fig. 7-J.

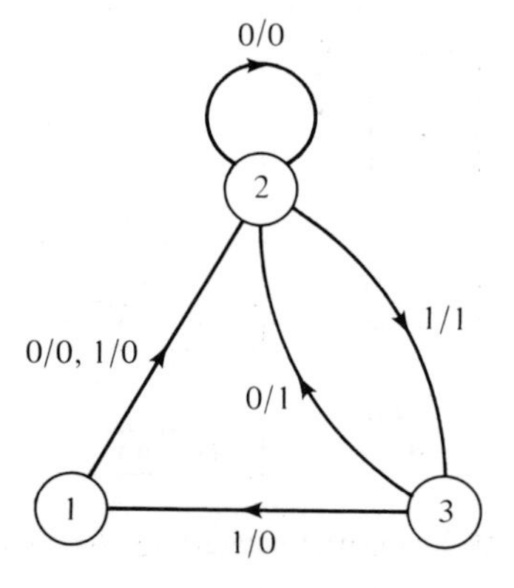

Fig. 7-J

4. Figure 7-K(a) shows the transition diagram of a finite-state machine that simulates the sequential network of Fig. 7-K(b), where the box is a combinational network. Synthesize this combinational network.

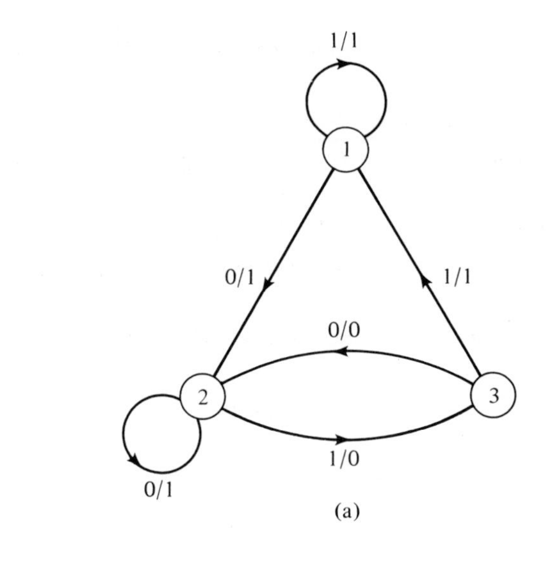

(a)

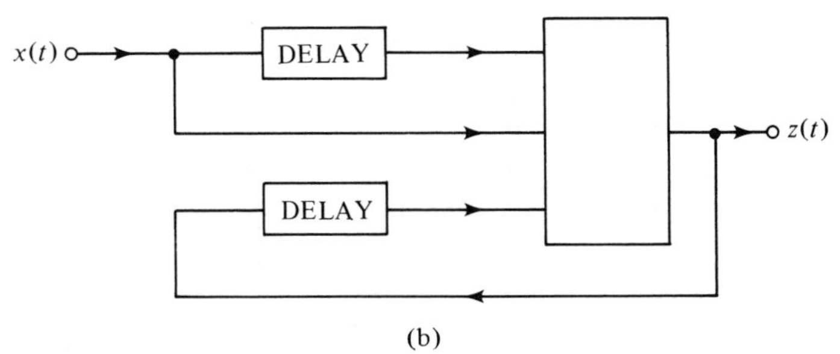

(b)

Fig. 7-K

REFERENCES

BOOTH, J. L., *Sequential Machines and Automata Theory*. New York: Wiley, 1967.

GILL, A., *Introduction to the Theory of Finite-State Machines*. New York: McGraw-Hill, 1962.

HARRISON, M. A., *Introduction to Switching and Automata Theory*. New York: McGraw-Hill, 1965.

HARTMANIS, J., and R. E. STEARNS, *Algebraic Structure Theory of Sequential Machines*. Englewood Cliffs, NJ: Prentice-Hall, 1966.

HENNIE, F. C., *Finite-State Models for Logical Machines*. New York: Wiley, 1968.

HOHN, F. E., *Applied Boolean Algebra*. New York: Macmillan, 1966.

KOHAVI, Z., *Switching and Finite Automata Theory.* New York: McGraw-Hill, 1970.

KORFHAGE, R. R., *Logic and Algorithms.* New York: Wiley, 1966.

MCCLUSKEY, E. J., *Introduction to the Theory of Switching Circuits.* New York: McGraw-Hill, 1965.

MILLER, R. E., *Switching Theory.* New York: Wiley, 1965.

WHITESITT, J. E., *Boolean Algebra and Its Applications.* Reading, MA: Addison-Wesley, 1961.

WOOD, P. E., *Switching Theory.* New York: McGraw-Hill, 1968.

8 LANGUAGES AND AUTOMATA

In this chapter we introduce the concept of a *language* (that is, a set of strings) over an alphabet and, in particular, the concept of a *phrase-structure language*, defined by means of a *grammar*. Special types of phrase-structure languages are defined, with main emphasis placed on *regular languages*. It is shown how regular languages can be recognized by *finite-state automata* (which are variants of the finite-state machines studied in Chapter 7), and it is demonstrated how every regular language can be expressed as a *regular set*. The chapter closes with a brief survey of other types of automata and a discussion of *Turing machines* and *computability*.

8-1. SETS OF STRINGS

In studying finite-state machines, as well as other mathematical models that will be introduced in this chapter, we constantly deal with *sequences* of symbols (for example, sequences of input symbols, output symbols, states, etc.). In this section we introduce notation and discuss some properties of such sequences.

For the purpose of this discussion, a set of elements is referred to as an *alphabet* and its elements as *symbols*. A *string over alphabet* $A = \{a_1, a_2, \ldots a_k\}$ is defined recursively as: A symbol a_i is a string ($i = 1, 2, \ldots, k$); if w_1 and w_2 are strings, so is w_1w_2. That is, a string over A is any ordered sequence of elements from A. For example, *abbca* is a string over $\{a, b, c, d\}$. A string composed of r successive as is often written as a^r.

A string is said to be *finite* if it consists of a finite number of symbols. The number of symbols in a finite string w is referred to as the *length* of w, and is denoted by $|w|$. Thus, if w_1 and w_2 are strings, then

$$|w_1 w_2| = |w_1| + |w_2|$$

The set of all finite strings over A is denoted by A^+.

It is often convenient to talk about the *empty string* (or *null string*), which is defined as the string of length 0 and is denoted by λ. For all strings w, we have, by definition:

$$w\lambda = \lambda w = w$$

The set of all strings over A, including the empty string, is denoted by A^*. Thus,

$$A^* = A^+ \cup \{\lambda\}$$

The string a^r is interpreted as λ when $r = 0$.

THEOREM 8-1

If A is finite, then A^* is denumerable.

Proof Let $A = \{a_1, a_2, \ldots, a_k\}$, and consider the function that transforms any string $a_{i_1} a_{i_2} \ldots a_{i_r} \in A^+$ into the base-$(k + 1)$ integer $i_1 i_2 \ldots i_r$. (For example, the string *abbca* over $\{a, b, c, d\}$ will be transformed into the base-5 integer 12231). Clearly, this function is a bijection from A^+ onto a subset of the set $\mathbb{N}$ of positive integers; hence, A^+ must be denumerable (see Sec. 3-6). Since A^+ is denumerable, so is $A^+ \cup \{\lambda\} = A^*$. $\square$

Under the reasonable assumption that the number of printable characters is finite, Theorem 8-1 implies that the number of distinct printable texts of finite length is denumerable. Thus, if T is any uncountable set (that is, any set whose cardinality is "greater" than that of $\mathbb{N}$), there is no way in which every element of T can be described by means of a finite text. For example, since the set A^* is denumerable, the set $\mathbf{2}^{A^*}$ is uncountable (see Sec. 3-6); hence, not every subset of A^* can be described by means of a finite text. As another example, consider the set of all functions $f\colon \mathbb{I} \longrightarrow \mathbb{I}$ (where $\mathbb{I}$ is the set of integers). Since this set is uncountable (see Sec. 3-6), there must exist some functions $f\colon \mathbb{I} \longrightarrow \mathbb{I}$ that cannot be described by means of finite texts and, hence, whose computation cannot be formulated with any conceivable programming language.

Subsets of A^* can be subjected to the usual operations of complementation, union, and intersection. In addition, the following operation can be defined: If T_1 and T_2 are subsets of A^*, then the *concatenation* of T_1 and T_2, denoted by $T_1 \cdot T_2$ (or simply $T_1 T_2$) is the subset of A^*, defined by

$$T_1 T_2 = \{t_1 t_2 \mid t_1 \in T_1, t_2 \in T_2\}$$

That is, T_1T_2 consists of all strings over A which are formed by letting a string from T_1 be followed by a string from T_2. For example, if $T_1 = \{0, 101\}$ and $T_2 = \{10, 11, 0110\}$, then

$$T_1T_2 = \{010, 011, 00110, 10110, 10111, 1010110\}$$

Note that concatenation is an associative operation: For all subsets T_1, T_2, T_3 of A^*,

$$T_1(T_2T_3) = (T_1T_2)T_3$$

hence, the parentheses are superfluous. Concatenation, however, is not commutative since, in general, $T_1T_2 \neq T_2T_1$. Notable exceptions are these identities, which hold for all $T \subset A^*$:

$$T\{\lambda\} = \{\lambda\}T = T$$
$$T\varnothing = \varnothing T = \varnothing$$

From the definition it also follows that concatenation is distributive over union: For all subsets T_1, T_2, T_3 of A^*,

$$T_1(T_2 \cup T_3) = (T_1T_2) \cup (T_1T_3)$$
$$(T_1 \cup T_2)T_3 = (T_1T_3) \cup (T_2T_3)$$

If T is any subset of A^*, we define $T^0 = \{\lambda\}$. For all $k \geq 1$, T^k is defined in the usual way. Thus, for any $k \geq 0$, A^k consists of all strings over A whose length is k. The *iteration* of T, denoted by T^*, is defined by

$$T^* = \bigcup_{k=0}^{\infty} T^k$$

In particular, A^* consists of all strings over A (including λ), which is precisely the definition given for A^* earlier in this section.

PROBLEMS

1. T_1 and T_2 are arbitrary subsets of A^*. Prove these identities:
(a) $(T_1 \cup T_2)^* = \{\lambda\} \cup \{T_{i_1}T_{i_2} \ldots T_{i_k} \mid i_1, i_2, \ldots, i_k \in \{1, 2\}, k = 1, 2, 3, \ldots\}$
(b) $T_1^*T_2 = T_2 \cup (T_1^*T_1T_2)$
(c) $T_1^*T_2 = (T_1^k)^*(T_2 \cup (T_1T_2) \cup (T_1^2T_2) \cup \ldots \cup (T_1^{k-1}T_2)) \qquad (k > 1)$
(d) $(T_1T_2 = T_2T_1) \Rightarrow ((T_1 \cup T_2)^* = T_1^*T_2^*)$

2. If $T_1 = \{\lambda, 11, 010\}$ and $T_2 = \{0, 01, 1001\}$, compute:
(a) T_1T_2 (b) T_2T_1 (c) $T_1^2 \cup T_2^2$

3. If T_1 and T_2 are as specified in Problem 2, compute the subset of strings of length 5 or less in:
(a) T_1^* (b) T_2^* (c) $(T_1T_2)^*$

4. Figure 8-A shows the transition diagram of a finite-state machine $\mathcal{M}$. Denote by T_{ij} $(i, j \in \{1, 2, 3\})$ the set of all output strings that $\mathcal{M}$ is capable of generating when it is taken from state i to state j. (For example, $T_{22} = \{\lambda, 0, 00, 11, 000, 011, 110, \ldots\}$.) Equate each T_{ij} to an expression composed only of subsets of $\{0, 1\}^*$ and the operations $\cup$, $\cdot$, and $*$. [For example, $T_{22} = (\{0\} \cup \{11\})^*$.]

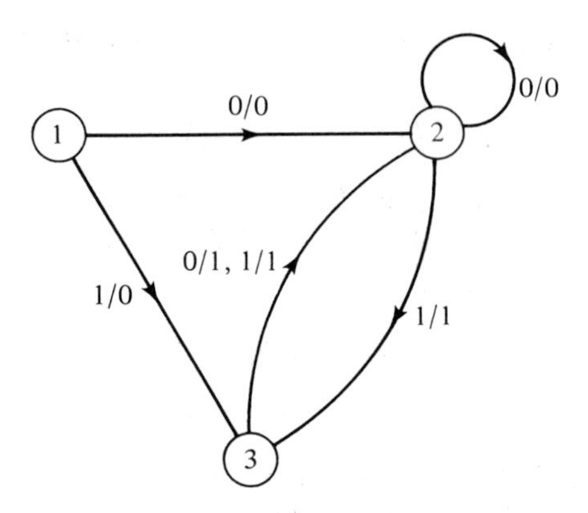

Fig. 8-A

8-2. LANGUAGES

In its everyday usage, the word "language" connotes the aggregate of written or spoken words and the rules by which they can be combined so as to communicate ideas. As used by computer scientists—and as used here—the word "language" has a much more precise and restricted connotation: Given a finite alphabet X, a *language over* X is any subset of X^*. For example, if $X = \{0, 1\}$, the set of strings

$$\{0, 1011, 01000\}$$

$$\{0, 1\} \cup (\{011\}^*\{1\})$$

$$\{w \mid w \in \{0, 1\}^* \text{ and contains a prime number of 1s}\}$$

are languages over X. (Languages thus defined are sometimes referred to as *formal languages*, to distinguish them from *natural languages* such as English, French, etc.)

One of the very basic problems in the study of languages is: Devise an algorithm (that is, a finite sequence of instructions that can be mechanically executed) capable of deciding—in a finite number of steps—whether or not any given string w belongs to the language L. Such an algorithm is called a *recognition algorithm* for L. A language for which there exists a recognition algorithm is said to be *recursive*.

Example 8-1

The language

$$L = \{w \mid w \in \{0, 1\}^* \text{ and contains the same number of 0s and 1s}\}$$

is a recursive language, for which the following is a possible recognition algorithm:

(i) Set m and n to 0. Set k to 1.
(ii) If $k > |w|$, then $w \in L$ if and only if $m = n$. Otherwise:
(iii) If the kth symbol of w is 0, increment m by 1. Otherwise, increment n by 1.
(iv) Increment k by 1 and return to step (ii). □

A task intimately related to that of recognizing a language is that of generating all its strings. A *generating algorithm* for language L is an algorithm that produces all and only those strings which belong to L; moreover, given any $w \in L$, w is producible by the algorithm in a finite number of steps. It is easy to see that, given any alphabet X, there is a generating algorithm for X^*. [If $\#X = k$, then strings over X can be generated as successive base-$(k + 1)$ integers, as is done in the proof of Theorem 8-1.] Given this algorithm, and a recognition algorithm for L over X, we can construct a generating algorithm for L in this manner:

(i) Generate a new string $w \in X^*$ (using the generating algorithm for X^*).
(ii) Decide whether or not w belongs to L (using the recognition algorithm for L). If $w \in L$, generate it.
(iii) Return to step (i).

In conclusion, for every recursive language there exists a generating algorithm.

For some languages there exists no recognition algorithm, but there may exist a *recognition quasi-algorithm*. A recognition quasi-algorithm for language L is an algorithm capable of deciding (in a finite number of steps) that $w \in L$ when this is indeed the case, but not necessarily capable of reaching any decision when $w \notin L$. A language for which there exists a recognition quasi-algorithm is said to be *recursively enumerable*. As it turns out, there do exist recursively enumerable languages that are not recursive.

Recognition algorithms and quasi-algorithms can be described by listing the actual instructions that make up the recognition process (using any convenient programming language). Alternatively, they can be described by an abstract "device," called an *automaton*, capable of carrying out this process by a finite sequence of "mechanical" actions. Such automata are introduced in subsequent sections.

In natural languages (where words play the role of alphabet symbols and sentences play the role of strings), the recognition of grammatically correct sentences is accomplished by the process of *parsing*. This process consists of replacing words with syntactic categories, then replacing groups of these categories with other syntactic categories, and continuing in this fashion until no additional replacements are possible. A sentence is ruled to be grammatical (although not necessarily semantically meaningful) if and only if the process terminates with a single category called "sentence." As an illustration, Fig. 8-1 shows the "parsing tree" of the English sentence

THIS OLD MAN MISSED THE BUS

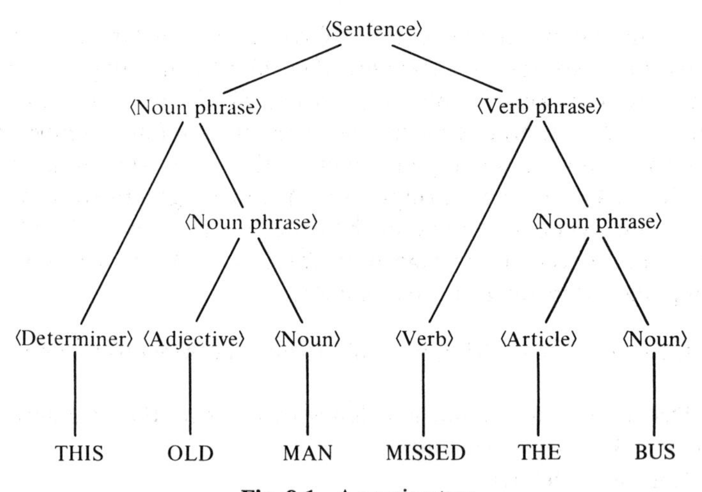

Fig. 8-1 A parsing tree

which, clearly, is a grammatically correct sentence. (In the figure, the words that describe syntactic categories are enclosed in angular brackets, in order to distinguish them from the language words.)

The parsing process is based on a set of rules that specify the manner in which words and categories can be replaced by categories. For example, the parsing tree of Fig. 8-1 makes use of the rules: THIS is a determiner; MAN and BUS are nouns; OLD is an adjective; THE is an article; MISSED is a verb; an adjective or an article followed by a noun is a noun phrase; a determiner followed by a noun phrase is a noun phrase; a verb followed by a noun phrase is a verb phrase; a noun phrase followed by a verb phrase is a sentence. A common way of listing these rules is:

$$\langle \text{Sentence} \rangle \longrightarrow \langle \text{Noun phrase} \rangle \langle \text{Verb phrase} \rangle$$
$$\langle \text{Noun phrase} \rangle \longrightarrow \langle \text{Determiner} \rangle \langle \text{Noun phrase} \rangle$$

⟨Noun phrase⟩ → ⟨Adjective⟩ ⟨Noun⟩
⟨Noun phrase⟩ → ⟨Article⟩⟨Noun⟩
⟨Verb phrase⟩ → ⟨Verb⟩ ⟨Noun phrase⟩
⟨Determiner⟩ → THIS
⟨Adjective⟩ → OLD
⟨Noun⟩ → MAN
⟨Noun⟩ → BUS
⟨Verb⟩ → MISSED
⟨Article⟩ → THE

Note that such a list can also be used to *generate* grammatically correct sentences, by starting with ⟨Sentence⟩ and successively performing any of the listed substitutions (right-hand categories or words replacing left-hand categories) until every category is replaced with a word. For example, the grammatically correct (but semantically nonsensical) sentence OLD BUS MISSED THE BUS can be generated in this manner (where ⇒ indicates a single substitution step):

⟨Sentence⟩ ⇒ ⟨Noun phrase⟩ ⟨Verb phrase⟩
⇒ ⟨Adjective⟩ ⟨Noun⟩ ⟨Verb phrase⟩
⇒ ⟨Adjective⟩ ⟨Noun⟩ ⟨Verb⟩ ⟨Noun phrase⟩
⇒ ⟨Adjective⟩ ⟨Noun⟩ ⟨Verb⟩ ⟨Article⟩ ⟨Noun⟩
⇒ OLD ⟨Noun⟩ ⟨Verb⟩ ⟨Article⟩ ⟨Noun⟩
⇒ OLD BUS ⟨Verb⟩ ⟨Article⟩ ⟨Noun⟩
⇒ OLD BUS MISSED ⟨Article⟩ ⟨Noun⟩
⇒ OLD BUS MISSED THE ⟨Noun⟩
⇒ OLD BUS MISSED THE BUS

A language (either natural or formal) is said to be *ambiguous* if it contains a sentence whose syntax is described by two or more distinct parsing trees. English is certainly an ambiguous language (consider, for example, the parsing of the sentence: "The horse flies like the devil"). In devising formal languages for computer use, it is essential to avoid ambiguities if a unique interpretation of all statements in the language is to result.

PROBLEMS

1. Show that if a language L has a generating algorithm, then it must be recursively enumerable.

2. Let L be a language over X. Show that if L and $X^*\text{-}L$ are both recursively enumerable, then L is recursive.

3. Show that the syntactic rules used to construct the parsing tree of Fig. 8-1 can be used to generate an infinite number of grammatically correct sentences.

4. Construct a set of grammatical rules from which we can generate all sentences of the form:

I RAN FASTER AND FASTER AND FASTER . . . AND FASTER

5. Show that the syntax of the sentence "The horse flies like the devil" can be described by two distinct parsing trees.

6. Consider a variant of FORTRAN, where a DO statement requires a blank (but not a comma) between the initial and final values of the loop index (for example, DO 5 I = 1 10). Is this variant ambiguous? Justify your answer.

8-3. PHRASE-STRUCTURE LANGUAGES

Languages whose strings are generable by a set of replacement rules such as exemplified in the preceding section, are called *phrase-structure languages.* The grammars that dominate such languages are called *phrase-structure grammars.* Besides replacement rules, which are commonly referred to as *productions*, a phrase-structure grammar is characterized by a set of elementary symbols called *terminals* (representing the vocabulary of the language), and a set of symbols called *nonterminals* (representing syntactic categories), of which the so-called *start symbol* (representing the category ⟨Sentence⟩) can generate all strings in the language.

Formally, a *phrase-structure grammar* (henceforth referred to simply as *grammar*) consists of:

(a) A finite nonempty set N of *nonterminals.*
(b) A finite nonempty set T of *terminals* ($T \cap N = \varnothing$).
(c) A finite nonempty set P of *productions*, each of the form[1] $\alpha \longrightarrow \beta$, where $\alpha \in (N \cup T)^+$, $\beta \in (N \cup T)^*$
(d) A *start symbol* $\sigma \in N$.

We denote a grammar $\mathcal{G}$ consisting of these entities by

$$\mathcal{G} = \langle N, T, P, \sigma \rangle$$

If $\gamma, \delta \in (N \cup T)^*$ and $(\alpha \longrightarrow \beta) \in P$, we say that $\gamma\alpha\delta$ *directly derives* $\gamma\beta\delta$ *in* $\mathcal{G}$, and write $\gamma\alpha\delta \underset{\mathcal{G}}{\Longrightarrow} \gamma\beta\delta$. If $\alpha_i \underset{\mathcal{G}}{\Longrightarrow} \alpha_{i+1}$ for $i = 1, 2, \ldots, r-1$, we say that α_i *derives* α_r *in* $\mathcal{G}$, and write

$$\alpha_1 \underset{\mathcal{G}}{\Longrightarrow} \alpha_2 \underset{\mathcal{G}}{\Longrightarrow} \alpha_3 \underset{\mathcal{G}}{\Longrightarrow} \cdots \underset{\mathcal{G}}{\Longrightarrow} \alpha_{r-1} \underset{\mathcal{G}}{\Longrightarrow} \alpha_r \tag{8-1}$$

Thus, α_1 derives α_r in $\mathcal{G}$ if α_r is producible from α_1 by employing a finite

[1]See Sec. 8-1 for the meaning of $^+$ and *.

number of productions from $\mathcal{G}$. An expression of the form (8-1) is called a *derivation in* $\mathcal{G}$, and can be abbreviated into the expression $\alpha_1 \overset{*}{\underset{\mathcal{G}}{\Rightarrow}} \alpha_r$. For completeness, we use the convention that, for any α, $\alpha \overset{*}{\underset{\mathcal{G}}{\Rightarrow}} \alpha$. (The $\mathcal{G}$ in $\underset{\mathcal{G}}{\Rightarrow}$ and $\overset{*}{\underset{\mathcal{G}}{\Rightarrow}}$ can be omitted whenever understood.)

The language generated by $\mathcal{G}$, denoted by $L(\mathcal{G})$, is defined by

$$L(\mathcal{G}) = \{w \mid w \in T^*, \sigma \overset{*}{\underset{\mathcal{G}}{\Rightarrow}} w\}$$

That is, $L(\mathcal{G})$ consists of all strings of terminals (or *terminal strings*) which can be derived in $\mathcal{G}$ from the start symbol σ.

Grammars $\mathcal{G}_1$ and $\mathcal{G}_2$ are said to be *equivalent* (written $\mathcal{G}_1 \sim \mathcal{G}_2$) if $L(\mathcal{G}_1) = L(\mathcal{G}_2)$.

Example 8-2

Consider the grammar $\mathcal{G} = \langle N, T, P, \sigma \rangle$, where $N = \{\sigma, A\}$, $T = \{0, 1\}$, and

$$P = \{\sigma \longrightarrow 0\sigma, \sigma \longrightarrow 1A, \sigma \longrightarrow 0, \sigma \longrightarrow 1, A \longrightarrow 0\sigma, A \longrightarrow 0\}$$

The following are some possible derivations in $\mathcal{G}$:

$$\sigma \Longrightarrow 0\sigma \Longrightarrow 00\sigma \Longrightarrow 000\sigma \Longrightarrow 0001A \Longrightarrow 00010\sigma \Longrightarrow 000101$$

$$\sigma \Longrightarrow 1A \Longrightarrow 10\sigma \Longrightarrow 101A \Longrightarrow 1010\sigma \Longrightarrow 10100$$

A close look at P reveals that

$$L(\mathcal{G}) = \{w \mid w \in \{0, 1\}^+, w \text{ contains no adjacent 1s}\}$$ □

The following are three special types of grammars that have been studied extensively because of their theoretical interest and practical ramifications (especially with regard to programming languages):

(a) *Context-sensitive grammars* are those in which all productions are of the form

$$\alpha_1 A \alpha_2 \longrightarrow \alpha_1 \beta \alpha_2, \text{ where } \alpha_1, \alpha_2 \in (N \cup T)^*, A \in N, \beta \in (N \cup T)^+$$

(The string β replaces A whenever A appears in the context of α_1 and α_2). Languages generated by such grammars are called *context-sensitive languages.*

(b) *Context-free grammars* are those in which all productions are of the form

$$A \longrightarrow \beta, \text{ where } A \in N, \beta \in (N \cup T)^+$$

Languages generated by such grammars are called *context-free languages.*

(c) *Regular grammars* are those in which all productions are of the form

$$A \longrightarrow aB \quad \text{or} \quad A \longrightarrow a, \text{ where } A, B \in N, a \in T$$

Languages generated by such grammars are called *regular languages.*

In all of these grammars, the production $\sigma \longrightarrow \lambda$ (where σ is the start symbol and λ is the empty string) is permitted, *provided σ does not appear on the right side of any production.*

Note that regular grammars are special cases of context-free grammars that, in turn, are special cases of context-sensitive grammars. Thus, every regular language is also a context-free and a context-sensitive language, and every context-free language is a context-sensitive language.

THEOREM 8-2

If L is a context-sensitive language, or a context-free language, or a regular language, then $L \cup \{\lambda\}$ is a context-sensitive language, or a context-free language, or a regular language, respectively.

Proof Let L be a context-sensitive, or a context-free, or a regular language, generated by $\mathcal{G} = \langle N, T, P, \sigma \rangle$. Construct a new grammar $\mathcal{G}' = \langle N', T', P', \sigma' \rangle$ (generating the language L'), where

$$N' = N \cup \{\sigma'\}$$
$$T' = T$$
$$P' = P \cup \{\sigma' \longrightarrow \alpha \mid (\sigma \longrightarrow \alpha) \in P\} \cup \{\sigma' \longrightarrow \lambda\}$$

That is, for each production $\sigma \longrightarrow \alpha$ in P, P' contains a production $\sigma' \longrightarrow \alpha$; in addition, P' contains all the P productions and the production $\sigma' \longrightarrow \lambda$. Since σ' (the start symbol of $\mathcal{G}'$) does not appear on the right side of any production, $\mathcal{G}'$ is of the same type as $\mathcal{G}$, and L' is of the same type as L. Now, suppose L has the derivation $\sigma \overset{*}{\underset{\mathcal{G}}{\Rightarrow}} w$. If the first production is $\sigma \underset{\mathcal{G}}{\Rightarrow} \alpha$, then $\sigma \underset{\mathcal{G}}{\Rightarrow} \alpha \overset{*}{\underset{\mathcal{G}}{\Rightarrow}} w$. Hence, L' must have the derivation $\sigma' \underset{\mathcal{G}'}{\Rightarrow} \alpha \overset{*}{\underset{\mathcal{G}'}{\Rightarrow}} w$, and $\sigma' \overset{*}{\underset{\mathcal{G}'}{\Rightarrow}} w$. Thus, $L \subset L'$.

Conversely, suppose L' has the derivation $\sigma' \overset{*}{\underset{\mathcal{G}'}{\Rightarrow}} w$ $(w \neq \lambda)$. If the first production is $\sigma' \underset{\mathcal{G}'}{\Rightarrow} \alpha$, then $\sigma' \underset{\mathcal{G}'}{\Rightarrow} \alpha \overset{*}{\underset{\mathcal{G}'}{\Rightarrow}} w$ (where α does not contain σ'). Hence, L' must have the derivation $\sigma \underset{\mathcal{G}}{\Rightarrow} \alpha \overset{*}{\underset{\mathcal{G}}{\Rightarrow}} w$, and $\sigma \overset{*}{\underset{\mathcal{G}}{\Rightarrow}} w$. Thus, $L' - \{\lambda\} \subset L$. In conclusion, $L' = L \cup \{\lambda\}$, and L' is of the same type as L. □

Example 8-3

In Example 8-2, a regular grammar $\mathcal{G} = \langle N, T, P, \sigma \rangle$ was specified which generates the language

$$L = \{w \mid w \in \{0, 1\}^+, w \text{ contains no adjacent 1s}\}$$

By Theorem 8-1, the language

$$L' = \{w \mid w \in \{0, 1\}^*, w \text{ contains no adjacent 1s}\} = L \cup \{\lambda\}$$

is generated by $\mathcal{G}' = \langle N', T', P', \sigma' \rangle$, where

$$N' = \{\sigma', \sigma, A\}$$
$$T' = \{0, 1\}$$
$$P' = \{\sigma' \longrightarrow \lambda, \sigma' \longrightarrow 0\sigma, \sigma' \longrightarrow 1A, \sigma' \longrightarrow 0, \sigma' \longrightarrow 1, \sigma \longrightarrow 0\sigma, \sigma \longrightarrow 1A, \sigma \longrightarrow 0, \sigma \longrightarrow 1, A \longrightarrow 0\sigma, A \longrightarrow 0\}$$

□

PROBLEMS

1. Consider the language L, generated by the grammar

$$\mathcal{G} = \langle \{\sigma\}, \{a, b, c, 0, 1\}, P, \sigma \rangle$$

where

$$P = \{\sigma \longrightarrow a, \sigma \longrightarrow b, \sigma \longrightarrow c, \sigma \longrightarrow \sigma a, \sigma \longrightarrow \sigma c, \sigma \longrightarrow \sigma 0, \sigma \longrightarrow \sigma 1\}$$

Wherever possible, give derivations for these strings in $\mathcal{G}$:
(a) a (b) $ab0$ (c) $a0c01$ (d) $0a$
(e) 11 (f) aaa

2. Consider the language L, generated by the grammar

$$\mathcal{G} = \langle \{\sigma, A, B\}, \{0, 1\}, P, \sigma \rangle$$

where

$$P = \{\sigma \longrightarrow 0B, \sigma \longrightarrow 1A, A \longrightarrow 0, A \longrightarrow 0\sigma, A \longrightarrow 1AA, B \longrightarrow 1, B \longrightarrow 1\sigma, B \longrightarrow 0BB\}$$

Construct all strings $w \in L$ such that $|w| \leq 6$. Can you express L in the form $L = \{w \mid \ldots\}$?

3. Construct grammars that generate the languages:
(a) $\{0\}^+$ (b) $\{0\}^*$ (c) $\{0, 1\}^+$ (d) $\{0, 1\}^*$
(e) $\{w \mid |w| = 3k, k = 1, 2, 3, \ldots\}$
(f) $\{w \mid |w| = 3k, k = 0, 1, 2, \ldots\}$
(g) $\{u^i \mid u \in \{0, 1\}^2, i = 1, 2, 3, \ldots\}$
(h) $\{u^i \mid u \in \{0, 1\}^2, i = 0, 1, 2, \ldots\}$

4. Using the form $L(\mathcal{G}) = \{w \mid \ldots\}$, describe the languages generated by the grammars

$$\mathcal{G} = \langle \{\sigma, A\}, \{0, 1\}, P, \sigma \rangle$$

where:
(a) $P = \{\sigma \longrightarrow A0,\ A \longrightarrow 1A,\ A0 \longrightarrow 10\}$
(b) $P = \{\sigma \longrightarrow 1\sigma,\ \sigma \longrightarrow 1A,\ A \longrightarrow 0A,\ A \longrightarrow 0\}$
(c) $P = \{\sigma \longrightarrow \sigma 0,\ \sigma \longrightarrow A1,\ A \longrightarrow 0A0,\ A \longrightarrow 1\}$
(d) $P = \{\sigma \longrightarrow 0\sigma 1,\ \sigma \longrightarrow 01\}$
In each case, indicate whether the grammar is context-sensitive, context-free, or regular.

5. Show that the grammar of Problem 4(a) is equivalent to a regular grammar (and, hence, that the corresponding language is actually a regular language).

6. L is the set of all strings composed of the 26 letters of the alphabet which start with UN and end with D. Find a grammar which generates L.

7. Find grammars that generate the decimal representations of:
(a) All integers (with leading zeros allowed).
(b) All even integers (with leading zeros allowed).
(c) All integers (with leading zeros not allowed).
(d) All even integers (with leading zeros not allowed).

8. Given a grammar $\mathcal{G} = \langle N, T, P, \sigma \rangle$, a relation $\underset{\mathcal{G}}{\Rightarrow}$ on $(N \cup T)^*$ can be defined as $\alpha \underset{\mathcal{G}}{\Rightarrow} \beta$ if and only if α directly derives β in $\mathcal{G}$. Similarly, a relation $\overset{*}{\underset{\mathcal{G}}{\Rightarrow}}$ on $(N \cup T)^*$ can be defined as $\alpha \overset{*}{\underset{\mathcal{G}}{\Rightarrow}} \beta$ if and only if α derives β in $\mathcal{G}$. Show that $\overset{*}{\underset{\mathcal{G}}{\Rightarrow}}$ is the reflexive transitive closure of $\underset{\mathcal{G}}{\Rightarrow}$ (see Sec. 2-5).

8-4. REGULAR LANGUAGES AND FINITE-STATE AUTOMATA

In the hierarchy of grammars defined in the preceding section, the regular grammar is the most restricted and the easiest to deal with. In the present section we describe devices, called *finite-state automata*, that are capable of executing algorithms for recognizing regular languages. We thus demonstrate that regular languages are recursive.

A finite-state automaton is a finite-state machine (see Sec. 7-9) with these special characteristics: It has a uniquely designated initial state; its output alphabet is $\{0, 1\}$; its output at any time depends only on the state at that time. Thus, the output function of such a machine can be specified simply by listing all those states that yield output 1. (These are called "accepting" or "final" states.) An input sequence is "recognized" (or "accepted") if it takes the machine from the initial state to any of the accepting states.

Formally, a *finite-state automaton* (abbreviated fsa) consists of:

(a) A finite nonempty set S called the *state set* (and consisting of *states*).
(b) A finite nonempty set X called the *input alphabet* (and consisting of *input symbols*).

(c) A function $f\colon S \times X \longrightarrow S$ called the *next-state function*.
(d) An element $\sigma_0 \in S$ called the *initial state*.
(e) A set $F \subset S$ of *accepting* (or *final*) *states*.

We denote a fsa $\mathfrak{M}$ consisting of these entities by

$$\mathfrak{M} = \langle S, X, f, \sigma_0, F \rangle$$

Following the convention established in Sec. 7-9, f also denotes the *extended next-state function* $S \times X^* \longrightarrow S$.

The transition diagram of a fsa is usually drawn with the edges labeled with input symbols only; the vertices of accepting states are designated by double circles, and an arrow head is attached to the initial state. As an example, Fig. 8-2 shows the transition diagram of a fsa

$$\mathfrak{M} = \langle \{\sigma_0, \sigma_1, \sigma_2\}, \{0, 1\}, f, \sigma_0, \{\sigma_0, \sigma_2\} \rangle$$

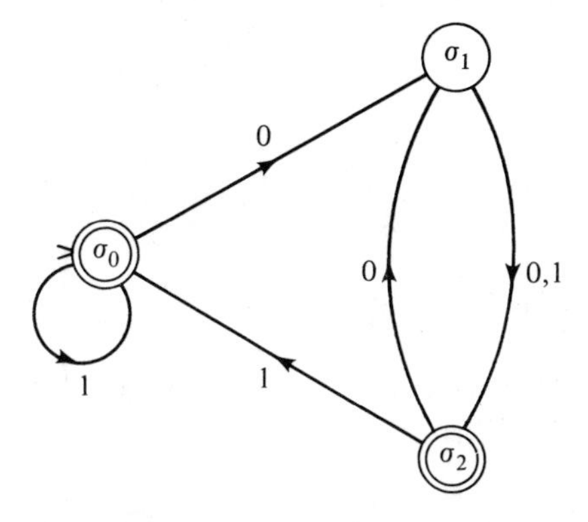

Fig. 8-2 A finite-state automaton

A string $w \in X^*$ is said to be *accepted* by $\mathfrak{M}$ if $f(\sigma_0, w) \in F$. The *accepted set* of $\mathfrak{M}$, denoted by $T(\mathfrak{M})$, is the set of all strings accepted by $\mathfrak{M}$; that is,

$$T(\mathfrak{M}) = \{w \mid f(\sigma_0, w) \in F\}$$

In terms of the transition diagram of $\mathfrak{M}$, $T(\mathfrak{M})$ is represented by all paths that initiate in vertex σ_0 and terminate in any of the doubly-circled vertices.[1] For example, the accepted set of the fsa of Fig. 8-2 contains the strings λ, 1, 00, 01, 11, 001, 011, 100, 101, 111 (and an infinite number of other strings).

THEOREM 8-3

The accepted set of every finite-state automaton is a regular language.

[1] In this chapter, a path of "length 0" (that is, a path consisting of l edges, where $l = 0$) is admitted as a legitimate path. Every vertex reaches itself via a path of length 0.

Proof Consider the fsa $\mathfrak{M} = \langle S, X, f, \sigma_0, F \rangle$. Construct the regular grammar $\mathcal{G} = \langle N, T, P, \sigma \rangle$, where

$$N = S$$

$$T = X$$

$$\sigma = \sigma_0$$

$$P = \{\sigma_i \longrightarrow \xi\sigma_j \mid f(\sigma_i, \xi) = \sigma_j\} \cup \{\sigma_i \longrightarrow \xi \mid f(\sigma_i, \xi) \in F\}$$

Now, $w = \xi_1\xi_2 \ldots \xi_r \in T(\mathfrak{M})$ $(w \neq \lambda)$ if and only if P contains the productions $\sigma_0 \longrightarrow \xi_1\sigma_1, \sigma_1 \longrightarrow \xi_2\sigma_2, \ldots, \sigma_{r-2} \longrightarrow \xi_{r-1}\sigma_{r-1}, \sigma_{r-1} \longrightarrow \xi_r$ (see Fig. 8-3); hence, if and only if $\mathcal{G}$ has the derivation

$$\sigma_0 \Rightarrow \xi_1\sigma_1 \Rightarrow \xi_1\xi_2\sigma_2 \Rightarrow \ldots \Rightarrow \xi_1\xi_2 \ldots \xi_{r-1}\sigma_{r-1} \Rightarrow \xi_1\xi_2 \ldots \xi_r = w$$

and, hence, if and only if $w \in L(\mathcal{G})$. Thus, $T(\mathfrak{M}) - \{\lambda\} = L(\mathcal{G})$; hence, $T(\mathfrak{M}) - \{\lambda\}$ is regular. By Theorem 8-2, if $T(\mathfrak{M}) - \{\lambda\}$ is regular, so is $T(\mathfrak{M})$. [If $\sigma_0 \notin F$, $T(\mathfrak{M}) = L(\mathcal{G})$; otherwise, $T(\mathfrak{M}) = L(\mathcal{G}')$, where $\mathcal{G}'$ is as specified in the proof of Theorem 8-2.] □

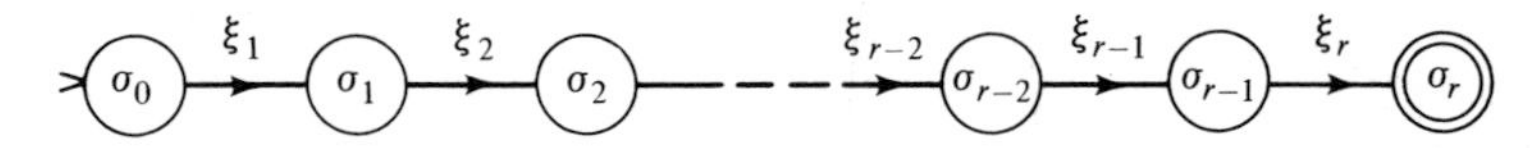

Fig. 8-3 For proof of Theorem 8-3

The proof of Theorem 8-3 suggests a procedure for constructing a grammar that generates the accepted set of any given fsa. The following illustrates this procedure.

Example 8-4

Figure 8-4 shows the transition diagram of a fsa $\mathfrak{M} = \langle \{\sigma_0, \sigma_1\}, \{0, 1\}, f, \sigma_0, \{\sigma_0\} \rangle$, where

$$T(\mathfrak{M}) = \{w \mid w \text{ contains an even number of 1s}\}$$

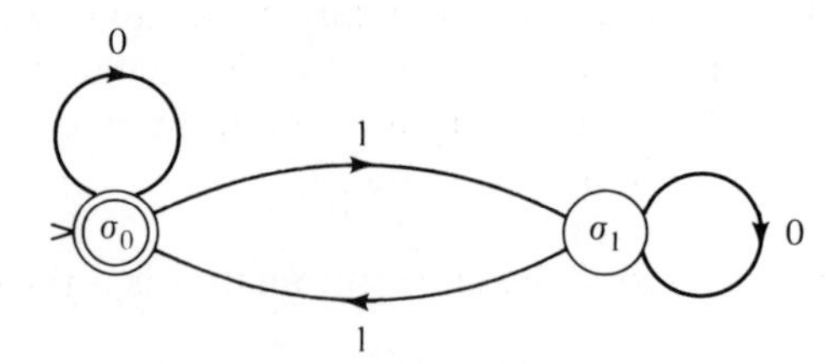

Fig. 8-4 Fsa for Example 8-4

The grammar generating $T(\mathfrak{M}) - \{\lambda\}$ is $\mathcal{G} = \langle\{\sigma_0, \sigma_1\}, \{0, 1\}, P, \sigma_0\rangle$, where

$$P = \{\sigma_0 \longrightarrow 0\sigma_0, \sigma_0 \longrightarrow 1\sigma_1, \sigma_0 \longrightarrow 0, \sigma_1 \longrightarrow 0\sigma_1, \sigma_1 \longrightarrow 1\sigma_0, \sigma_1 \longrightarrow 1\}$$

The grammar generating $T(\mathfrak{M})$ is $\mathcal{G}' = \langle\{\sigma, \sigma_0, \sigma_1\}, \{0, 1\}, P', \sigma\rangle$, where

$$P' = \{\sigma \longrightarrow \lambda, \sigma \longrightarrow 0\sigma_0, \sigma \longrightarrow 1\sigma_1, \sigma \longrightarrow 0, \sigma_0 \longrightarrow 0\sigma_0,$$
$$\sigma_0 \longrightarrow 1\sigma_1, \sigma_0 \longrightarrow 0, \sigma_1 \longrightarrow 0\sigma_1, \sigma_1 \longrightarrow 1\sigma_0, \sigma_1 \longrightarrow 1\} \qquad \square$$

Of greater interest to us than Theorem 8-3 is its converse—namely, that for every regular language there exists a fsa which accepts this language. Before proving such a result, however, it is necessary to introduce a generalization of a fsa, called a *nondeterministic finite-state automaton* (abbreviated nfsa). A nfsa $\mathfrak{M} = \langle S, X, f, \sigma_0, F\rangle$ is defined in the same manner as a fsa, except that the next-state function has the power set of S (rather than the set S) as its range; that is,

$$f\colon \quad S \times X \longrightarrow \mathbf{2}^S$$

Thus, for any $\sigma \in S$ and $\xi \in X$, $f(\sigma, \xi)$ is a *set* of states (which may be empty), rather than a single state. In the transition diagram of $\mathfrak{M}$ this is reflected by the fact that, for any $\xi \in X$, a vertex can originate more than one edge labeled ξ (or none at all). As an example, Fig. 8-5 shows the transition diagram of a nfsa

$$\mathfrak{M} = \langle\{\sigma_0, \sigma_1, \sigma_2, \sigma_3\}, \{0, 1\}, f, \sigma_0, \{\sigma_0, \sigma_2\}\rangle$$

For nfsa's, as for fsa's, the next-state function f can be extended to the domain $S \times X^*$ in the usual manner. In addition, it is also convenient to

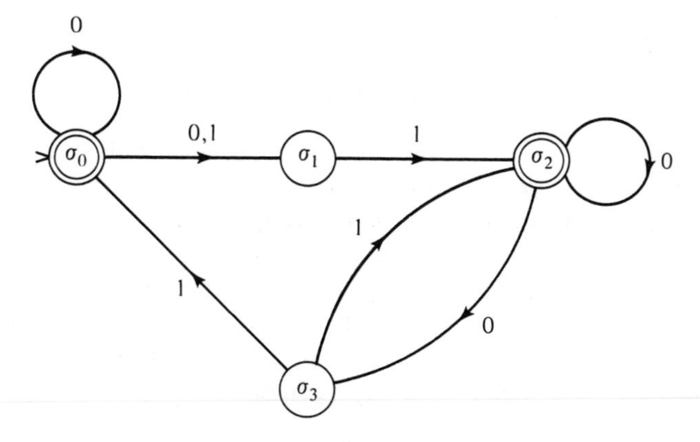

Fig. 8-5 A nondeterministic finite-state automaton

extend f to the domains $\mathbf{2}^S \times X$ and $\mathbf{2}^S \times X^*$ such that, for any $\{\sigma_1, \sigma_2, \ldots, \sigma_r\} \subset S$, $\xi \in X$ and $w \in X^*$,

$$f(\{\sigma_1, \sigma_2, \ldots, \sigma_r\}, \xi) = \bigcup_{i=1}^r f(\sigma_i, \xi)$$
$$f(\{\sigma_1, \sigma_2, \ldots, \sigma_r\}, w) = \bigcup_{i=1}^r f(\sigma_i, w)$$

Since any two of these functions coincide at the intersection of their domains, the same symbol f can be used for all of them. Note that every fsa is a nfsa; hence, the preceding notation can also be applied to fsa's.

A string $w \in X^*$ is said to be *accepted* by the nfsa $\mathfrak{M}$ if $[f(\sigma_0, w)] \cap F \neq \varnothing$; that is, if in the transition diagram of $\mathfrak{M}$ there is *at least one* path that initiates in vertex σ_0 and terminates in a doubly circled vertex. The *accepted set* of $\mathfrak{M}$, accordingly, is given by

$$T(\mathfrak{M}) = \{w \mid [f(\sigma_0, w)] \cap F \neq \varnothing\}$$

For example, the accepted set of the nfsa of Fig. 8-5 contains the strings λ, 0, 00, 01, 11, 000, 001, 010, 011, 110 (and an infinite number of other strings).

THEOREM 8-4

Given any nondeterministic finite-state automaton $\mathfrak{M}$, there exists a finite-state automaton $\tilde{\mathfrak{M}}$ such that $T(\mathfrak{M}) = T(\tilde{\mathfrak{M}})$.

Proof Let $\mathfrak{M} = \langle S, X, f, \sigma_0, F\rangle$ be a nfsa. Construct a fsa $\tilde{\mathfrak{M}} = \langle \tilde{S}, X, \tilde{f}, \tilde{\sigma}_0, \tilde{F}\rangle$ as follows:

$$\tilde{S} = \mathbf{2}^S$$

(that is, states in $\tilde{\mathfrak{M}}$ represent sets of states in $\mathfrak{M}$); for any $\{\sigma_1, \sigma_2, \ldots, \sigma_r\} \in \tilde{S}$ and $\xi \in X$, $\tilde{f}(\{\sigma_1, \sigma_2, \ldots, \sigma_r\}, \xi) = \{\sigma'_1, \sigma'_2, \ldots, \sigma'_s\}$ if and only if $f(\{\sigma_1, \sigma_2, \ldots, \sigma_r\}, \xi) = \{\sigma'_1, \sigma'_2, \ldots, \sigma'_s\}$; for all $\xi \in X$, $\tilde{f}(\varnothing, \xi) = \varnothing$;

$$\tilde{\sigma}_0 = \{\sigma_0\}$$
$$\tilde{F} = \{\{\sigma_1, \sigma_2, \ldots, \sigma_r\} \mid \{\sigma_1, \sigma_2, \ldots, \sigma_r\} \cap F \neq \varnothing\}$$

(that is, state $\{\sigma_1, \sigma_2, \ldots, \sigma_r\}$ in $\tilde{S}$ is accepting if and only if at least one of the states $\sigma_1, \sigma_2, \ldots, \sigma_r$ in S is accepting).

By induction on $|w|$ we now prove that, for any $w \in X^*$, $f(\sigma_0, w) = \{\sigma_1, \sigma_2, \ldots, \sigma_r\}$ if and only if $\tilde{f}(\{\sigma_0\}, w) = \{\sigma_1, \sigma_2, \ldots, \sigma_r\}$. (*Basis*). When $|w| = 0$, $f(\sigma_0, w) = \tilde{f}(\{\sigma_0\}, w) = \{\sigma_0\}$. (*Induction step*). The induction hypothesis is that, for any w such that $|w| = k$, $f(\sigma_0, w) = \{\sigma_1, \sigma_2, \ldots, \sigma_r\}$ if and only if $\tilde{f}(\{\sigma_0\}, w) = \{\sigma_1, \sigma_2, \ldots, \sigma_r\}$. Now, consider any string $w\xi \in X^{k+1}$, where $|w| = k$ and $\xi \in X$. By induction hypothesis,

$$f(\sigma_0, w\xi) = f(f(\sigma_0, w), \xi) = f(\{\sigma_1, \sigma_2, \ldots, \sigma_r\}, \xi)$$

if and only if

$$\tilde{f}(\{\sigma_0\}, w\xi) = \tilde{f}(\tilde{f}(\{\sigma_0\}, w), \xi) = \tilde{f}(\{\sigma_1, \sigma_2, \ldots, \sigma_r\}, \xi)$$

But, by definition of $\tilde{f}, f(\{\sigma_1, \sigma_2, \ldots, \sigma_r\}, \xi) = \{\sigma'_1, \sigma'_2, \ldots, \sigma'_s\}$ if and only if $\tilde{f}(\{\sigma_1, \sigma_2, \ldots, \sigma_r\}, \xi) = \{\sigma'_1, \sigma'_2, \ldots, \sigma'_s\}$. Hence, $f(\sigma_0, w\xi) = \{\sigma'_1, \sigma'_2, \ldots, \sigma'_s)$ if and only if $\tilde{f}(\{\sigma_0\}, w\xi) = \{\sigma'_1, \sigma'_2, \ldots, \sigma'_s\}$. This completes the induction.

Now, let w be any string in X^*, and let $f(\sigma_0, w) = \tilde{f}(\{\sigma_0\}, w) = \{\sigma_1, \sigma_2, \ldots, \sigma_r\}$. Then $w \in T(\mathfrak{M})$ if and only if $\{\sigma_1, \sigma_2, \ldots, \sigma_r\} \cap F \neq \varnothing$; hence, if and only if $\{\sigma_1, \sigma_2, \ldots, \sigma_r\} \in \tilde{F}$ and, hence, if and only if $w \in T(\tilde{\mathfrak{M}})$. □

The proof of Theorem 8-4 shows how, given any nfsa $\mathfrak{M}$, we can construct a fsa $\mathfrak{M}$ that has the same accepted set as $\mathfrak{M}$. As an illustration, Fig. 8-6 shows the fsa that has the same accepted set as the nfsa of Fig. 8-5. Note that, in the construction of $\tilde{\mathfrak{M}}$, not all elements of $\tilde{S}$ may be needed; all those states which are not reachable from $\{\sigma_0\}$ (such as the states $\{\sigma_3\}$, $\{\sigma_0, \sigma_3\}$, $\{\sigma_1, \sigma_2, \sigma_3\}$, etc.) are clearly irrelevant insofar as $T(\tilde{\mathfrak{M}})$ is concerned, and can be ignored.

Theorem 8-4 shows that the class of languages accepted by nfsa's is the same as the class of languages accepted by fsa's and, hence, that nfsa's are by no means "more powerful" than fsa's.

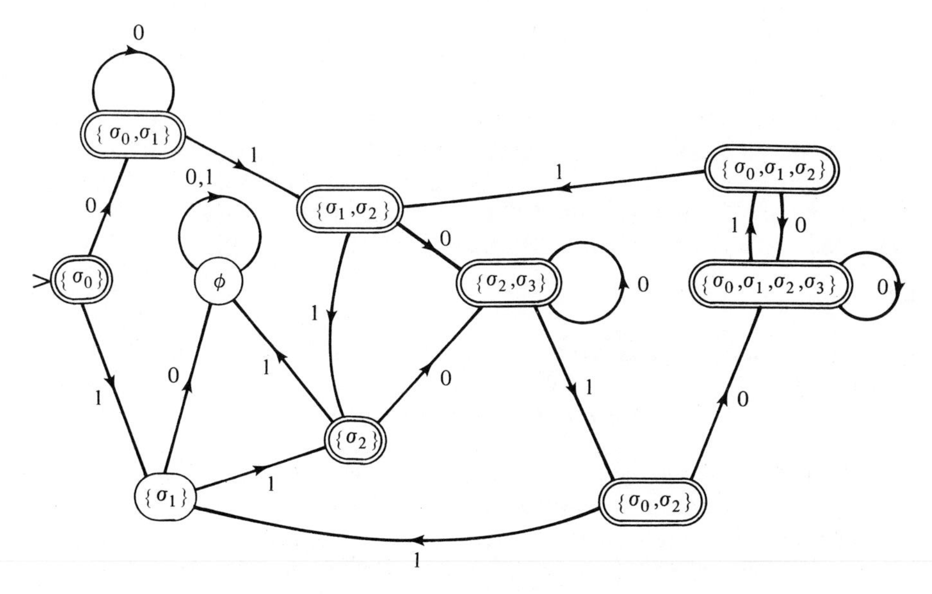

Fig. 8-6 Fsa with same accepted set as nfsa of Fig. 8-5

THEOREM 8-5

Given any regular language, there exists a finite-state automaton whose accepted set is this language.

Proof Consider the regular grammar $\mathcal{G} = \langle N, T, P, \sigma \rangle$. Construct the nfsa $\mathfrak{M} = \langle S, X, f, \sigma_0, F \rangle$, where

$$S = N \cup \{\tau\} \quad (\tau \notin N)$$
$$X = T$$

f is defined by:

$$(A \longrightarrow aB) \in P \text{ implies } B \in f(A, a)$$
$$(A \longrightarrow a) \in P \text{ implies } \tau \in f(A, a)$$
$$f(\tau, a) = \varnothing \text{ for all } a \in X$$

Also:

$$\sigma_0 = \sigma$$
$$F = \begin{cases} \{\tau\} & \text{if } (\sigma \longrightarrow \lambda) \notin P \\ \{\sigma_0, \tau\} & \text{otherwise} \end{cases}$$

Now, $\lambda \in L(\mathcal{G})$ if and only if $(\sigma \longrightarrow \lambda) \in P$; hence, if and only if $\sigma_0 \in F$ and, hence, if and only if $\lambda \in T(\mathfrak{M})$. Also, $w = a_1 a_2 \ldots a_r \in L(\mathcal{G})$ if and only if $L(\mathcal{G})$ has a derivation

$$\sigma \Longrightarrow a_1 A_1 \Longrightarrow a_1 a_2 A_2 \Longrightarrow \ldots \Longrightarrow a_1 a_2 \ldots a_{r-1} A_{r-1} \Longrightarrow a_1 a_2 \ldots a_r = w$$

hence, if and only if P has the productions $\sigma \longrightarrow a_1 A_1$, $A_1 \longrightarrow a_2 A_2, \ldots,$ $A_{r-2} \longrightarrow a_{r-1} A_{r-1}$, $A_{r-1} \longrightarrow a_r$; hence, if and only if $A_1 \in f(\sigma, a_1)$, $A_2 \in f(A_1, a_2), \ldots, A_{r-1} \in f(A_{r-2}, a_{r-1})$, $\tau \in f(A_{r-1}, a_r)$ (see Fig. 8-7), hence, if and only if $w \in T(\mathfrak{M})$. Thus, $T(\mathfrak{M}) = L(\mathcal{G})$. By Theorem 8-4, a fsa $\tilde{\mathfrak{M}}$ exists such that $T(\tilde{\mathfrak{M}}) = T(\mathfrak{M})$ and, hence, such that $T(\tilde{\mathfrak{M}}) = L(\mathcal{G})$. □

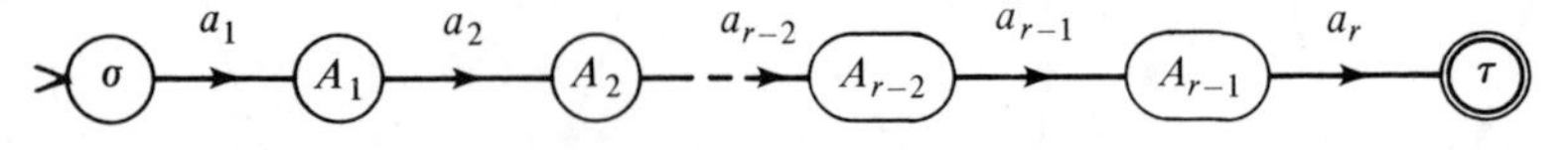

Fig. 8-7 For proof of Theorem 8-5

The proof of Theorem 8-5, in conjunction with the proof of Theorem 8-4, provides a procedure for constructing a fsa whose accepted set is any given regular language. It thus demonstrates the recursiveness of regular languages. Combining Theorems 8-3 and 8-5, we can conclude that the class of all sets accepted by fsa's is the same as the class of all regular languages.

Example 8-5

Consider the regular language $L(\mathcal{G})$, where $\mathcal{G} = \langle\{\sigma, A, B\}, \{0, 1\}, P, \sigma\rangle$ and

$$P = \{\sigma \longrightarrow 1\sigma, \sigma \longrightarrow 0A, \sigma \longrightarrow 0B, \sigma \longrightarrow 0, A \longrightarrow 0\sigma,$$
$$A \longrightarrow 0A, B \longrightarrow 0\sigma, B \longrightarrow 1A, B \longrightarrow 0B, B \longrightarrow 0\}$$

The nfsa $\mathfrak{M}$ whose accepted set is $L(\mathcal{G})$ is shown in Fig. 8-8. The fsa $\tilde{\mathfrak{M}}$ whose accepted set equals $T(\mathfrak{M})$, and, hence, equals $L(\mathcal{G})$, is shown in Fig. 8-9. □

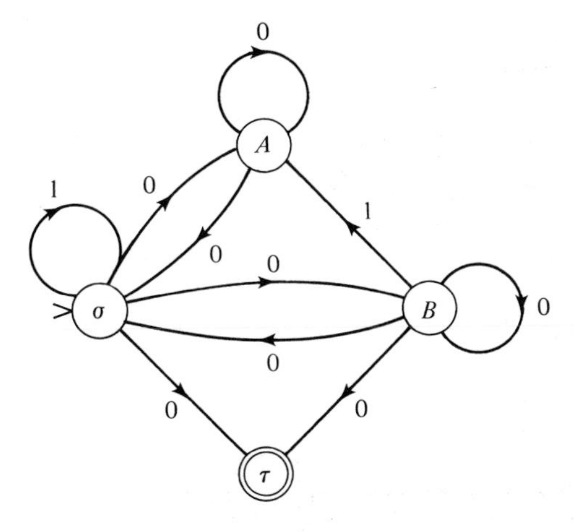

Fig. 8-8 Nfsa for Example 8-5

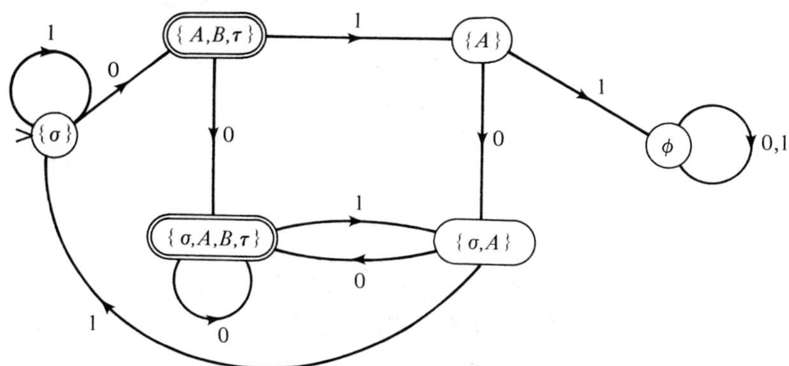

Fig. 8-9 Fsa for Example 8-5

PROBLEMS

1. A fsa $\mathfrak{M} = \langle S, X, f, \sigma_0, F\rangle$ is said to be *connected* if, for every $\sigma \in S$, there is $w \in X^*$ such that $f(\sigma_0, w) = \sigma$. Show that for every fsa $\mathfrak{M}$ there is a connected fsa $\bar{\mathfrak{M}}$ such that $T(\mathfrak{M}) = T(\bar{\mathfrak{M}})$.

2. Figure 8-B shows the transition diagrams of three fsa's. Find grammars which generate their accepted sets.

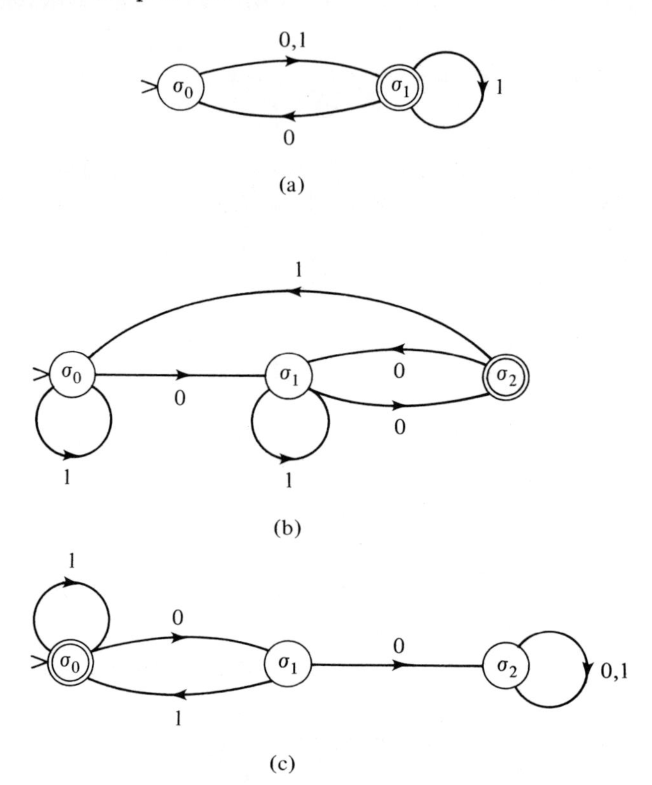

Fig. 8-B

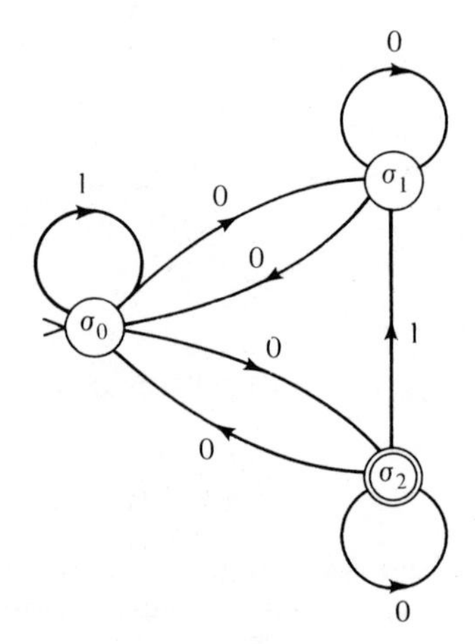

Fig. 8-C

3. Figure 8-C shows the transition diagram of a nfsa $\mathfrak{M}$.
   (a) Find a fsa $\tilde{\mathfrak{M}}$ such that $T(\tilde{\mathfrak{M}}) = T(\mathfrak{M})$.
   (b) Find a grammar which generates $T(\mathfrak{M})$.

4. Consider the regular grammars $\mathcal{G} = \langle\{\sigma, A\}, \{0, 1\}, P, \sigma\rangle$, where
   (a) $P = \{\sigma \longrightarrow 0A,\ A \longrightarrow 0A,\ A \longrightarrow 1\sigma,\ A \longrightarrow 0\}$
   (b) $P = \{\sigma \longrightarrow 1\sigma,\ \sigma \longrightarrow 1A,\ A \longrightarrow 0A,\ A \longrightarrow 0\}$
   (c) $P = \{\sigma \longrightarrow \lambda,\ \sigma \longrightarrow 0A,\ \sigma \longrightarrow 1A,\ A \longrightarrow 1A,\ A \longrightarrow 1\}$
   For each of these cases, construct a fsa $\mathfrak{M}$ such that $T(\mathfrak{M}) = L(\mathcal{G})$.

5. L is a regular language over X. Show that the language

$$L_0 = \{w \mid wu \in L \text{ for some } u \in X^*\}$$

   is also regular.

6. Prove that if L is a regular language, so is the language $\{w^R \mid w \in L\}$, where w^R denotes the string w written in reverse. (*Hint*: See what happens when all the arrows are reversed in the transition diagram of the fsa which accepts L).

7. Prove that a regular language L accepted by an n-state fsa is nonempty if and only if there is a string $w \in L$ such that $|w| \leq n - 1$.

8-5. REGULAR SETS

Given any sets of strings T_1 and T_2 over an alphabet X, we can form the new sets $T_1 \cup T_2$, T_1T_2, and T_1^* by performing the operations of union, concatenation, and iteration, respectively, as indicated in Sec. 8-1. These three operations will be referred to as the *regular operations*. A *regular set over X* is any subset of X^* that can be created by a finite number of regular operations on strings of length at most 1. More precisely, a regular set over the alphabet $X = \{\xi_1, \xi_2, \ldots, \xi_l\}$ is defined recursively as:

(*Basis*). $\varnothing, \{\lambda\}, \{\xi_1\}, \{\xi_2\}, \ldots, \{\xi_l\}$ are regular sets.
(*Induction step*). If R_1 and R_2 are regular sets, so are (R_1), $R_1 \cup R_2$, R_1R_2, and R_1^*. [Proliferation of parentheses is avoided by assuming that iteration has precedence over concatenation and union, and that concatenation has precedence over union. For example, $R_1 \cup (R_2(R_3^*))$ is written as $R_1 \cup R_2R_3^*$.]

Examples of regular sets over $\{0, 1\}$ are:

$$\{1, 0010010\}$$
$$(\{0\} \cup \{1, 011\}^*)\{11\}$$
$$(\{10\}^*\{0101\})^* \cup \{10, 010\}^*$$

THEOREM 8-6
The accepted set of every finite-state automaton is a regular set.

Proof Consider the fsa $\mathfrak{M} = \langle S, X, f, \sigma_1, F \rangle$, where $S = \{\sigma_1, \sigma_2, \ldots, \sigma_n\}$ and $F = \{\sigma_{v_1}, \sigma_{v_2}, \ldots, \sigma_{v_h}\}$. For $i = 1, 2, \ldots, n$, $j = 1, 2, \ldots, n$, and $k = 0, 1, \ldots, n$, define the sets R_{ij}^k recursively as follows:

(*Basis*). $$R_{ij}^0 = \{\xi \mid \xi \in X, f(\sigma_i, \xi) = \sigma_j\}$$

(*Induction step*).

$$R_{ij}^k = R_{ij}^{k-1} \cup R_{ik}^{k-1}(R_{kk}^{k-1})^* R_{kj}^{k-1} \qquad (k \geq 1)$$

By the recursive definition of regular sets it follows that the R_{ij}^k are regular sets.

We now prove by induction on k that

$$R_{ij}^k = \{w \mid w \in X^+, f(\sigma_i, w) = \sigma_j, \text{ and } w \text{ does not cause } \mathfrak{M} \text{ to pass through any of the states in } \{\sigma_{k+1}, \sigma_{k+2}, \ldots, \sigma_n\} \text{ as intermediate states}\} \tag{8-2}$$

(*Basis*). For $k = 0$,

$$\begin{aligned} R_{ij}^k = R_{ij}^0 &= \{\xi \mid \xi \in X, f(\sigma_i, \xi) = \sigma_j\} \\ &= \{w \mid w \in X^+, f(\sigma_i, w) = \sigma_j, \text{ and } w \text{ does not cause } \mathfrak{M} \text{ to pass through any of the states in } \{\sigma_1, \sigma_2, \ldots, \sigma_n\} \text{ as intermediate states}\} \end{aligned}$$

(*Induction step*). The induction hypothesis is that, for any $p \geq 1$,

$$R_{ij}^{p-1} = \{w \mid w \in X^+, f(\sigma_i, w) = \sigma_j, \text{ and } w \text{ does not cause } \mathfrak{M} \text{ to pass through any of the states in } \{\sigma_p, \sigma_{p+1}, \ldots, \sigma_n\} \text{ as intermediate states}\}$$

Now, referring to Fig. 8-10, the set of paths represented by

$$T = \{w \mid w \in X^+, f(\sigma_i, w) = \sigma_j, \text{ and } w \text{ does not cause } \mathfrak{M} \text{ to pass through any of the states in } \{\sigma_{p+1}, \sigma_{p+2}, \ldots, \sigma_n\} \text{ as intermediate states}\}$$

can be partitioned into paths that pass through σ_p and those that do not pass through σ_p. Those that do, can be formed from subpaths which connect σ_i to σ_p, followed by subpaths which connect σ_p to σ_p, and finally followed by subpaths which connect σ_p to σ_j. By induction hypothesis, then:

$$T = R_{ij}^{p-1} \cup R_{ip}^{p-1}(R_{pp}^{p-1})^* R_{pj}^{p-1} = R_{ij}^p$$

which completes the induction.

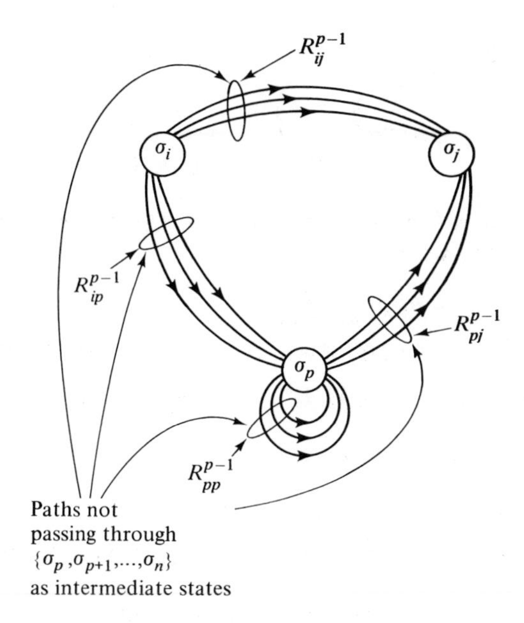

Fig. 8-10 For proof of Theorem 8-6

Letting $i = 1$, $j = \nu_j$ and $k = n$ in (8-2), we have

$$\begin{aligned} R^n_{1\nu_j} &= \{w \mid w \in X^+,\ f(\sigma_1, w) = \sigma_{\nu_j}, \text{ and } w \\ &\qquad \text{does not cause } \mathfrak{M} \text{ to pass through any} \\ &\qquad \text{of the states in } \varnothing \text{ as intermediate states}\} \\ &= \{w \mid w \in X^+,\ f(\sigma_1, w) = \sigma_{\nu_j}\} \end{aligned}$$

Since $\{\sigma_{\nu_1}, \sigma_{\nu_2}, \ldots, \sigma_{\nu_h}\} = F$,

$$T(\mathfrak{M}) - \{\lambda\} = \bigcup_{j=1}^{h} R^n_{1\nu_j}$$

and, hence,

$$T(\mathfrak{M}) = \begin{cases} \bigcup_{j=1}^{h} R^n_{1\nu_j} & \text{if } \sigma_1 \notin F \\ (\bigcup_{j=1}^{h} R^n_{1\nu_j}) \cup \{\lambda\} & \text{if } \sigma_1 \in F \end{cases} \tag{8-3}$$

Since the $R^n_{1\nu_j}$ are regular sets, so is $T(\mathfrak{M})$. □

The proof of Theorem 8-6 suggests an algorithm for expressing the accepted set of any fsa as a regular set. This algorithm is illustrated by:

Example 8-6

Figure 8-11 shows the transition diagram of a fsa

$$\mathfrak{M} = \langle \{\sigma_1, \sigma_2\}, \{0, 1\}, f, \sigma_1, \{\sigma_2\} \rangle$$

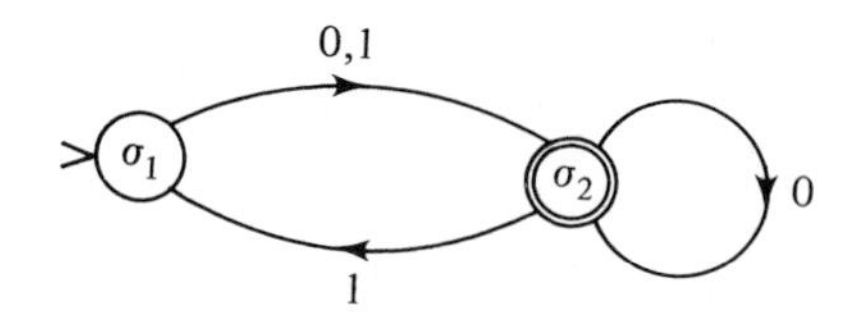

Fig. 8-11 Fsa for Example 8-6

In this case $n = 2$, $h = 1$, $v_1 = 2$, and, by (8-3), $T(\mathfrak{M}) = R_{12}^2$. Using the recursive definition of R_{ij}^k, we have:

$$R_{12}^2 = R_{12}^1 \cup R_{12}^1(R_{22}^1)^* R_{22}^1 \tag{8-4}$$

$$R_{12}^1 = R_{12}^0 \cup R_{11}^0(R_{11}^0)^* R_{12}^0, \; R_{22}^1 = R_{22}^0 \cup R_{21}^0(R_{11}^0)^* R_{12}^0 \tag{8-5}$$

$$R_{11}^0 = \varnothing, R_{12}^0 = \{0, 1\}, R_{21}^0 = \{1\}, R_{22}^0 = \{0\} \tag{8-6}$$

Substituting (8-6) in (8-5), we have:

$$R_{12}^1 = \{0, 1\} \cup \varnothing \varnothing^* \{0, 1\} = \{0, 1\} \tag{8-7}$$

$$R_{22}^1 = \{0\} \cup \{1\} \varnothing^* \{0, 1\} = \{0\} \cup \{1\}\{0, 1\} = \{0, 10, 11\} \tag{8-8}$$

Substitution of (8-7) and (8-8) in (8-4) finally yields

$$\begin{aligned} R_{12}^2 = T(\mathfrak{M}) &= \{0, 1\} \cup \{0, 1\}\{0, 10, 11\}^* \{0, 10, 11\} \\ &= \{0, 1\}\{0, 10, 11\}^* \end{aligned}$$

□

THEOREM 8-7

Given any finite-state automata $\mathfrak{M}_1$ and $\mathfrak{M}_2$ with input alphabet X, there are finite-state automata that accept: (a) $X^* - T(\mathfrak{M}_1)$, (b) $T(\mathfrak{M}_1) \cup T(\mathfrak{M}_2)$, (c) $T(\mathfrak{M}_1) \cap T(\mathfrak{M}_2)$.

Proof Let $\mathfrak{M}_1 = \langle S_1, X, f_1, \sigma_1 F_1 \rangle$ and $\mathfrak{M}_2 = \langle S_2, X, f_2, \sigma_2, F_2 \rangle$.

(a) Construct the fsa $\mathfrak{M} = \langle S, X, f, \sigma, F \rangle$, where $S = S_1, f = f_1, \sigma = \sigma_1$ and $F = S - F_1$. Clearly, for all $w \in X^*$, $w \in T(\mathfrak{M})$ if and only if $w \notin T(\mathfrak{M}_1)$. Hence, $T(\mathfrak{M}) = x^* - T(\mathfrak{M}_1)$.

(b) Construct the fsa $\mathfrak{M} = \langle S, X, f, \sigma, F \rangle$, where $S = S_1 \times S_2$; for all $(\sigma', \sigma'') \in S$ and $\xi \in X$,

$$f((\sigma', \sigma''), \xi) = (f_1(\sigma', \xi), f_2(\sigma'', \xi))$$

$\sigma = (\sigma_1, \sigma_2)$; and

$$F = \{(\sigma' \sigma''\} \,|\, \sigma' \in F_1 \textit{ or } \sigma'' \in F_2\}$$

By induction on $|w|$, for any $w \in X^*$ we have

$$f((\sigma_1, \sigma_2), w) = (f_1(\sigma_1, w), f_2(\sigma_2\, w))$$

Now, $w \in T(\mathfrak{M})$ if and only if $f(\sigma, w) \in F$; hence, if and only if $f((\sigma_1, \sigma_2), w) \in F$; hence, if and only if $(f(\sigma_1, w), f_2(\sigma_2, w)) \in F$; hence, if and only if $f_1(\sigma_1, w) \in F_1$ *or* $f_2(\sigma_2, w) \in F_2$; and, hence, if and only if $w \in T(\mathfrak{M}_1)$ *or* $w \in T(\mathfrak{M}_2)$. Thus, $T(\mathfrak{M}) = T(\mathfrak{M}_1) \cup T(\mathfrak{M}_2)$.

(c) Same proof as in part (b), but change *or* into *and* and $\cup$ into $\cap$. □

Theorem 8-7 shows that the set of all subsets of X^* accepted by fsa's is closed under the operations of complementation, union, and intersection. This set, therefore, forms an algebra of sets (or a Boolean algebra). The proof of Theorem 8-7 suggests algorithms for constructing fsa's that accept the

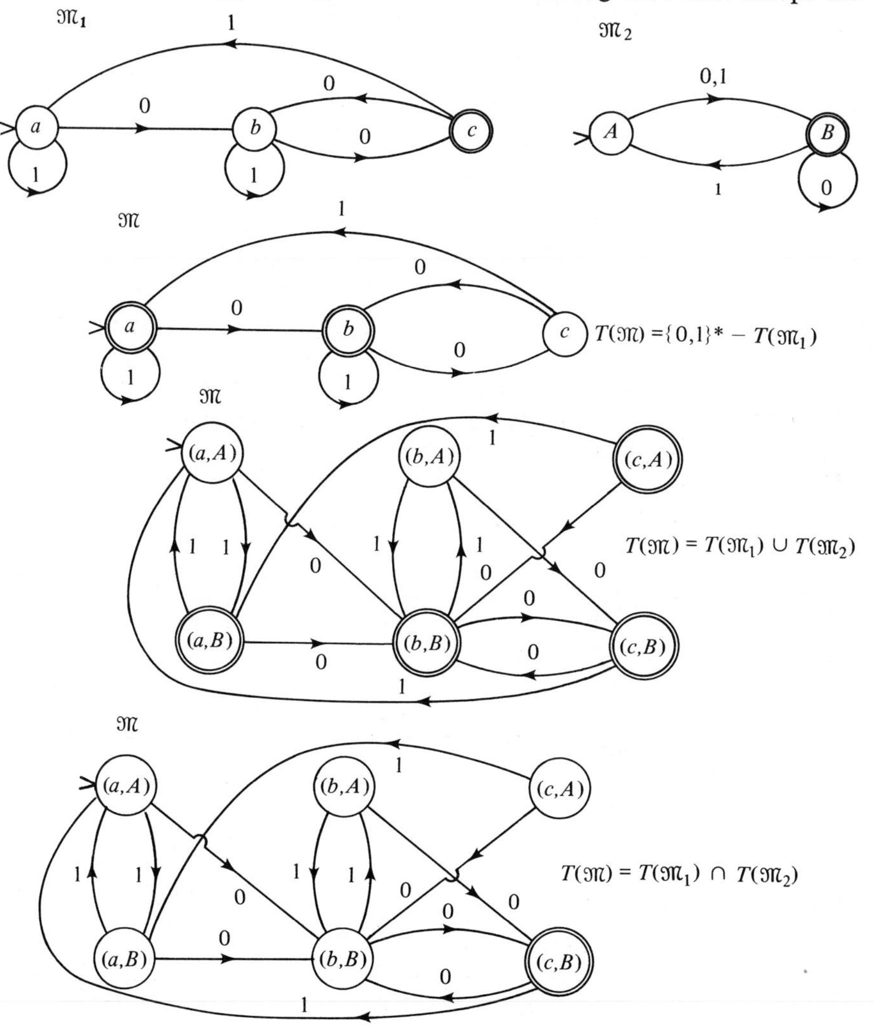

Fig. 8-12 Illustration of Theorem 8-7

complements, unions, and intersections of sets accepted by given fsa's. These algorithms are illustrated in Fig. 8-12.

THEOREM 8-8

Given any finite-state automata $\mathfrak{M}_1$ and $\mathfrak{M}_2$ with input alphabet X, there are finite-state automata that accept: (a) $T(\mathfrak{M}_1)T(\mathfrak{M}_2)$, (b) $[T(\mathfrak{M}_1)]^*$.

Proof Let $\mathfrak{M}_1 = \langle S_1, X, f_1, \sigma_1\ F_1\rangle$ and $\mathfrak{M}_2 = \langle S_2, X, f_2, \sigma_2, F_2\rangle$.

(a) Construct the nondeterministic fsa $\mathfrak{M} = \langle S, X, f, \sigma_0, F\rangle$, where $S = S_1 \cup S_2$; for all $\sigma \in S$ and $\xi \in X$,

$$f(\sigma, \xi) = \begin{cases} \{f_1(\sigma, \xi)\} & \text{if } \sigma \in S_1 - F_1 \\ \{f_1(\sigma, \xi), f_2(\sigma_2, \xi)\} & \text{if } \sigma \in F_1 \\ \{f_2(\sigma, \xi)\} & \text{if } \sigma \in F_2 \end{cases}$$

$\sigma_0 = \sigma_1$; and

$$F = \begin{cases} F_2 & \text{if } \sigma_2 \notin F_2 \\ F_1 \cup F_2 & \text{if } \sigma_2 \in F_2 \end{cases}$$

From Fig. 8-13 it can be deduced that $w_1 = \xi_{i1}\xi_{i2} \ldots \xi_{ir} \in T(\mathfrak{M}_1)$ and $w_2 = \xi_{j1}\xi_{j2} \ldots \xi_{js} \in T(\mathfrak{M}_2)$ $(w_2 \neq \lambda)$ if and only if $w_1w_2 = \xi_{i1} \ldots \xi_{ir}\xi_{j1} \ldots \xi_{js} \in T(\mathfrak{M})$. When $w_2 = \lambda \in T(\mathfrak{M}_2)$, then $w_1 \in T(\mathfrak{M}_1)$ if and only if $w_1 \in T(\mathfrak{M})$. Hence, $T(\mathfrak{M}) = T(\mathfrak{M}_1)T(\mathfrak{M}_2)$. By Theorem 8-4, then, there is a fsa $\tilde{\mathfrak{M}}$ such that $T(\tilde{\mathfrak{M}}) = T(\mathfrak{M}) = T(\mathfrak{M}_1)T(\mathfrak{M}_2)$.

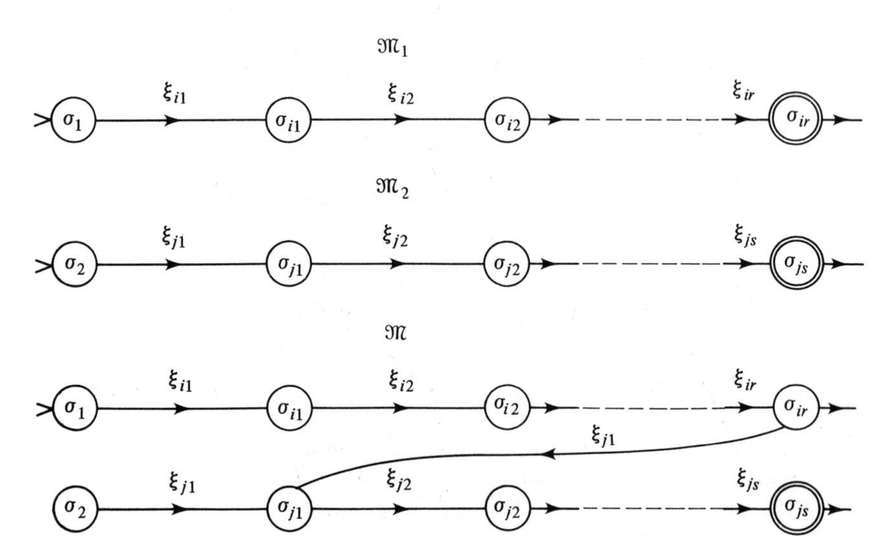

Fig. 8-13 For proof of Theorem 8-8(a)

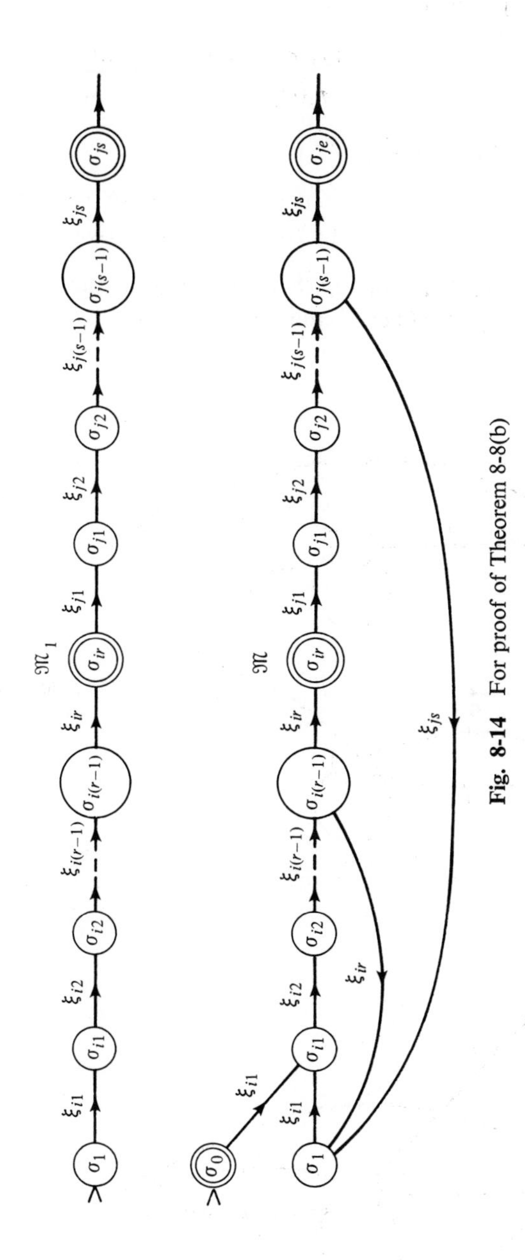

Fig. 8-14 For proof of Theorem 8-8(b)

(b) Construct the nondeterministic fsa $\mathfrak{M} = \langle S, X, f, \sigma_0, F \rangle$, where $S = S_1 \cup \{\sigma_0\}$; for all $\xi \in X$,

$$f(\sigma_0, \xi) = \begin{cases} \{f_1(\sigma_1, \xi), \sigma_1\} & \text{if } f_1(\sigma_1, \xi) \in F_1 \\ \{f_1(\sigma_1, \xi)\} & \text{otherwise} \end{cases}$$

and for all $\sigma \in S_1$ and $\xi \in X$,

$$f(\sigma, \xi) = \begin{cases} \{f_1(\sigma, \xi), \sigma_1\} & \text{if } f(\sigma, \xi) \in F_1 \\ \{f_1(\sigma, \xi)\} & \text{otherwise} \end{cases}$$

and $F = F_1 \cup \{\sigma_0\}$. From Fig. 8-14 it can be deduced that $\lambda \in T(\mathfrak{M})$, and that $w_1 w_2 \ldots w_k \in T(\mathfrak{M})$ if and only if $w_1, w_2, \ldots, w_k \in T(\mathfrak{M}_1)$. Hence,

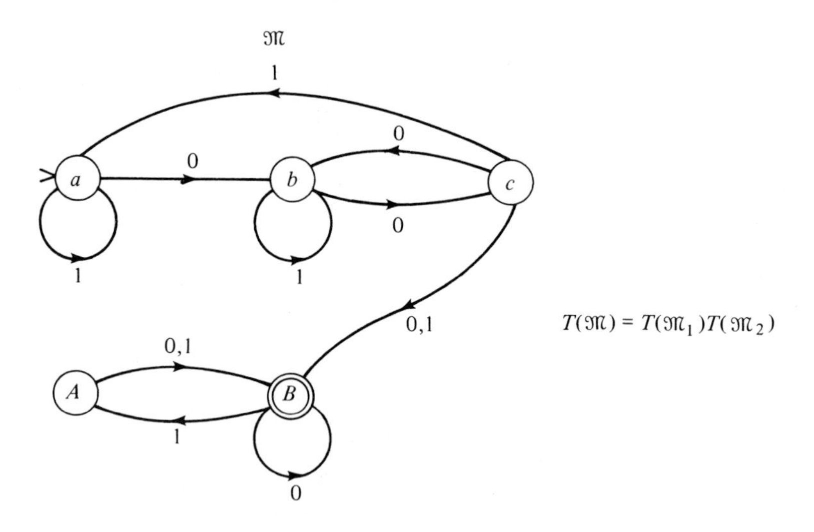

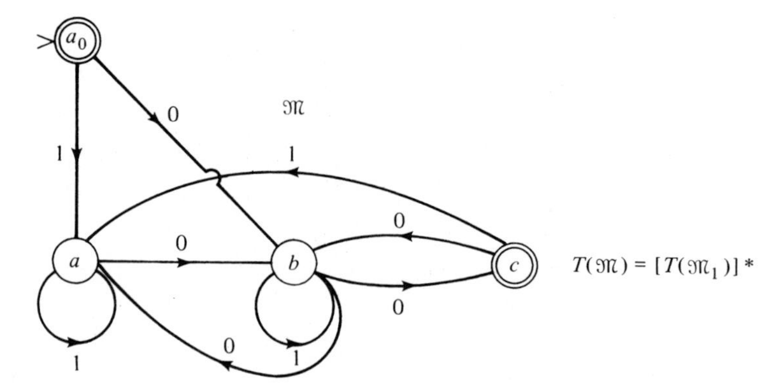

Fig. 8-15 Illustrations of Theorem 8-8

$T(\mathfrak{M}) = [T(\mathfrak{M}_1)]^*$. Again, by Theorem 8-4, there is a fsa $\tilde{\mathfrak{M}}$ such that $T(\tilde{\mathfrak{M}}) = T(\mathfrak{M}) = [T(\mathfrak{M}_1)]^*$.

The proof of Theorem 8-8 suggests algorithms for constructing nfsa's that accept the concatenations and iterations of sets accepted by given fsa's. These algorithms are illustrated in Fig. 8-15, where $\mathfrak{M}_1$ and $\mathfrak{M}_2$ refer to the fsa's of Fig. 8-12. The conversion of the nfsa's shown in the figure to fsa's (using the algorithm in the proof of Theorem 8-4 is left to the reader.

From Theorems 8-7 and 8-8 it can be deduced that the set of all subsets of X^* accepted by fsa's is closed under the regular operations (that is, union, concatenation, and iteration). This important conclusion leads to:

THEOREM 8-9

For every regular set there is a finite-state automaton whose accepted set is this regular set.

Proof The theorem is proved by induction on the shape of the given regular set R over $X = \{\xi_1, \xi_2, \ldots, \xi_l\}$.

(*Basis*). The fsa's $\mathfrak{M}$ whose accepted sets are $R = \varnothing$, $R = \{\lambda\}$, and $R = \{\xi_i\}$ ($i = 1, 2, \ldots, l$) are shown in Fig. 8-16.

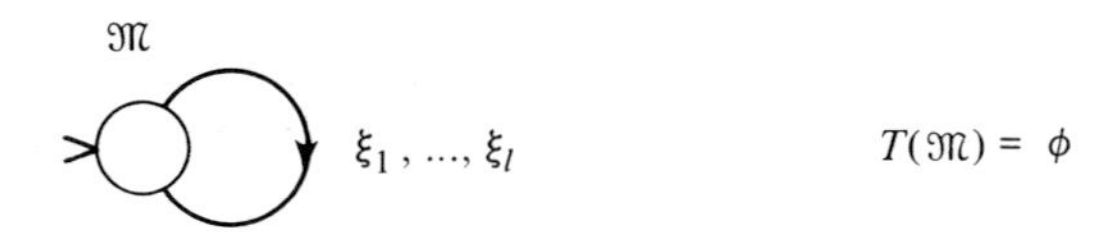

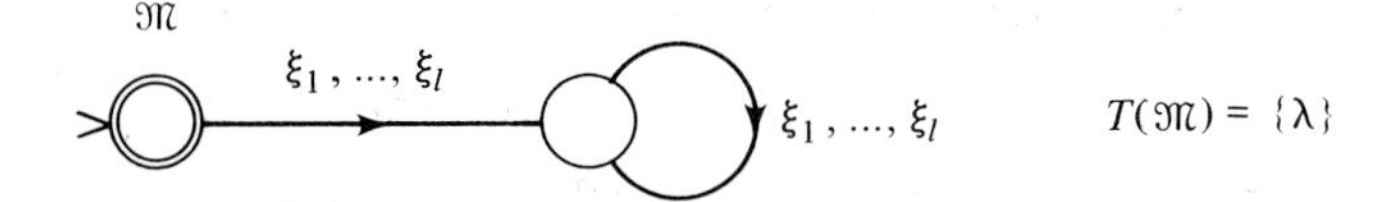

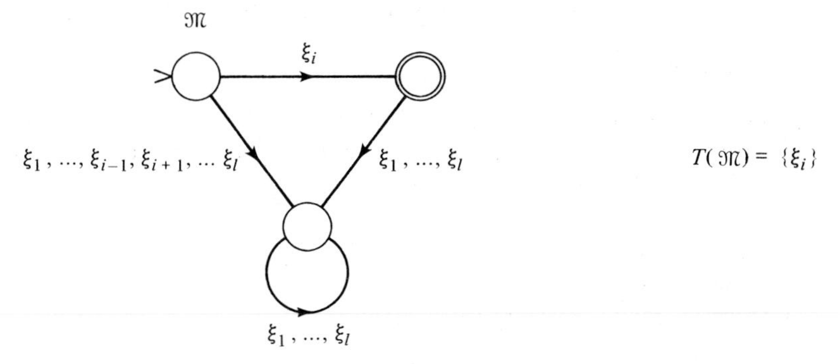

Fig. 8-16 For proof of Theorem 8-9

(*Induction step*). By Theorems 8-7 and 8-8, if R_1 and R_2 are sets accepted by fsa's, so are $R_1 \cup R_2$, R_1R_2, and R_1^*. $\square$

The process by which a fsa can be constructed for a given regular set R is a recursive one, utilizing the algorithms outlined in the proofs of Theorems 8-7, 8-8, and 8-9. For example, if $R = (\{0\}^* \cup \{10\})^*$ (over $\{0, 1\}$), we first construct fsa's that accept $R_1 = \{0\}$ and $R_2 = \{1\}$, then the fsa that accepts $R_3 = R_1^*$, then the fsa that accepts $R_4 = R_2R_1 = \{10\}$, then the fsa that accepts $R_5 = R_3 \cup R_4 = \{0\}^* \cup \{10\}$, and, finally, the fsa that accepts $R_6 = R_5^* = (\{0\}^* \cup \{10\})^* = R$. (It might be simpler at each stage to construct a nfsa rather than a fsa; the last nfsa can then be converted to a fsa that accepts R.)

Combining Theorems 8-6 and 8-9, we can conclude that the class of all sets accepted by fsa's is precisely the class of all regular sets. In view of Theorems 8-3 and 8-5 we can finally conclude that *every regular set is a regular language and conversely*. Thus, a regular language can always be specified through the conveniently compact form of a regular set.

PROBLEMS

1. Given a regular set R, describe a simple procedure for deciding whether:
(a) $R = \varnothing$.
(b) R is finite.

2. Find the regular sets accepted by the fsa's shown in Fig. 8-B.

3. For the nfsa's shown in Fig. 8-15, find fsa's $\tilde{\mathfrak{M}}$ such that $T(\tilde{\mathfrak{M}}) = T(\mathfrak{M})$.

4. Construct fsa's $\mathfrak{M}$ with input alphabet $\{0, 1\}$, such that:
(a) $T(\mathfrak{M}) = \{w \mid w$ contains exactly one 1$\}$
(b) $T(\mathfrak{M}) = \{w \mid w$ contains an odd number of 0s$\}$
(c) $T(\mathfrak{M}) = \{w \mid w$ has no 0s or w has no 1s$\}$
(d) $T(\mathfrak{M}) = \{w \mid w$ contains an even number of 0s followed by an even number of 1s$\}$
Express each $T(\mathfrak{M})$ as a regular set.

5. Given the finite set of strings $W = \{w_1, w_2, \ldots, w_k\} \subset X^*$, describe the construction of a fsa $\mathfrak{M}$ such that $T(\mathfrak{M}) = W$. Illustrate the construction with the sets $W = \{0, 10, 0100, 1100\}$ (over $\{0, 1\}$).

6. Construct a fsa $\mathfrak{M}$ such that:
(a) $T(\mathfrak{M}) = \{0\}^*$
(b) $T(\mathfrak{M}) = \{01\}^*$
(c) $T(\mathfrak{M}) = \{00\}^* \cup \{11\}^*$
(d) $T(\mathfrak{M}) = \{1\} \cup [(\{0\}\{1\}^*\{0\})^*(\{0\}\{1\}^*\{0\})]$

7. Given the regular sets $R_1 = \{1, 01\}^*$ and $R_2 = \{\lambda\} \cup \{0\}\{1\}^*$, construct fsa's that accept the sets:

(a) R_1 (b) R_2 (c) $\{0, 1\}^* - R_1$
(d) $R_1 \cup R_2$ (e) $R_1 \cap R_2$ (f) R_1R_2
(g) R_2R_1 (h) R_1^* (i) R_2^*
(j) $[(\{0, 1\}^* - R_2) \cup R_1]R_2$

8. Express the following sets of strings as regular sets (using union, concatenation, and iteration only):
(a) $\{0, 1\}^* - \{1\}^*$ (b) $\{0, 1\}^* - \{1, 01\}^*$
(c) $(\{1\}\{0\}^*\{1\}) \cap (\{1\}^*\{00\}^*\{1\}^*)$

9. Given the regular grammar $\mathcal{G} = \langle\{\sigma, A\}, \{0, 1\}, P, \sigma\rangle$, where

$$P = \{\sigma \longrightarrow 0\sigma, \sigma \longrightarrow 1A, \sigma \longrightarrow 1, A \longrightarrow 0\sigma, A \longrightarrow 1\sigma\}$$

find a regular set R such that $R = L(\mathcal{G})$.

10. Formulate an algorithm for deciding whether two regular sets are equal. Try your algorithm with:

$$R_1 = \{0, 1\}\{0\}^*$$
$$R_2 = \{0, 1\} \cup (\{0, 1\}\{0\}^*\{0\})$$

8-6. RECOGNIZERS AND TURING MACHINES

The finite-state automaton introduced in Sec. 8-4 is an example of a more general "device" shown in Fig. 8-17. This general model, which we call a *recognizer*, consists of the following components:

(a) an *input tape* on which the string to be recognized is recorded. Each input symbol occupies one *cell* on the tape. In some cases *endmarkers* are inserted before and after the string to designate its boundaries. The tape may be finite or infinite in one or both directions.

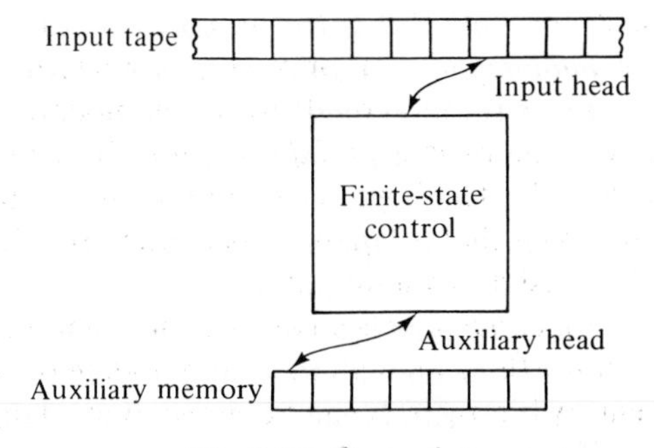

Fig. 8-17 Recognizer

(b) an *auxiliary memory*, also composed of cells, on which strings can be recorded in the course of recognition. It serves as the recognizer's "scratch pad."

(c) a *finite-state control*, which is a mechanism endowed with a set of *states* and capable of scanning input tape in one direction (a *one-way* recognizer) or in both directions (a *two-way* recognizer). Scanning is done via the *input head*, which may be capable of recording new symbols on the tape as well as reading symbols already recorded. The control is also capable of reading symbols from and recording symbols on the auxiliary memory via an *auxiliary head*. The precise action taken by the control and its state at any instant of time are determined by the scanned symbols and the control's state at the preceding instant of time. The action and new state may or may not be unique (in which cases the recognizer is said to be *deterministic* or *nondeterministic*, respectively).

Initially, the finite-state control is in its *initial state*, the input head confronts the leftmost symbol of the input string, and the auxiliary head confronts some specified cell in the auxiliary memory. Subsequently, the heads move from one cell to the next, leaving in their trail an assortment of symbols—as dictated by the control. The input string is said to be *accepted* by the recognizer if, upon confronting the string's rightmost symbol (after a finite number of moves), the recognizer enters any one of a set of states designated as *accepting states*. (In cases where the recognizer is nondeterministic, this entry may correspond to one or more possible sequences of action permitted by the control.) The set of all strings accepted by the recognizer constitutes its *accepted set*.

Various restrictions on the general structure described above result in a variety of recognizers of special interest.

(1) A *finite-state automaton* is simply a one-way, deterministic recognizer whose input tape is finite (long enough to accommodate the input string) and whose auxiliary memory is absent. We have already proved that a language is regular if and only if it is the accepted set of such a recognizer.

(2) A *pushdown automaton* is a one-way, nondeterministic recognizer whose input tape is finite (long enough to accommodate the input string) and whose auxiliary memory is a *pushdown store*. In a pushdown store, a symbol can be added or deleted from one end only (like plates in a cafeteria's dishwell). It can be shown that a language is context-free if and only if it is the accepted set of a pushdown automaton.

(3) A *linear-bounded automaton* is a two-way, nondeterministic recognizer whose input tape is finite (long enough to accommodate the input string) and whose auxiliary memory is absent. It can be shown that a language is context-sensitive if and only if it is the accepted set of a linear-bounded automaton.

(4) A *Turing machine* (named after the English mathematician Alan M. Turing, 1912–1954) is a two-way, nondeterministic recognizer whose input tape is infinite in one direction and whose auxiliary memory is absent. It can be shown that a language is phrase-structure if and only if it is accepted by a Turing machine.

Besides serving as recognizers, Turing machines (abbreviated tm) can also serve as mechanisms for computing functions. We say that a Turing machine *computes* the function $f: A \longrightarrow B$ if, when presented with an input tape on which any $a \in A$ is recorded, it leaves on the input tape, after a finite number of moves, the value of $f(a)$ [the representation of a and $f(a)$ on tape may follow any convenient convention]. If there exists a Turing machine that computes f, we say that f is a *recursive function.*

There is overwhelming evidence that any function that can be computed by a physical device of any conceivable complexity (for example, a modern digital computer) can also be computed by a Turing machine. This assertion, known as *Church's thesis* (after the American mathematician Alonzo Church) cannot be proved; however, mathematicians' faith in it is so strong that they recognize a function f as *computable* if and only if there exists a tm that computes f.

A tm of special interest is the so-called *universal Turing machine* which is capable of simulating the operation of any tm. Roughly, a universal tm $\mathfrak{M}_U$ operates as follows: when the tape of $\mathfrak{M}_U$ is initially provided with the encoded description $w_{\mathfrak{M}}$ of any tm $\mathfrak{M}$ and with any string w, $\mathfrak{M}_U$ is capable of moving back and forth between $w_{\mathfrak{M}}$ and w and mimicking the actions that $\mathfrak{M}$ would have taken if it were presented with the input string w. In this manner, a universal tm can be used to compute any computable function f (much like a stored-program general-purpose digital computer), simply by presenting the universal tm with the description of the tm that computes f and with the argument of f.

In this and previous chapters we used the concept of "algorithm" rather informally—implying by it a scheme for accomplishing a task through a finite sequence of "executable" operations. We are now in a position to define this concept precisely: An *algorithm* is a Turing machine that is guaranteed to halt for all input strings.

A problem is said to be *recursively unsolvable* if there exists no algorithm for solving it. One of the most famous problems is the *halting problem for Turing machines*, which is that of deciding whether or not an arbitrary tm $\mathfrak{M}$, presented with an arbitrary input string w, will ever halt. It is possible to demonstrate that this problem is recursively unsolvable—that is, there exists no tm which always halts and which, when presented with the descriptions of $\mathfrak{M}$ and w, decides whether or not $\mathfrak{M}$, when presented with w, eventually halts.

Another well-known recursively unsolvable problem is *Post's corre-*

spondence problem (named after the American mathematician Emil L. Post, 1897–1954), formulated as follows: Given an alphabet X, where $\# X \geq 2$, and two arbitrary l-tuples $(u_1, u_2, \ldots, u_l)$ and $(v_1, v_2, \ldots, v_l)$ of nonempty strings over X, decide whether or not there exists a nonempty set of subscripts $\{i_1, i_2, \ldots, i_k\}$ $(1 \leq i_\nu \leq l)$ such that $u_{i_1}u_{i_2} \ldots u_{i_k} = v_{i_1}v_{i_2} \ldots v_{i_k}$. (Although this problem might be solvable for some special cases, there exists no algorithm that solves it for all possible alphabets and l-tuples.)

Many problems can be shown to be recursively unsolvable by demonstrating that they are equivalent to the tm halting problem or to Post's correspondence problem. For example, we can show that the problem of deciding whether or not a grammar $\mathcal{G}$ is ambiguous [that is, whether or not there is $w \in L(\mathcal{G})$ for which there are two distinct parsing trees] is recursively unsolvable, since this problem is equivalent to Post's correspondence problem.

PROBLEMS

1. Show that the following functions are computable, by constructing Turing machines that compute them. (Assume that n is recorded on tape as a string of n 1s.)

(a) $g(n) = n + 1 \quad (n \in \mathbb{N})$

(b) $g(n_1, n_2) = \begin{cases} n_1 - n_2 & (n_1 > n_2) \\ \text{undefined} & (n_1 \leq n_2) \end{cases} \quad (n_1, n_2 \in \mathbb{N})$

(c) $g(n_1, n_2) = n_1 n_2 \quad (n_1, n_2 \in \mathbb{N})$

2. Show that the following problems are recursively unsolvable.

(a) Does the (arbitrary) tm $\mathfrak{M}$ ever print the symbol Z when presented with the (arbitrary) input string w?

(b) Does the (arbitrary) tm $\mathfrak{M}$ ever halt when presented with an all-blank tape?

3. Solve Post's correspondence problem for the special cases where $X = \{0, 1\}$ and where the two l-tuples are:

(a) (0111, 0, 10), (1, 001, 1)

(b) (01, 010, 100), (010, 100, 00)

(c) (0, 00, 010, 100), (00, 010, 100, 0100)

4. Show that Post's correspondence problem is solvable for all cases where $\# X = 1$.

REFERENCES

Arbib, M. A., *Theory of Abstract Automata*. Englewood Cliffs, NJ: Prentice-Hall, 1969.

Booth, J. L., *Sequential Machines and Automata Theory*. New York: Wiley, 1967.

Davis, M., *Computability and Unsolvability*. New York: McGraw-Hill, 1958.

GINSBURG, S., *The Mathematical Theory of Context-Free Languages*. New York: McGraw-Hill, 1966.

GINZBURG, A., *Algebraic Theory of Automata*. New York: Academic Press, 1968.

HARRISON, M. A., *Introduction to Switching and Automata Theory*. New York: McGraw-Hill, 1965.

HOPCROFT, J. E., and J. D. ULLMAN, *Formal Languages and Their Relation to Automata*. Reading, MA: Addison-Wesley, 1969.

KAIN, R. Y., *Automata Theory, Machines and Languages*. New York: McGraw-Hill, 1972.

KORFHAGE, R. R., *Logic and Algorithms*. New York: Wiley, 1966.

MINSKY, M. L., *Computation: Finite and Infinite Machines*. Englewood Cliffs, NJ: Prentice-Hall, 1967.

NELSON, R. J., *Introduction to Automata*. New York: Wiley, 1968.

ROGERS, H., *Theory of Recursive Functions and Effective Computability*. New York: McGraw-Hill, 1967.

9 GROUPS

In this chapter we return to abstract mathematics and to the study of various algebraic systems. In particular, we focus our attention on *binary algebras*—algebraic systems with a single binary operation. From the most primitive such algebras—the *semigroups*—we move to *monoids*, and finally to the highly structured *groups* that are of utmost importance in both theory and applications. After deriving basic properties of groups, we introduce the concepts of *subgroups* and *cosets*.

To the computer scientist, the familiarity with groups is important because they are basic to the study of error detecting and correcting codes, various aspects of automata theory, and many important enumeration problems.

9-1. BINARY ALGEBRAS

In this chapter we deal with algebraic systems called *binary algebras*, for which a single binary operation is defined. This operation is generally denoted by $*$, and the corresponding algebraic system by $\mathcal{S} = \langle S; * \rangle$.

An element $e_l \in S$ is said to be a *left identity* of the binary algebra $\langle S; * \rangle$ if, for all $a \in S$, $e_l * a = a$. An element $e_r \in S$ is a *right identity* of $\langle S; * \rangle$ if, for all $a \in S$, $a * e_r = a$. An element $e \in S$ is an *identity* of $\langle S; * \rangle$ if, for all $a \in S$, $e * a = a * e = a$.

THEOREM 9-1

If a binary algebra has a left identity and a right identity, then these identities are unique and are equal to the unique identity of the algebra.

Proof Let e_l and e'_l be left identities of $\langle S; *\rangle$, and e_r and e'_r right identities of $\langle S; *\rangle$. Then:

$$e_l * e_r = e'_l * e_r = e_r = e_l = e'_l$$
$$e_l * e_r = e_l * e'_r = e_l = e_r = e'_r$$

Hence, the left and right identities are unique and equal to each other. Letting $e_l = e_r = e$, we have, for all $a \in S$:

$$e * a = a * e = a$$

and, hence, e is the (unique) identity of $\langle S; *\rangle$. □

An element $z_l \in S$ is said to be a *left zero* of $\langle S; *\rangle$ if, for all $a \in S$, $z_l * a = z_l$. An element $z_r \in S$ is a *right zero* of $\langle S; *\rangle$ if, for all $a \in S$, $a * z_r = z_r$. An element $z \in S$ is a *zero* of $\langle S; *\rangle$ if, for all $a \in S$, $z * a = a * z = a$.

THEOREM 9-2

If a binary algebra has a left zero and a right zero, then these zeros are unique and equal to the unique zero of the algebra.

Proof In a manner analogous to that used in Theorem 9-1 (replacing e with z), we can show that z_l and z_r are unique and equal to each other. Letting $z_l = z_r = z$, we have, for all $a \in S$,

$$z * a = a * z = z$$

hence, z is the (unique) zero of $\langle S; *\rangle$. □

Let $\langle S; *\rangle$ be a binary algebra, and let T_1 and T_2 be any subsets of S. The following is a commonly used notation:

$$T_1 * T_2 = \{t_1 * t_2 \mid t_1 \in T_1, t_2 \in T_2\}$$

(This is a generalization of the notation introduced in Sec. 8-1 for the case where T_1 and T_2 are sets of strings and $*$ denotes string concatenation.) When $T_1 = \{a\}$ (that is, when T_1 consists of a single element only), $T_1 * T_2$ can be written simply as $a * T_2$. Similarly, when $T_2 = \{a\}$, $T_1 * T_2$ can be written as $T_1 * a$. When the operation $*$ is associative, the above notation can be generalized to any number of subsets of S:

$$T_1 * T_2 * \ldots * T_r = \{t_1 * t_2 * \ldots * t_r \mid t_1 \in T_1, t_2 \in T_2, \ldots, t_r \in T_r\}$$

PROBLEMS

1. Consider the binary algebra $\mathcal{S} = \langle S; * \rangle$, where $S = \{a, b, c\}$ and $*$ is defined by

$*$	a	b	c
a	a	a	a
b	a	c	b
c	b	a	c

Does $\mathcal{S}$ have a right identity? Left identity? Right zero? Left zero?

2. Exhibit a binary algebra that has more than one left identity.

3. Let $\mathcal{S} = \langle S; * \rangle$ be the binary algebra of Problem 1. Enumerate all pairs of subsets T_1, T_2 of S such that $T_1 * T_2 = T_2 * T_1$.

4. Let $\mathcal{S} = \langle S; * \rangle$ be a binary algebra, and consider the binary algebra $\mathbf{2}^{\mathcal{S}} = \langle \mathbf{2}^S; * \rangle$, where $\mathbf{2}^S$ is the power set of S and $*$ is as defined at the end of Sec. 9-1.

(a) Show that $\mathbf{2}^{\mathcal{S}}$ has a zero.

(b) Show that if $\mathcal{S}$ has an identity, so does $\mathbf{2}^{\mathcal{S}}$.

(c) Show that if $*$ is commutative and associative in $\mathcal{S}$, then it has the same properties in $\mathbf{2}^{\mathcal{S}}$.

9-2. SEMIGROUPS AND MONOIDS

Perhaps the simplest binary algebra conceivable is the one called *semigroup*, for which only the associative law is postulated. Formally, a semigroup $\mathcal{S} = \langle S; * \rangle$ is a binary algebra for which the following law is satisfied:

Associativity. For all $a, b, c \in S$, $a * (b * c) = (a * b) * c$ (This law permits an unambiguous interpretation of the parentheses-free expression $a_1 * a_2 * \ldots * a_r$.)

An example of a semigroup is the binary algebra $\langle A^+; \cdot \rangle$, where A^+ is the set of all nonempty strings over the alphabet A, and where $\cdot$ denotes the string concatenation operation (defined by $w_1 \cdot w_2 = w_1 w_2$). Other examples are the algebraic systems $\langle \mathbb{N}; + \rangle$ and $\langle \mathbb{N}; \cdot \rangle$, where $+$ and $\cdot$ are the ordinary addition and multiplication operations.

A semigroup with an identity is called a *monoid.* Thus, a monoid $\mathcal{S} = \langle S; * \rangle$ is a binary algebra for which the following laws are satisfied:

(a) *Associativity.* For all $a, b, c \in S$,

$$a * (b * c) = (a * b) * c$$

(b) *Identity*. There exists an element $e \in S$, such that, for all $a \in S$,

$$e * a = a * e = a$$

A monoid $\mathcal{S} = \langle S; * \rangle$ is said to be *commutative* if, in addition to (a) and (b), it satisfies this law:

(c) *Commutativity*. For all $a, b \in S$,

$$a * b = b * a$$

A number of monoids have already been encountered in preceding chapters. For example: $\langle \mathbf{2}^U; \cup \rangle$ and $\langle \mathbf{2}^U; \cap \rangle$, where the identities are $\varnothing$ and U, respectively; $\langle \mathbb{Z}; + \rangle$ and $\langle \mathbb{Z}; \cdot \rangle$, where the identities are 0 and 1, respectively; $\langle R_A; \circ \rangle$, where R_A is the set of all relations on A and $\circ$ is the composition operation, and where the identity is I_A (see Sec. 2-3); $\langle A^*; \cdot \rangle$ where A^* is the set of all strings over A (including λ) and $\cdot$ is the string concatenation operation, and where the identity is λ. (Note that the last two examples represent noncommutative monoids.)

The *powers* of an element a in a monoid $\langle S; * \rangle$ with identity e are defined recursively as:

(*Basis*). $a^0 = e$

(*Induction step*). $a^{i+1} = a^i * a$ $(i = 0, 1, 2, \ldots)$

Following Theorem 4-2, for all nonnegative integers i and j, we have

$$a^i * a^j = a^{i+j}, \qquad (a^i)^j = a^{ij}$$

An element a in a monoid $\langle S; * \rangle$ is called an *idempotent* if $a * a = a$. By induction, it can be readily established that if a is an idempotent, then, for all integers $r \geq 1$, $a^r = a$. Clearly, every monoid contains at least one idempotent—the identity.

A monoid $\mathcal{S} = \langle S; * \rangle$ is said to be *cyclic* if it contains an element g (called the *generator* of $\mathcal{S}$) such that each element $a \in S$ can be written in the form g^i $(i \geq 0)$.

THEOREM 9-3

Every cyclic monoid is commutative.

Proof Let the cyclic monoid be $\langle S; * \rangle$, with the generator g. If $a, b \in S$, we can write

$$a * b = g^i * g^j = g^{i+j} = g^{j+i} = g^j * g^i = b * a \qquad \square$$

The binary algebra $\langle \mathbb{N}; + \rangle$ is an example of a cyclic monoid; its identity is the integer 0 and its generator is the integer 1.

Let $\langle S; *\rangle$ be a finite cyclic monoid with identity e and generator g, and consider the infinite sequence $e, g, g^2, g^3, \ldots$ (which must contain all elements of S). If n is the least integer such that g^n is an element that already appeared in the sequence, say, as g^m $(m < n)$, then the sequence can be written as

$$e, g, g^2, \ldots, g^m, g^{m+1}, \ldots, g^{n-1}, g^m, g^{m+1}, \ldots$$
$$g^{n-1}, g^m, g^{m+1}, \ldots, g^{n-1}, \ldots$$

Thus, S has exactly n elements, namely,

$$S = \{e, g, g^2, \ldots, g^{n-1}\}$$

If $n - m = l$, then, for all $i \geq m$ and $\nu \geq 0$, $g^{i+\nu l} = g^i$. Setting $i = kl$, where kl is the least multiple of l such that $kl \geq m$, and setting $\nu = k$, we have

$$g^{kl+kl} = g^{kl} * g^{kl} = g^{kl}$$

hence, g^{kl} is an idempotent. When $m \neq 0$, $g^{kl} \neq e$; hence, S contains at least one idempotent other than e.

THEOREM 9-4

Let $\mathcal{S} = \langle S; *\rangle$ be a finite monoid. Then, for each $a \in S$, there is an integer $j \geq 1$ such that a^j is an idempotent.

Proof For any $a \in S$, consider the binary algebra $\mathcal{S}_a = \langle S_a; *\rangle$, where

$$S_a = \{e, a, a^2, a^3, \ldots\}$$

Clearly, $\mathcal{S}_a$ is a finite cyclic monoid with generator a and, hence, has at least one idempotent a^{kl}, where k and l are as defined in the preceding discussion. □

Example 9-1

Consider the monoid $\langle\{1, \alpha, \beta, \gamma, \delta\}; \cdot\rangle$, where $\cdot$ is defined in Table 9-1. This monoid is cyclic, with the generator γ, since:

$$\gamma^0 = 1,\ \gamma^1 = \gamma,\ \gamma^2 = \gamma\cdot\gamma = \beta$$
$$\gamma^3 = \gamma\cdot\beta = \alpha,\ \gamma^4 = \gamma\cdot\alpha = \delta$$

In this case (since $\gamma^5 = \gamma\cdot\delta = \beta = \gamma^2$), we have $m = 2$ and $l = n - m = 5 - 2 = 3$. Hence, $\gamma^{i+3\nu} = \gamma^i$ for all $i \geq 2$ and $\nu \geq 0$. In particular, $\gamma^{3+3} = \gamma^3$; hence, $\gamma^3 = \alpha$ is idempotent (and, indeed, from the table we have $\alpha^2 = \alpha$).

Table 9-1 THE · OPERATION FOR EXAMPLE 9-1

$\cdot$	1	α	β	γ	δ
1	1	α	β	γ	δ
α	α	α	β	δ	δ
β	β	β	δ	α	α
γ	γ	δ	α	β	β
δ	δ	δ	α	β	β

To illustrate Theorem 9-4, consider the monoid $\langle\{1, \delta, \delta^2, \delta^3, \ldots\}; \cdot\rangle$. Since

$$\delta^2 = \beta,\ \delta^3 = \alpha,\ \delta^4 = \delta$$

we have $\delta^{i+3v} = \delta^i$ for all $i \geq 1$ and $v \geq 0$. In particular, $\delta^{3+3} = \delta^3$; hence, δ^3 is an idempotent. □

PROBLEMS

1. Exhibit a semigroup that has a left identity and a right zero, but is not a monoid.
2. Do the monoids $\langle 2^U; \cup\rangle$, $\langle 2^U; \cap\rangle$, $\langle \mathbb{Z}; +\rangle$, $\langle \mathbb{Z}; \cdot\rangle$, $\langle R_A; \cdot\rangle$ and $\langle A^*; \cdot\rangle$ possess zeros? If so, what are they?
3. Show that the binary algebra $\langle M_n; \times\rangle$, where M_n is the set of all $n \times n$ stochastic matrices and $\times$ is the matrix multiplication operation (see Problem 4, Sec. 4-1), is a monoid. Is it a commutative monoid?
4. Consider the binary algebra $\mathcal{S} = \langle\{\alpha, \beta, \gamma, \delta\}; \cdot\rangle$, where $\cdot$ is defined by

$\cdot$	α	β	γ	δ
α	α	β	γ	δ
β	β	γ	δ	α
γ	γ	δ	α	β
δ	δ	α	β	γ

   (a) Show that $\mathcal{S}$ is a cyclic monoid and list its generators.
   (b) If g is a generator of $\mathcal{S}$, express each element of $\mathcal{S}$ as a power of g.
   (c) List all the idempotents of $\mathcal{S}$.
   (d) Show that every element of $\mathcal{S}$ raised to some power is an idempotent.

5. Consider the binary algebra $\mathcal{S} = \langle\{\alpha, \beta, \gamma, \delta\}; \cdot\rangle$, where $\cdot$ is defined by

$\cdot$	α	β	γ	δ
α	γ	β	α	δ
β	β	β	β	β
γ	α	β	γ	δ
δ	δ	β	δ	β

(a) Is $\mathcal{S}$ a cyclic monoid? Prove your answer.
(b) Show that every element of $\mathcal{S}$ raised to some power is an idempotent.

6. Let $\mathfrak{M} = \langle S, X, f, \sigma_0, F\rangle$ be a fsa, and define a relation ρ on X^* as follows: $w_1 \rho w_2$ if and only if $f(\sigma_0, w_1) = f(\sigma_0, w_2)$.
(a) Show that ρ is a right congruence relation on the monoid $\langle X^*; \cdot\rangle$, partitioning X^* into a finite number of equivalence classes (see Sec. 4-8).
(b) Show that $T(\mathfrak{M})$ can be expressed as the union of some of the equivalence classes obtained in part (a).

9-3. GROUPS

Let $\mathcal{S} = \langle S; *\rangle$ be a monoid with identity e. An element $a \in S$ is said to be *left-invertible* if there exists an element $a_l^{-1} \in S$ (called a *left inverse* of a) such that $a_l^{-1} * a = e$. The element $a \in S$ is *right-invertible* if there exists an element $a_r^{-1} \in S$ (called a *right inverse* of a) such that $a * a_r^{-1} = e$. The element $a \in S$ is *invertible* if there exists an element $a^{-1} \in S$ (called an *inverse* of a) such that $a^{1-} * a = a * a^{-1} = e$.

THEOREM 9-5
If an element a of a monoid has a left inverse and a right inverse, then these inverses are unique and equal to the unique inverse of a.

Proof If a_l^{-1} and a_r^{-1} are the left inverse and right inverse, respectively, of a, then

$$\begin{aligned} a_l^{-1} * a * a_r^{-1} &= a_l^{-1} * e = a_l^{-1} \\ &= e * a_r^{-1} = a_r^{-1} \end{aligned}$$

Hence, $a_l^{-1} = a_r^{-1} = a^{-1}$ is an inverse of a. If b is also an inverse of a, then $b * a = e$, and

$$b = b * e = b * a * a^{-1} = e * a^{-1} = a^{-1}$$

which proves the uniqueness of a^{-1}. □

A monoid in which every element is invertible is called a *group*. Thus, a group $\mathcal{G} = \langle G; *\rangle$ is a binary algebra for which these laws are satisfied:

(a) *Associativity.* For all $a, b, c \in G$,

$$a * (b * c) = (a * b) * c$$

(b) *Identity.* There exists an element $e \in G$ such that, for all $a \in G$,

$$e * a = a * e = a$$

(c) *Inverse.* For every $a \in G$ there exists an element $a^{-1} \in G$ such that

$$a^{-1} * a = a * a^{-1} = e$$

A group $\mathcal{G} = \langle G; * \rangle$ is said to be *Abelian* (after the Norwegian mathematician Niels H. Abel, 1802–1829) or *commutative* if, in addition to (a), (b), and (c), it satisfies this law:

(d) *Commutativity.* For all $a, b \in G$,

$$a * b = b * a$$

When the group operation is $+$ (addition), the group is said to be an *additive group*. In this case $a + b$ is called the *sum* of a and b, the identity is denoted by 0 ("zero"), and the inverse of a is written as $-a$ [$a + (-b)$ is usually written as $a - b$]. When the group operation is $\cdot$ (multiplication), the group is said to be a *multiplicative group*. In this case $a \cdot b$ (or simply ab) is called the *product* of a and b, and the identity is denoted by 1 ("one"). The most familiar additive and multiplicative groups are $\langle \mathbb{I}; + \rangle$ and $\langle \mathbb{Q}, \cdot \rangle$, respectively (both of which are Abelian). (The binary algebras $\langle \mathbb{Z}; + \rangle$ and $\langle \mathbb{I}; \cdot \rangle$, on the other hand, are not groups, since they do not obey the inverse law.)

The cardinality of a group is sometimes called the *order* of the group; this practice, however, is not followed in this book.

Example 9-2

Let $P = \{1, \alpha, \beta, \gamma, \delta, \epsilon\}$ be the set of all six permutations of order 3 on $\{A, B, C\}$ (see Sec. 3-5). Specifically, we denote:

$$1 = \begin{pmatrix} A & B & C \\ A & B & C \end{pmatrix}, \quad \alpha = \begin{pmatrix} A & B & C \\ A & C & B \end{pmatrix}, \quad \beta = \begin{pmatrix} A & B & C \\ B & A & C \end{pmatrix},$$

$$\gamma = \begin{pmatrix} A & B & C \\ B & C & A \end{pmatrix}, \quad \delta = \begin{pmatrix} A & B & C \\ C & A & B \end{pmatrix}, \quad \epsilon = \begin{pmatrix} A & B & C \\ C & B & A \end{pmatrix}$$

We can now define a binary algebra $\langle P; \cdot \rangle$, where $p \cdot q$ ($p, q \in P$) is the composition of p and q—that is, the permutation resulting from permutation

q followed by permutation p. For example, $\beta \cdot \delta = \epsilon$, since δ applied to (A, B, C) results in (C, A, B), and β applied to (C, A, B) results in (C, B, A). Table 9-2 lists the values of $p \cdot q$ for all $p, q \in P$.

That $\langle P; \cdot \rangle$ is associative can be verified from the table by showing that, for all $p, q, r \in P, p \cdot (q \cdot r) = (p \cdot q) \cdot r$. For example:

$$\alpha \cdot (\delta \cdot \epsilon) = \alpha \cdot \beta = \delta, \qquad (\alpha \cdot \delta) \cdot \epsilon = \beta \cdot \epsilon = \delta$$

Clearly, $\langle P; \cdot \rangle$ has an identity—namely 1. Also, every element in P is invertible, since every row and column in the table has a 1 entry ($1^{-1} = 1$, $\alpha^{-1} = \alpha$, $\beta^{-1} = \beta$, $\gamma^{-1} = \delta$, $\delta^{-1} = \gamma$, $\epsilon^{-1} = \epsilon$). Thus, $\langle P; \cdot \rangle$ is a group. (From the fact that the table is not symmetric about its principal diagonal it is apparent that $\langle P; \cdot \rangle$ is not Abelian.) □

Table 9-2 THE $\cdot$ OPERATION FOR EXAMPLE 9-2

$\cdot$	1	α	β	γ	δ	ϵ
1	1	α	β	γ	δ	ϵ
α	α	1	δ	ϵ	β	γ
β	β	γ	1	α	ϵ	δ
γ	γ	β	ϵ	δ	1	α
δ	δ	ϵ	α	1	γ	β
ϵ	ϵ	δ	γ	β	α	1

Example 9-2 is an illustration of a *permutation group*, that is, a group whose elements are permutations and whose operation is that of composition. The set of all $n!$ permutations of order n, together with the composition operation, always forms a group, since it obeys the associative, identity, and inverse laws (see Sec. 3-5). This group is called the *symmetric group of degree n*. Thus, Table 9-2 defines the symmetric group of degree 3. Example 9-3 describes a permutation group that is not symmetric—that is, which does not include *all* possible permutations of a given order.

Example 9-3

Figure 9-1 shows a square whose corners—starting with the upper left corner and proceeding clockwise—are represented by the 4-tuple (A, B, C, D). This square can be rotated in various ways, such that the corners (scanned in the same order) represent various rearrangements of the 4-tuple (A, B, C, D). Specifically, these operations correspond to the eight permutations of order 4 on $\{A, B, C, D\}$, given by:

$$1 = \begin{pmatrix} A & B & C & D \\ A & B & C & D \end{pmatrix} \qquad \text{(No change)}$$

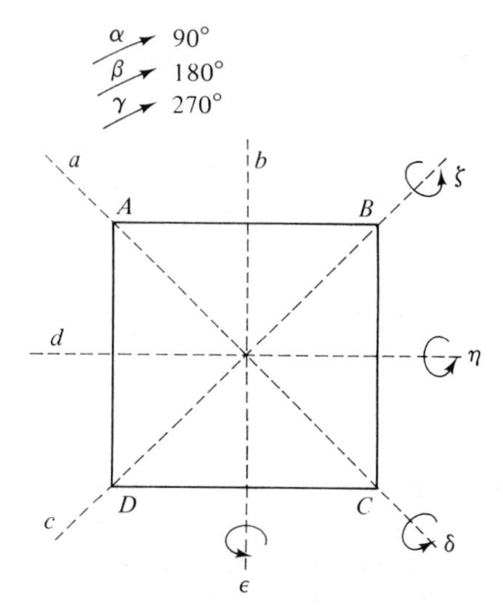

Fig. 9-1. For Example 9-3

$$\alpha = \begin{pmatrix} A & B & C & D \\ D & A & B & C \end{pmatrix} \quad \text{(90° clockwise rotation)}$$

$$\beta = \begin{pmatrix} A & B & C & D \\ C & D & A & B \end{pmatrix} \quad \text{(180° clockwise rotation)}$$

$$\gamma = \begin{pmatrix} A & B & C & D \\ B & C & D & A \end{pmatrix} \quad \text{(270° clockwise rotation)}$$

$$\delta = \begin{pmatrix} A & B & C & D \\ A & D & C & B \end{pmatrix} \quad \text{(Flip about axis } a\text{)}$$

$$\epsilon = \begin{pmatrix} A & B & C & D \\ B & A & D & C \end{pmatrix} \quad \text{(Flip about axis } b\text{)}$$

$$\zeta = \begin{pmatrix} A & B & C & D \\ C & B & A & D \end{pmatrix} \quad \text{(Flip about axis } c\text{)}$$

$$\eta = \begin{pmatrix} A & B & C & D \\ D & C & B & A \end{pmatrix} \quad \text{(Flip about axis } d\text{)}$$

While the number of possible permutations is $4! = 24$, the eight permutations listed above are all those permutations that can be achieved by means of the labeled square. Table 9-3 defines the composition operation $\cdot$ for the set $P = \{1, \alpha, \beta, \gamma, \delta, \epsilon, \zeta, \eta\}$. For example $\gamma \cdot \zeta$ is the permutation ζ, which transforms (A, B, C, D) into (C, B, A, D), followed by γ, which transforms (C, B, A, D) into (D, C, B, A); hence, $\gamma \cdot \zeta = \eta$.

Table 9-3 THE $\cdot$ OPERATION FOR EXAMPLE 9-3

$\cdot$	1	α	β	γ	δ	ϵ	ζ	η
1	1	α	β	γ	δ	ϵ	ζ	η
α	α	β	γ	1	η	δ	ϵ	ζ
β	β	γ	1	α	ζ	η	δ	ϵ
γ	γ	1	α	β	ϵ	ζ	η	δ
δ	δ	ϵ	ζ	η	1	α	β	γ
ϵ	ϵ	ζ	η	δ	γ	1	α	β
ζ	ζ	η	δ	ϵ	β	γ	1	α
η	η	δ	ϵ	ζ	α	β	γ	1

From Table 9-3 it is clear that P is closed under $\cdot$, and $\langle P; \cdot\rangle$ is a bona fide binary algebra. By exhaustive checking it can be verified that $\langle P; \cdot\rangle$ is associative. It is also apparent that $\langle P; \cdot\rangle$ has an identity (namely, 1) and that every element in P is invertible. Thus, $\langle P; \cdot\rangle$ is a group. (This group is called the *group of symmetries of the square*.) $\square$

A group $\mathcal{G} = \langle G; *\rangle$ is called a *cyclic group*, if it contains an element g such that every element $a \in G$ can be written in the form g^i $(i \in \mathbb{I})$. If this is the case, g is called a *generator* of $\mathcal{G}$, and $\mathcal{G}$ is said to be *generated* by g. If g is a generator of $\mathcal{G}$, the domain of $\mathcal{G}$ is often denoted by (g). By Theorem 9-3, *every cyclic group must be Abelian.*

PROBLEMS

1. Which of the following binary algebras $\mathcal{S} = \langle S; *\rangle$ forms a group? In each case where $\mathcal{S}$ is a group, indicate the identity and determine the inverse of each element.

(a) $S = \{1, 10\}$; $*$ is multiplication modulo 11.
(b) $S = \{1, 3, 4, 5, 9\}$; $*$ is multiplication modulo 11.
(c) $S = \mathbb{Q}$; $*$ is ordinary addition.
(d) $S = \mathbb{Q}$; $*$ is ordinary multiplication.
(e) $S = \mathbb{I}$; $*$ is ordinary subtraction.
(f) $S = \{\alpha, \beta, \gamma, \delta\}$; $*$ is the operation $\circ$ defined below.
(g) $S = \{\alpha, \beta, \gamma, \delta\}$; $*$ is the operation $\square$ defined below.

$\circ$	α	β	γ	δ
α	β	δ	α	γ
β	δ	γ	β	α
γ	α	β	γ	δ
δ	γ	α	δ	β

$\square$	α	β	γ	δ
α	α	β	γ	δ
β	β	α	δ	γ
γ	γ	δ	α	α
δ	δ	γ	β	β

2. Show that, up to isomorphism (that is, except, possibly, for element names), there is only one group with three elements and only two groups with four elements, and that both are Abelian.
3. With reference to the square of Fig. 9-1, show that the set of rotations (that is, the permutations 1, α, β, and γ in Example 9-3) together with the composition operation form an Abelian group.
4. Using Example 9-3 for guidance, give a geometrical interpretation to the permutation group of Example 9-2.
5. In the group of Example 9-2, solve the following equations:
   (a) $\gamma * x = \beta$ (b) $\delta * x^2 = \gamma$ (c) $x^3 = 1$
6. Let $\mathcal{G} = \langle G; * \rangle$ be a group with identity e, where the cardinality of G is even. Show that there is an element $a \in G$ such that $a \neq e$ and $a^2 = e$.

9-4. SOME GROUP PROPERTIES

In this section we discuss some important properties of groups.

THEOREM 9-6

If $\langle G; * \rangle$ is a group, then for any $a, b \in G$: (a) There is a unique element $x \in G$ such that $a * x = b$. (b) There is a unique element $y \in G$ such that $y * a = b$.

Proof (a) There is at least one x that satisfies $a * x = b$, namely, $x = a^{-1} * b$. [Check: $a * (a^{-1} * b) = (a * a^{-1}) * b = e * b = b$.] Now, if x is any element in G such that $a * x = b$, then

$$x = e * x = (a^{-1} * a) * x = a^{-1} * (a * x) = a^{-1} * b$$

Hence, $x = a^{-1} * b$ is the *only* element satisfying $a * x = b$.

(b) Analogous steps demonstrate that $y = b * a^{-1}$ is the only element in G that satisfies $y * a = b$. □

In terms of the table defining the operation $*$, Theorem 9-6 implies that, in any given row and in any given column of the table, every element of G appears exactly once.

THEOREM 9-7 (*Cancellation law*)

If $\langle G; * \rangle$ is a group, then for any $a, b, c \in G$: (a) $a * b = a * c$ implies $b = c$. (b) $b * a = b * c$ implies $a = c$.

Proof The theorem is a direct corollary of Theorem 9-6. □

THEOREM 9-8

If $\langle G; * \rangle$ is a group, then for any $a, b \in G$, $(a * b)^{-1} = b^{-1} * a^{-1}$.

Proof Since $(a * b) * (a * b)^{-1} = e$, and

$$(a * b) * (b^{-1} * a^{-1}) = a * (b * b^{-1}) * a^{-1}$$
$$= a * a^{-1} = e$$

we have, by Theorem 9-6 (or, alternatively, by Theorem 9-5), $(a * b)^{-1} = b^{-1} * a^{-1}$. □

Theorem 9-8 can be readily extended, by induction, to this result. For any $a_1, a_2, \ldots, a_r \in G$,

$$(a_1 * a_2 * \ldots * a_r)^{-1} = a_r^{-1} * a_{r-1}^{-1} * \ldots * a_1^{-1} \tag{9-1}$$

Note that, when $\langle G; *\rangle$ is Abelian, (9-1) can be written as

$$(a_1 * a_2 * \ldots * a_r)^{-1} = a_1^{-1} * a_2^{-1} * \ldots * a_r^{-1}$$

In groups, we can define negative powers of elements, as well as positive powers: a^{-r} is defined as $a^{-1} * a^{-1} * \ldots * a^{-1}$ (r times), or, recursively:

(*Basis*). $a^0 = e$

(*Induction step*). $a^{-i-1} = a^{-i} * a^{-1}$ $(i = 0, 1, 2, \ldots)$

For example, in the group of Example 9-3

$$\alpha^{-3} = \alpha^{-1} \cdot \alpha^{-1} \cdot \alpha^{-1} = \gamma \cdot \gamma \cdot \gamma = \beta \cdot \gamma = \alpha$$

Using a proof analogous to that used for Theorem 4-2, we have for *all* integers i and j (positive or negative):

$$a^i * a^j = a^{i+j}, \qquad (a^i)^j = a^{ij}$$

An element a of a group $\langle G; *\rangle$ is said to have a *finite order* if a positive integer r exists such that $a^r = e$. The least such integer is called the *order* of a. If no such integer exists, a is said to have *infinite order*. Clearly, the order of the identity e is always 1

THEOREM 9-9

If an element a of a group is of (finite) order r, then $a^k = e$ if and only if k is a multiple of r.

Proof If $k = \nu r$ for some integer ν, then

$$a^k = a^{\nu r} = (a^r)^\nu = e^\nu = e$$

Conversely, suppose $a^k = e$. By the division theorem, we can write $k = \nu r$

$+ \mu$ $(0 \leq \mu < r)$. Now,

$$a^{\mu} = a^{k-\nu r} = a^k * a^{-\nu r} = e * e^{-\nu} = e * e = e$$

Since $0 \leq \mu < r$ and since, by definition, r is the least positive integer such that $a^r = e$, we must have $\mu = 0$; hence, $k = \nu r$. □

Thus, if $a^r = e$ and if for no divisor d of r $(1 < d < r)$ do we have $a^d = e$, then r is the order of a. For example, if $a^8 = e$, but $a^2 \neq e$ and $a^4 \neq e$, then 8 must be the order of a.

THEOREM 9-10
A group element and its inverse have the same order.

Proof If a has a finite order r, then $a^r = e$ and, hence,

$$(a^{-1})^r = (a^r)^{-1} = e^{-1} = e$$

Thus, if r' is the order of a^{-1}, we have $r' \leq r$. Now,

$$a^{r'} = ((a^{-1})^{r'})^{-1} = e^{-1} = e$$

and, hence, $r \leq r'$. In conclusion, $r' = r$. □

THEOREM 9-11
In a finite group $\mathcal{G} = \langle G; * \rangle$, every element has a finite order. Moreover, the order of every element is at most $\#G$.

Proof Let a be any element in G. In the sequence $a, a^2, a^3, \ldots, a^{(\#G)+1}$ at least two elements are equal, say, $a^r = a^p$, where $1 \leq p < r \leq (\#G) + 1$. Now,

$$e = a^0 = a^{r-r} = a^r * a^{-r} = a^r * a^{-p} = a^{r-p}$$

and, hence, a is of order at most $r - p \leq \#G$. □

THEOREM 9-12
Let $\mathcal{G} = \langle G; * \rangle$ be an Abelian group, and let $r_1, r_2, \ldots, r_k$ be the orders of the elements $a_1, a_2, \ldots, a_k$, respectively $(k \geq 2)$. If $r_1, r_2, \ldots, r_k$ are relatively prime in pairs, then the order of the element $a_1 * a_2 * \ldots * a_k$ is $r_1 r_2 \ldots r_k$.

Proof (by induction on k) (*Basis*). Clearly,

$$(a_1 * a_2)^{r_1 r_2} = (a_1^{r_1})^{r_2} * (a_2^{r_2})^{r_1} = e * e = e$$

Suppose $(a_1 * a_2)^r = e$. Since $\mathcal{G}$ is Abelian, we can write

$$e = (a_1 * a_2)^{rr_1} = (a_1^{r_1})^r * a_2^{rr_1} = e * a_2^{rr_1} = a_2^{rr_1}$$

Hence, by Theorem 9-9, r_2 divides rr_1. Since r_1 and r_2 are relatively prime, this implies that r_2 divides r. Similarly, it can be shown that r_1 divides r. Thus, since r_1 and r_2 are relatively prime, $r_1 r_2$ divides r; hence, $r_1 r_2$ is the order of $a_1 * a_2$.

(*Induction step*). Hypothesize that the theorem is true for $k \geq 2$, and consider the element $a_1 * a_2 * \ldots a_k * a_{k+1}$, where $r_1, r_2, \ldots, r_{k+1}$ (the orders of $a_1, a_2, \ldots, a_{k+1}$, respectively) are relatively prime in pairs. By induction hypothesis, the order of $a_1 * a_2 * \ldots * a_k$ is $r_1 r_2 \ldots r_k$. Since $r_1, r_2, \ldots, r_{k+1}$ are relatively prime in pairs, the integers $r_1 r_2 \ldots r_k$ and r_{k+1} are relatively prime, and, again by the induction hypothesis, the order of $(a_1 * a_2 * \ldots * a_k) * a_{k+1}$ is $(r_1 r_2 \ldots r_k) r_{k+1}$. □

Let $\mathcal{G} = \langle G; * \rangle$ be a finite group, where $G = \{a_1, a_2, \ldots, a_n\}$, and where r_i is the order of a_i $(i = 1, 2, \ldots, n)$. The greatest of these n orders is called the *exponent* of $\mathcal{G}$.

THEOREM 9-13

Let $\mathcal{G} = \langle G; * \rangle$ be a finite Abelian group. Then the order of each element of G divides the exponent of $\mathcal{G}$.

Proof Let $\hat{r}$ denote the exponent of $\mathcal{G}$, and let a be the element of G whose order is $\hat{r}$. Let b be any other element, say, of order r, and hypothesize that r does not divide $\hat{r}$. Thus, there exists some prime number p such that, with some $j \geq 0$ and $k > j$, we can write $\hat{r} = p^j \alpha$ and $r = p^k \beta$, where p and α, and also p and β, are relatively prime. Hence, the order of a^{p^j} is α and the order of b^β is p^k. Since α and p^k are relatively prime, the order of $a^{p^j} * b^\beta$, by Theorem 9-12, is $p^k \alpha$. But $p^k \alpha > p^j \alpha = \hat{r}$, which implies that the order of some element in G is greater than the exponent of $\mathcal{G}$—a contradiction. In conclusion, r must divide $\hat{r}$. □

PROBLEMS

1. Let $\mathcal{S} = \langle S; * \rangle$ be a finite commutative monoid where, for all $a, b \in S$, $a * b = a * c$ implies $b = c$. Show that $\mathcal{S}$ is an Abelian group.
2. Let $\mathcal{G} = \langle G; * \rangle$ be a group with identity e.
   (a) Show that if, for some $a \in G$, $a^2 = e$, then $a = e$.
   (b) Show that if, for all $a \in G$, $a^2 = e$, then $\mathcal{G}$ must be Abelian.
3. Let a be an element of a group $\langle G; * \rangle$. Prove, by induction, that for all integers i and j (whether positive or negative):
   (a) $a^i * a^j = a^{i+j}$ (b) $(a^i)^j = a^{ij}$

4. Verify that $\langle \mathbb{Z}_m; * \rangle$, where $\mathbb{Z}_m = \{0, 1, \ldots, m-1\}$ and $*$ is addition modulo m, is a group. Show that d is a divisor of m if and only if d is the order of some element of $\mathbb{Z}_m$.

5. Find the order of every element in the group $\langle \{0, 1, \ldots, 5\}; * \rangle$, where $*$ is addition modulo 6. Use this group to illustrate Theorems 9-10, 9-11, 9-12, and 9-13.

6. Let $\langle B; {}^-, \vee, \wedge \rangle$ be a Boolean algebra, and consider the binary algebra $\langle B; * \rangle$ where, for all $a, b \in B$,

$$a * b = (a \wedge \bar{b}) \vee (\bar{a} \wedge b)$$

Show that $\langle B; * \rangle$ is an Abelian group where every nonidentity element has the order 2.

9-5. SUBGROUPS

Given a group $\mathcal{G} = \langle G; * \rangle$, a subsystem $\tilde{\mathcal{G}} = \langle \tilde{G}; * \rangle$ of $\mathcal{G}$ is any algebraic system such that $\tilde{G}$ is a nonempty subset of G closed under $*$ (see Sec. 4-2). If $\tilde{\mathcal{G}}$ is a group (that is, if $\tilde{\mathcal{G}}$ satisfies the associative, identity, and inverse laws with respect to $*$), it is called a *subgroup* of $\mathcal{G}$.

If $\tilde{G} = G$ or if $\tilde{G} = \{e\}$ (where e is the identity of $\mathcal{G}$), then $\tilde{\mathcal{G}} = \langle \tilde{G}; * \rangle$ is, trivially, a subgroup of $\mathcal{G}$. In these cases $\tilde{\mathcal{G}}$ is called a *trivial subgroup* of $\mathcal{G}$. If $\tilde{G}$ is a proper subset of G, $\tilde{\mathcal{G}}$ is said to be a *proper subgroup* of $\mathcal{G}$. If $\tilde{\mathcal{G}}$ is a proper subgroup of $\mathcal{G}$, and if no proper subgroup $\tilde{\mathcal{G}}' = \langle \tilde{G}', * \rangle$ of $\mathcal{G}$ exists such that $\tilde{G}$ is a proper subset of $\tilde{G}'$, then $\tilde{\mathcal{G}}$ is called a *maximal subgroup* of $\mathcal{G}$. For example, the set of all multiples of 6 forms a proper subgroup of the group $\langle \mathbb{I}; + \rangle$. It is not, however, a maximal subgroup, since the set of all multiples of 3 also forms a subgroup of $\langle \mathbb{I}; + \rangle$.

Suppose the identities of $\mathcal{G} = \langle G; * \rangle$ and its subgroup $\tilde{\mathcal{G}} = \langle \tilde{G}; * \rangle$ are e and $\tilde{e}$, respectively. Then $\tilde{e} * e = \tilde{e} = e$; hence, $\mathcal{G}$ and $\tilde{\mathcal{G}}$ have the same identity. Suppose the inverses of a in $\mathcal{G}$ and $\tilde{\mathcal{G}}$ are b and $\tilde{b}$, respectively. Then $a * b = a * \tilde{b} = e = \tilde{e}$ and, by Theorem 9-7, $b = \tilde{b}$. Thus, the inverse of any element in $\tilde{\mathcal{G}}$ is the same as its inverse in $\mathcal{G}$.

THEOREM 9-14

Let $\mathcal{G} = \langle G; * \rangle$ be a group, and $\tilde{\mathcal{G}} = \langle \tilde{G}; * \rangle$ be a subsystem of $\mathcal{G}$. Then $\tilde{\mathcal{G}}$ is a subgroup of $\mathcal{G}$ if and only if it satisfies the inverse law.

Proof. Clearly, associativity is automatically inherited by $\tilde{\mathcal{G}}$ from $\mathcal{G}$. If $\tilde{\mathcal{G}}$ satisfies the inverse law, then, for any $a \in \tilde{G}$, we have $a^{-1} \in G$, and (since $\tilde{G}$ is closed under $*$), $a * a^{-1} = e \in \tilde{G}$. Thus, $\tilde{\mathcal{G}}$ satisfies the identity law and, hence, is a group. Conversely, if $\tilde{\mathcal{G}}$ is a subgroup, it must, by definition, satisfy the inverse law. □

Thus, if $\tilde{\mathcal{G}} = \langle \tilde{G}, * \rangle$ is a group, to establish whether a nonempty set $\tilde{G}$ of G forms a subgroup of $\mathcal{G}$, it is sufficient to establish:

(a) *Closure.* For any $a, b \in \tilde{G}$, $a * b$ is also in $\tilde{G}$.
(b) *Inverse.* For any $a \in \tilde{G}$, a^{-1} is also in $\tilde{G}$.

If $\mathcal{G}$ is Abelian, the commutativity of $\mathcal{G}$ is inherited by $\tilde{\mathcal{G}}$; hence, a subgroup of any Abelian group must also be Abelian.

THEOREM 9-15
Let $\mathcal{G} = \langle G; * \rangle$ be a *finite* group, and $\tilde{\mathcal{G}} = \langle \tilde{G}; * \rangle$ be a subsystem of $\mathcal{G}$. Then $\tilde{\mathcal{G}}$ is a subgroup of $\mathcal{G}$.

Proof In view of Theorem 9-14, we need to prove only that $\tilde{\mathcal{G}}$ satisfies the inverse law. Now, let a be any element of $\tilde{G}$. By Theorem 9-11, a has a finite order, say, r. Since $\tilde{G}$ is closed under $*$, the elements $a, a^2, \ldots, a^r$ are all in $\tilde{G}$. In particular, $a^{r-1} = a^r * a^{-1} = e * a^{-1} = a^{-1}$ is in $\tilde{G}$; hence, a is invertible in $\tilde{\mathcal{G}}$. □

Thus, if $\mathcal{G} = \langle G; * \rangle$ is a finite group, to establish whether a nonempty subset $\tilde{G}$ of G forms a subgroup of $\mathcal{G}$, it is sufficient to establish closure of $\tilde{G}$ under $*$.

Example 9-4

In Example 9-2 $\{1, \alpha\}$ is closed under $\cdot$ (see Table 9-2); hence, $\langle \{1, \alpha\}; \cdot \rangle$ is a subgroup of $\langle \{1, \alpha, \beta, \gamma, \delta, \epsilon\}; \cdot \rangle$. Similarly, since $\{1, \gamma, \delta\}$ is closed under $\cdot$, $\langle \{1, \gamma, \delta\}; \cdot \rangle$ is a subgroup of $\langle \{1, \alpha, \beta, \gamma, \delta, \epsilon\}; \cdot \rangle$. It can be verified that both of these subgroups are maximal. Tables 9-4 and 9-5 define $\cdot$ for the two subgroups. (What other subgroups can you identify?) □

Table 9-4 THE $\cdot$ OPERATION FOR EXAMPLE 9-4

$\cdot$	1	α
1	1	α
α	α	1

Table 9-5 THE $\cdot$ OPERATION FOR EXAMPLE 9-4

$\cdot$	1	γ	δ
1	1	γ	δ
γ	γ	δ	1
δ	δ	1	γ

PROBLEMS

1. Identity all the subgroups of the group $\langle \{1, \alpha, \beta, \gamma, \delta, \epsilon, \zeta, \eta\}; \cdot \rangle$ of Example 9-3.
2. Identify all the subgroups of the group $\langle \mathbb{Z}_{12}; * \rangle$, where $*$ is addition modulo 12.

3. Show that the subgroups $\langle\{1, \alpha\}; \cdot\rangle$ and $\langle\{1, \gamma, \delta\}; \cdot\rangle$ of Example 9-4 are maximal.

4. Let $\mathcal{G} = \langle G; *\rangle$ be a group and $\tilde{G}$ a nonempty subset of G. Show that $\langle\tilde{G}; *\rangle$ is a subgroup of $\mathcal{G}$ if and only if $a, b \in \tilde{G}$ implies $a^{-1} * b \in G$.

5. Let $\mathcal{G} = \langle G; *\rangle$ be a group, and define the subset $\tilde{G}$ of G as

$$\tilde{G} = \{a \mid a * b = b * a \text{ for all } b \in G\}$$

Show that $\langle\tilde{G}; *\rangle$ is a subgroup of $\mathcal{G}$. (This subgroup is called the *center* of $\mathcal{G}$.)

9-6. COSETS

Let $\mathcal{G} = \langle G; *\rangle$ be a group, and let $\langle C_0; *\rangle$ be any subgroup of $\mathcal{G}$. If a is any element of G, then the set

$$C_0 * a = \{b * a \mid b \in C_0\}$$

(see the notation introduced at the end of Sec. 9-1) is called a *right coset of* C_0 *in* $\mathcal{G}$. The set

$$a * C_0 = \{a * b \mid b \in C_0\}$$

is called a *left coset of* C_0 *in* $\mathcal{G}$. If, for every $a \in G$, we have $C_0 * a = a * C_0$, then $\langle C_0; *\rangle$ is called a *normal subgroup* of $\mathcal{G}$, and right and left cosets are simply called *cosets*. Clearly, if $\mathcal{G}$ is Abelian, then every subgroup of $\mathcal{G}$ is normal.

Example 9-5

Consider the group $\mathcal{G} = \langle\{1, \alpha, \beta, \gamma, \delta, \epsilon\}; \cdot\rangle$ of Example 9-2 and its subgroup $\langle\{1, \alpha\}; \cdot\rangle$. The right cosets of $\{1, \alpha\}$ in $\mathcal{G}$ are:

$$\begin{aligned}
\{1, \alpha\}\cdot 1 &= \{1\cdot 1, \alpha\cdot 1\} = \{1, \alpha\} \\
\{1, \alpha\}\cdot \alpha &= \{1\cdot \alpha, \alpha\cdot \alpha\} = \{\alpha, 1\} \\
\{1, \alpha\}\cdot \beta &= \{1\cdot \beta, \alpha\cdot \beta\} = \{\beta, \delta\} \\
\{1, \alpha\}\cdot \gamma &= \{1\cdot \gamma, \alpha\cdot \gamma\} = \{\gamma, \epsilon\} \\
\{1, \alpha\}\cdot \delta &= \{1\cdot \delta, \alpha\cdot \delta\} = \{\delta, \beta\} \\
\{1, \alpha\}\cdot \epsilon &= \{1\cdot \epsilon, \alpha\cdot \epsilon\} = \{\epsilon, \gamma\}
\end{aligned}$$

Thus, $\{1, \alpha\}$ has three distinct right cosets in $\mathcal{G}$, namely,

$$\{1, \alpha\}, \{\beta, \delta\}, \{\gamma, \epsilon\}$$

The left cosets of $\{1, \alpha\}$ in $\mathcal{G}$ are:

$$1\cdot\{1, \alpha\} = \{1\cdot 1, 1\cdot\alpha\} = \{1, \alpha\}$$

$$\alpha\cdot\{1, \alpha\} = \{\alpha\cdot 1, \alpha\cdot\alpha\} = \{\alpha, 1\}$$

$$\beta\cdot\{1, \alpha\} = \{\beta\cdot 1, \beta\cdot\alpha\} = \{\beta, \gamma\}$$

$$\gamma\cdot\{1, \alpha\} = \{\gamma\cdot 1, \gamma\cdot\alpha\} = \{\gamma, \beta\}$$

$$\delta\cdot\{1, \alpha\} = \{\delta\cdot 1, \delta\cdot\alpha\} = \{\delta, \epsilon\}$$

$$\epsilon\cdot\{1, \alpha\} = \{\epsilon\cdot 1, \epsilon\cdot\alpha\} = \{\epsilon, \delta\}$$

Thus, $\{1, \alpha\}$ has three distinct left cosets in $\mathcal{G}$, namely,

$$\{1, \alpha\}, \{\beta, \gamma\}, \{\delta, \epsilon\}$$

The set $\{1, \alpha\}$, then, is not a normal subgroup of $\mathcal{G}$.

Now, consider the subgroup $\langle\{1, \gamma, \delta\}; \cdot\rangle$ of $\mathcal{G}$. The right cosets of $\{1, \gamma, \delta\}$ in $\mathcal{G}$ are:

$$\{1, \gamma, \delta\}\cdot 1 = \{1\cdot 1, \gamma\cdot 1, \delta\cdot 1\} = \{1, \gamma, \delta\}$$

$$\{1, \gamma, \delta\}\cdot\alpha = \{1\cdot\alpha, \gamma\cdot\alpha, \delta\cdot\alpha\} = \{\alpha, \beta, \epsilon\}$$

$$\{1, \gamma, \delta\}\cdot\beta = \{1\cdot\beta, \gamma\cdot\beta, \delta\cdot\beta\} = \{\beta, \epsilon, \alpha\}$$

$$\{1, \gamma, \delta\}\cdot\gamma = \{1\cdot\gamma, \gamma\cdot\gamma, \delta\cdot\gamma\} = \{\gamma, \delta, 1\}$$

$$\{1, \gamma, \delta\}\cdot\delta = \{1\cdot\delta, \gamma\cdot\delta, \delta\cdot\delta\} = \{\delta, 1, \gamma\}$$

$$\{1, \gamma, \delta\}\cdot\epsilon = \{1\cdot\epsilon, \gamma\cdot\epsilon, \delta\cdot\epsilon\} = \{\epsilon, \alpha, \beta\}$$

Thus, $\{1, \gamma, \delta\}$ has two distinct right cosets in $\mathcal{G}$, namely,

$$\{1, \gamma, \delta\}, \{\alpha, \beta, \epsilon\}$$

As can be readily verified, for every $a \in \{1, \alpha, \beta, \gamma, \delta, \epsilon\}$ we have $\{1, \gamma, \delta\}\cdot a = a\cdot\{1, \gamma, \delta\}$; hence, $\langle\{1, \gamma, \delta\}; \cdot\rangle$ is a normal subgroup of $\mathcal{G}$. □

THEOREM 9-16

Let $\langle C_0; *\rangle$ be a subgroup of the group $\mathcal{G} = \langle G; *\rangle$. Then:

(a) $b \in C_0 * a$ if and only if $b * a^{-1} \in C_0$.

(b) $b \in a * C_0$ if and only if $a^{-1} * b \in C_0$.

Proof. (a) $b \in C_0 * a$ if and only if there is some $c \in C_0$ such that $b = c * a$; hence, if and only if $b * a^{-1} = c$ and, hence, if and only if $b * a^{-1} \in C_0$.

(b) The proof is analogous to that of part (a). □

Note that when $\mathcal{G}$ is Abelian and additive, statements (a) and (b) above reduce to this single statement:

$$b \in C_0 + a = a + C_0 \text{ if and only if } b - a \in C_0$$

THEOREM 9-17

Let $\langle C_0; *\rangle$ be a subgroup of the group $\mathcal{G} = \langle G; *\rangle$, and let a and b be any elements of G. Then:

(a) Either $C_0 * a = C_0 * b$, or $(C_0 * a) \cap (C_0 * b) = \varnothing$.

(b) Either $a * C_0 = b * C_0$, or $(a * C_0) \cap (b * C_0) = \varnothing$.

Proof (a) If $(C_0 * a) \cap (C_0 * b) \neq \varnothing$, let c be any element in $(C_0 * a) \cap (C_0 * b)$. Hence, for some $c_1, c_2 \in C_0$, $c = c_1 * a = c_2 * b$, and $a = c_1^{-1} * c_2 * b$. Now, let a' be *any* element of $C_0 * a$. Then, for some $c' \in C_0$, $a' = c' * a = c' * c_1^{-1} * c_2 * b$. Since c', c_1, and c_2 are all in C_0 (which is closed under $*$ and satisfies the inverse law), $c' * c_1^{-1} * c_2$ is also in C_0; hence, $a' = (c' * c_1^{-1} * c_2) * b$ is in $C_0 * b$. Thus, every element of $C_0 * a$ is also an element of $C_0 * b$. Selecting an arbitrary element b' of $C_0 * b$, the argument can be repeated to show that every element of $C_0 * b$ is also an element of $C_0 * a$. In conclusion, $C_0 * a = C_0 * b$.

(b) The proof is analogous to that of part (a). □

THEOREM 9-18

Let $\langle C_0; *\rangle$ be a subgroup of the group $\mathcal{G} = \langle G; *\rangle$. Then:

(a) If $b \in C_0 * a$, then $C_0 * b = C_0 * a$.

(b) If $b \in a * C_0$, then $b * C_0 = a * C_0$.

Proof (a) Since $C_0 * b$ must contain $e * b = b$, $(C_0 * b) \cap (C_0 * a) \neq \varnothing$; hence, by Theorem 9-17, $C_0 * b = C_0 * a$.

(b) The proof is analogous to that of part (a). □

THEOREM 9-19

Let $\langle C_0, *\rangle$ be a subgroup of the group $\mathcal{G} = \langle G; *\rangle$. Then:

(a) The distinct right cosets of C_0 in $\mathcal{G}$ form a partition of G.

(b) The distinct left cosets of C_0 in $\mathcal{G}$ form a partition of G.

Proof (a) By Theorem 9-17(a), the distinct right cosets of C_0 in $\mathcal{G}$ are disjoint. Moreover, every element $a \in G$ is in some right coset—namely, $C_0 * a$ (which contains $e * a = a$).

(b) The proof is analogous to that of part (a). □

Example 9-6

In Example 9-5, the right cosets of $\{1, \alpha\}$ in $\mathcal{G} = \langle\{1, \alpha, \beta, \gamma, \delta, \epsilon\}; \cdot\rangle$ form the partition

$$\{\{1, \alpha\}, \{\beta, \delta\}, \{\gamma, \epsilon\}\}$$

and the left cosets of $\{1, \alpha\}$ in $\mathcal{G}$ form the partition

$$\{\{1, \alpha\}, \{\beta, \gamma\}, \{\delta, \epsilon\}\}$$

The left (or right) cosets of $\{1, \gamma, \delta\}$ in $\mathcal{G}$ form the partition

$$\{\{1, \gamma, \delta\}, \{\alpha, \beta, \epsilon\}\}$$

Note that, since $\{1, \gamma, \delta\}\cdot 1 = \{1, \gamma, \delta\}$, we have

$$\{1, \gamma, \delta\}\cdot 1 = \{1, \gamma, \delta\}\cdot\gamma = \{1, \gamma, \delta\}\cdot\delta$$

(in conformance with Theorem 9-18). Similarly, since $\{1, \gamma, \delta\}\cdot\alpha = \{\alpha, \beta, \epsilon\}$, we have

$$\{1, \gamma, \delta\}\cdot\alpha = \{1, \gamma, \delta\}\cdot\beta = \{1, \gamma, \delta\}\cdot\epsilon$$ □

The partition defined in Theorem 9-19(a) [or (b)] is called the *right* (or *left*) *coset partition relative to* C_0 *in* $\mathcal{G}$. This partition is the equivalence partition induced on G by the relation ρ, where $a_1 \rho a_2$ if and only if a_1 and a_2 are in the same right (or left) coset of C_0 in $\mathcal{G}$. When $\langle C_0; *\rangle$ is a normal subgroup of $\mathcal{G}$, this partition is simply called the *coset partition relative to* C_0 *in* $\mathcal{G}$. The preceding theorems suggest the following recursive procedure for constructing coset partitions:

ALGORITHM 9-1

$\langle C_0; *\rangle$ is a subgroup of the group $\mathcal{G} = \langle G; *\rangle$. To construct the (distinct) right or left cosets $C_0, C_1, C_2, \ldots$ of C_0 in $\mathcal{G}$:

(i) (*Basis*). $C_0 = \{e, c_1, c_2, c_3, \ldots\}$ is given.

(ii) (*Induction step*). Select any $a \in G$ such that $a \notin C_0 \cup C_1 \cup \ldots \cup C_i$. Then the $(i + 1)$st right coset is

$$C_{i+1} = \{a, c_1 * a, c_2 * a, c_3 * a, \ldots\} \qquad (i = 0, 1, 2, \ldots)$$

The $(i + 1)$st left coset is

$$C_{i+1} = \{a, a * c_1, a * c_2, a * c_3, \ldots\} \qquad (i = 0, 1, 2, \ldots)$$ □

Note that all cosets have the same cardinality. Also, the number of right cosets and left cosets (if finite) is the same.

Example 9-7

Consider the (Abelian) group $\langle \mathbb{I}; +\rangle$ and its subgroup $\langle C_0; +\rangle$, where C_0 is the set of all multiples of 4. Then:

$$C_0 = \{0, 4, -4, 8, -8, 12, -12, \ldots\}$$
$$C_1 = \{1, 5, -3, 9, -7, 13, -11, \ldots\}$$

$$C_2 = \{2, 6, -2, 10, -6, 14, -10, \ldots\}$$
$$C_3 = \{3, 7, -1, 11, -5, 15, -9, \ldots\}$$

By Theorem 9-16, integers i and j are in the same coset if and only if $i - j \in C_0$ and, hence, if and only if $i - j$ is a multiple of 4. □

The number of left (or right) cosets of C_0 in $\mathcal{G} = \langle G; *\rangle$ is called the *index* of C_0 in $\mathcal{G}$. In Example 9-6, the index of $\{1, \alpha\}$ in $\mathcal{G}$ is 3, and that of $\{1, \gamma, \delta\}$ is 2. A direct corollary of the preceding results is the well-known *Lagrange's theorem* (after the French mathematician Joseph L. Lagrange, 1736–1813):

THEOREM 9-20

If $\mathcal{G} = \langle G; *\rangle$ is a finite group with the subgroup $\langle C_0; *\rangle$, and d is the index of C_0 in $\mathcal{G}$, then

$$\#G = d(\#C_0)$$

From this it follows that $\#C_0$ must be a divisor of $\#G$; hence, if $\#G$ is prime, the only subgroups of $\mathcal{G}$ are $\langle\{e\}; *\rangle$ and $\mathcal{G}$.

We are now in a position to strengthen Theorem 9-11.

THEOREM 9-21

In a finite group $\mathcal{G} = \langle G; *\rangle$, the order of any element is a divisor of $\#G$.

Proof If $a \in G$ is of order r, then

$$\langle\{e, a, a^2, \ldots, a^{r-1}\}; *\rangle$$

forms a subgroup of $\mathcal{G}$. By Theorem 9-20, then, r divides $\#G$. □

A direct corollary of Theorem 9-21 is that if $\#G$ is prime, then the order of any nonidentity element in G is precisely $\#G$. In addition, this result can be proved:

THEOREM 9-22

Let $\mathcal{G} = \langle G; *\rangle$ be a finite Abelian group, where the order of each element is a power of a prime number p. Then $\#G$ is also a power of p.

PROBLEMS

1. Prove part (b) of Theorems 9-16, 9-17, 9-18, and 9-19.
2. Let $\mathcal{G} = \langle \mathbb{Z}_6; *\rangle$, where $*$ is addition modulo 6. Construct all subgroups of $\mathcal{G}$, and the cosets of each subgroup in $\mathcal{G}$.
3. Let G be the set of all rational numbers of the form $2^i 3^j$ ($i, j \in \mathbb{I}$), and $\tilde{G}$ the set of all rational numbers of the form 2^i ($i \in \mathbb{I}$). Show that $\mathcal{G} = \langle G; \cdot\rangle$ (where $\cdot$ is

ordinary multiplication) is a group, and that $\tilde{\mathcal{G}} = \langle \tilde{G}; \cdot \rangle$ is a subgroup of $\mathcal{G}$. What are the cosets of $\tilde{G}$ in $\mathcal{G}$?

4. Let $\tilde{\mathcal{G}}$ be the center of group $\mathcal{G}$. (See Problem 5, Sec. 9-5.) Show that $\tilde{\mathcal{G}}$ is a normal subgroup of $\mathcal{G}$.

5. Let $\mathcal{G} = \langle G; * \rangle$ be a group, and $\tilde{\mathcal{G}} = \langle \tilde{G}; * \rangle$ be a subgroup of $\mathcal{G}$. Define the subset H of G as follows:

$$H = \{a \mid a * \tilde{G} = \tilde{G} * a\}$$

Show that:
(a) $\mathcal{H} = \langle H; * \rangle$ is a subgroup of $\mathcal{G}$.
(b) $\tilde{\mathcal{G}}$ is a normal subgroup of $\mathcal{H}$.
(c) If $\mathcal{K} = \langle K; * \rangle$ is any subgroup of $\mathcal{G}$ such that $\tilde{G} \subseteq K$, and such that, for all $a \in K$, $\tilde{G} * a = a * \tilde{G}$, then $\mathcal{K}$ is a subgroup of $\mathcal{H}$.

6. Prove that if $\langle C_0; * \rangle$ is a subgroup of $\mathcal{G}$ and if the index of C_0 in $\mathcal{G}$ is 2, then $\langle C_0; * \rangle$ must be a normal subgroup of $\mathcal{G}$.

7. Let $\mathcal{G} = \langle G; * \rangle$ be a finite group where $\# G = mk$, and $\tilde{\mathcal{G}} = \langle \tilde{G}; * \rangle$ be a subgroup of $\mathcal{G}$ where $\# G = m$. Show that, for any $a \in G$, there is an integer ν, where $1 \leq \nu \leq k$, such that $a^\nu \in \tilde{G}$. (*Hint*: Consider the sets $\tilde{G}$, $a * \tilde{G}$, $a^2 * \tilde{G}, \ldots$)

8. Let $\mathcal{G} = \langle G; * \rangle$ be a group and $\langle C_0; * \rangle$ be a normal subgroup of $\mathcal{G}$. Show that the relation ρ, which induces the coset partition relative to C_0 in $\mathcal{G}$, is a congruence relation on $\mathcal{G}$.

REFERENCES

BURNSIDE, W., *Theory of Groups of Finite Order*. New York: Dover, 1955.

CARMICHAEL, R. D., *Introduction to the Theory of Groups of Finite Order*. New York: Dover, 1956.

CHEVALLEY, C., *Fundamental Concepts of Algebra*. New York: Academic Press, 1956.

DICKSON, L. E., *Linear Groups*. New York: Dover, 1958.

HALL, M., *Theory of Groups*. New York: Macmillan, 1959.

HERSTEIN, I., *Topics in Algebra*. Waltham, MA: Xerox, 1964.

LEDERMAN, W., *Introduction to the Theory of Finite Groups*. New York: Interscience, 1953.

LJAPIN, S., *Theory of Semigroups*. Providence, RI: American Mathematical Society, 1963.

MACLANE, S., and G. BIRKHOFF, *Algebra*. New York: Macmillan, 1967.

VAN DER WAERDEN, B. L., *Modern Algebra*. Ungar, 1931.

WIELANDT, H., *Finite Permutation Groups*. New York: Academic Press, 1964.

10 RINGS AND FIELDS

Our discussion of algebraic systems continues with the study of *rings*—highly structured systems with two binary operations. The properties of a special type of subring—the *ideal*—are developed, and it is shown that ideals in rings play a role which is analogous to that played by normal subgroups in groups. Additional structure is imposed on rings to result in an algebraic system known as a *field*, which is perhaps the most commonly encountered in practice. At this point we turn to study properties of rings formed from *polynomials* whose coefficients are elements of rings or fields. In particular, we explore the ideals of such rings, which helps us derive the important *Euclidean algorithm* and *prime factorization theorem* for polynomials. A discussion of *extension fields* leads us to the study of finite fields and to the fundamental result that every finite field is isomorphic to a *Galois field* whose cardinality is a power of a prime number. The chapter concludes with a cursory introduction to *vector spaces* and *matrices* over fields; as these subjects are beyond the scope of this book, only the most basic notions are exposed and results are stated mostly without proofs.

Like the knowledge of group theory, the knowledge of the theory of rings, ideals, fields, polynomials, and vector spaces is essential in the study of error detecting and correcting codes and their physical implementations. These, in turn, are of interest to the computer scientist because of their applicability to fault-combatting schemes in computing systems, the design of counters and arithmetic units, the generation of "random" sequences, and numerous other digital applications.

10-1. RINGS

In this chapter we deal with algebraic systems with *two* binary operations. We start with a discussion of an algebraic system called a *ring*, where the two operations are commonly denoted by $+$ (addition) and $\cdot$ (multiplication).[1] A ring is characterized by the fact that it is an Abelian group under addition, and that its multiplication operation is associative and distributive over addition. Thus, formally:

The algebraic system $\mathcal{R} = \langle R; +, \cdot \rangle$ is a *ring* if it is an Abelian group with respect to $+$, and if it satisfies these laws:

(a) *Associativity.* For all $a, b, c \in R$,

$$a(bc) = (ab)c$$

(b) *Distributivity.* For all $a, b, c \in R$,

$$a(b + c) = ab + ac, \qquad (b + c)a = ba + ca$$

The distributivity law can be readily extended by induction to the following: For all $a, b_1, b_2, \ldots, b_k \in R$,

$$a(b_1 + b_2 + \ldots + b_k) = ab_1 + ab_2 + \ldots + ab_k$$
$$(b_1 + b_2 + \ldots + b_k)a = b_1a + b_2a + \ldots + b_ka$$

A ring $\mathcal{R} = \langle R; +, \cdot \rangle$ is said to be *commutative* if, in addition to (a) and (b), it satisfies this law:

(c) *Commutativity.* For all $a, b \in R$,

$$ab = ba$$

A ring $\mathcal{R} = \langle R; +, \cdot \rangle$ may or may not satisfy these laws:

(d) *Identity.* There exists an element $1 \in R$ such that, for all $a \in R$,

$$1a = a1 = a$$

(1 is sometimes called the *multiplicative identity* of $\mathcal{R}$, to distinguish it from the *additive identity* 0.)

(e) *Cancellation.* If $a \neq 0$, then, for all $b, c \in R$,

$$ab = ac \text{ implies } b = c$$
$$ba = ca \text{ implies } b = c$$

[1]The conventional notation for additive and multiplicative groups is retained for rings; that is, the additive inverse of a is denoted by $-a$ [with $a + (-b)$ commonly written as $a - b$], and the multiplicative inverse by a^{-1}. The product $a \cdot b$ is usually written as ab. The nth power of a under $+$ is denoted by na, and under $\cdot$ by a^n. With the absence of parentheses we assume that exponentiation has precedence over multiplication which, in turn, has precedence over addition.

Perhaps the most familiar ring is $\langle \mathbb{I}; +, \cdot \rangle$ (where $+$ and $\cdot$ are ordinary addition and multiplication);[1] it is a commutative ring with multiplicative identity, satisfying the cancellation law. For any positive integer m, the algebraic system $\mathcal{Z}_m = \langle \mathbb{Z}_m; +, \cdot \rangle$, where $+$ and $\cdot$ are addition and multiplication modulo m,[2] is a ring. For example, $\langle \mathbb{Z}_3; +, \cdot \rangle$ is a commutative ring with multiplicative identity, satisfying the cancellation law (see Table 10-1). The system $\langle \mathbb{Z}_4; +, \cdot \rangle$ is a commutative ring with multiplicative identity, which does not satisfy the cancellation law (as seen from Table 10-2, $2a = 2c$ does not imply $a = c$). The system $\langle \mathbb{R}[x]; +, \cdot \rangle$, where $\mathbb{R}[x]$ is the set of all polynomials in x with real coefficients, and $+$ and $\cdot$ are ordinary polynomial addition and multiplication, is a commutative ring with multiplicative identity (namely, the polynomial 1), satisfying the cancellation law.

Table 10-1 OPERATION TABLES FOR $\langle \mathbb{Z}_3; +, \cdot \rangle$

$+$	0	1	2
0	0	1	2
1	1	2	0
2	2	0	1

$\cdot$	0	1	2
0	0	0	0
1	0	1	2
2	0	2	1

Table 10-2 OPERATION TABLES FOR $\langle \mathbb{Z}_4; +, \cdot \rangle$

$+$	0	1	2	3
0	0	1	2	3
1	1	2	3	0
2	2	3	0	1
3	3	0	1	2

$\cdot$	0	1	2	3
0	0	0	0	0
1	0	1	2	3
2	0	2	0	2
3	0	3	2	1

Comparing the definition of a ring with the definition of an integral domain (see Sec. 4-4), it is seen that *an integral domain is equivalent to a commutative ring with multiplicative identity, satisfying the cancellation law.*

THEOREM 10-1

Let $\mathcal{R} = \langle R; +, \cdot \rangle$ be a ring. Then, for all $a, b \in R$,

(a) $a0 = 0a = 0$

(b) $(-a)b = a(-b) = -(ab)$

[1] Henceforth, the $+$ and $\cdot$ in $\langle \mathbb{I}; +, \cdot \rangle$ are understood to denote ordinary addition and multiplication.

[2] Henceforth, the $+$ and $\cdot$ in $\langle \mathbb{Z}_m; +, \cdot \rangle$ are understood to denote addition and multiplication modulo m. (To compute modulo m, first compute as per rules of $\langle \mathbb{I}; +, \cdot \rangle$ and then replace the result with its remainder upon division by m.)

Proof (a) $0 = a0 - a0 = a(0 + 0) - a0$

$$= a0 + a0 - a0 = a0$$

Similarly, it can be shown that $0 = 0a$.

(b) $(-a)b = ab + (-a)b - (ab) = (a + (-a))b - (ab)$

$$= (a - a)b - (ab) = 0b - (ab) = 0 - (ab)$$

$$= -(ab)$$

Similarly, it can be shown that $a(-b) = -(ab)$. □

We say that a ring $\mathcal{R} = \langle R; +, \cdot \rangle$ has *divisors of zero* if it has *nonzero* elements a and b such that $ab = 0$. For example, $\langle \mathbb{Z}_4; +, \cdot \rangle$ has divisors of zero (since $2 \cdot 2 = 0$).

THEOREM 10-2

A ring $\mathcal{R} = \langle R; +, \cdot \rangle$ has no divisors of zero if and only if $\mathcal{R}$ satisfies the cancellation law.

Proof Suppose $\mathcal{R}$ has no divisors of zero, and let $a, b, c \in R$, with $a \neq 0$, be such that $ab = ac$. Hence, $ab - ac = 0$, or $a(b - c) = 0$, which implies (since there are no divisors of zero) that $b - c = 0$ and, hence, $b = c$. Thus, $\mathcal{R}$ satisfies the cancellation law.

Conversely, suppose $\mathcal{R}$ satisfies the cancellation law, and let $a, b \in R$ be such that $ab = 0$. Hypothesizing that $a \neq 0$, we have $ab - a0 = 0$, or $ab = a0$, which implies (by the cancellation law) that $b = 0$. Similarly, if $b \neq 0$, then $a = 0$. Thus, $\mathcal{R}$ has no divisors of zero. □

Given a ring $\mathcal{R} = \langle R; +, \cdot \rangle$, a subsystem $\tilde{\mathcal{R}} = \langle \tilde{R}; +, \cdot \rangle$ of $\mathcal{R}$ is any algebraic system such that $\tilde{R}$ is a nonempty subset of R and is closed under both $+$ and $\cdot$ (see Sec. 4-2). If $\tilde{\mathcal{R}}$ is a ring, it is said to be a *subring* of $\mathcal{R}$. If, in addition, $\tilde{R}$ is a proper subset of R, then $\tilde{\mathcal{R}}$ is said to be a *proper subring* of $\mathcal{R}$.

THEOREM 10-3

Let $\mathcal{R} = \langle R; +, \cdot \rangle$ be a ring, and $\tilde{\mathcal{R}} = \langle \tilde{R}; +, \cdot \rangle$ be a subsystem of $\mathcal{R}$. Then $\tilde{\mathcal{R}}$ is a subring of $\mathcal{R}$ if and only if it satisfies the inverse law with respect to $+$.

Proof By Theorem 9-14, $\tilde{\mathcal{R}}$ is an additive group if and only if it satisfies the inverse law with respect to $+$. In addition, commutativity, associativity, and distributivity (with respect to either $+$ or $\cdot$) are automatically inherited by $\tilde{\mathcal{R}}$ from $\mathcal{R}$. □

Thus, if $\mathcal{R} = \langle R; +, \cdot \rangle$ is a ring, to establish whether a nonempty subset $\tilde{R}$ of R forms a subring of $\mathcal{R}$, it is sufficient to establish:

(a) *Closure*. For any $a, b \in \tilde{R}$, $a + b$ and ab are in $\tilde{R}$.
(b) *Inverse*. For any $a \in \tilde{R}$, $-a$ is also in $\tilde{R}$.

If $\mathcal{R}$ is commutative, the commutativity property is inherited by $\tilde{\mathcal{R}}$; hence, any subring of a commutative ring must also be commutative.

As an example, the set of all multiples of 6 forms a commutative subring of $\langle \mathbb{I}; +, \cdot \rangle$; the set $\{0, 2\}$ forms a commutative subring of $\langle \mathbb{Z}_4; +, \cdot \rangle$ (since $-2 = 2$).

PROBLEMS

1. Verify that $\langle \mathbb{I}; +, \cdot \rangle$ and $\langle \mathbb{Z}_m; +, \cdot \rangle$ are rings.
2. Let $\mathcal{R} = \langle R; +, \cdot \rangle$ be a ring, and let a, b, and c be arbitrary elements in R. Show that:
   (a) If $ab = ba$, then $a(-b) = (-b)a$, $a(nb) = (nb)a$ (with any nonnegative integer n), and $a(b^{-1}) = (b^{-1})a$.
   (b) If $ab = ba$ and $ac = ca$, then $a(b + c) = (b + c)a$, and $a(bc) = (bc)a$.
3. Let $\mathcal{R} = \langle R; +, \cdot \rangle$ be a ring, and define a subset G of R as

$$G = \{a \mid a^{-1} \text{ is defined in } \mathcal{R}\}$$

   Show that $\langle G; \cdot \rangle$ is a group.
4. Consider the ring $\langle \mathbb{Z}; +, \cdot \rangle$ (where $+$ and $\cdot$ are ordinary addition and multiplication) and the arbitrary ring $\mathcal{R} = \langle R; +, \cdot \rangle$. Define the algebraic system $\mathcal{S} = \langle \mathbb{Z} \times R; +, \cdot \rangle$ where, for all $i, j \in \mathbb{Z}$, and $r, s \in R$,

$$(i, r) + (j, s) = (i + j, r + s)$$
$$(i, r) \cdot (j, s) = (ij, is + jr + rs)$$

   Show that $\mathcal{S}$ is a ring.
5. Consider the ring $\langle \{5v \mid v \in \mathbb{I}\}; +, \cdot \rangle$ (where $+$ and $\cdot$ are ordinary addition and multiplication). Is it an integral domain?
6. Let $\mathcal{R} = \langle R; +, \cdot \rangle$ be a ring where, for all $a \in R$, $a^2 = a$. (Such a ring is called a *Boolean ring*.)
   (a) Show that $\mathcal{R}$ must be commutative.
   (b) Show that, for all $a \in R$, $a + a = 0$.
   (c) Show that, if $\# R > 2$, then $\mathcal{R}$ cannot be an integral domain.
7. Let R be the set of all 2×2 matrices whose entries are real numbers.
   (a) Show that $\mathcal{R} = \langle R; +, \cdot \rangle$, where $+$ and $\cdot$ are matrix addition and multiplication, is a ring.

(b) Show that if $\tilde{R}$ is the set of all 2×2 matrices of the form

$$\begin{bmatrix} a & b \\ 0 & c \end{bmatrix}$$

(where a, b, and c are real numbers), then $\langle \tilde{R}; +, \cdot \rangle$ is a subring of $\mathfrak{R}$.

8. Let $\mathfrak{R} = \langle R; +, \cdot \rangle$ be a ring, where R is the set of all (continuous) functions whose domain is the set of real numbers between -1 and $+1$ (inclusively) and whose range is $\mathbb{R}$, and where $+$ and $\cdot$ are defined as:

$$(f + g)(x) = f(x) + g(x)$$
$$(f \cdot g)(x) = f(x) \cdot g(x)$$

(the $+$ and $\cdot$ on the right denoting ordinary addition and multiplication). Show that $\mathfrak{R}$ has divisors of zero.

10-2. IDEALS

Let $\mathfrak{R} = \langle R; +, \cdot \rangle$ be a ring. A subring $\mathfrak{D} = \langle D; +, \cdot \rangle$ of $\mathfrak{R}$ is called an *ideal* of $\mathfrak{R}$ if, for all $a \in R$ and $d \in D$, ad and da are in D.

If $D = R$, or if $D = \{0\}$, then $\mathfrak{D} = \langle D; +, \cdot \rangle$ is, trivially, an ideal of $\mathfrak{R}$. In these cases $\mathfrak{D}$ is called a *trivial ideal.* If D is a proper subset of R, the ideal $\langle D; +, \cdot \rangle$ is said to be a *proper ideal* of $\mathfrak{R}$. If $\mathfrak{D}$ is a proper ideal of $\mathfrak{R}$, and if no proper ideal $\mathfrak{D}' = \langle D'; +, \cdot \rangle$ of $\mathfrak{R}$ exists such that D is a proper subset of D', then $\mathfrak{D}$ is called a *maximal ideal* of $\mathfrak{R}$.

If $\mathfrak{D} = \langle D; +, \cdot \rangle$ is an ideal such that $D = Rg$ for some $g \in D$,[1] then $\mathfrak{D}$ is called a *principal ideal.* In this case g is called the *generator* of $\mathfrak{D}$, $\mathfrak{D}$ is said to be *generated* by g, and D is denoted by (g). A ring in which every ideal is a principal ideal is called a *principal ideal ring*.

Example 10-1

Consider the subring of the ring $\mathfrak{I} = \langle \mathbb{I}; +, \cdot \rangle$, given by

$$\mathfrak{D}_m = \langle \{mi \mid i \in \mathbb{I}\}; +, \cdot \rangle$$

where m is some (fixed) nonnegative integer. Clearly, $\mathfrak{D}_m$ is an ideal of $\mathfrak{I}$; moreover, since $\{mi \mid i \in \mathbb{I}\} = \mathbb{I}m$, $\mathfrak{D}_m$ is precisely the principal ideal $\langle (m); +, \cdot \rangle$. In fact, every ideal of $\mathfrak{I}$ is a principal ideal $\langle (m); +, \cdot \rangle$ for some nonnegative integer m. This fact can be shown as follows: When the ideal is $\langle \{0\}; +, \cdot \rangle$, then, trivially, $\{0\} = (0)$. When the ideal is $\mathfrak{D} = \langle D; +, \cdot \rangle$

[1] The notation Ta and $T + a$ (where T is a set and a is an element) are the additive and multiplicative versions of the notation $T * a$ introduced in Sec. 9-1.

where $D \neq \{0\}$, let m be the least positive integer in D. For any $n \in D$ we can write $n = mi + r$, where $i \in \mathbb{I}$ and $0 \leq r < m$. Since $mi \in D$, so is $n - mi = r$. But, by assumption, m is the least positive integer in D; hence, we must have $r = 0$ and $n = mi$. Thus, $D = (m)$. In conclusion, then, $\mathcal{I}$ is a principal ideal ring. □

By defintion, a ring $\mathcal{R} = \langle R; +, \cdot \rangle$ is commutative with respect to $+$; thus, every ideal $\mathfrak{D} = \langle D; +, \cdot \rangle$ is a normal additive subgroup of $\mathcal{R}$. By Problem 8, Sec. 9-6, the relation ρ, which induces the coset partition relative to D in $\mathcal{R}$, satisfies the substitution property with respect to $+$. That ρ also satisfies the substitution property with respect to $\cdot$ can be shown as follows: Consider the elements $a, b \in R$, belonging to cosets $D + c_1$ and $D + c_2$, respectively. For some $d_1, d_2 \in D$,

$$ab = (d_1 + c_1)(d_2 + c_2) = d_1 d_2 + d_1 c_2 + c_1 d_2 + c_1 c_2$$

Since $\mathfrak{D}$ is an ideal, $d_1 d_2$, $d_1 c_2$, and $c_1 d_2$ are in D, and so is $d_1 d_2 + d_1 c_2 + c_1 d_2$. Hence, $ab \in D + c_1 c_2$. Similarly, if $a', b' \in R$ belong to $D + c_1$ and $D + c_2$, respectively, then $a'b' \in D + c_1 c_2$. Thus, $a \rho a'$ and $b \rho b'$ imply $(ab)\rho(a'b')$ which implies, in turn, that ρ satisfies the substitution property with respect to $\cdot$. In conclusion, then, ρ is a congruence relation on $\mathcal{R}$. Combining this result with Theorem 4-14, we have:

THEOREM 10-4

Let $\mathfrak{D} = \langle D; +, \cdot \rangle$ be an ideal in the ring $\mathcal{R} = \langle R; +, \cdot \rangle$, and let ρ be the relation that induces the (additive) coset partition relative to D. Then $\mathcal{R}/\rho$ is an epimorphic image of $\mathcal{R}$.

From Theorem 4-11 it follows that $\mathcal{R}/\rho$ (as well as any other epimorphic image of $\mathcal{R}$) is a ring. This ring is called a *quotient ring* (or a *factor ring*) of $\mathcal{R}$ and is often denoted by $\mathcal{R}/D$ (rather than $\mathcal{R}/\rho$).

Consider now an arbitrary epimorphism h from the ring $\mathcal{R}_1 = \langle R_1; +, \cdot \rangle$ onto the ring $\mathcal{R}_2 = \langle R_2; +, \cdot \rangle$. Clearly, h is also an epimorphism from the additive group of $\mathcal{R}_1$ onto the additive group of $\mathcal{R}_2$. It can be readily shown that the set

$$K = \{a \in R_1 \mid h(a) = 0\}$$

(which is called the *kernel* of h) forms an additive (normal) subgroup of $\mathcal{R}_1$ and, hence (by Theorem 10-3), $\langle K; +, \cdot \rangle$ is a subring of $\mathcal{R}_1$. Moreover, for any $a \in R_1$ and $c \in K$,

$$h(ac) = h(a)h(c) = h(a)0 = 0$$

and, hence, $ac \in K$. Similarly, ca $\in K$. We can thus state:

THEOREM 10-5

Let K be the kernel of the epimorphism h from ring $\mathcal{R}_1 = \langle R_1; +, \cdot\rangle$ onto ring $\mathcal{R}_2 = \langle R_2; +, \cdot\rangle$. Then $\langle K; +, \cdot\rangle$ is an ideal of $\mathcal{R}_1$.

Combining Theorems 10-4 and 10-5, we have:

THEOREM 10-6

The ring $\mathcal{R}_2 = \langle R_2; +, \cdot\rangle$ is an epimorphic image of the ring $\mathcal{R}_1 = \langle R_1; +, \cdot\rangle$ if and only if $\mathcal{R}_2$ is isomorphic to a quotient group $\mathcal{R}_1/D$, where $\langle D; +, \cdot\rangle$ is some ideal of $\mathcal{R}_1$.

Example 10-2

Consider the ring $\mathcal{I} = \langle \mathbb{I}; +, \cdot\rangle$ and its ideal

$$\mathfrak{D}_m = \langle D_m; +, \cdot\rangle = \langle (m); +, \cdot\rangle$$

(for some positive integer m). By Theorem 9-16, i_1 and i_2 are in the same (additive) coset relative to D_m if and only if $i_1 - i_2 \in D_m$ and, hence, if and only if, for some $\nu \in \mathbb{I}$, $i_1 - i_2 = \nu m$. Thus, i_1 and i_2 are in the same coset if and only if $\text{res}_m(i_1) = \text{res}_m(i_2)$. The coset partition relative to D_m is, therefore,

$$\{D_m, D_m + 1, D_m + 2, \ldots, D_m + m - 1\}$$

The quotient ring $\mathcal{I}/D_m$ is an epimorphic image of $\mathcal{I}$, under the epimorphism

$$h\colon\ \mathbb{I} \longrightarrow \{D_m, D_m + 1, D_m + 2, \ldots, D_m + m - 1\}$$

where $\qquad h(i) = D_m + \text{res}_m(i)$

(The kernel of h is D_m, which is also the zero of $\mathcal{I}/D_m$.) From Example 4-4 it is apparent that $\mathcal{I}/D_m$ is isomorphic to $\mathcal{Z}_m = \langle \mathbb{Z}_m; +, \cdot\rangle$. □

The proofs of the following results will be left to the reader:

THEOREM 10-7

Let $\mathfrak{D}_1 = \langle D_1; +, \cdot\rangle$ and $\mathfrak{D}_2 = \langle D_2; +, \cdot\rangle$, where $D_1 \subset D_2$, be ideals of the ring $\mathcal{R} = \langle R; +, \cdot\rangle$. Then $\mathcal{R}/D_2$ is an epimorphic image of $\mathcal{R}/D_1$. $\mathcal{R}/D_2$ is isomorphic to $\mathcal{R}/D_1$ if and only if $D_1 = D_2$.

THEOREM 10-8

Let $\mathfrak{D} = \langle D; +, \cdot\rangle$ be an ideal of the ring $\mathcal{R} = \langle R; +, \cdot\rangle$. If $\mathcal{R}/D$ has an epimorphic image $\mathcal{R}'$, then $\mathcal{R}$ must have an ideal $\mathfrak{D}' = \langle D'; +, \cdot\rangle$ such that $\mathcal{R}/D'$ is isomorphic to $\mathcal{R}'$ and $D \subset D'$. $\mathcal{R}'$ is isomorphic to $\mathcal{R}/D$ if and only if $D = D'$.

When the only ideals of the ring $\mathcal{R} = \langle R; +, \cdot\rangle$ are the trivial ones, the only quotient rings of $\mathcal{R}$ are the one consisting of the zero only, and the one that is isomorphic to $\mathcal{R}$. These quotient rings are sometimes referred to as the *trivial epimorphic images* of $\mathcal{R}$.

Let $\mathfrak{D} = \langle D; +, \cdot\rangle$ be an ideal of the ring $\mathcal{R} = \langle R; +, \cdot\rangle$. $\mathfrak{D}$ is said to be a *prime ideal* if, for any $a, b \in R$, $ab \in D$ implies $a \in D$ or $b \in D$. For example, the ideal $\langle\{5i \mid i \in \mathbb{I}\}; +, \cdot\rangle$ of $\mathcal{I} = \langle \mathbb{I}; +, \cdot\rangle$ is a prime ideal, since $ab = 5i$ implies (by the prime factorization theorem) that either a or b is a multiple of 5 and, hence, in the ideal. The ideal $\langle\{6i \mid i \in \mathbb{I}\}; +, \cdot\rangle$, however, is not a prime ideal, since, for example, $6\cdot 1 = 2\cdot 3$ and neither 2 nor 3 is a multiple of 6.

THEOREM 10-9

Let $\mathfrak{D} = \langle D; +, \cdot\rangle$ be an ideal of a commutative ring $\mathcal{R} = \langle R; +, \cdot\rangle$ with multiplicative identity. Then $\mathcal{R}/D$ is an integral domain if and only if $\mathfrak{D}$ is a prime ideal.

Proof By Theorem 4-11, $\mathcal{R}/D$ is a commutative ring with multiplicative identity. It will, therefore, suffice to prove that $\mathcal{R}/D$ satisfies the cancellation law (or, equivalently, has no divisors of zero) if and only if $\mathfrak{D}$ is a prime ideal.

Let C_1 and C_2 be any elements of $\mathcal{R}/D$. Then we can write $C_1 = D + a_1$, $C_2 = D + a_2$, and, hence,

$$C_1C_2 = (D + a_1)(D + a_2) = D(D + a_1 + a_2) + a_1a_2 = D + a_1a_2$$

Now, $\mathfrak{D}$ is prime if and only if $a_1a_2 \in D$ implies $a_1 \in D$ or $a_2 \in D$; hence, if and only if $C_1C_2 = D$ implies $C_1 = D$ or $C_2 = D$ and, (since D is the zero of $\mathcal{R}/D$), if and only if $\mathcal{R}/D$ has no divisors of zero. □

THEOREM 10-10

The ring $\mathcal{Z}_m = \langle \mathbb{Z}_m; +, \cdot\rangle$ is an integral domain if and only if m is a prime number.

Proof From Example 10-2 it follows that $\mathcal{Z}_m$ is isomorphic to $\mathcal{I}/D_m$, where $\mathcal{I} = \langle \mathbb{I}; +, \cdot\rangle$ and $D_m = (m)$. When m is prime, for any $ab \in D_m$ we can write $ab = mi$ such that either a or b is a multiple of m and, hence, for any $ab \in D_m$ either $a \in D_m$ or $b \in D_m$. Thus, when m is prime, $\langle D_m; +, \cdot\rangle$ is a prime ideal and, by Theorem 10-9, $\mathcal{I}/D_m$ is an integral domain.

When m is not a prime number, say, $m = m_1m_2$ $(1 < m_1 < m, 1 < m_2 < m)$, then for some $ab \in D_m$ we can write $ab = m1 = m_1m_2$ such that neither a nor b is a multiple of m. Thus, $\langle D_m; +, \cdot\rangle$ is not a prime ideal and (again by Theorem 10-9), $\mathcal{I}/D_m$ is not an integral domain. □

PROBLEMS

1. Find all the subrings, ideals, and epimorphic images of the rings $\langle \mathbb{Z}_m; +, \cdot \rangle$, where:
   (a) $m = 6$ (b) $m = 8$ (c) $m = 11$ (d) $m = 12$

2. Let $\mathfrak{D}_1 = \langle D_1; +, \cdot \rangle$ and $\mathfrak{D}_2 = \langle D_2; +, \cdot \rangle$ be ideals of $\mathfrak{R}$, a commutative ring with multiplicative identity. Let

$$S = \{a_1b_1 + a_2b_2 + \ldots + a_kb_k \mid a_1, a_2, \ldots, a_k \in D_1, b_1, b_2, \ldots, b_k \in D_2, k \geq 1\}$$

   Show that $\langle S; +, \cdot \rangle$ is an ideal of $\mathfrak{R}$.

3. What are the ideals of $\mathfrak{I} \times \mathfrak{I}$ (where $\mathfrak{I} = \langle \mathbb{I}; +, \cdot \rangle$)?

4. Show that, if $\mathfrak{D}_1 = \langle D_1; +, \cdot \rangle$ and $\mathfrak{D}_2 = \langle D_2; +, \cdot \rangle$ are ideals of $\mathfrak{R}$, so are:
   (a) $\langle D_1 + D_2; +, \cdot \rangle$ (b) $\langle D_1 \cap D_2; +, \cdot \rangle$

5. Let $\mathfrak{R}$ be a ring and $S = \{D_1, D_2, D_3, \ldots\}$ be the set of domains of all the ideals of $\mathfrak{R}$. For all $D_i, D_j \in S$, define

$$D_i + D_j = D_i \vee D_j$$
$$D_i \cap D_j = D_1 \wedge D_j$$

   Show that $\langle S; \vee, \wedge \rangle$ forms a lattice under the partial ordering $\subset$.

6. Show that every prime ideal of $\mathfrak{I} = \langle \mathbb{I}; +, \cdot \rangle$ is maximal.

7. Show that every maximal ideal must be a prime ideal.

8. Prove Theorems 10-7 and 10-8.

10-3. FIELDS

If the algebraic system $\mathfrak{F} = \langle F; +, \cdot \rangle$ is a commutative ring with multiplicative identity, in which every nonzero element has an inverse with respect to $\cdot$, then it is called a *field*. Thus:

The algebraic system $\mathfrak{F} = \langle F; +, \cdot \rangle$ is a *field* if it is a commutative ring and if it satisfies these laws:

(a) *Identity*. There exists an element $1 \in F$ such that, for all $a \in F$,

$$1a = a1 = a$$

(b) *Inverse*. For every $a \in F$ such that $a \neq 0$, there exists an element $a^{-1} \in F$ such that

$$a^{-1}a = aa^{-1} = 1$$

(a^{-1} is sometimes called the *multiplicative inverse* of a, to distinguish it from the *additive inverse* $-a$.)

It is understood that the additive and multiplicative identities of $\mathfrak{F}$ are distinct, so that the cardinality of any field must be at least 2.

Perhaps the most familiar fields are $\langle \mathbb{R}; +, \cdot \rangle$ (where $\mathbb{R}$ is the set of real numbers) and $\langle \mathbb{Q}; +, \cdot \rangle$ (where $\mathbb{Q}$ is the set of rational numbers). The ring $\langle \mathbb{Z}_3; +, \cdot \rangle$ (see Table 10-1) is also a field, since it is commutative, has a multiplicative identity (namely, 1), and each of its nonzero elements has a multiplicative inverse ($1^{-1} = 1, 2^{-1} = 2$).

THEOREM 10-11

Every field satisfies the cancellation law (and, hence, has no divisors of zero).

Proof Let the field be $\mathfrak{F} = \langle F; +, \cdot \rangle$, and let $a \neq 0$, b, and c be elements in F. If $ab = ac$, then $a^{-1}ab = a^{-1}ac$ and, hence, $b = c$. □

From Theorem 10-11 we immediately have:

THEOREM 10-12

Every field is an integral domain.

The converse of this theorem, while generally false (for example, $\langle \mathbb{I}; +, \cdot \rangle$ is not a field), does hold when the field is finite:

THEOREM 10-13

Every finite integral domain is a field.

Proof Let $\mathfrak{F} = \langle F; +, \cdot \rangle$ be a finite integral domain. If a and b are any elements in F such that $a \neq b$, then, for all nonzero $c \in F$ (by the cancellation law), $ac \neq bc$. Hence, $Fc = F$ and for some $d \in F$ we have $dc = 1$. Thus, every nonzero $c \in F$ has a multiplicative inverse in $\mathfrak{F}$, which implies that $\mathfrak{F}$ is a field. □

The following theorem offers an alternative characterization of a field.

THEOREM 10-14

Let $\mathfrak{F} = \langle F; +, \cdot \rangle$ be a commutative ring with multiplicative identity. Then $\mathfrak{F}$ is a field if and only if every epimorphic image of $\mathfrak{F}$ is trivial.

Proof Assume that every epimorphic image of $\mathfrak{F}$ is trivial. Let c be any nonzero element in F, and consider the set Fc. Since, for all $a \in F$ and all $a'c \in Fc$, $aa'c$ and $caa' = aa'c$ are in Fc, $\langle Fc; +, \cdot \rangle$ is an ideal of $\mathfrak{F}$. Since every epimorphic image of $\mathfrak{F}$ is trivial, either $Fc = \{0\}$ or $Fc = F$; but, since

$c \neq 0$, $Fc \neq \{0\}$ and, hence, we must have $Fc = F$. Thus, $1 \in Fc$ and there must be some $a \in F$ such that $ac = 1$. In conclusion, every nonzero $c \in F$ has a multiplicative inverse; hence, $\mathfrak{F}$ is a field.

Conversely, assume that $\mathfrak{F}$ is a field, and let $\mathfrak{D} = \langle D; +, \cdot \rangle$ be any ideal of $\mathfrak{F}$ such that $D \neq \{0\}$. If c is any nonzero element of D, for any $a \in F$ we have $(ac^{-1})c = a \in D$ and, hence, $D = F$. Thus, $\mathfrak{D}$ must be a trivial ideal; hence, every epimorphic image of $\mathfrak{F}$ must also be trivial. □

Theorem 10-14 implies that, unlike quotient rings, quotient fields are completely uninteresting.

THEOREM 10-15

Let $\mathfrak{D} = \langle D; +, \cdot \rangle$ be an ideal of a commutative ring $\mathfrak{R} = \langle R; +, \cdot \rangle$ with multiplicative identity. Then $\mathfrak{R}/D$ is a field if and only if $\mathfrak{D}$ is a maximal ideal.

Proof Assume that $\mathfrak{D}$ is *not* a maximal ideal. Then there exists an ideal $\mathfrak{D}' = \langle D'; +, \cdot \rangle$ of $\mathfrak{R}$ such that D is a proper subset of D'. By Theorem 10-7, $\mathfrak{R}/D'$ is a nontrivial epimorphic image of $\mathfrak{R}/D$. By Theorem 10-14, then, $\mathfrak{R}/D$ is *not* a field.

Conversely, assume that $\mathfrak{R}/D$ is *not* a field. By Theorem 10-14, $\mathfrak{R}/D$ has a nontrivial epimorphic image $\mathfrak{R}'$. By Theorem 10-8, $\mathfrak{R}$ has an ideal $\mathfrak{D}' = \langle D'; + \cdot \rangle$ such that $\mathfrak{R}/D'$ is isomorphic to $\mathfrak{R}'$ and $D \subset D'$. Since $\mathfrak{R}'$ is nontrivial, $D' \neq D$, hence, $\mathfrak{D}$ cannot be a maximal ideal of $\mathfrak{R}$.

In conclusion, $\mathfrak{R}/D$ is *not* a field if and only if $\mathfrak{D}$ is *not* a maximal ideal—which implies the theorem. □

Example 10-3

Consider the ring $\mathfrak{I} = \langle \mathbb{I}; +, \cdot \rangle$ and its proper ideal

$$\mathfrak{D}_p = \langle D_p; +, \cdot \rangle = \langle (p); +, \cdot \rangle$$

where p is a prime number. Suppose there exists another ideal $\mathfrak{D}' = \langle D'; +, \cdot \rangle$ such that D_p is a proper subset of D'. From Example 10-1 it follows that $D' = (m)$ for some nonnegative integer m; hence, $p = mi$ for some integer i. But, since p is prime, we must have $i = 1, p = m, (p) = (m)$ and, hence, $D_p = D'$. Thus, D_p cannot be a proper subset of D'; hence, $\mathfrak{D}_p$ is a maximal ideal of $\mathfrak{I}$. From Theorem 10-15 it follows, then, that $\mathfrak{I}/D_p$ is a field. Since $\mathfrak{Z}_p = \langle \mathbb{Z}_p; +, \cdot \rangle$ is isomorphic to $\mathfrak{I}/D_p$ (see Example 10-2), it follows that $\mathfrak{Z}_p$ is a field. □

The definition of a field implies that, if $\mathfrak{F} = \langle F; +, \cdot \rangle$ is a field, then $\langle F; + \rangle$ is an additive Abelian group (referred to as the *additive group* of $\mathfrak{F}$). Since, for every nonzero $a, b \in F$, ab is nonzero (see Theorem 10-11), it follows that $\langle F - \{0\}; \cdot \rangle$ forms a multiplicative Abelian group (referred to as the *multiplicative* group of $\mathfrak{F}$. A nonzero element $a \in F$, therefore, has two orders associated with it: an *additive order* (which equals the least positive integer r such that $ra = 0$), and a *multiplicative order* (which equals the least positive integer r such that $a^r = 1$).

THEOREM 10-16

Let $\mathfrak{F} = \langle F; +, \cdot \rangle$ be a field. Then all nonzero elements in F have the same additive order.

Proof Consider any nonzero elements $a, b \in F$, and suppose r is the additive order of a. Then

$$rb = r(aa^{-1}b) = (ra)a^{-1}b = 0(a^{-1}b) = 0$$

Similarly, if r' is the additive order of b, then $r'a = 0$. Hence, the additive orders of a and b must be the same. □

The unique additive order of all nonzero elements of F is called the *characteristic* of $\mathfrak{F} = \langle F; +, \cdot \rangle$. For example, the characteristic of the field $\mathfrak{Z}_p = \langle \mathbb{Z}_p; +, \cdot \rangle$ (where p is prime) is p, since the additive order of any $a \in \mathbb{Z}_p$ is precisely p (see Theorem 9-21).

THEOREM 10-17

The characteristic of every finite field is a prime number.

Proof Let $\mathfrak{F} = \langle F; +, \cdot \rangle$ be a finite field with characteristic r (which, by Theorem 9-11, must be finite). Hypothesize that r is not prime and that $r = mn$, where $m < r$ and $n < r$. Hence,

$$0 = r1 = (mn)1 = m(n1)$$

Since r is the additive order of $n1 \in F$, we must have $r < m$, a contradiction. In conclusion, r must be prime. □

THEOREM 10-18

Let $\mathfrak{F} = \langle F; +, \cdot \rangle$ be a finite field with the prime characteristic p. Then $\#F = p^k$ for some positive integer k.

Proof By Theorem 10-17, the additive order of each element $a \in F$ is a power of p (it is $p^1 = p$ when $a \neq 0$, and $p^0 = 1$ when $a = 0$). The present theorem then follows directly from Theorem 9-22. □

One field whose characteristic and cardinality are both the same prime number p is $\mathfrak{Z}_p = \langle \mathbb{Z}_p; +, \cdot \rangle$ (see Example 10-3). The following two theorems state some interesting properties of this field.

THEOREM 10-19

Let a and b be any elements of $\mathbb{Z}_p = \{0, 1, \ldots, p-1\}$, and let s be any multiple of p. Then, in $\mathbf{Z}_p = \langle \mathbb{Z}_p; +, \cdot \rangle$,

$$(a+b)^s = a^s + b^s$$

Proof Let $s = pq$. Using the binomial expansion,

$$(a+b)^s = a^s + \binom{s}{1} a^{s-1}b + \binom{s}{2} a^{s-2}b^2 + \ldots + \binom{s}{s-1} ab^{s-1} + b^s$$

where, for $i = 1, 2, \ldots, s-1$,

$$\binom{s}{i} = \frac{s!}{i!(s-i)!} = \left[\frac{(s-i)!q}{i!(s-i)!}\right]p = n_i p \qquad (n_i \in \mathbb{N})$$

Since the characteristic of $\mathbf{Z}_p$ is p, for $i = 1, 2, \ldots, s-1$ we must have

$$(n_i p)a^{s-i}b^i = 0$$

and, hence, $(a+b)^s = a^s + b^s$. □

By induction, Theorem 10-19 can be readily generalized to the following: If $a_1, a_2, \ldots, a_n \in \mathbb{Z}_p$, and if s is a multiple of p, then

$$(a_1 + a_2 + \ldots + a_n)^s = a_1^s + a_2^s + \ldots + a_n^s$$

THEOREM 10-20

Let a be any element in $\mathbb{Z}_p = \{0, 1, \ldots, p-1\}$, and let k be any positive integer. Then, in $\mathbf{Z}_p = \langle \mathbb{Z}_p; +, \cdot \rangle$,

$$a^{p^k} = a$$

Proof From the cancellation law it follows that, for any $a, b_1, b_2 \in \mathbb{Z}_p$, $ab_1 = ab_2$ if and only if $b_1 = b_2$. Thus,

$$\{1, 2, 3, \ldots, p-1\} = \{a, 2a, 3a, \ldots, (p-1)a\}$$

and we can write

$$\begin{aligned} 1 \cdot 2 \cdot 3 \cdot \ldots \cdot (p-1) &= a \cdot 2a \cdot 3a \cdot \ldots \cdot (p-1)a \\ &= a^{p-1}[1 \cdot 2 \cdot 3 \cdot \ldots \cdot (p-1)] \end{aligned}$$

which implies $1 = a^{p-1}$ and, hence, $a^p = a$.

The theorem can now be proved by induction on k. (*Basis*). When $k = 1$,

$a^p = a$, as has just been shown. (*Induction step*). Hypothesizing that $a^{p^k} = a$, we have

$$a^{p^{k+1}} = (a^{p^k})^p = a^p = a \qquad \square$$

PROBLEMS

1. Construct a field of three elements.
2. The algebraic system $\mathfrak{F} = \langle F; +, \cdot \rangle$ is defined by:

$+$	α	β	γ	δ
α	α	β	γ	δ
β	β	α	δ	γ
γ	γ	δ	α	β
δ	δ	γ	β	α

$\cdot$	α	β	γ	δ
α	α	α	α	α
β	α	β	γ	δ
γ	α	γ	δ	β
δ	α	δ	β	γ

   (a) Show that $\mathfrak{F}$ is a field.
   (b) Solve the following equations in $\mathfrak{F}$:

$$x + \gamma y = \alpha$$
$$\gamma x + y = \beta$$

3. Show that the algebraic system $\langle \mathbb{C}; +, \cdot \rangle$, where $\mathbb{C}$ is the set of complex numbers and $+$ and $\cdot$ denote complex addition and multiplication, is a field.
4. Determine whether the algebraic system $\mathfrak{F} = \langle F; +, \cdot \rangle$ is a field, if $+$ and $\cdot$ are ordinary addition and subtraction, and if:
   (a) $F = \mathbb{I}$ (b) $F = \{q_1 + q_2\sqrt{3} \mid q_1, q_2 \in \mathbb{Q}\}$
5. Let $\mathfrak{F} = \langle F; +, \cdot \rangle$ be a field, where ab^{-1}, is denoted by $\frac{a}{b}$. Show that if

$$\frac{a_1}{b_1} = \frac{a_2}{b_2} = \ldots = \frac{a_n}{b_n} \qquad (a_1, \ldots, a_n, b_1, \ldots, b_n \in F)$$

   then

$$\frac{c_1 a_1^k + c_2 a_2^k + \ldots + c_n a_n^k}{c_1 b_1^k + c_2 b_2^k + \ldots + c_n b_n^k} = \frac{a_1^k}{b_1^k}$$

   where $c_1, c_2, \ldots, c_n$ are arbitrary elements of F (not all zero), and where k is an arbitrary positive integer.
6. Let $\mathfrak{F} + \langle F; +, \cdot \rangle$ be a field and $\mathfrak{R} = \langle R; +, \cdot \rangle$ a subring of $\mathfrak{F}$. Show that $\mathfrak{R}$ is an integral domain.
7. Prove that a direct product of two fields is never a field.
8. A *skew field* $\mathfrak{F} = \langle F; +, \cdot \rangle$ is a ring (not necessarily commutative) with multiplicative identity, in which every nonzero element has a multiplicative inverse.

Show that if F is the set of all 2×2 matrices of the form

$$\begin{bmatrix} a + b\sqrt{-1} & c + d\sqrt{-1} \\ -c + d\sqrt{-1} & a - b\sqrt{-1} \end{bmatrix} \qquad (a, b, c, d \in \mathbb{R})$$

and if $+$ and $\cdot$ are matrix addition and multiplication, then $\langle F; +, \cdot \rangle$ is a skew field.

10-4. POLYNOMIALS OVER RINGS

A *polynomial over a ring* $\mathcal{R} = \langle R; +, \cdot \rangle$ is defined as an expression of the form

$$a_0 + a_1x + a_2x^2 + \ldots + a_nx^n \qquad (n \geq 0)$$

where $a_0, a_1, \ldots, a_n \in R$ and $a_n \neq 0$ if $n > 0$.[1] The a_i are the *coefficients,* the symbol x is the *indeterminate,* and the nonnegative integer n is the *degree* of the polynomial. The term a_nx^n is called the *leading term,* and a_n the *leading coefficient* of the polynomial. A polynomial of degree 0 (that is, an element of R) is called a *constant polynomial.* The constant polynomial 0 is called the *zero polynomial.* A polynomial whose leading coefficient is 1 is called a *monic polynomial.*

Note that, basically, a polynomial with the indeterminate x is not a function of x, but an abstract form in which the x^i appear for the sole purpose of imposing some ordering on the coefficients. Nevertheless, we represent such polynomials with the functional notation $a(x)$, $b(x)$, etc. The degree of a polynomial $a(x)$ will be denoted by $\deg(a(x))$.

Let

$$a(x) = a_0 + a_1x + a_2x^2 + \ldots + a_nx^n$$
$$b(x) = b_0 + b_1x + b_2x^2 + \ldots + b_mx^m$$

be polynomials over $\mathcal{R} = \langle R; +, \cdot \rangle$. The *sum* of $a(x)$ and $b(x)$ is a polynomial over $\mathcal{R}$, defined by

$$a(x) + b(x) = c_0 + c_1x + c_2x^2 + \ldots + c_rx^r$$

where r is the greater of the two degrees n and m, and where

$$c_i = a_i + b_i \qquad (i = 0, 1, \ldots, r)$$

(In this definition, a_i is taken as 0 if $i > n$, and b_i is taken as 0 if $i > m$.)

[1]A term $+ (-a_i)x^i$ in a polynomial can be written as $-a_ix^i$, a term $1x^i$ can be written as x^i, and a term $0x^i$ can be deleted altogether. a_0 is equivalent to a_0x^0.

The *product* of $a(x)$ and $b(x)$ is a polynomial over $\mathcal{R}$ defined by

$$a(x)\cdot b(x) = c_0 + c_1x + c_2x^2 + \ldots + c_{n+m}x^{n+m}$$

where

$$c_i = a_0b_i + a_1b_{i-1} + a_2b_{i-2} + \ldots + a_ib_0 \qquad (i = 0, 1, \ldots, n + m)$$

(Here, again, $a_i = 0$ if $i > n$, and $b_i = 0$ if $i > m$.) It is seen that addition and multiplication of $a(x)$ and $b(x)$ can be performed by adding and multiplying $a(x)$ and $b(x)$ as if their terms were functions of x, and then "collecting terms" associated with the same powers of x. This process must surely be familiar to the reader from his high-school algebra.

Note that the evaluation of $a_0 + b_0$ and $a_0 \cdot b_0$ ($a_0, b_0 \in R$) is the same whether a_0 and b_0 are interpreted as elements of R or as polynomials over $\mathcal{R} = \langle R; +, \cdot \rangle$. For that reason, we can, without any ambiguity, use the addition and multiplication symbols of $\mathcal{R}$ ($+$ and $\cdot$) for the addition and multiplication of polynomials over $\mathcal{R}$.

Example 10-4

Consider the polynomials

$$a(x) = 1 + 2x + x^3$$
$$b(x) = 3 + x^2 + 2x^3$$

over the ring $\mathcal{Z}_4 = \langle \mathbb{Z}_4; +, \cdot \rangle$. In this case:

$$\begin{aligned}
a(x) + b(x) &= (1 + 2x + x^3) + (3 + x^2 + 2x^3) \\
&= (1 + 3) + (2 + 0)x + (0 + 1)x^2 + (1 + 2)x^3 \\
&= 2x + x^2 + 3x^3 \\
a(x)b(x) &= (1 + 2x + x^3)(3 + x^2 + 2x^3) \\
&= 3 \qquad + x^2 + 2x^3 \\
&\qquad + 2x \qquad + 2x^3 \\
&\qquad\qquad + 3x^3 + x^5 + 2x^6 \\
&= 3 + 2x + x^2 + 3x^3 + x^5 + 2x^6
\end{aligned}$$

□

The set of polynomials (with indeterminate x) over the ring $\mathcal{R} = \langle R; +, \cdot \rangle$ is denoted by $R[x]$. This set, clearly, is closed under the $+$ and $\cdot$ operations defined above; hence, we can talk about the algebraic system $\mathcal{R}[x] = \langle R[x]; +, \cdot \rangle$. The associativity and commutativity of $\mathcal{R}$ with respect to $+$ imply the associativity and commutativity of $\mathcal{R}[x]$ with respect to $+$. Moreover, $\mathcal{R}[x]$ has a zero (namely, the zero polynomial), and for every

$a(x) = a_0 + a_1x + \ldots + a_nx^n \in R[x]$ there is an additive inverse $-a(x) \in R[x]$ (namely, $-a_0 - a_1x - \ldots - a_nx^n$). Thus, $\mathcal{R}[x]$ is an Abelian group with respect to $+$. The associativity of $\mathcal{R}$ with respect to $\cdot$, and the distributivity of $\mathcal{R}$ with respect to $\cdot$ over $+$, imply the same properties in $\mathcal{R}[x]$. Thus, $\mathcal{R}[x]$ is a ring. It is called the *ring of polynomials over* $\mathcal{R}$.

It can be readily verified that $\mathcal{R}[x]$ is commutative if and only if $\mathcal{R}$ is commutative, and that $\mathcal{R}[x]$ has a multiplicative identity if and only if $\mathcal{R}$ possesses one (specifically, if 1 is the multiplicative identity of $\mathcal{R}$, then the polynomial 1 is the multiplicative identity of $\mathcal{R}[x]$).

THEOREM 10-21

The ring $\mathcal{R}[x]$ is an integral domain if and only if the ring $\mathcal{R}$ is an integral domain.

Proof It suffices to show that $\mathcal{R}[x]$ has no divisors of zero if and only if $\mathcal{R} = \langle R; +, \cdot\rangle$ has no divisors of zero. Let $a(x) = a_0 + a_1x + \ldots + a_nx^n$ and $b(x) = b_0 + b_1x + \ldots + b_mx^m$ be two polynomials in $R[x]$. Then $a(x)b(x) = c_0 + c_1x + \ldots + c_{n+m}x^{n+m}$, where $c_{n+m} = a_nb_m$. If $\mathcal{R}$ has no divisors of zero, $c_{n+m} = 0$ implies $a_n = 0$ or $b_m = 0$; hence, $a(x)b(x) = 0$ implies $a(x) = 0$ or $b(x) = 0$. Thus, $\mathcal{R}[x]$ has no divisors of zero.

If $\mathcal{R}$ has divisors of zero, then for some nonzero $a_0, b_0 \in R$, we have $a_0b_0 = 0$. Hence, there exist two nonzero (constant) polynomials $a_0, b_0 \in R[x]$ such that $a_0b_0 = 0$. $\mathcal{R}[x]$, therefore, has divisors of zero. □

For example, $\mathcal{Z}_3 = \langle \mathbb{Z}_3; +, \cdot\rangle$ is an integral domain (in fact, a field) and, hence, $\mathcal{Z}_3[x]$ is an integral domain. On the other hand, $\mathcal{Z}_4 = \langle \mathbb{Z}_4; +, \cdot\rangle$ is not an integral domain ($2\cdot 2 = 0$) and, hence, $\mathcal{Z}_4[x]$ is not an integral domain [for example, $(2 + 2x)(2 + 2x + 2x^2) = 0$].

PROBLEMS

1. Provide the details of the proof that $\mathcal{R}[x]$ is a ring. Also prove that $\mathcal{R}[x]$ is commutative with multiplicative identity if $\mathcal{R}$ is commutative with multiplicative identity.

2. Let $a(x) = -4 + 5x + 3x^3$ and $b(x) = 3 - x + 4x^3$ be polynomials over $\mathcal{Z}_7 = \langle \mathbb{Z}_7; +, \cdot\rangle$. Evaluate $a(x) + b(x)$ and $a(x)b(x)$.

3. Let $\mathcal{Z}_6[x]$ be the ring of polynomials over $\langle \mathbb{Z}_6; +, \cdot\rangle$. List all polynomials $a(x)$ and $b(x)$ in this ring which are of degree 3 or less, and such that $a(x)b(x) = 0$.

4. The *derivative* $a'(x)$ [or $a(x)'$] of a polynomial $a(x)$ over a ring $\mathcal{R} = \langle R; +, \cdot\rangle$ is defined as follows:

If $$a(x) = a_0 + a_1x + a_2x^2 + \ldots + a_nx^n$$

then $$a'(x) = a_1 + 2a_2x + 3a_3x^2 + \ldots + na_nx^{n-1}$$

[where $\nu a_i = a_i + a_i + \ldots + a_i(\nu \text{ times})$]. Show that:
(a) $(ca(x))' = ca'(x) \quad (c \in R)$
(b) $(a(x) + b(x))' = a'(x) + b'(x)$
(c) $(a(x)b(x))' = a(x)b'(x) + a'(x)b(x)$
(d) $((a(x))^n)' = n(a(x))^{n-1}a'(x)$

10-5. POLYNOMIALS OVER FIELDS

A *polynomial over a field* $\mathfrak{F} = \langle F; +, \cdot \rangle$ is defined as an expression of the form

$$a_0 + a_1x + a_2x^2 + \ldots + a_nx^n \qquad (n \geq 0)$$

where $a_0, a_1, \ldots, a_n \in F$ and $a_n \neq 0$ if $n > 0$. Every polynomial over a field is, of course, also a polynomial over a ring; hence, all the terminology and results presented in Sec. 10-4 for polynomials over a ring are directly applicable to polynomials over a field. Following the notation established in that section, the ring of polynomials over $\mathfrak{F}$ is denoted by $\mathfrak{F}[x] = \langle F[x]; +, \cdot \rangle$. Since $\mathfrak{F}$ is commutative and has multiplicative identity, $\mathfrak{F}[x]$ must also be commutative and have multiplicative identity (namely, the polynomial 1).

Given the polynomials $a(x)$ and $b(x)$ over a field $\mathfrak{F}$, $b(x)$ is said to *divide* $a(x)$ [or be a *divisor* or a *factor* of $a(x)$] if, for some polynomial $q(x) \in F[x]$, we have $a(x) = q(x)b(x)$. In this case $a(x)$ is said to be a *multiple* of $b(x)$. A polynomial $a(x) \in F[x]$ is called a *reducible polynomial* (over $\mathfrak{F}$) if, for some nonconstant $q(x), b(x) \in F[x]$, we have $a(x) = q(x)b(x)$; otherwise, $a(x)$ is said to be an *irreducible polynomial* (over $\mathfrak{F}$). Note that, over any field, all polynomials of degree 0 or 1 are irreducible.

THEOREM 10-22 (*The division theorem for polynomials*).
Let $a(x)$ and $b(x)$ be any polynomials over a field $\mathfrak{F}[x] = \langle F[x]; +, \cdot \rangle$, and assume $b(x) \neq 0$. Then there exist unique polynomials $q(x), r(x) \in F[x]$, such that

$$a(x) = q(x)b(x) + r(x) \qquad (10\text{-}1)$$

where $\deg(r(x)) < \deg(b(x))$, or $r(x) = 0$ if $b(x)$ is constant.

Proof The theorem, apart from the uniqueness clause, is proved by induction on $\deg(a(x))$. (*Basis*). Suppose $\deg(a(x)) = 0$. If $\deg(b(x)) > 0$, we can write $a(x) = 0b(x) + r(x)$, where $r(x) = a(x)$; hence, $\deg(r(x)) < \deg(b(x))$. If $\deg(b(x)) = 0$ [that is, $b(x)$ is constant], let $a(x) = a_0$ and $b(x) = b_0$ ($a_0, b_0 \in F$). Then we can write $a(x) = (a_0b_0^{-1})b(x) + r(x)$, where $r(x) = 0$.

(*Induction step*). Let $\deg(a(x)) = n$ and $\deg(b(x)) = m$. If $m > n$, we can write $a(x) = 0b(x) + r(x)$, where $r(x) = a(x)$; hence, $\deg(r(x)) < \deg(b(x))$. If $n \geq m$, let $a(x) = a_0 + a_1x + \ldots + a_nx^n$, $b(x) = b_0 + b_1x + \ldots + b_mx^m$, and define

$$\tilde{a}(x) = a(x) - (a_n b_m^{-1} x^{n-m})b(x)$$

Since $\deg(\tilde{a}(x)) < n$, induction hypothesis permits us to write

$$\tilde{a}(x) = \tilde{q}(x)b(x) + \tilde{r}(x)$$

where $\deg(\tilde{r}(x)) < \deg(b(x))$, or $\tilde{r}(x) = 0$ if $b(x)$ is constant. We can thus write

$$a(x) = (\tilde{q}(x) + a_n b^{m-1} x^{n-m})b(x) + \tilde{r}(x)$$

which completes the induction.

To prove uniqueness, assume that, in addition to (10-1), we have some $\bar{q}(x), \bar{r}(x) \in F[x]$, such that

$$a(x) = \bar{q}(x)b(x) + \bar{r}(x)$$

where $\deg(r(x)) < \deg(b(x))$ or $r(x) = 0$. Hence,

$$q(x)b(x) + r(x) = \bar{q}(x)b(x) + \bar{r}(x)$$

and

$$(q(x) - \bar{q}(x))b(x) = r(x) - \bar{r}(x)$$

If $q(x) \neq \bar{q}(x)$, then $q(x) - \bar{q}(x) \neq 0$ and, hence, $\deg(r(x) - \bar{r}(x)) \geq \deg(b(x))$. This implies either $\deg(r(x)) \geq \deg(b(x))$, or $\deg(\bar{r}(x)) \geq \deg(b(x))$. This, in turn, is possible only if $r(x) = \bar{r}(x) = 0$, $b(x)$ is constant, and $(q(x) - \bar{q}(x))b(x) = 0$. Since $b(x) \neq 0$, this implies that (see Theorem (10-12) $q(x) = \bar{q}(x)$, a contradiction. In conclusion, $q(x) = \bar{q}(x)$ and, hence, $r(x) = \bar{r}(x)$. □

In Theorem 10-22, the polynomials $q(x)$ and $r(x)$ are referred to as the *quotient* [sometimes written as $a(x)/b(x)$] and *remainder*, respectively, of $a(x)$ upon division by $b(x)$. These polynomials can be obtained through the familiar process of long division, where all operations are performed as per rules of the field $\mathfrak{F}$.

Example 10-5

Consider the polynomials

$$a(x) = 2 + 2x + x^2 + x^4 + 2x^5, \; b(x) = 1 + 2x^2$$

over the field $\mathcal{Z}_3 = \langle \mathbb{Z}_3; +, \cdot \rangle$. The quotient and remainder of $a(x)$ upon division by $b(x)$ can be obtained as shown below. (Note that, modulo 3, $-1 = 2$, $-2 = 1$, and $2 \cdot 2 = 1$). Thus, we can write

$$2 + 2x + x^2 + x^4 + 2x^5 = (1 + x + 2x^2 + x^3)(1 + 2x^2) + (1 + x)$$

$$\begin{array}{rrl}
 & x^3 + 2x^2 + x + 1 & \longleftarrow \text{quotient} \\
b(x) \longrightarrow 2x^2 + 1 & \overline{)\,2x^5 + x^4 \qquad + x^2 + 2x + 2} & \longleftarrow a(x) \\
 & \underline{2x^5 \qquad + x^3} & \\
 & x^4 + 2x^3 & \\
 & \underline{x^4 \qquad + 2x^2} & \\
 & 2x^3 + 2x^2 & \\
 & \underline{2x^3 \qquad + x} & \\
 & 2x^2 + x & \\
 & \underline{2x^2 \qquad + 1} & \\
 & x + 1 & \longleftarrow \text{remainder}
\end{array}$$

□

Given a polynomial $a(x)$ over the field $\mathfrak{F} = \langle F; +, \cdot \rangle$, and any element $c \in F$, the *value of* $a(x)$ *at* c will be defined as the element of F obtained when every x in $a(x)$ is replaced by c, and all the additions and multiplications in $a(x)$ are performed as per rules of $\mathfrak{F}$. Thus, for the purpose of computing the value of $a(x)$ at c, $a(x)$ can be regarded as a function $a: F \longrightarrow F$, evaluated for the argument c. The element $c \in F$ is called a *root* of $a(x)$ if $a(c) = 0$.

THEOREM 10-23

Let $a(x)$ be a polynomial over a field $\mathfrak{F} = \langle F; +, \cdot \rangle$. Then $x - c$ divides $a(x)$ if and only if c is a root of $a(x)$.

Proof Using the division theorem, we can write

$$a(x) = q(x)(x - c) + r(x)$$

where $r(x) = 0$ or $\deg(r(x)) < \deg(x - c) = 1$. Hence, $r(x) = r \in F$. Now,

$$a(c) = q(c)0 + r = r$$

Hence, $a(c) = 0$ if and only if $r = 0$; hence, if and only if $a(x) = q(x)(x - c)$.

□

THEOREM 10-24

Let $a(x)$ be a polynomial of degree $n > 0$ over a field $\mathfrak{F} = \langle F; +, \cdot \rangle$. Then $a(x)$ has at most n roots in F.

Proof (by induction on n). (*Basis*). When $n = 1$, $a(x)$ has the form $a_0 + a_1x$ and there is only one root, namely, $-a_0a_1^{-1}$. (*Induction step*). Let $a(x)$ have the degree $n > 1$. If $a(x)$ has no roots, then the theorem is trivially valid. If it does have a root, say, c, then—by Theorem 10-23—we can write $a(x) = (x - c)b(x)$, where $\deg(b(x)) = n - 1$. Now, $x - c$ has one root and, by induction hypothesis, $b(x)$ has at most $n - 1$ roots. Thus, $a(x)$ has at most n roots. □

PROBLEMS

1. The following are polynomials over the field $\mathcal{Z}_2 = \langle \mathbb{Z}_2; +, \cdot \rangle$. Factor each one of them into irreducible polynomials over $\mathcal{Z}_2$.
   (a) $1 + x^4$ (b) $1 + x^3 + x^4$
   (c) $1 + x + x^2 + x^3 + x^4 + x^5$ (d) $1 + x^6$
   (e) $1 + x^7$ (f) $1 + x^4 + x^8$

2. Let $a(x)$ be a polynomial over $\mathfrak{F} = \langle F; +, \cdot \rangle$ of degree 2 or 3. Show that $a(x)$ is irreducible over $\mathfrak{F}$ if and only if $a(c) = 0$ for all $c \in F$.

3. Let $a(x) = 1 + x + x^2 + \ldots + x^n$ be a polynomial over $\mathcal{Z}_2 = \langle \mathbb{Z}_2; +, \cdot \rangle$. Show that $1 + x$ is a factor of $a(x)$ if and only if n is odd.

4. Let $a(x) = 1 + x + x^2 + \ldots + x^n$ be a polynomial over $\mathcal{Z}_3 = \langle \mathbb{Z}_3; +, \cdot \rangle$.
   (a) Show that $1 + x$ is a factor of $a(x)$ if and only if n is odd.
   (b) Show that $2 + x$ is a factor of $a(x)$ if and only if $n = 3k - 1$ ($k = 1, 2, 3, \ldots$).

5. Let $a(x) = 1 + x + x^3 + x^4$ be a polynomial over $\mathcal{Z}_2 = \langle \mathbb{Z}_2; +, \cdot \rangle$. Without actually computing $a(x)/b(x)$, determine whether the following polynomials $b(x)$ over $\mathcal{Z}_2$ are divisors of $a(x)$:
   (a) $b(x) = 1 + x$ (b) $b(x) = 1 + x + x^2$
   (c) $b(x) = 1 + x + x^3$

6. Show that the polynomial $1 + x^2$ is irreducible over the field of real numbers $\langle \mathbb{R}; +, \cdot \rangle$.

7. Show that if

$$a(x) = a_0 + a_1x + a_2x^2 + \ldots + a_nx^n$$

is an irreducible polynomial over the field $\mathfrak{F}$, so is

$$b(x) = a_n + a_{n-1}x + a_{n-2}x^2 + \ldots + a_0x^n$$

[*Hint*: $b(x) = x^na(1/x)$.]

10-6. POLYNOMIAL IDEALS

In this section we study the subsets of the ring $\mathfrak{F}[x] = \langle F[x]; +, \cdot \rangle$ (where $\mathfrak{F} = \langle F; +, \cdot \rangle$ is a field) which form ideals. These ideals are called *polynomial ideals over* $\mathfrak{F}$. From the defintion of an ideal it follows that, if

$\mathfrak{D} = \langle D; +, \cdot \rangle$ is a polynomial ideal over $\mathfrak{F}$, then for all $a(x) \in F[x]$ and $d(x) \in D$, $a(x)d(x)$ is in D. If $\mathfrak{D}$ is a principal ideal, then $D = (g(x))$ for some $g(x) \in F[x]$.

THEOREM 10-25
Every polynomial ideal over a field $\mathfrak{F}$ is a principal ideal.

Proof Let $\mathfrak{D} = \langle D; +, \cdot \rangle$ be an ideal of $\mathfrak{F}[x] = \langle F[x]; +, \cdot \rangle$. If $D = \{0\}$, then, trivially, $\mathfrak{D}$ is a principal ideal (with the generator 0). If $D \neq \{0\}$, let $g(x)$ be any nonzero element *of least degree* in D. Clearly, D contains all multiples of $g(x)$. That D contains *only* multiples of $g(x)$ can be shown as follows: Let $a(x)$ be any element of D. By the division theorem, we can write $a(x) = q(x)g(x) + r(x)$, where $r(x) = 0$ or $\deg(r(x)) < \deg(g(x))$. Since $r(x) = a(x) - q(x)g(x)$ is in D, and since $g(x)$ is a least degree nonzero polynomial in D, we must have $r(x) = 0$ and, hence, $a(x) = q(x)g(x)$. In conclusion, then, $D = (g(x))$. □

Theorem 10-25 implies that, like $\mathfrak{I} = \langle \mathbb{I}; +, \cdot \rangle$ (see Example 10-1), $\mathfrak{F}[x] = \langle F[x]; + \times \cdot \rangle$ is a principal ideal ring.

Any ideal $\langle (g(x)); +, \cdot \rangle$ of the ring $\mathfrak{F}[x] = \langle F[x]; +, \cdot \rangle$ is, of course, an additive normal subgroup of $\mathfrak{F}$ (see Sec. 10-2). By Theorem 9-16, the polynomials $a_1(x), a_2(x) \in F[x]$ are in the same coset of $(g(x))$ in the additive group of $\mathfrak{F}$, if and only if $a_1(x) - a_2(x) = q(x)g(x)$ for some $q(x) \in F[x]$; hence, if and only if their remainders, upon division by $g(x)$, are identical. Thus, each coset of $(g(x))$ consists of all polynomials expressible in the form $q(x)g(x) + r(x)$, where $r(x)$ is fixed for the coset and is of degree less than $\deg(g(x))$. [Each coset, therefore, can be written as $(g(x)) + r(x)$.] These cosets are referred to as *polynomial residue classes modulo* $g(x)$. Clearly, there are as many residue classes modulo $g(x)$ as there are polynomials in $F[x]$ of degree less than $\deg(g(x))$.

Example 10-6

Consider the ring of polynomials over the field $\mathfrak{Z}_2 = \langle \mathbb{Z}_2; +, \cdot \rangle$. The ideal $(\langle 1 + x + x^3); +, \cdot \rangle$ consists of all multiples of $1 + x + x^3$. Its cosets are given by:

$$(1 + x + x^3), (1 + x + x^3) + 1, (1 + x + x^3) + x,$$
$$(1 + x + x^3) + 1 + x, (1 + x + x^3) + x^2,$$
$$(1 + x + x^3) + 1 + x^2, (1 + x + x^3) + x + x^2,$$
$$(1 + x + x^3) + 1 + x + x^2$$

[where $(1 + x + x^3)$ denotes the domain of the ideal itself]. □

For each ideal $\langle(g(x)); +, \cdot\rangle$ of $\mathfrak{F}[x]$ we can define a quotient ring $\mathfrak{F}[x]/(g(x))$ which, according to Theorem 10-4, is an epimorphic image of $\mathfrak{F}[x]$. The elements of this ring are the polynomial residue classes modulo $g(x)$, and its cardinality equals the number of polynomials over $\mathfrak{F}$ whose degree is less than that of $g(x)$. It is convenient to denote each residue class, and, hence, each element of the quotient ring, by $[r(x)]$ where $r(x)$ is the remainder of any class element upon division by $g(x)$. Thus, the domain of $\mathfrak{F}[x]/(g(x))$ is given by

$$\{[r(x)] \mid \deg(r(x)) < \deg(g(x))\}$$

Operations in the quotient ring $\mathfrak{F}[x]/(g(x))$ can be carried out by first adding and mutliplying as per rules of $\mathfrak{F}[x]$, and then replacing the result with its remainder upon division by $g(x)$.

Example 10-7

Consider the ideal $\langle(1 + x + x^3); +, \cdot\rangle$ of Example 10-6. The domain of the quotient ring $\mathfrak{Z}_2[x]/(1 + x + x^3)$ is given by

$$\{[0], [1], [x], [1 + x], [x^2], [1 + x^2], [x + x^2], [1 + x + x^2]\}$$

For example, to compute $[1 + x^2][1 + x + x^2] + [1 + x]$ in this ring, first compute (as per rules of $\mathfrak{Z}_2[x]$):

$$\begin{aligned}(1 + x^2)(1 + x + x^2) + (1 + x) &= x^3 + x^4 \\ &= (1 + x)(1 + x + x^3) + (1 + x^2)\end{aligned}$$

Hence,

$$[1 + x^2][1 + x + x^2] + [1 + x] = [1 + x^2]$$ □

THEOREM 10-26

Let $\langle(g(x)); +, \cdot\rangle$ be an ideal of $\mathfrak{F}[x] = \langle F[x]; +, \cdot\rangle$. Then $\mathfrak{F}[x]/(g(x))$ is a field if and only if $g(x)$ is irreducible over $\mathfrak{F}$.

Proof By Theorem 10-15, it sufficies to prove that $\langle(g(x)); +, \cdot\rangle$ is a maximal ideal if and only if $g(x)$ is irreducible over $\mathfrak{F}$. Suppose $g(x) = q(x)a(x)$ for some $a(x) \in F[x]$. Clearly, $(g(x)) \subset (a(x))$. Now, $\langle(g(x)); +, \cdot\rangle$ is maximal if and only if $(a(x)) = (g(x))$ or $(a(x)) = F[x]$; hence, if and only if $q(x) = 1$ or $a(x) = 1$, and, hence, if and only if $g(x)$ is irreducible. □

Example 10-8

Consider the ring of polynomials over the field $\mathfrak{Z}_2 = \langle\mathbb{Z}_2; +, \cdot\rangle$ and its ideal $\langle(1 + x + x^2); +, \cdot\rangle$. The quotient ring in this case is

$$\mathfrak{Z}_2(x)/(1 + x + x^2) = \langle\{[0], [1], [x], [1 + x]\}; +, \cdot\rangle$$

Table 10-3 OPERATION TABLES FOR $\mathbb{Z}_2[x]/(1 + x + x^2)$

$+$	$[0]$	$[1]$	$[x]$	$[1 + x]$
$[0]$	$[0]$	$[1]$	$[x]$	$[1 + x]$
$[1]$	$[1]$	$[0]$	$[1 + x]$	$[x]$
$[x]$	$[x]$	$[1 + x]$	$[0]$	$[1]$
$[1 + x]$	$[1 + x]$	$[x]$	$[1]$	$[0]$

$\cdot$	$[0]$	$[1]$	$[x]$	$[1 + x]$
$[0]$	$[0]$	$[0]$	$[0]$	$[0]$
$[1]$	$[0]$	$[1]$	$[x]$	$[1 + x]$
$[x]$	$[0]$	$[x]$	$[1 + x]$	$[1]$
$[1 + x]$	$[0]$	$[1 + x]$	$[1]$	$[x]$

where $+$ and $\cdot$ are as defined in Table 10-3. Since $1 + x + x^2$ is irreducible over $\mathbb{Z}_2$ (verify!), $\mathbb{Z}_2[x]/(1 + x + x^2)$ is a field. □

The elements $[a(x)]$ of $\mathfrak{F}[x]/(g(x))$ are often written simply as $a(x)$. With this notation, the domain of $\mathfrak{F}[x]/(g(x))$ simply consists of all polynomials over $\mathfrak{F}$ of degree less than $\deg(g(x))$, and all operations are carried out "modulo $g(x)$" [that is, all operations are performed as per rules of $\mathfrak{F}[x]$ and results are ultimately replaced by their remainders upon division by $g(x)$]. For example, if the elements $[x]$ and $[1 + x]$ in Example 10-8 are denoted by x and $1 + x$, respectively, then the element $[x][1 + x]$ is denoted by 1, since

$$x(1 + x) = x^2 + x = 1(1 + x + x^2) + 1$$

If $a(x)$ is a divisor of each polynomial in some nonempty set $P = \{a_1(x), a_2(x), \ldots, a_k(x)\}$ of polynomials over a field $\mathfrak{F} = \langle F; +, \cdot \rangle$, it is said to be a *common divisor* of P. A common divisor of P that is divisible by every other common divisor of P is called a *greatest common divisor* of P, and is denoted by $\gcd(a_1(x), a_2(x), \ldots, a_k(x))$. Clearly, if P has two or more greatest common divisors, they must divide each other and, hence, must differ only by a constant factor. We refer to the unique *monic* greatest common divisor of P (that is, the one whose leading coefficient is 1) as *the* greatest common divisor of P.

THEOREM 10-27

Let $a(x)$ and $b(x)$ be polynomials over a field $\mathfrak{F} = \langle F; +, \cdot \rangle$. Then there exists a polynomial $d(x) \in F[x]$ such that $d(x) = \gcd(a(x), b(x))$. Moreover, there exist polynomials $m(x), n(x) \in F[x]$ such that

$$d(x) = m(x)a(x) + n(x)b(x) \tag{10-2}$$

Proof Let S denote the set of all polynomials in $F[x]$ of the form $a_1(x)a(x) + b_1(x)b(x)$. It can be readily shown that $\langle S; +, \cdot \rangle$ is an ideal of $\mathfrak{F}[x]$ and, hence, that $S = (d(x))$ for some $d(x) \in F[x]$ which divides every element in S. Since $a(x)$, $b(x)$, and $d(x)$ are all in S, $d(x)$ is a common divisor of $a(x)$ and $b(x)$; moreover, it is expressible in the form (10-2) with some appropriate

$m(x), n(x) \in F[x]$. From (1) it also follows that every divisor of $a(x)$ and $b(x)$ is also a divisor of $d(x)$; hence, $d(x) = \gcd(a(x), b(x))$. □

The computation of $\gcd(a_1(x), a_2(x))$ is facilitated by the polynomial version of the *Euclidean algorithm* described in Sec. 3-11. It consists of forming, with the aid of the division theorem, this series of identities:

$$\begin{aligned}
a_1(x) &= q_1(x)a_2(x) + a_3(x) && [\deg(a_3(x)) < \deg(a_2(x))] \\
a_2(x) &= q_2(x)a_3(x) + a_4(x) && [\deg(a_4(x)) < \deg(a_3(x))] \\
a_3(x) &= q_3(x)a_4(x) + a_5(x) && [\deg(a_5(x)) < \deg(a_4(x))] \\
&\ \,\vdots \\
a_{h-2}(x) &= q_{h-2}(x)a_{h-1}(x) + a_h(x) && [\deg(a_h(x)) < \deg(a_{h-1}(x))] \\
a_{h-1}(x) &= q_{h-1}(x)a_h(x) + 0
\end{aligned} \tag{10-3}$$

Since $\deg(a_i(x)) < \deg(a_{i-1}(x))$ for $i = 3, 4, 5, \ldots$, it is guaranteed that eventually a zero remainder will be obtained, as indicated in the last identity. Now, any divisor of $a_2(x)$ and $a_3(x)$ is also a divisor of $a_1(x)$, and any divisor of $a_1(x)$ and $a_2(x)$ is also a divisor of $a_3(x)$. Thus, the common divisors of $\{a_1(x), a_2(x)\}$ are the same as the common divisors of $\{a_2(x), a_3(x)\}$ and, hence, $\gcd(a_1(x), a_2(x)) = \gcd(a_2(x), a_3(x))$. Proceeding down the series of identities (10-3), we have by the same argument:

$$\begin{aligned}
\gcd(a_1(x), a_2(x)) &= \gcd(a_2(x), a_3(x)) = \ldots = \gcd(a_{h-2}(x), a_{h-1}(x)) \\
&= \gcd(a_{h-1}(x), a_h(x)) = a_h(x)
\end{aligned}$$

In conclusion, $\gcd(a_1(x), a_2(x))$ equals the last nonzero remainder in the series (10-3).

Example 10-9

The identities (10-3) for the polynomials

$$a_1(x) = 1 + x^2 + x^3 + x^4, \quad a_2(x) = 1 + x^4$$

over $\mathbf{Z}_2 = \langle \mathbb{Z}_2; +, \cdot \rangle$ are:

$$\begin{aligned}
1 + x^2 + x^3 + x^4 &= 1(1 + x^4) + x^2 + x^3 \\
1 + x^4 &= (1 + x)(x^2 + x^3) + 1 + x^2 \\
x^2 + x^3 &= (1 + x)(1 + x^2) + 1 + x \\
1 + x^2 &= (1 + x)(1 + x) + 0
\end{aligned}$$

Hence,
$$\gcd(1 + x^2 + x^3 + x^4, 1 + x^4) = 1 + x$$

A convenient way for computing the preceding quotients and remainders is the "divide and invert" method illustrated in Table 4.

Table 10-4 FOR EXAMPLE 10-9

$$
\begin{array}{rl}
 & 1 \longleftarrow q_1(x) \\
a_2(x) \longrightarrow x^4 + 1 & \overline{)\, x^4 + x^3 + x^2 + 1} \longleftarrow a_1(x) \\
 & \underline{x^4 + 1} \quad x + 1 \longleftarrow q_2(x) \\
 & a_3(x) \longrightarrow x^3 + x^2 \,\overline{)\, x^4 + 1} \longleftarrow a_2(x) \\
 & \underline{x^4 + x^3} \\
 & x^3 + 1 \\
 & \underline{x^3 + x^2} \quad x + 1 \longleftarrow q_3(x) \\
 & a_4(x) \longrightarrow x^2 + 1 \,\overline{)\, x^3 + x^2} \longleftarrow a_3(x) \\
 & \underline{x^3 + x} \\
 & x^2 + x \\
 & \underline{x^2 + 1} \quad x + 1 \longleftarrow q_4(x) \\
 & a_5(x) \longrightarrow x + 1 \,\overline{)\, x^2 + 1} \longleftarrow a_4(x) \\
 & \underline{x^2 + x} \\
 & x + 1 \\
 & \underline{x + 1} \\
 & 0
\end{array}
$$

The series of identities (10-3) can also serve to express the greatest common divisor in the form (10-2) introduced in Theorem 10-27:[1]

$$
\begin{aligned}
1 + x &= (x^2 + x^3) - (1 + x)(1 + x^2) \\
&= (x^2 + x^3) - (1 + x)((1 + x^4) - (1 + x)(x^2 + x^3)) \\
&= (1 + x)(1 + x^4) + x^2(x^2 + x^3) \\
&= (1 + x)(1 + x^4) + x^2((1 + x^2 + x^3 + x^4) - 1(1 + x^4)) \\
&= (1 + x + x^2)(1 + x^4) + x^2(1 + x^2 + x^3 + x^4)
\end{aligned}
$$

Hence,

$$
\begin{aligned}
\gcd(1 + x^2 + x^3 + x^4, 1 + x^4) = x^2&(1 + x^2 + x^3 + x^4) \\
&+ (1 + x + x^2)(1 + x^4) \qquad \square
\end{aligned}
$$

THEOREM 10-28

Let $p(x)$ be an irreducible polynomial over a field $\mathfrak{F} = \langle F; +, \cdot \rangle$. If $p(x)$ divides $a(x)b(x)$ [$a(x), b(x) \in F[x]$], then $p(x)$ must either divide $a(x)$ or divide $b(x)$.

Proof Let $a(x)b(x) = q(x)p(x)$. Hypothesizing that $p(x)$ does not divide $a(x)$, we have $\gcd(p(x), a(x)) = 1$; hence, $1 = m(x)p(x) + n(x)a(x)$ for some

[1] Recall that, in $\mathbb{Z}_2$, $-1 = 1$.

$m(x), n(x) \in F[x]$. Consequently, we can write

$$\begin{aligned} b(x) &= m(x)p(x)b(x) + n(x)a(x)b(x) \\ &= m(x)p(x)b(x) + n(x)q(x)p(x) \\ &= p(x)(m(x)b(x) + n(x)q(x)) \end{aligned}$$

which implies that $p(x)$ divides $b(x)$. Thus, $p(x)$ must either divide $a(x)$ or divide $b(x)$. □

THEOREM 10-29 (*The prime factorization theorem for polynomials*)
Every nonconstant polynomial $a(x)$ over a field $\mathfrak{F} = \langle F; +, \cdot \rangle$ can be written in the *prime factorization* form

$$a(x) = cp_1(x)p_2(x) \ldots p_r(x)$$

where $c \in F$ and the $p_i(x)$ are irreducible monic polynomials over $\mathfrak{F}$. This form is unique, except for the ordering of the factors.

Proof [by induction on $\deg(a(x))$] (*Basis*). When $\deg(a(x)) = 1$, the theorem is true, since $a(x)$ is irreducible. (*Induction step*). Suppose $a(x)$ has two prime factorization forms:

$$a(x) = cp_1(x)p_2(x) \ldots p_r(x) = dq_1(x)q_2(x) \ldots q_s(x)$$

where $c, d \in F$ and the $p_i(x)$ and $q_i(x)$ are irreducible and monic. By Theorem 10-28, since $p_1(x)$ divides $q_1(x)q_2(x) \ldots q_s(x)$, it must divide one of the $q_i(x)$, say, $q_j(x)$. Since $q_j(x)$ is irreducible and since both $p_1(x)$ and $q_j(x)$ are monic, $p_1(x) = q_j(x)$ and we can write

$$\begin{aligned} a_1(x) = a(x)/p_1(x) &= cp_2(x) \ldots p_r(x) \\ &= dq_1(x) \ldots q_{j-1}(x)q_{j+1}(x) \ldots q_s(x) \end{aligned}$$

By induction hypothesis, the prime factorization of $a_1(x)$ is unique except for factor ordering; that is, for every $p_i(x)$ there is an equal $q_k(x)$, and conversely. Hence, the prime factorization of $a(x)$ must also be unique except for factor ordering. □

Consider the polynomial

$$a(x) = a_0 + a_1x + a_2x^2 + \ldots + a_nx^n$$

over the field $\mathfrak{F} = \langle F; +, \cdot \rangle$. The *derivative* of $a(x)$ is a polynomial over $\mathfrak{F}$, denoted by $a'(x)$ [or $a(x)'$] and defined by

$$a'(x) = a_1 + 2a_2x + 3a_3x^2 + \ldots + na_nx^{n-1}$$

(This, of course, is a generalization of the concept of derivative as applied to polynomials over the real field $\langle \mathbb{R}; +, \cdot \rangle$; in the general case, however, there is no geometric interpretation that relates derivatives to "slopes.") It can be readily verified (see Problem 4, Sec. 10-4) that, for all $a(x), b(x) \in F[x]$,

$$(a(x) + b(x))' = a'(x) + b'(x)$$
$$(a(x)b(x))' = a'(x)b(x) + a(x)b'(x)$$

THEOREM 10-30

Let $\mathfrak{F} = \langle F; +, \cdot \rangle$ be a field, and let $p(x)$ be an irreducible polynomial over $\mathfrak{F}$ which divides $a(x) \in F[x]$. Then $(p(x))^2$ divides $a(x)$ if and only if $p(x)$ divides $a'(x)$.

Proof Since $p(x)$ divides $a(x)$, we can write $a(x) = p(x)q(x)$ and, hence, $a'(x) = p'(x)q(x) + p(x)q'(x)$. Thus, if $(p(x))^2$ divides $a(x)$, then $p(x)$ divides $q(x)$ and, hence, divides $a'(x)$. Conversely, if $p(x)$ divides $a'(x)$, then $p(x)$ divides $p'(x)q(x)$ and, hence (by Theorem 10-28), $p(x)$ divides either $p'(x)$ or $q(x)$. But $\deg(p'(x)) < \deg(p(x))$. Hence, $p(x)$ divides $q(x)$, and $(p(x))^2$ divides $p(x)q(x) = a(x)$. □

PROBLEMS

1. Let $a(x)$ and $b(x)$ be fixed polynomials over the field $\mathfrak{F} = \langle F; +, \cdot \rangle$. Show that if S is the set of all polynomials of the form $a_1(x)a(x) + b_1(x)b(x)$, where $a_1(x), b_1(x) \in F[x]$, then $\langle S; +, \cdot \rangle$ is an ideal of $\mathfrak{F}[x]$.

2. Construct the operation tables for the quotient rings $\mathbb{Z}_2[x]/(g(x))$, where:
   (a) $g(x) = 1 + x + x^3$
   (b) $g(x) = 1 + x + x^2 + x^3 + x^4$

3. The entries of the following matrix $\mathbf{M}$ are elements of the field $\mathbb{Z}_2[x]/(1 + x + x^2)$. Evaluate the determinant of $\mathbf{M}$.

$$\mathbf{M} = \begin{bmatrix} 1 & 1+x & x \\ x & 0 & x \\ 1+x & 1 & 1+x \end{bmatrix}$$

4. (a) Prove that, for any polynomials $a(x)$, $b(x)$, and $c(x)$ over a field $\mathfrak{F}$,

$$\gcd(a(x), b(x), c(x)) = \gcd(\gcd(a(x), b(x)), c(x))$$

(b) Suggest an algorithm for computing $\gcd(a_1(x), a_2(x), \ldots, a_n(x))$ for any polynomials $a_1(x), a_2(x), \ldots, a_n(x)$ over $\mathfrak{F}$.

5. The following are polynomials over $\mathbf{Z}_3 = \langle \mathbb{Z}_3; +, \cdot \rangle$:

$$a(x) = 1 + x + x^5 + x^7$$
$$b(x) = 1 + x + 2x^5 + x^6 + 2x^7$$

(a) Find $\gcd(a(x), b(x))$.
(b) Express $\gcd(a(x), b(x))$ in the form $m(x)a(x) + n(x)b(x)$.

6. Let $a(x)$ and $b(x)$ be polynomials over a field $\mathfrak{F} = \langle F; +, \cdot \rangle$, such that $\gcd(a(x), b(x)) = 1$. Show that any polynomial $c(x) \in F[x]$ can be written in the form $m(x)a(x) + n(x)b(x)$, with some $m(x), n(x) \in F[x]$.

7. Use Theorem 10-28 to determine whether the polynomial $1 + 2x + x^2$ divides the polynomial $1 + x^2 + 2x^3 + 2x^4$ over $\mathbf{Z}_3 = \langle \mathbb{Z}_3; +, \cdot \rangle$.

10-7. EXTENSION RINGS AND FIELDS

Given a field $\mathfrak{F} = \langle F; +, \cdot \rangle$, a subsystem $\tilde{\mathfrak{F}} = \langle \tilde{F}; +, \cdot \rangle$ of $\mathfrak{F}$ is any algebraic system such that $\tilde{F}$ is a nonempty subset of F and is closed under both $+$ and $\cdot$ (see Sec. 4-2). If $\tilde{\mathfrak{F}}$ is a field, it is called a *subfield* of $\mathfrak{F}$. If, in addition, $\tilde{F}$ is a proper subset of F, then $\tilde{\mathfrak{F}}$ is said to be a *proper subfield* of $\mathfrak{F}$.

THEOREM 10-31

Let $\mathfrak{F} = \langle F; +, \cdot \rangle$ be a field and $\tilde{\mathfrak{F}} = \langle \tilde{F}; +, \cdot \rangle$ be a subsystem of $\mathfrak{F}$. Then $\tilde{\mathfrak{F}}$ is a subfield if and only if $\langle \tilde{F}; + \rangle$ and $\langle \tilde{F} - \{0\}; \cdot \rangle$ satisfy the inverse law with respect to $+$ and $\cdot$, respectively.

Proof Clearly, if $\tilde{\mathfrak{F}}$ is a subfield of $\mathfrak{F}$, then $\langle \tilde{F}; + \rangle$ and $\langle \tilde{F} - \{0\}; \cdot \rangle$ satisfy the inverse law with respect to $+$ and $\cdot$, respectively. Conversely, if $\langle \tilde{F}; + \rangle$ satisfies the inverse law with respect to $+$, it must (by Theorem 10-3) be a subring of $\mathfrak{F}$—in fact, a commutative subring, since commutativity is inherited by $\tilde{\mathfrak{F}}$ from $\mathfrak{F}$. If, in addition, $\langle \tilde{F} - \{0\}; \cdot \rangle$ satisfies the inverse law with respect to $\cdot$, it must contain a multiplicative identity, and hence $\tilde{\mathfrak{F}}$ must be a field. □

Thus, if $\mathfrak{F} = \langle F; +, \cdot \rangle$ is a field, to establish whether a nonempty subset $\tilde{F}$ of F forms a subfield of $\mathfrak{F}$, it is sufficient to establish:

(a) *Closure*. For any $a, b \in \tilde{F}$, $a - b$ and ab are in $\tilde{F}$.
(b) *Inverse*. For any $a \in \tilde{F}$, $-a$ is in $\tilde{F}$; and for any $a \in \tilde{F} - \{0\}$, a^{-1} is in $\tilde{F}$.

THEOREM 10-32

Let $\mathfrak{F} = \langle F; +, \cdot \rangle$ be a *finite* field and $\tilde{\mathfrak{F}} = \langle \tilde{F}; +, \cdot \rangle$ be a subsystem of $\mathfrak{F}$. Then $\tilde{\mathfrak{F}}$ is a subfield of $\mathfrak{F}$.

Proof When $\mathfrak{F}$ is finite, then, by Theorem 9-15, $\langle \tilde{F}; + \rangle$ and $\langle \tilde{F} - \{0\}; \cdot \rangle$ automatically satisfy the inverse law with respect to $+$ and $\cdot$, respectively; hence, $\tilde{\mathfrak{F}}$ is a subfield. □

Thus, if $\mathfrak{F} = \langle F; +, \cdot \rangle$ is a finite field, to establish whether a nonempty subset $\tilde{F}$ of F forms a subfield of $\mathfrak{F}$, it is sufficient to establish closure under $+$ and $\cdot$.

Let the field $\mathfrak{F} = \langle F; +, \cdot \rangle$ be a subsystem of the algebraic system $\mathcal{E} = \langle E; +, \cdot \rangle$. If $\mathcal{E}$ is a ring, then $\mathcal{E}$ is called an *extension ring* of $\mathfrak{F}$; if it is a field, it is called an *extension field* of $\mathfrak{F}$. In this context, $\mathfrak{F}$ is called the *base field* of $\mathcal{E}$.

Given a base field $\mathfrak{F} = \langle F; +, \cdot \rangle$, an extension ring of $\mathfrak{F}$ can be constructed as follows: Select a nonconstant polynomial $w(x)$ over $\mathfrak{F}$, and define a new element ξ (not in F) which, by definition, has the property that $w(\xi) = 0$. Now, define E as the set of all elements obtained by successive applications of the $+$ and $\cdot$ operations of $\mathfrak{F}$ to the elements of $F \cup \{\xi\}$. Thus, every element of E would have the form $a_0 + a_1\xi + a_2\xi^2 + \ldots + a_m\xi^m$; hence, E is precisely the set of all polynomials over $\mathfrak{F}$ with the indeterminate ξ (that is, $E = F[\xi]$). Since E is closed under $+$ and $\cdot$, it forms an algebraic system $\mathcal{E} = \langle E; +, \cdot \rangle$ of which $\mathfrak{F}$ is a subsystem. Using the division theorem, any element $a(\xi) \in E$ can be expressed as

$$a(\xi) = q(\xi)w(\xi) + r(\xi)$$

where $r(\xi) = 0$ or $\deg(r(\xi)) < \deg(w(\xi))$. Since, by definition, $w(\xi) = 0$, we have $a(\xi) = r(\xi)$; hence, $a_1(\xi) = a_2(\xi)$ if and only if $a_1(\xi)$ and $a_2(\xi)$ are congruent modulo $w(\xi)$. Comparing $\mathcal{E}$ with the quotient ring $\mathfrak{F}[x]/(w(x))$ as defined in Sec. 10-6, it is seen that the two systems are identical, with elements of $\mathcal{E}$ playing the same role as that played in $\mathfrak{F}[x]/(w(x))$ by residue classes modulo $w(x)$. More precisely, $\mathcal{E}$ is isomorphic to $\mathfrak{F}[x]/(w(x))$, with the image of any $a(\xi) \in E$ being the residue class modulo $w(x)$ to which $a(x)$ belongs. In conclusion, then, $\mathcal{E}$ is an extension ring of $\mathfrak{F}$. By Theorem 10-26, if $w(x)$ is irreducible over $\mathfrak{F}$, then $\mathcal{E}$ is an extension *field* of $\mathfrak{F}$.

It should be emphasized that, unlike x, ξ does not represent an indeterminate, but a bona fide element of E. Since, by defintion, $w(\xi) = 0$, the procedure just described for constructing $\mathcal{E} = \langle E; +, \cdot \rangle$ guarantees that E will contain a root for the polynomial $w(x)$; in fact, $\mathcal{E}$ as constructed above is the *smallest* ring (or field) in which $w(x)$ has a root. It is common to say that $\mathcal{E}$ is produced by *adjoining the root ξ of $w(x)$ to the field* $\mathfrak{F}$. Using this terminology, we can summarize the foregoing as:

THEOREM 10-33

The algebraic system $\mathcal{E} = \langle E; +, \cdot \rangle$ produced by adjoining the root ξ of the polynomial $w(x) \in F[x]$ to the field $\mathcal{F} = \langle F; +, \cdot \rangle$, is an extension ring of $\mathcal{F}$ [or an extension field of $\mathcal{F}$ if $w(x)$ is irreducible over $\mathcal{F}$] which is isomorphic to $\mathcal{F}[x]/(w(x))$. Under this isomorphism, the image of $a(\xi) \in E$ is the residue class modulo $w(x)$ in $F[x]$ to which $a(x)$ belongs.

Example 10-10

Consider the base field $\mathcal{Z}_2 = \langle \mathbb{Z}_2; +, \cdot \rangle$ and its extension ring $\mathcal{E} = \langle E; +, \cdot \rangle$ produced by adjoining the root ξ of the polynomial $1 + x + x^2 \in \mathbb{Z}_2[x]$ to the field $\mathcal{Z}_2$. In this case, E consists of all polynomials over $\mathcal{Z}_2$ with indeterminate ξ and degree less than 2. That is,

$$E = \{0, 1, \xi, 1 + \xi\}$$

Any other polynomial over $\mathcal{Z}_2$ with the indeterminate ξ must equal one of the four elements of E. For example,

$$1 + \xi + \xi^2 + \xi^6 + \xi^7 = (\xi^2 + \xi^3 + \xi^5)(1 + \xi + \xi^2) + 1 + \xi$$

Since, by definition, $1 + \xi + \xi^2 = 0$, we have

$$1 + \xi + \xi^2 + \xi^6 + \xi^7 = 1 + \xi$$

The operation tables which define $\mathcal{E}$ are shown in Table 10-5. For example:

$$\xi + (1 + \xi) = 1 + (1 + 1)\xi = 1 + 0 = 1$$

$$\begin{aligned}(1 + \xi)(1 + \xi) &= 1 + (1 + 1)\xi + \xi^2 = 1 + \xi^2 \\ &= 1(1 + \xi + \xi^2) + \xi = \xi\end{aligned}$$

Comparison of Table 10-5 with Table 10-3 reveals the isomorphism of $\mathcal{E}$ to $\mathcal{Z}_2[x]/(1 + x + x^2)$, under which the image of $a_0 + a_1\xi$ ($a_0, a_1 \in \{0, 1\}$) is $[a_0 + a_1 x]$. Since $1 + x + x^2$ is irreducible over $\mathcal{Z}_2$, $\mathcal{E}$ is a field. □

Table 10-5 OPERATION TABLES FOR EXAMPLE 10-10

$+$	0	1	ξ	$1+\xi$
0	0	1	ξ	$1+\xi$
1	1	0	$1+\xi$	ξ
ξ	ξ	$1+\xi$	0	1
$1+\xi$	$1+\xi$	ξ	1	0

$\cdot$	0	1	ξ	$1+\xi$
0	0	0	0	0
1	0	1	ξ	$1+\xi$
ξ	0	ξ	$1+\xi$	1
$1+\xi$	0	$1+\xi$	1	ξ

Example 10-11

Consider the base field $\langle \mathbb{R}; +, \cdot \rangle$ (where $\mathbb{R}$ is the set of real numbers and $+$ and $\cdot$ are ordinary addition and multiplication) and its extension ring $\mathfrak{C} = \langle \mathbb{C}; +, \cdot \rangle$ produced by adjoining the root i of the polynomial $1 + x^2 \in \mathbb{R}[x]$ to $\langle \mathbb{R}; +, \cdot \rangle$. (Here we use the symbol i rather than ξ for the root, to conform with the conventional notation for $\sqrt{-1}$.) In this case $\mathbb{C}$ consists of all polynomials over $\langle \mathbb{R}; +, \cdot \rangle$ with indeterminate i and degree less than 2; that is,

$$\mathbb{C} = \{a + bi \mid a, b \in \mathbb{R}\}$$

(which can be recognized as the set of complex numbers).

Addition and multiplication in $\mathfrak{C}$ (noting that $1 + i^2 = 0$) are defined by:

$$(a_1 + b_1 i) + (a_2 + b_2 i) = (a_1 + a_2) + (b_1 + b_2)i$$

$$\begin{aligned}(a_1 + b_1 i)(a_2 + b_2 i) &= a_1 a_2 + (a_1 b_2 + (b_1 a_2)i + b_1 b_2 i^2 \\ &= b_1 b_2(1 + i^2) + a_1 a_2 - b_1 b_2 + (a_1 b_2 + b_1 a_2)i \\ &= a_1 a_2 - b_1 b_2 + (a_1 b_2 + b_1 a_2)i\end{aligned}$$

Since $1 + x^2$ is irreducible over $\langle \mathbb{R}; +, \cdot \rangle$ (see Theorem 10-23), $\mathfrak{C} = \langle \mathbb{C}; +, \cdot \rangle$ is a field (in fact, the familiar field of complex numbers). □

THEOREM 10-34

Let $\mathfrak{F} = \langle F; +, \cdot \rangle$ be a field, and $w(x)$ any nonconstant polynomial over $\mathfrak{F}$. Then there always exists an extension field of $\mathfrak{F}$ in which $w(x)$ contains a root.

Proof If $w(x)$ is irreducible over $\mathfrak{F}$, the theorem follows immediately from Theorem 10-33. Otherwise, we can write $w(x) = w_1(x)q(x)$, where $w_1(x)$ is irreducible, and produce an extension field of $\mathfrak{F}$ by adjoining to it the root ξ of $w_1(x)$. Since ξ is also a root of $w(x)$, the theorem follows. □

An extension field of $\mathfrak{F} = \langle F; +, \cdot \rangle$ in which $w(x) \in F[x]$ contains a root ξ is, by Theorem 10-23, a field in which $w(x)$ can be factored in the form $w(x) = (x - \xi)w_1(x)$. A generalization of such an extension field is the so-called "root field" of $\mathfrak{F}$. Specifically, $\mathcal{E} = \langle E; +, \cdot \rangle$ is called a *root field of* $\mathfrak{F}$ *for* $w(x)$ if, in $\mathcal{E}$, $w(x)$ can be written in the form

$$w(x) = c(x - \xi_1)(x - \xi_2) \ldots (x - \xi_n)$$

where $c \in F, \xi_1, \xi_2, \ldots, \xi_n \in E$, and $n = \deg(w(x))$. The following verifies that a root field exists of every base field $\mathfrak{F} = \langle F; +, \cdot \rangle$ and for every nonconstant polynomial over $\mathfrak{F}$.

ALGORITHM 10-1

To construct a root field of a field $\mathcal{F} = \langle F; +, \cdot \rangle$ for a nonconstant polynomial $w(x) \in F[x]$:

(i) Denote $\mathcal{E}_1 = \mathcal{F}$ and $w_1(x) = w(x)$. Set $i = 1$.

(ii) If, in $\mathcal{E}_i = \langle E_i; +, \cdot \rangle$, every factor of $w_i(x)$ is of degree 1, then, in $\mathcal{E}_i$, every factor of $w(x)$ is of degree 1; hence, $\mathcal{E}_i$ is a root field of $\mathcal{F}$ for $w(x)$. Otherwise:

(iii) Let $w_i(x) = p_i(x)q_i(x)$, where $p_i(x)$ is of degree greater than 1 and is irreducible over $\mathcal{E}_i$. Let $\mathcal{E}_{i+1} = \langle E_{i+1}; +, \cdot \rangle$ be the extension field produced by adjoining the root ξ_i of $p_i(x)$ to $\mathcal{E}_i$. In $\mathcal{E}_{i+1}$, then, $w_i(x)$ can be written in the form $w_i(x) = (x - \xi_i)w_{i+1}(x)$.

(iv) Increment i by 1 and return to step (ii). □

Thus, the construction of a root field of $\mathcal{F}$ for $w(x)$ consists of successive construction of fields $\mathcal{E}_1 = \mathcal{F}, \mathcal{E}_2, \mathcal{E}_3, \ldots$, each $\mathcal{E}_i$ $(i > 1)$ being an extension field of $\mathcal{E}_{i-1}$ and produced by adjoining to $\mathcal{E}_{i-1}$ a root of $w(x)$ which $\mathcal{E}_{i-1}$ does not contain. If $\deg(w(x)) = n$, at most $n - 1$ such extensions will be necessary before the construction of the root field is completed.

Example 10-12

To construct a root field of $\mathbf{Z}_2 = \langle \mathbb{Z}_2; +, \cdot \rangle$ for $w(x) = 1 + x^2 + x^3 + x^4$, we first check whether or not $w(x)$ has any factor of degree 1—that is, whether or not $w(x)$ has a root in $\mathbf{Z}_2$. Since $w(1) = 0$, the answer is in the affirmative, and we can write $w(x) = (x - 1)u_1(x)$. Division of $w(x)$ by $x - 1$ yields $u_1(x) = 1 + x + x^3$. Since $u_1(0) = u_1(1) = 1$, $u_1(x)$ has no roots in $\mathbf{Z}_2$; hence, $x - 1$ is the only factor of degree 1 of $w(x)$. Since $1 + x + x^3$ is irreducible over $\mathbf{Z}_2$, our next step is to produce an extension field $\mathcal{E} = \langle E; +, \cdot \rangle$ of $\mathbf{Z}_2$ by adjoining to $\mathbf{Z}_2$ a root of $u(x) = 1 + x + x^3$.

The field $\mathcal{E}$ has the domain

$$E = \{0, 1, \xi, 1 + \xi, \xi^2, 1 + \xi^2, \xi + \xi^2, 1 + \xi + \xi^2\}$$

and its operations can be readily performed by recalling that it is isomorphic to $\mathbf{Z}_2[x]/(1 + x + x^3)$. Since $\xi \in E$ is a root of $u_1(x)$, $u_1(x)$ can be factored in $\mathcal{E}$ in the form $u_1(x) = (x - \xi)u_2(x)$. Division of $u_1(x)$ by $x - \xi$ (see Table 10-6) yields $u_2(x) = 1 + \xi^2 + \xi x + x^2$. To check whether $u_2(x)$ has a root in $\mathcal{E}$, we evaluate $u_2(c)$ for all $c \in E$. In particular, we find out that

$$\begin{aligned} u_2(\xi^2) &= 1 + \xi^2 + \xi^3 + \xi^4 = (1 + \xi)(1 + \xi + \xi^3) \\ &= (1 + \xi)0 = 0 \end{aligned}$$

and, hence, that ξ^2 is a root of $u_2(x)$ in $\mathcal{E}$. Thus, we can write $u_2(x) = (x - \xi^2)u_3(x)$. Division of $u_2(x)$ by $x - \xi^2$ (see Table 10-6) yields $u_3(x) = x -$

Table 10-6 FOR EXAMPLE 10-12

$$
\begin{array}{ll}
 & x^2 + \xi x + (1 + \xi^2) \\
x - \xi & \overline{)x^3 + x + 1} \\
 & x^2 - \xi x^2 \\
 & \xi x^2 + x + 1 \\
 & \xi x^2 + \xi^2 x \\
 & (1 + \xi^2)x + 1 \\
 & (1 + \xi^2)x - \xi(1 - \xi^2) \\
 & 1 + \xi + \xi^3 = 0 \\
 & \\
 & x + (\xi - \xi^2) \\
x - \xi^2 & \overline{)x^2 + \xi x + (1 + \xi^2)} \\
 & x^2 - \xi^2 x \\
 & (\xi - \xi^2)x + (1 + \xi^2) \\
 & (\xi - \xi^2)x - \xi^2(\xi - \xi^2) \\
 & 1 + \xi^2 + \xi^3 + \xi^4 = 1 + \xi^2 + 1 + \xi + \xi + \xi^2 = 0
\end{array}
$$

$\xi(1 - \xi)$. Thus, $\mathcal{E}$ is the root field of $\mathbf{Z}_2$ for $w(x)$. The factored form of $w(x)$ in $\mathcal{E}$ is given by

$$
\begin{aligned}
w(x) &= 1 + x^2 + x^3 + x^4 \\
&= (x - 1)(x - \xi)(x - \xi^2)(x - \xi(1 - \xi))
\end{aligned}
\qquad \square
$$

If $\mathcal{E} = \langle E; +, \cdot \rangle$ is a root field of $\mathcal{F} = \langle F; +, \cdot \rangle$ for $w(x) \in F[x]$, and if there exists no other root field $\tilde{\mathcal{E}} = \langle \tilde{E}; +, \cdot \rangle$ of $\mathcal{F}$ for $w(x)$ such that $\tilde{E}$ is a proper subset of E, then $\mathcal{E}$ is called a *splitting field of* $\mathcal{F}$ *for* $w(x)$. Thus, a splitting field of $\mathcal{F}$ for $w(x)$ is the "smallest" existing root field of $\mathcal{F}$ for $w(x)$. In this connection, we state this result without proof:

THEOREM 10-35

A splitting field exists of every field $\mathcal{F} = \langle F; +, \cdot \rangle$ and for every non-constant polynomial over $\mathcal{F}$. Moreover, any two such splitting fields are isomorphic, with each element of F being its own image.

PROBLEMS

1. Construct extension fields $\mathcal{E} = \langle E; +, \cdot \rangle$ of $\mathbf{Z}_2 = \langle \mathbb{Z}_2; +, \cdot \rangle$, such that:
(a) $\#E = 8$ (b) $\#E = 16$

2. Construct the operation tables of a field of cardinality 9.

3. Construct a field that contains a root of the polynomial $w(x)$ over $\mathbf{Z}_2 = \langle \mathbb{Z}_2; +, \cdot \rangle$, where:
(a) $w(x) = 1 + x + x^3$ (b) $w(x) = 1 + x^2 + x^6$
(c) $w(x) = 1 + x^2 + x^3 + x^4$ (d) $w(x) = 1 + x^2 + x^4$

4. Construct an extension field of $\langle\mathbb{Q}; +, \cdot\rangle$ (where $\mathbb{Q}$ is the set of rational numbers and $+$ and $\cdot$ are ordinary addition and multiplication) by adjoining to $\langle\mathbb{Q}; +, \cdot\rangle$ the root of the polynomial $3 - x^5 \in \mathbb{Q}[x]$.

5. Construct a root field of $\mathcal{Z}_2 = \langle\mathbb{Z}_2; +, \cdot\rangle$ for $w(x) = 1 + x + x^2 + x^4$. Is this field a splitting field?

6. Find a splitting field of $\mathcal{Z}_2 = \langle\mathbb{Z}_2; +, \cdot\rangle$ for the polynomial $(1 + x^{16})(1 + x + x^2) \in \mathcal{Z}_2[x]$.

7. Construct a field $\mathcal{E} = \langle E; +, \cdot\rangle$ such that *every* polynomial $a(x) \in \mathcal{Z}_2[x]$ $(\mathcal{Z}_2 = \langle\mathbb{Z}_2; +, \cdot\rangle)$ of degree 1, 2, or 3 can be written in $\mathcal{E}$ in the form

$$a(x) = c(x - \xi_1)(x - \xi_2) \ldots (x - \xi_n)$$

where $c \in \mathbb{Z}_2, \xi_1, \xi_2, \ldots, \xi_n \in E$, and $1 \leq n \leq 3$.

10-8. GALOIS FIELDS

In this section we deal with *finite* fields only. Theorem 10-18 has already stated that the cardinality of every finite field must be a power of a prime number. The following result states that for every power of a prime number there is a field of that cardinality.

THEOREM 10-36

For every prime number p and a positive integer k there exists a field $\mathcal{F} = \langle F; +, \cdot\rangle$ such that $\#F = p^k$.

Proof Consider the polynomial $m(x) = x^{p^k} - x$ over the field $\mathcal{Z}_p = \langle\mathbb{Z}_p; +, \cdot\rangle$. From Theorem 10-35 we know that $\mathcal{Z}_p$ has a splitting field $\mathcal{F} = \langle F; +, \cdot\rangle$ for $m(x)$ and, hence, that we can factor $m(x)$ in $\mathcal{F}$ in the form

$$m(x) = x^{p^k} - x = (x - \xi_1)(x - \xi_2) \ldots (x - \xi_{p^k})$$

That all the ξ_i are distinct can be shown as follows: Since, in $\mathcal{Z}_p$, $p^k = 0$, the derivative of $m(x)$ (see Sec. 10-6) is given by

$$m'(x) = p^k x^{p^k - 1} - 1 = -1$$

and, hence, no factor $x - \xi_i$ divides $m'(x)$. By Theorem 10-30, then, $(x - \xi_i)^2$ does not divide $m(x)$ and thus no two ξ_i can be identical.

Now, consider the subset $\tilde{F} = \{\xi_1, \xi_2, \ldots, \xi_{p^k}\}$ of F. Using the fact that $(\xi_i + \xi_j)^{p^k} = \xi_i^{p^k} + \xi_j^{p^k}$ and that $\xi_i^{p^k} = \xi_i$ (see Theorems 10-19 and 10-20), we have:

$$\begin{aligned} m(\xi_i + \xi_j) &= (\xi_i + \xi_j)^{p^k} - (\xi_i + \xi_j) \\ &= (\xi_i^{p^k} - \xi_i) + (\xi_j^{p^k} - \xi_j) = 0 + 0 = 0 \end{aligned}$$

$$m(\xi_i\xi_j) = (\xi_i\xi_j)^{p^k} - (\xi_i\xi_j)$$
$$= \xi_j^{p^k}(\xi_i^{p^k} - \xi_i) + \xi_i(\xi_j^{p^k} - \xi_j) = 0 + 0 = 0$$

Hence, $\xi_i + \xi_j$ and $\xi_i\xi_j$ are roots of $m(x)$ in $\mathfrak{F}$ and, therefore, are elements of $\tilde{F}$. Thus, $\tilde{F}$ is closed under $+$ and $\cdot$ and, by Theorem 10-32, $\tilde{\mathfrak{F}} = \langle \tilde{F}; +, \cdot \rangle$ is a subfield of $\mathfrak{F}$. Since $\tilde{\mathfrak{F}}$ is a root field and $\mathfrak{F}$ a splitting field of $\mathbb{Z}_p$ for $m(x)$, we have $\#\tilde{F} = p^k \geq \#F$. But $\tilde{F} \subset F$. Hence, $\#F = p^k$ and thus, $\mathfrak{F} = \langle F; +, \cdot \rangle$ is the sought field of cardinality p^k. □

The field $\mathfrak{F} = \langle F; +, \cdot \rangle$ described in the preceding proof (that is, the splitting field of $\mathbb{Z}_p$ for $x^{p^k} - x$) is referred to as the *Galois field of characteristic* p^k (after the French mathematician Evariste Galois, 1811–1832) and is denoted by $GF(p^k)$. [In particular, the field $\mathbb{Z}_p = \langle \mathbb{Z}_p; +, \cdot \rangle$ is referred to as the Galois field of characteristic p and denoted by $GF(p)$.]

THEOREM 10-37

Every finite field is isomorphic to some Galois field.

Proof Let $\mathfrak{F} = \langle F; +, \cdot \rangle$ be a finite field. By Theorem 10-18, $\#F = p^k$ for some prime number p and positive integer k. Thus, $F - \{0\}$ forms a multiplicative group of cardinality $p^k - 1$. By Theorem 9-21, the order of every $a \in F - \{0\}$ is a divisor of $p^k - 1$; hence, every $a \in F - \{0\}$ is a root of the polynomial $x^{p^k-1} - 1$ over $\mathfrak{F}$. Thus, every $a \in F$ (including $a = 0$) is a root of $m(x) = x^{p^k} - x$. Since $m(x)$ has at most p^k roots (see Theorem 10-24), $\mathfrak{F}$ must be a root field of $\mathbb{Z}_p$ for $m(x)$. Moreover, since no proper subfield of $\mathfrak{F}$ contains all p^k roots of $m(x)$, $\mathfrak{F}$ is a splitting field of $\mathbb{Z}_p$ for $m(x)$ and, hence, a Galois field of characteristic p^k. Since, by Theorem 10-35, all such splitting fields are isomorphic, it follows that all finite fields of cardinality p^k are isomorphic to $GF(p^k)$. □

Theorem 10-37 is highly significant, inasmuch as it establishes that, except for element labeling, there is no difference between a finite field and a Galois field of the same cardinality. Thus, the study of finite fields can be reduced to that of Galois fields.

The construction of a Galois field of cardinality p^k can be carried out by successively adjoining roots of $m(x) = x^{p^k} - x$ to $\mathbb{Z}_p$ until, for the first time, a root field for $m(x)$ is obtained (see Algorithm 10-1). An easier method is to pick up an irreducible polynomial $p(x)$ of degree k over $\mathbb{Z}_p$ and construct the quotient ring $\mathbb{Z}_p[x]/(p(x))$. By Theorem 10-26, this quotient ring is a field. Moreover, since there are p^k polynomials over $\mathbb{Z}_p$ of degree less than k, the cardinality of this field is precisely p^k. As will be shown presently, an irreducible polynomial of degree k over $\mathbb{Z}_p$ exists for every positive integer k and prime number p; hence, this method can be used to construct any field of cardinality p^k.

Example 10-13

To construct a field of order $64 = 2^6$, we look for an irreducible polynomial of degree 6 over $\mathbb{Z}_2$. Such a polynomial can be found by trial and error to be $1 + x + x^6$. The desired field is then $\mathbb{Z}_2[x]/(1 + x + x^6)$, whose domain is given by

$$\{[a(x)] \mid a(x) \in \mathbb{Z}_2[x], \deg(a(x)) < 6\} \qquad \square$$

THEOREM 10-38

The polynomial $x^{p^k} - x$ is divisible by every irreducible polynomial over $\mathbb{Z}_p = \langle \mathbb{Z}_p; +, \cdot \rangle$ whose degree is k, but by no irreducible polynomial over $\mathbb{Z}_p$ whose degree is greater than k.

Proof Let $p(x)$ be an irreducible polynomial of degree k over $\mathbb{Z}_p$, and consider the field $\mathfrak{F} = \langle F; +, \cdot \rangle$ produced by adjoining the root ξ of $p(x)$ to $\mathbb{Z}_p$. Then, in $\mathfrak{F}$, $a(\xi) = 0$ if and only if $a(\xi)$ is a multiple of $p(\xi)$. Since $\#F = p^k$, $\mathfrak{F}$ is isomorphic to $GF(p^k)$ and, hence, every element of F is a root of $m(x) = x^{p^k} - x$ in $\mathfrak{F}$. In particular, $m(\xi) = 0$; hence, $m(\xi)$ is a multiple of $p(\xi)$. In conclusion, $m(x)$ is divisible by $p(x)$.

Now, let $p(x)$ be an irreducible polynomial of degree $l > k$ over $\mathbb{Z}_p$, and hypothesize that $p(x)$ divides $m(x) = x^{p^k} - x$. Consider the field $\mathfrak{F} = \langle F; +, \cdot \rangle$ produced by adjoining the root ξ of $p(x)$ to $\mathbb{Z}_p$, where

$$F = \{a(\xi) \mid a(\xi) = a_0 + a_1\xi + \ldots + a_{l-1}\xi^{l-1}, a_i \in \mathbb{Z}_p \quad (i = 0, 1, \ldots, l-1)\}$$

Using Theorem 10-19 and 10-20, we have

$$\begin{aligned}(a(\xi))^{p^k} &= a_0^{p^k} + a_1^{p^k}\xi^{p^k} + \ldots + a_{l-1}^{p^k}\xi^{(l-1)p^k} \\ &= a_0 + a_1\xi^{p^k} + \ldots + a_{l-1}\xi^{(l-1)p^k}\end{aligned}$$

If $p(x)$ divides $m(x)$, ξ [which is a root of $p(x)$] must also be a root of $m(x)$ and, hence, $\xi^{p^k} = \xi$. Thus we have:

$$a(\xi))^{p^k} = a_0 + a_1\xi + \ldots + a_{l-1}\xi^{l-1} = a(\xi)$$

which implies that every $a(\xi) \in F$ is a root of $x^{p^k} - x$. But this, in turn, implies that $m(x)$ has p^l distinct roots—an impossibility, since $p^l > p^k$. By contradiction, then, $p(x)$ cannot divide $m(x)$. $\square$

THEOREM 10-39

For every prime number p and positive integer k, there exists an irreducible polynomial over $\mathbb{Z}_p = \langle \mathbb{Z}_p; +, \cdot \rangle$ of degree k.

Proof Suppose $m(x) = x^{p^k} - x$ is factored into irreducible factors over $\mathbb{Z}_p$. All these factors are distinct (see proof of Theorem 10-36), and none of them is of degree greater than k (see Theorem 10-38). Now, by Theorem 10-38, for every $i > 0$, the polynomial $x^{p^i} - x$ is divisible by every irreducible polynomial of degree i; hence, the sum of the degrees of all such polynomials is at most p^i. Consequently, the sum of the degrees of all irreducible polynomials of degrees less than k is at most

$$\sum_{i=0}^{k-1} p^i = 1 + p + p^2 + \ldots + p^{k-1} = \frac{p^k - 1}{p - 1} < p^k$$

(where $+$ denotes ordinary addition). Hence, at least one of the p^k irreducible factors of $m(x)$ must be of degree precisely k. □

There is no easy way of discovering irreducible polynomials over a specified field $\mathbb{Z}_p$ and of a specified degree k (although, by Theorem 10-39, discovery is always guaranteed by an exhaustive search). Lists of such polynomials can be found in various books.[1]

We close this section with a characterization of the multiplicative group of a finite field.

THEOREM 10-40

If $\mathfrak{F} = \langle F; +, \cdot \rangle$ is a finite field, then its multiplicative group $\langle F - \{0\}; \cdot \rangle$ is cyclic.

Proof Let $\#F = p^k$, and let $\hat{r}$ denote the exponent of $\langle F - \{0\}; \cdot \rangle$. By Theorem 9-13, the (multiplicative) order of every element $a \in F - \{0\}$ must divide $\hat{r}$; hence, every $a \in F - \{0\}$ is a root of the polynomial $x^{\hat{r}} - 1$ over $\mathfrak{F}$. Since a polynomial of degree $\hat{r}$ can have at most $\hat{r}$ roots, $\#(F - \{0\}) = p^k - 1 \leq \hat{r}$. On the other hand (by Theorem 9-11), the order of any element of $F - \{0\}$ cannot exceed $\#(F - \{0\})$ and, hence, $\hat{r} \leq p^k - 1$. In conclusion, $\hat{r} = p^k - 1$, and $F - \{0\} = (g)$, where g is an element in $F - \{0\}$ whose order is $p^k - 1$. □

An element that generates the multiplicative group of a finite field $\mathfrak{F} = \langle F; +, \cdot \rangle$ [that is, any element whose multiplicative order is $(\#F) - 1$] is called a *primitive element* of $\mathfrak{F}$. If in the field $\mathfrak{F}[x]/(p(x))$ [where $p(x)$ is an irreducible polynomial over $\mathfrak{F}$] the element $[x]$ generates the multiplicative group, then $p(x)$ is called a *primitive polynomial* (over $\mathfrak{F}$). In this case, the domain of the group is given by

$$\{[1], [x], [x]^2, [x]^3, \ldots, [x]^{(\#F)-2}\}$$

[1]For example: A. Gill, *Linear Sequential Circuits*. New York: McGraw-Hill Book Co., 1966. W. W. Peterson, *Error-Correcting Codes*. Cambridge, MA: MIT Press, 1961.

Example 10-14

Consider the field $\mathbb{Z}_2[x]/(1 + x + x^3)$. Using the simplified notation [where $[a(x)]$ is written as $a(x)$], the domain of this field is given by

$$\{0, 1, x, 1 + x, x^2, 1 + x^2, x + x^2, 1 + x + x^2\}$$

Now, in this field,

$$x^0 = 1, \quad x^1 = x, \quad x^2 = x^2,$$
$$x^3 = 1 + x, \quad x^4 = x + x^2, \quad x^5 = 1 + x + x^2,$$
$$x^6 = 1 + x^2$$

(and $x^7 = 1$). Thus, $1 + x + x^3$ is a primitive polynomial over $\mathbb{Z}_2$. □

PROBLEMS

1. (a) Produce $GF(8)$ by constructing a splitting field of $\mathbb{Z}_2$ for $x^8 - x$.
(b) Construct the field $\mathbb{Z}_2[x]/(1 + x + x^3)$.
(c) Show that the fields produced in parts (a) and (b) are isomorphic.

2. (a) Construct the operation tables for the field $\mathfrak{F}_1 = \mathbb{Z}_3[x]/(1 + x + 2x^2)$.
(b) Construct the operation tables for the field $\mathbb{Z}_2 = \mathbb{Z}_3[x]/(1 + x^2)$.
(c) Find a primitive element g_1 of $\mathfrak{F}_1$, and express every element of $\mathfrak{F}_2$ in the form g_1^i.
(d) Find a primitive element g_2 of $\mathfrak{F}_2$, and express every element of $\mathfrak{F}_1$ in the form g_2^i.
(e) Which of the polynomials $1 + x + 2x^2$ and $1 + x^2$ (over $\mathbb{Z}_3 = \langle \mathbb{Z}_3; +, \cdot \rangle$) is primitive?
(f) Show that $\mathfrak{F}_1$ and $\mathfrak{F}_2$ are isomorphic.

3. (a) Construct the field $\mathfrak{F} = \mathbb{Z}_2[x]/(1 + x + x^2 + x^3 + x^4)$.
(b) Find a primitive element of $\mathfrak{F}$.
(c) Show that $1 + x + x^2 + x^3 + x^4$ is not a primitive polynomial.

4. Let $\mathfrak{F} = \langle F; +, \cdot \rangle$ be a field of cardinality p^k. For any $a \in F$, the *minimum polynomial* of a is the (unique) least-degree monic polynomial $u(x)$ such that $u(a) = 0$.
(a) Show that $u(x)$ is irreducible over $\mathfrak{F}$ and of degree at most k.
(b) Find the minimum polynomial of every element of the field $\mathfrak{F}$ constructed in Problem 3.

5. Using Theorem 10-38, find the prime factorization of the polynomial $1 + x^7$ over $\mathbb{Z}_2 = \langle \mathbb{Z}_2; +, \cdot \rangle$.

6. (a) Show that any subfield of $GF(p^k)$ has the cardinality p^l, where l is a divisor of k.
(b) Show that if l divides k, then $p^l - 1$ divides $p^k - 1$.
(c) Show that if l divides k, then $GF(p^k)$ has a subfield of cardinality p^l.

10-9. VECTOR SPACES

An additive Abelian group $\mathcal{V} = \langle V; +\rangle$ is called a *vector space over field* $\mathcal{F} = \langle F; +, \cdot\rangle$, if it satisfies these laws:

(a) *Closure.* For all $a \in F$ and $\mathbf{v} \in V$, $a\mathbf{v}$ is defined and is in V.
(b) *Associativity.* For all $a, b \in F$, and $\mathbf{v} \in V$,

$$(ab)\mathbf{v} = a(b\mathbf{v})$$

(c) *Distributivity.* For all $a, b \in F$, and $\mathbf{v}, \mathbf{w} \in V$,

$$a(\mathbf{v} + \mathbf{w}) = a\mathbf{v} + a\mathbf{w}$$
$$(a + b)\mathbf{v} = a\mathbf{v} + b\mathbf{v}$$

(d) *Identity.* For all $\mathbf{v} \in V$,

$$1\mathbf{v} = \mathbf{v}$$

The elements of a vector space $\mathcal{V} = \langle V; +\rangle$ over $\mathcal{F} = \langle F; +, \cdot\rangle$ are called *vectors over* $\mathcal{F}$ and are commonly printed boldface. The zero of $\mathcal{V}$ (printed $\mathbf{0}$, to distinguish it from the zero of $\mathcal{F}$, which is printed 0) is called the *null vector* of $\mathcal{V}$. To distinguish the elements of F from those of V, it is common to refer to the former as *scalars.*

THEOREM 10-41
Let $\mathcal{V} = \langle V; +\rangle$ be a vector space over field $\mathcal{F} = \langle F; +, \cdot\rangle$. Then, for all $\mathbf{v} \in V$ and $a \in F$,
(a) $0\mathbf{v} = \mathbf{0}$ (b) $-\mathbf{v} = (-1)\mathbf{v}$ (c) $a\mathbf{0} = \mathbf{0}$

Proof (a) From the identity and distributivity laws, we have

$$\mathbf{v} = 1\mathbf{v} = (1 + 0)\mathbf{v} = 1\mathbf{v} + 0\mathbf{v} = \mathbf{v} + 0\mathbf{v}$$

Adding $-\mathbf{v}$ to each side yields $0\mathbf{v} = \mathbf{0}$.
(b) Using the same laws:

$$\mathbf{v} + (-1)\mathbf{v} = 1\mathbf{v} + (-1)\mathbf{v} = (1 - 1)\mathbf{v} = 0\mathbf{v} = \mathbf{0}$$

which implies that $(-1)\mathbf{v}$ is the additive inverse of $\mathbf{v}$ and, hence, $-\mathbf{v} = (-1)\mathbf{v}$.
(c) Using the identity, distributivity, and associativity laws, we have

$$a\mathbf{0} = a(\mathbf{v} + (-\mathbf{v})) = a\mathbf{v} + a(-\mathbf{v}) = a\mathbf{v} + a((-1)\mathbf{v})$$
$$= a\mathbf{v} + (-a)\mathbf{v} = (a - a)\mathbf{v} = 0\mathbf{v} = \mathbf{0} \qquad \square$$

Example 10-15

Consider the field $\mathfrak{F} = \langle F; +, \cdot \rangle$ and the set F^n (that is, the set of all n-tuples of elements from F). If, for all $(a_1, a_2, \ldots, a_n), (b_1, b_2, \ldots, b_n) \in F^n$ and $c \in F$, we define

$$(a_1, a_2, \ldots, a_n) + (b_1, b_2, \ldots, b_n) = (a_1 + b_1, a_2 + b_2, \ldots, a_n + b_n)$$
$$c(a_1, a_2, \ldots, a_n) = (ca_1, ca_2, \ldots, ca_n)$$

then $\mathfrak{F}^n = \langle F^n; + \rangle$ is a vector space over $\mathfrak{F}$. (Verify!) The null vector of $\mathfrak{F}^n$ is the n-tuple $(0, 0, \ldots, 0)$ and, for every $(a_1, a_2, \ldots, a_n) \in F^n$,

$$-(a_1, a_2, \ldots, a_n) = (-a_1, -a_2, \ldots, -a_n)$$ □

Let $\mathcal{V} = \langle V; + \rangle$ be a vector space over field $\mathfrak{F} = \langle F; +, \cdot \rangle$, and let U be a nonempty subset of V. Then the subsystem $\mathcal{U} = \langle U; + \rangle$ of $\mathcal{V}$ is called a *subspace* of $\mathcal{V}$ if $\mathcal{U}$ is a vector space over $\mathfrak{F}$.

THEOREM 10-42

Let $\mathcal{V} = \langle V; + \rangle$ be a vector space over field $\mathfrak{F} = \langle F; +, \cdot \rangle$, and let U be a nonempty subset of V. Then $\mathcal{U} = \langle U; + \rangle$ is a subspace of $\mathcal{V}$ if and only if, for all $\mathbf{v}, \mathbf{w} \in U$ and $a, b \in F$,

(a) $\mathbf{v} + \mathbf{w} \in U$ and $a\mathbf{v} \in U$

or (b) $a\mathbf{v} + b\mathbf{w} \in U$

Proof (a) If $\mathcal{V}$ is a vector space, then it is an additive group and, hence, for all $\mathbf{v}, \mathbf{w} \in U$, $\mathbf{v} + \mathbf{w} \in U$. By closure we must also have, for all $a \in F$ and $\mathbf{v} \in U$, $a\mathbf{v} \in U$. Conversely, if for all $\mathbf{v}, \mathbf{w} \in U$ we have $\mathbf{v} + \mathbf{w} \in U$, then U is closed under addition. If, for all $a \in F$ and $\mathbf{v} \in U$ we have $a\mathbf{v} \in U$, then, for all $\mathbf{v} \in U$, $(-1)\mathbf{v} = -\mathbf{v} \in U$. By Theorem 9-14, then, $\langle U; + \rangle$ is an Abelian group. Associativity, distributivity, and identity are inherited by $\mathcal{U}$ from $\mathcal{V}$; hence, $\mathcal{U}$ is a vector space.

(b) Setting $a = b = 1$, condition (b) becomes $\mathbf{v} + \mathbf{w} \in U$; setting $b = 0$, it becomes $a\mathbf{v} \in U$. Hence, condition (b) implies condition (a). That (a) implies (b) is obvious. □

Example 10-16

Consider the vector space $\mathcal{R}^3 = \langle \mathbb{R}^3; + \rangle$ over the field $\langle \mathbb{R}; +, \cdot \rangle$. [The vectors of $\mathcal{R}^3$ are all 3-tuples (a_1, a_2, a_3), where the a_i are real numbers.] Now, consider the subset of $\mathcal{R}^3$, given by

$$U = \{a_1, a_2, 0)| \; a_1, a_2 \in \mathbb{R}\}$$

For all $(a_1, a_2, 0), (b_1, b_2, 0) \in U$, and $a, b \in \mathbb{R}$, we have

$$\begin{aligned} a(a_1, a_2, 0) + b(b_1, b_2, 0) &= (aa_1, aa_2, 0) + (bb_1, bb_2, 0) \\ &= (aa_1 + bb_1, aa_2 + bb_2, 0) \end{aligned}$$

which is in U. By Theorem 10-42(b), therefore, $\langle U; +\rangle$ is a subspace of $\mathcal{R}^3$. □

A *linear combination* of n vectors $\mathbf{v}_1, \mathbf{v}_2, \ldots, \mathbf{v}_n$ over $\mathfrak{F} = \langle F; +, \cdot\rangle$ is a sum of the form

$$a_1\mathbf{v}_1 + a_2\mathbf{v}_2 + \ldots + a_n\mathbf{v}_n \qquad (a_1, a_2, \ldots, a_n \in F)$$

The vectors $\mathbf{v}_1, \mathbf{v}_2, \ldots, \mathbf{v}_n$ are said to be *linearly independent* if

$$a_1\mathbf{v}_1 + a_2\mathbf{v}_2 + \ldots + a_n\mathbf{v}_n = \mathbf{0}$$
$$\text{implies } a_1 = a_2 = \ldots = a_n = 0$$

Otherwise, $\mathbf{v}_1, \mathbf{v}_2, \ldots, \mathbf{v}_n$ are said to be *linearly dependent*. Note that if $\mathbf{v}_1, \mathbf{v}_2, \ldots, \mathbf{v}_n$ are linearly dependent, we can write $a_1\mathbf{v}_1 + a_2\mathbf{v}_2 + \ldots + a_n\mathbf{v}_n = \mathbf{0}$ with some $a_i \neq 0$ and, hence, can express some $\mathbf{v}_i$ as a linear combination of all the remaining $n - 1$ vectors:

$$\begin{aligned} \mathbf{v}_i = &-a_i^{-1}a_1\mathbf{v}_1 - a_i^{-1}a_2\mathbf{v}_2 - \ldots - a_i^{-1}a_{i-1}\mathbf{v}_{i-1} \\ &-a_i^{-1}a_{i+1}\mathbf{v}_{i+1} - \ldots - a_i^{-1}a_n\mathbf{v}_n \end{aligned}$$

The vectors $\mathbf{v}_1, \mathbf{v}_2, \ldots, \mathbf{v}_n$ are said to *span* the vector space $\mathcal{V} = \langle V; +\rangle$ if every vector in V can be expressed as a linear combination of $\mathbf{v}_1, \mathbf{v}_2, \ldots, \mathbf{v}_n$. If $\mathbf{v}_1, \mathbf{v}_2, \ldots, \mathbf{v}_n$ are linearly independent as well as span $\mathcal{V}$, then the set $\{\mathbf{v}_1, \mathbf{v}_2, \ldots, \mathbf{v}_n\}$ is called a *basis* of $\mathcal{V}$. It can be shown that if $\mathcal{V}$ has a finite basis, then all the bases of $\mathcal{V}$ have the same cardinality. This unique cardinality is called the *dimension* of $\mathcal{V}$. A vector space of dimension n is said to be *n-dimensional*.

Example 10-17

The vector space $\mathfrak{F}^n = \langle F; +\rangle$ of Example 10-15 is spanned by the n vectors

$$\begin{gathered} (1, 0, 0, \ldots, 0) \\ (0, 1, 0, \ldots, 0) \\ (0, 0, 1, \ldots, 0) \\ \cdot \\ \cdot \\ \cdot \\ (0, 0, 0, \ldots, 1) \end{gathered}$$

Since these vectors are linearly independent (why?), they constitute a basis of $\mathfrak{F}^n$. Thus, $\mathfrak{F}^n$ is n-dimensional. □

Example 10-18

Let $g(x)$ be a polynomial of degree n over the field $\mathfrak{F} = \langle F; +, \cdot \rangle$. Consider the quotient ring $\mathfrak{F}[x]/(g(x))$, where the elements $[a(x)]$ [$a(x)$ being a polynomial over $\mathfrak{F}$ of degree less than n] are denoted simply by $a(x)$. Since, in this ring,

$$\begin{aligned} &(a_0 + a_1x + \ldots + a_{n-1}x^{n-1}) + (b_0 + b_1x + \ldots + b_{n-1}x^{n-1}) \\ &= (a_0 + b_0) + (a_1 + b_1)x + \ldots + (a_{n-1} + b_{n-1})x^{n-1} \\ &c(a_0 + a_1x + \ldots + a_{n-1}x^{n-1}) \\ &= ca_0 + ca_1x + \ldots + ca_{n-1}x^{n-1} \end{aligned}$$

the ring can be regarded as a vector space over $\mathfrak{F}$. This vector space is spanned by the vectors $1, x, x^2, \ldots, x^{n-1}$, which are linearly independent. (Why?) Thus, $\mathfrak{F}[x]/(g(x))$ is n-dimensional with a basis $\{1, x, x^2, \ldots, x^{n-1}\}$. □

Let $\mathfrak{V} = \langle V; + \rangle$ be an n-dimensional vector space over $\mathfrak{F} = \langle F; +, \cdot \rangle$. Given any m vectors $\mathbf{v}_1, \mathbf{v}_2, \ldots, \mathbf{v}_m \in V$, the set U of all vectors spanned by these m vectors forms a subspace $\mathfrak{U} = \langle U; + \rangle$ of $\mathfrak{V}$ (see Theorem 10-42). A basis for $\mathfrak{V}$ can be constructed by selecting from the set $\{\mathbf{v}_1, \mathbf{v}_2, \ldots, \mathbf{v}_m\}$ a largest subset (the selection is generally not unique) in which all vectors are linearly independent.

PROBLEMS

1. Show that, in every vector space over field $\mathfrak{F} = \langle F; +, \cdot \rangle$, $a\mathbf{v} = \mathbf{0}$ implies $a = 0$ or $\mathbf{v} = \mathbf{0}$ ($a \in F$, $\mathbf{v} \in V$).
2. Prove that $\mathfrak{F}^n$ and $\mathfrak{F}[x]/(g(x))$ of Examples 10-15 and 10-18 are vector spaces over $\mathfrak{F}$.
3. Let V be the set of all functions $f(x)$ whose domain is the set of all real numbers between 0 and 1 (inclusively) and whose range is $\mathbb{R}$. Show that $\mathfrak{V} = \langle V; + \rangle$ is a vector space over $\langle \mathbb{R}; +, \cdot \rangle$, and determine which of the following subsets of V forms a subspace of $\mathfrak{V}$.
   (a) $\{a_0 + a_1x + a_2x^2 + a_3x^3 \mid a_0, a_1, a_2, a_3 \in \mathbb{R}\}$
   (b) $\{a_0 + a_1x + a_2x^2 + x^3 \mid a_0, a_1, a_2 \in \mathbb{R}\}$
   (c) $\{f(x) \mid 2f(0) = f(1)\}$
   (d) $\{f(x) \mid f(x) = f(1 - x) \text{ for all } 0 \leq x \leq 1\}$
4. Consider the vector space $\mathfrak{F}^3 = \langle F^3; + \rangle$ (see Example 10-15), where $\mathfrak{F} = \langle \mathbb{Z}_2; +, \cdot \rangle$. Which of the following subsets of F^3 forms a subspace of $\mathfrak{F}^3$?
   (a) $\{(a_1, a_2, a_3) \mid a_1 = 1\}$

(b) $\{(a_1, a_2, a_3) \mid a_1 = 0 \text{ or } a_3 = 0\}$
(c) $\{(a_1, a_2, a_3) \mid a_1 + a_2 = 0\}$

5. Consider the vector space $\mathfrak{F}^n = \langle F^n; +\rangle$ over $\mathfrak{F} = \langle \mathbb{Z}_p; +, \cdot\rangle$, where p is a prime number. Show that every additive subgroup of $\mathfrak{F}^n$ is a subspace of $\mathfrak{F}^n$.

6. Prove that the set of vectors $\{(1, 0, \ldots, 0), (0, 1, \ldots, 0), (0, 0, \ldots, 1)\}$ is a basis of the vector space $\mathfrak{F}^n$ of Example 10-15. Construct another basis for $\mathfrak{F}^n$.

7. Prove that $\{1, x, \ldots, x^{n-1}\}$ is a basis of $\mathfrak{F}[x]/(g(x))$, where $g(x)$ is a polynomial over $\mathfrak{F}$ of degree n (see Example 10-18).

8. (a) Why is $\{(1, 1, 0), (1, 0, 1), (0, 1, 1)\}$ not a basis of $\mathfrak{F}^3$ of Problem 4?
(b) Find bases for all subspaces specified in Problem 4.

9. Consider the vector space $\mathfrak{F}^4 = \langle F^4; +\rangle$ over $\mathfrak{F} = \langle \mathbb{Z}_3; +, \cdot\rangle$.
(a) Find the vector space $\mathfrak{U}$ spanned by the set of $\mathfrak{F}^4$ vectors

$$\{(0, 1, 0, 0), (2, 0, 0, 1), (1, 2, 0, 2), (0, 2, 0, 2)\}$$

(b) Find a basis for $\mathfrak{U}$ of part (a).

10-10. MATRICES OVER FIELDS

An *$m \times n$ matrix over field* $\mathfrak{F} = \langle F; +, \cdot\rangle$ is an array of mn elements of F, arranged in m rows and n columns. The element located in the ith row and jth column of matrix $\mathbf{M}$ is referred to as the (i, j) *entry* of $\mathbf{M}$. An $m \times n$ matrix whose (i, j) entry has the generic label m_{ij}, is sometimes denoted by $[m_{ij}]_{m\times n}$. Thus,

$$\mathbf{M} = [m_{ij}]_{m\times n} = \begin{bmatrix} m_{11} & m_{12} & \cdots & m_{1n} \\ m_{21} & m_{22} & \cdots & m_{2n} \\ \cdot & & & \\ \cdot & & & \\ \cdot & & & \\ m_{m1} & m_{m2} & \cdots & m_{mn} \end{bmatrix}$$

The *transpose* of a matrix $\mathbf{M} = [m_{ij}]_{m\times n}$ is the matrix $\mathbf{M}^{TR} = [m_{ji}]_{n\times m}$.

An $n \times n$ matrix is called a *square* matrix. The (i, i) entries of a square matrix constitute its *principal diagonal*. An $n \times n$ square matrix whose principal diagonal consists entirely of 1s and where all other entries are 0, is called an *identity matrix* and is denoted by $\mathbf{I}_n$ (or simply by $\mathbf{I}$, when its size is irrelevant). A matrix that consists entirely of 0 entries is called a *null matrix* and is denoted by $\mathbf{0}$.

The rows (or columns) of a matrix $\mathbf{M}$ over field $\mathfrak{F} = \langle F; +, \cdot\rangle$, having n columns (or rows), can be regarded as vectors from the n-dimensional vector space $\mathfrak{F}^n = \langle F^n; +\rangle$ defined in Example 10-15. For this reason, the rows (or columns) of $\mathbf{M}$ are often referred to as *row* (or *column*) *vectors* of **M**. A matrix composed of a single row (or column) that consists of n entries, is referred to as an *n-dimensional row* (or *column*) *vector*. (For that reason, matrices, like vectors, are commonly denoted by boldface symbols.)

The subspace $\mathcal{U} = \langle U; +\rangle$ spanned by the row (or column) vectors of **M** is called the *row* (or *column*) *space* of **M**; its dimension is called the *row* (or *column*) *rank* of **M**. It can be shown that the row and column ranks of a matrix are always equal and, hence, we can talk simply about the *rank* of a matrix. An $n \times n$ square matrix of rank n is called a *nonsingular matrix*; otherwise, it is a *singular matrix*.

Example 10-19

The following is a 4×5 matrix over $\mathcal{Z}_2 = \langle \mathbb{Z}_2; +, \cdot\rangle$:

$$\mathbf{M} = \begin{bmatrix} 0 & 1 & 0 & 0 & 0 \\ 1 & 0 & 1 & 0 & 0 \\ 0 & 1 & 0 & 1 & 0 \\ 1 & 0 & 1 & 1 & 0 \end{bmatrix}$$

The second row vector of **M** differs from the first; hence, the first two row vectors are linearly independent. The third row vector differs from the first two and also from the sum (modulo 2) of the first two; hence, the first three row vectors are linearly independent. The fourth row vector equals the sum of the first three. Thus, the first three row vectors of **M** are linearly independent and span the row space of **M**. The rank of **M** is, therefore, 3.

As a check, we can note that the second column vector of **M** differs from the first, the third equals the first, the fourth differs from the first two and their sum, and the fifth is a null vector. Thus, the first, second, and fourth column vectors of **M** are linearly independent and span the column space of **M**. □

Matrices over a field can be transformed into others by a sequence of so-called *elementary row* (or *column*) *operations* of these types:

(a) Interchange of any two row (or column) vectors.
(b) Multiplication of any row (or column) vector by a nonzero scalar.
(c) Addition of a scalar-multiple of any row (or column) vector to any other row (or column) vector.

If matrix $\mathbf{M}_2$ is obtained from $\mathbf{M}_1$ by a sequence of elementary row (or column) operations, then $\mathbf{M}_1$ and $\mathbf{M}_2$ are said to be *row-* (or *column-*) *equivalent*. From the definition of elementary row (or column) operations it can be seen that the space spanned by the row (or column) vectors of a matrix remains invariant under such operations. Thus, if $\mathbf{M}_1$ and $\mathbf{M}_2$ are row- (or column-) equivalent, they have the same row (or column) space.

When elementary row (or column) operations are judiciously applied, a matrix $\mathbf{M}$ can be transformed into a matrix $\hat{\mathbf{M}}$ which is called the *row-* (or *column-*) *reduced echelon canonical form* of $\mathbf{M}$, and which exhibits the following properties: The first nonzero entry of every nonzero row (or column) is 1; every column (or row) containing such a 1 has 0s elsewhere; the first 1 in every nonzero row (or column) is to the right of (or below) the first 1 in the preceding row (or column); zero rows (or columns) appear below (or to the right of) all nonzero rows (or columns). The properties of the row- (or column-) reduced echelon canonical form imply that the non-null row (or column) vectors of $\hat{\mathbf{M}}$ are linearly independent. Thus, these row (or column) vectors can serve as a basis of the row (or column) space of $\hat{\mathbf{M}}$ and, hence, as a basis of the row (or column) space of $\mathbf{M}$.

Example 10-20

The following illustrates how a matrix over $\mathcal{Z}_2 = \langle \mathbb{Z}_2; +, \cdot \rangle$ can be transformed into its row-reduced echelon canonical form. The given matrix is

$$\mathbf{M} = \begin{bmatrix} 1 & 1 & 1 & 1 & 0 & 1 \\ 1 & 1 & 0 & 0 & 1 & 1 \\ 0 & 0 & 1 & 1 & 1 & 1 \\ 1 & 1 & 1 & 1 & 0 & 0 \\ 0 & 0 & 0 & 0 & 0 & 1 \end{bmatrix}$$

Adding row 1 to rows 2 and 4 (in order to eliminate all 1s from column 1 except for the first), we obtain

$$\mathbf{M}_1 = \begin{bmatrix} 1 & 1 & 1 & 1 & 0 & 1 \\ 0 & 0 & 1 & 1 & 1 & 0 \\ 0 & 0 & 1 & 1 & 1 & 1 \\ 0 & 0 & 0 & 0 & 0 & 1 \\ 0 & 0 & 0 & 0 & 0 & 1 \end{bmatrix}$$

Adding row 2 to rows 1 and 3 (in order to eliminate all 1s from column 3 except for the second), we obtain

$$\mathbf{M}_2 = \begin{bmatrix} 1 & 1 & 0 & 0 & 1 & 1 \\ 0 & 0 & 1 & 1 & 1 & 0 \\ 0 & 0 & 0 & 0 & 0 & 1 \\ 0 & 0 & 0 & 0 & 0 & 1 \\ 0 & 0 & 0 & 0 & 0 & 1 \end{bmatrix}$$

Finally, adding row 3 to rows 1, 4, and 5 (in order to eliminate all 1s from column 6 except for the third), we obtain

$$\hat{\mathbf{M}} = \begin{bmatrix} 1 & 1 & 0 & 0 & 1 & 0 \\ 0 & 0 & 1 & 1 & 1 & 0 \\ 0 & 0 & 0 & 0 & 0 & 1 \\ 0 & 0 & 0 & 0 & 0 & 0 \\ 0 & 0 & 0 & 0 & 0 & 0 \end{bmatrix}$$

which is of the desired form. Clearly, the row space of $\hat{\mathbf{M}}$ is the same as that of the matrix $\tilde{\mathbf{M}}$ obtained by deleting all zero rows from $\hat{\mathbf{M}}$:

$$\tilde{\mathbf{M}} = \begin{bmatrix} 1 & 1 & 0 & 0 & 1 & 0 \\ 0 & 0 & 1 & 1 & 1 & 0 \\ 0 & 0 & 0 & 0 & 0 & 1 \end{bmatrix}$$

The row vectors of $\tilde{\mathbf{M}}$ constitute a basis for the row space of **M**. (Thus, the row space of **M** is 3-dimensional). □

When the row (or column) vectors of **M** are linearly independent, its row- (or column-) reduced echelon canonical form consists of an identity matrix with possibly some additional columns (or rows) interspersed among the columns (or rows) of this identity matrix. For example, $\tilde{\mathbf{M}}$ of Example 10-20 consists of a 3×3 identity matrix with three additional interspersed columns (namely, the second, fourth, and fifth).

Let $\mathbf{M} = [m_{ij}]_{r \times s}$ and $\mathbf{N} = [n_{ij}]_{s \times t}$ be two matrices over field $\mathfrak{F} = \langle F; +, \cdot \rangle$ (with the number of columns in **M** being equal to the number of rows in **N**). Then the product **MN** of **M** and **N** is a matrix $\mathbf{L} = [l_{ij}]_{r \times t}$ over $\mathfrak{F}$, where, for $i = 1, 2, \ldots, r$ and $j = 1, 2, \ldots, t$,

$$l_{ij} = \textstyle\sum_{k=1}^{s} m_{ik} n_{kj}$$

(with addition and multiplication as per rules of $\mathfrak{F}$). The product $a\mathbf{M}$, where $a \in F$, is the matrix $[am_{ij}]_{r \times s}$ ($-1\mathbf{M}$ is written as $-\mathbf{M}$). Note that, since matrix multiplication is not commutative, **N** *premultiplied* by **M** (that is, **MN**) is not necessarily the same as **N** *postmultiplied* by **M** (that is, **NM**).

Given an $n \times n$ matrix **M** over $\mathfrak{F}$, the *inverse* of **M** is an $n \times n$ matrix $\mathbf{M}^{-1}$ over $\mathfrak{F}$ such that

$$\mathbf{M}\mathbf{M}^{-1} = \mathbf{M}^{-1}\mathbf{M} = \mathbf{I}_n$$

It can be shown that **M** has an inverse (and a unique one) if and only if **M** is nonsingular. $\mathbf{M}^{-1}$ can be constructed by performing on $\mathbf{I}_n$ all the row operations performed on $\mathbf{M}$ in the course of transforming $\mathbf{M}$ into its row-reduced echelon canonical form (that is, into $\mathbf{I}_n$).

Example 10-21

The following illustrates how the inverse of a matrix over $\mathbb{Z}_2 = \langle \mathbb{Z}; +, \cdot \rangle$ can be constructed. The given matrix is

$$\mathbf{M} = \begin{bmatrix} 1 & 1 & 1 \\ 1 & 0 & 1 \\ 0 & 1 & 1 \end{bmatrix}$$

We start by forming the composite matrix $(\mathbf{M} \mid \mathbf{I}_3)$:

$$\left[\begin{array}{ccc|ccc} 1 & 1 & 1 & 1 & 0 & 0 \\ 1 & 0 & 1 & 0 & 1 & 0 \\ 0 & 1 & 1 & 0 & 0 & 1 \end{array}\right]$$

Adding row 1 to row 2, we obtain

$$\left[\begin{array}{ccc|ccc} 1 & 1 & 1 & 1 & 0 & 0 \\ 0 & 1 & 0 & 1 & 1 & 0 \\ 0 & 1 & 1 & 0 & 0 & 1 \end{array}\right]$$

Adding row 2 to rows 1 and 3, we obtain

$$\left[\begin{array}{ccc|ccc} 1 & 0 & 1 & 0 & 1 & 0 \\ 0 & 1 & 0 & 1 & 1 & 0 \\ 0 & 0 & 1 & 1 & 1 & 1 \end{array}\right]$$

Finally, adding row 3 to row 1, we obtain

$$\left[\begin{array}{ccc|ccc} 1 & 0 & 0 & 1 & 0 & 1 \\ 0 & 1 & 0 & 1 & 1 & 0 \\ 0 & 0 & 1 & 1 & 1 & 1 \end{array}\right]$$

Thus,

$$\mathbf{M}^{-1} = \begin{bmatrix} 1 & 0 & 1 \\ 1 & 1 & 0 \\ 1 & 1 & 1 \end{bmatrix}$$

(Check!) □

Let $\mathbf{M}$ be an $m \times n$ matrix over field $\mathfrak{F} = \langle F; +, \cdot \rangle$, and consider the set N of all n-dimensional row vectors $\mathbf{v}$ over $\mathfrak{F}$ such that

$$\mathbf{M}\mathbf{v}^{TR} = \mathbf{0}$$

($\mathbf{0}$ being the column vector consisting of m zeros). If $\mathbf{v}_1$ and $\mathbf{v}_2$ are in N, then, for all $a_1, a_2 \in F$,

$$\mathbf{M}(a_1\mathbf{v}_1 + a_2\mathbf{v}_2)^{TR} = a_1\mathbf{M}\mathbf{v}_1^{TR} + a_2\mathbf{M}\mathbf{v}_2^{TR} = a_1\mathbf{0} + a_2\mathbf{0} = \mathbf{0}$$

and, hence, $a_1\mathbf{v}_1 + a_2\mathbf{v}_2$ is in N. By Theorem 10-42, then, N is a subspace of $\mathfrak{F}^n = \langle F^n; + \rangle$. This subspace is called the *null space* of $\mathbf{M}$. It can be shown that if the rank of $\mathbf{M}$ is k, then the dimension of its null space (called the *nullity* of $\mathbf{M}$) is $n - k$.

PROBLEMS

1. Show that the following two matrices over $\langle \mathbb{R}; +, \cdot \rangle$ are equivalent.

$$\mathbf{M}_1 = \begin{bmatrix} 5 & 2 & 7 \\ -3 & 4 & 1 \\ -1 & -2 & 3 \end{bmatrix} \qquad \mathbf{M}_2 = \begin{bmatrix} 1 & 0 & 1 \\ 0 & 1 & 1 \\ 0 & 0 & 0 \end{bmatrix}$$

2. Find the row-reduced echelon canonical forms and the ranks of the following matrices:

(a)
$$\mathbf{M} = \begin{bmatrix} 0 & 0 & 2 & 2 & 0 & 2 \\ 2 & 2 & 6 & 8 & 4 & 8 \\ 1 & 1 & 5 & 6 & 2 & 5 \\ 1 & 1 & 3 & 4 & 2 & 7 \end{bmatrix} \qquad (\text{over } \langle \mathbb{R}; +, \cdot \rangle)$$

(b)
$$\mathbf{M} = \begin{bmatrix} 1 & 1 & 0 & 1 \\ 1 & 1 & 1 & 0 \\ 0 & 1 & 1 & 0 \\ 0 & 1 & 0 & 1 \end{bmatrix} \qquad (\text{over } \langle \mathbb{Z}_2; +, \cdot \rangle)$$

3. The matrix $\mathbf{M}$ given below is over $\langle \mathbb{R}; +, \cdot \rangle$:

$$\mathbf{M} = \begin{bmatrix} 1 & 2 & 4 & 5 & 7 \\ 1 & 2 & 3 & 4 & 5 \\ -1 & -2 & 0 & 2 & 1 \end{bmatrix}$$

(a) Find a basis for the row space of $\mathbf{M}$.
(b) Express each row vector of $\mathbf{M}$ as a linear combination of basis vectors.

4. Determine the number of distinct 2×4 row-reduced echelon canonical matrices of rank 2 over $\mathbf{Z}_2 = \langle \mathbb{Z}_2; +, \cdot \rangle$.

5. Let

$$\mathbf{M} = \begin{bmatrix} \mathbf{I}_k & \mathbf{Q} \\ \mathbf{0} & \mathbf{I}_{n-k} \end{bmatrix}$$

be an $n \times n$ matrix over $\mathfrak{F}$ [where $\mathbf{0}$ is an $(n - k) \times k$ null matrix and $\mathbf{Q}$ is an arbitrary $k \times (n - k)$ matrix]. Show that

$$\mathbf{M}^{-1} = \begin{bmatrix} \mathbf{I}_k & -\mathbf{Q} \\ \mathbf{0} & \mathbf{I}_{n-k} \end{bmatrix}$$

6. The following is a nonsingular matrix over $\mathbf{Z}_3 = \langle \mathbb{Z}_3; +, \cdot \rangle$. Find its inverse.

$$\mathbf{M} = \begin{bmatrix} 2 & 1 & 0 \\ 1 & 1 & 2 \\ 2 & 2 & 2 \end{bmatrix}$$

7. The following matrix is over $\mathbf{Z}_2 = \langle \mathbb{Z}_2; +, \cdot \rangle$:

$$\mathbf{P} = \begin{bmatrix} 1 & 1 & 1 & 1 & 0 & 1 \\ 1 & 1 & 0 & 0 & 1 & 1 \\ 0 & 0 & 1 & 1 & 1 & 1 \end{bmatrix}$$

Find a matrix $\mathbf{Q}$ such that $\mathbf{PQ} = \mathbf{0}$. ($\mathbf{Q}$ is called the *right inverse* of $\mathbf{P}$.) (*Hint*: $\mathbf{P}$ is the 3×3 matrix $\mathbf{M}$ of Example 10-21, with three additional columns interspersed.)

8. Find a basis for the null space of the matrix $\mathbf{P}$ of Problem 7.

9. Show that the relation of row (or column) equivalence on the set of $m \times n$ matrices over $\mathfrak{F}$ is an equivalence relation.

REFERENCES

ARTIN, E., *Galois Theory*. Notre Dame, IN: University of Notre Dame Press, 1959.

GAAL, L., *Classical Galois Theory with Examples*. Chicago: Markham, 1971.

HERSTEIN, I., *Topics in Algebra*. Waltham, MA: Xerox, 1964.

MACLANE, S., and G. BIRKHOFF, *Algebra*. New York: Macmillan, 1967.

MCCOY, N. H., *Rings and Ideals*. Mathematical Association of America (distributed by Wiley), 1948.

NERING, E. D., *Linear Algebra and Matrix Theory*. New York: Wiley, 1963.

VAN DER WAERDEN, B. L., *Modern Algebra*. Ungar, 1931.

11 CODES

The mathematical theories expounded in the Chapters 9 and 10 are used extensively in the implementation of error-correcting and error-detecting codes and in the design of counters and "pseudo-random" sequences. In this chapter we focus our attention on *linear codes*, with special emphasis on *Hamming codes*.

11-1. THE COMMUNICATION CHANNEL MODEL

Roughly speaking, "codes" are means by which items of information can be represented so as to make them amenable to an economical and reliable transmittal from one physical location to another. For purpose of discussing codes, the physical devices and environmental factors relevant to transmittal are idealized into a so-called *communication channel model* which is shown diagramatically in Fig. 11-1. The model consists of a *source*, capable of transmitting a variety of items of information called *messages*, and a *destination* that receives and utilizes them. The transmitted messages can be continuous with time (such as variations in air pressure produced by human voice) or discrete with time (such as patterns of holes on a punched card). These messages are applied to an *encoder* (such as a microphone or a card reader) whose function is to transform each message into a form (such as a wave or a series of pulses of electric current) suitable for transmittal to the destination. The output of the encoder, consisting of the *transmitted encoded messages*, is applied to the *channel* that is the physical means by which the messages are to be conveyed from source to destination. The channel, in actuality, can be a transmission device (such as a telephone cable) or a storage medium (such as a magnetic tape). In all practical cases, imperfections in the channel cause

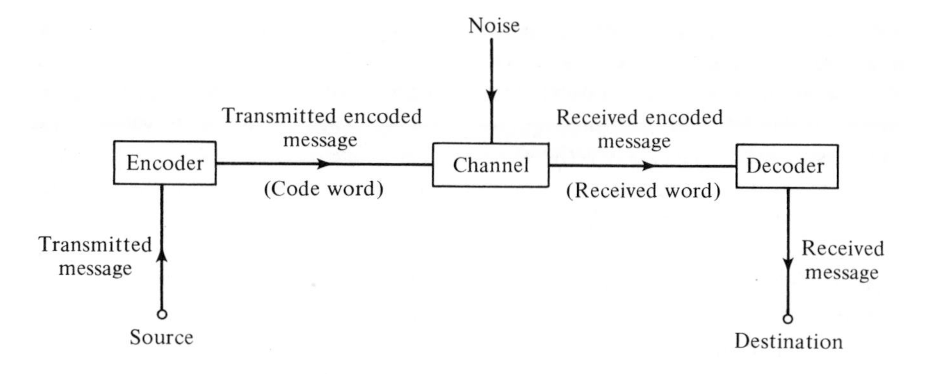

Fig. 11-1 Communication channel model

its output—the *received encoded messages*—to differ to lesser or greater extent from its input. The cumulative effect of these imperfections is known as *noise*. The received encoded messages are applied to a *decoder* whose function, ideally, is to reproduce the originally transmitted messages from the received encoded ones. Due to noise, this goal is not always achievable, with the result that the *received messages*—that is, the messages as they appear at the destination—do not always match the transmitted ones. The central problem in the theory of codes is: For a given channel, how can we design an encoder and a decoder so as to minimize the probability of a received message differing from the transmitted one?

Before attempting to tackle this question, we make some simplifying assumptions with respect to our communication channel model. First, we assume that the messages are discrete—that is, each message is an element out of a finite *message set*

$$M = \{m_1, m_2, \ldots, m_r\}$$

Second, we assume that the channel can accept, at any instant of time, only one of a finite number of *channel symbols* that belong to the channel alphabet

$$X = \{\xi_1, \xi_2, \ldots, \xi_l\}$$

($X = \{0, 1\}$ is the most common alphabet.) The encoder transforms each transmitted message m_i into a finite-length string of channel symbols, called a *code word*, which constitutes the encoded message and which is applied, one symbol at a time, to the channel. The set of r code words corresponding to the r elements of M is referred to as the *code* of M. (When $X = \{0, 1\}$, this code is said to be a *binary code*.) We assume that all r code words are distinct and have the same length n. (A code that conforms with these assumptions is sometimes called a *block code*.) Thus, the encoder can be regarded as a function from M into X^n, under which every message m_i is *encoded* into some n-tuple of elements from X.[1] When the code word $\xi_{i_1}\xi_{i_2}\ldots\xi_{i_n}$ ($\xi_{i_1}, \xi_{i_2}, \ldots, \xi_{i_n} \in X$) is applied to the channel, the corresponding output string—called the *received word*—is an n-tuple $\xi'_{i_1}\xi'_{i_2}\ldots\xi'_{i_n}$ ($\xi'_{i_1}, \xi'_{i_2}, \ldots, \xi'_{i_n} \in X$) which, due to noise, is not necessarily the same as the input code word. The received word is applied to the decoder whose function is to determine the original message m_i and convey this m_i to the destination.

The heart of the decoder is the so-called *decoding table* that, for each i, lists all those received words which should yield the decision that the transmitted message was m_i. Some decoding tables may also list those received words that should be declared as erroneous, without any decision being made as to which m_i they correspond to. The operation whereby a decoder produces the correct m_i, despite injurious noise effects, is referred to as *error correction.* The operation whereby the decoder merely recognizes these effects, without attempting to counter them, is referred to as *error detection.* Error detection, although less effective than error correction, is not useless; in some applications, the detection of an error automatically results in the retransmittal of the message and in a new attempt (this time, possibly, successful) to decode it.

Example 11-1

Consider the message set $M = \{m_1, m_2, m_3, m_4\}$ operating in conjunction with a channel whose alphabet is $X = \{0, 1\}$. Let m_1, m_2, m_3, and m_4 be encoded into the 5-tuples 00110, 01101, 10011, and 11000, respectively. A possible decoding table for this code is shown in Table 11-1, where each of the $2^5 = 32$ possible received words is in exactly one of four sets (columns)

Table 11-1 DECODING TABLE FOR EXAMPLE 11-1

Message	m_1	m_2	m_3	m_4
Code word	00110	01101	10011	11000
Code words with one error	00111 00100 00010 01110 10110	01100 01111 01001 00101 11101	10010 10001 10111 11011 00011	11001 11010 11100 10000 01000
Code words with two errors	00000 10100	10101 00001	01011 11111	11110 01010

[1]In the discussion of codes, the notation $(a_1, a_2, \ldots, a_n)$ for an n-tuple will often be replaced with the string notation $a_1a_2\ldots a_n$.

of such words. Every received word that belongs to the column headed "m_i," is decoded by the decoder as m_i.

Note that column "m_i" ($i = 1, 2, 3, 4$) includes the code word assigned to m_i as well as all five possible received words obtained by altering exactly one symbol (from 0 to 1 or from 1 to 0) in the code word assigned to m_i. Thus, every code word which, due to noise, is subjected to a single error (that is, to an alteration of exactly one of its symbols) is correctly decoded by the decoder. The chosen code is thus a *single-error correcting code.* It can also be noted that some code words subjected to two errors are correctly decoded —for example, 00110, corrupted into 00000, is correctly interpreted as m_1. However, this is not always the case—for example, 00110, corrupted into 11110, is incorrectly interpreted as m_4. The chosen code, therefore, cannot be classified as a *double-error correcting code.*

We can modify the decoding table by shifting all those received words which, in Table 11-1, appear in the last two rows, to a new column headed "Error." Any received word that falls into the new category is not decoded into any of the four messages, but simply declared as erroneous. When associated with such a decoding table, the chosen code is said to be single-error correcting and *double-error detecting.* □

The evaluation of a code effectiveness in combating errors depends critically on our knowledge of the statistical properties of the noise. In many practical situations, we can assume a *binary symmetric channel* where the alphabet is $\{0, 1\}$ and where the probability of any symbol being altered by noise is a fixed constant ψ ($0 \leq \psi \leq 1$). Thus, the probability of any given code word of length n being subjected to i errors ($0 \leq i \leq n$) is $\psi^i(1 - \psi)^{n-i}$. Since, in all practical cases, $\psi < 1 - \psi$ (the error probability usually being much smaller than 1), the probability of undergoing i errors decreases with i. This implies that, given a received word, the transmitted code word assigned to it is, most likely, the one which differs from it in the fewest positions. In the decoding table, therefore, a received word should appear in the same column as the code word that is "closest" to it in terms of the number of differing positions. (Decoding with such a table is called a *maximum-likelihood decoding.*) The decoding table in Table 11-1 satisfies this requirement.

The number of positions in which two words differ is referred to as the *distance* between these words. Thus, i errors result in distance i between a code word and the corresponding received word. The least distance between any two distinct code words in a code is referred to as the *minimum distance* of the code. If a code is constructed in such a way that its minimum distance is at least $l + 1$, then any pattern of l or fewer errors can be detected (since such a pattern can never result in another code word). Also, any code whose minimum distance is at least $2l + 1$ can be used to correct any pattern of l or fewer errors, employing maximum-likelihood decoding: Every received word is listed in the decoding table with the (unique) code word that differs from it in at most l positions. For example, the minimum distance of the code of Example 11-1 is 3; hence, this code can be used as a single-error correcting and a double-error detecting code.

PROBLEMS

1. Construct a maximum-likelihood decoding table for the code consisting of the code words 0000, 1001, 0110, 1010. What errors are corrected and detected by this code?

2. Construct a double-error correcting binary block code for a message set that consists of 4 messages (using shortest possible code words). What errors are detected by this code?

3. Consider a binary code of cardinality 2^k ($k \geq 1$), where the code words are all k-digit binary strings. By adding one digit to each code word, make the code single-error detecting.

4. Let $d(w_i, w_j)$ denote the distance between code words w_i and w_j. Show that, for any code words w_1, w_2, w_3,

$$d(w_1, w_2) \leq d(w_1, w_3) + d(w_2, w_3)$$

5. Show that a block code is both l-error correcting and k-error detecting ($k \geq l$) if its minimum distance is $l + k + 1$.

6. Assume that the code and decoding table of Example 11-1 are used in conjunction with a binary symmetric channel where the probability of any symbol being altered by noise is ψ. In terms of ψ, find the probability of a message being correctly decoded by the decoder.

11-2. LINEAR CODES

In the remainder of this chapter we restrict ourselves to the case where the channel alphabet is $\mathbb{Z}_p = \{0, 1, 2, \ldots, p - 1\}$, where p is a prime number. Correspondingly, code words and received words of length n are regarded as vectors from the n-dimensional vector space $\mathcal{Z}_p^n = \langle \mathbb{Z}_p^n; + \rangle$ over the field $\mathcal{Z}_p = \langle \mathbb{Z}_p; +, \cdot \rangle$ (see Example 10-15), and are referred to as *code vectors* and *received vectors*, respectively. For example, the code words and received words of Example 11-1 are regarded as vectors from the 5-dimensional vector space $\mathcal{Z}_2^5 = \langle \mathbb{Z}_2^5; + \rangle$ over $\mathcal{Z}_2 = \langle \mathbb{Z}_2; +, \cdot \rangle$, and are referred to as 5-dimensional code vectors and received vectors, respectively. Consistent with these conventions, all additions and multiplications involving elements of $\mathbb{Z}_p$ are assumed to be performed as per rules of $\mathcal{Z}_p$ (that is, modulo p).

An (n, k) *linear code over* $\mathbb{Z}_p$ (where $n \geq k$) is a code whose code vectors form a k-dimensional subspace of the n-dimensional vector space $\mathcal{Z}_p^n = \langle \mathbb{Z}_p^n; + \rangle$. If the basis of this code is $\{\mathbf{u}_1, \mathbf{u}_2, \ldots, \mathbf{u}_k\}$ (where each $\mathbf{u}_i$ is an n-dimensional code vector), then an n-dimensional vector $\mathbf{v}$ is in the code if and only if it equals some linear combination of $\mathbf{u}_1, \mathbf{u}_2, \ldots, \mathbf{u}_k$ (see Sec. 10-9). Since each coefficient in the sum $a_1\mathbf{u}_1 + a_2\mathbf{u}_2 + \ldots + a_k\mathbf{u}_k$ ($a_1, a_2, \ldots, a_k \in \mathbb{Z}_p$) can assume p distinct values, the number of distinct linear combinations of $\mathbf{u}_1, \mathbf{u}_2, \ldots, \mathbf{u}_k$ is p^k. Thus, the cardinality of an (n, k) linear code over $\mathcal{Z}_p$ is p^k. Linear codes over $\mathbb{Z}_2$ are referred to as linear *binary* codes.

Example 11-2

Consider the code whose code vectors are:

$$\begin{aligned}
\mathbf{u}_1 &= 00000\\
\mathbf{u}_2 &= 00101\\
\mathbf{u}_3 &= 01010\\
\mathbf{u}_4 &= 01111\\
\mathbf{u}_5 &= 10011\\
\mathbf{u}_6 &= 10110\\
\mathbf{u}_7 &= 11001\\
\mathbf{u}_8 &= 11100
\end{aligned}$$

It can be readily verified that $\mathbf{u}_2$, $\mathbf{u}_3$, and $\mathbf{u}_5$ form a basis for this code and, hence, that this code is a (5, 3) linear code over $\mathbb{Z}_2$. □

Clearly, any basis of a linear code completely specifies the code. Thus, a convenient way of describing an (n, k) linear code over $\mathbb{Z}_p$ is by means of a $k \times n$ matrix over $\mathcal{Z}_p$, whose k rows are the k code vectors that constitute a basis of the code. This matrix, denoted by $\mathbf{G}$, is referred to as a *generator matrix* of the code (and the code is said to be *generated* by this matrix). Given $\mathbf{G}$, the code it generates can be produced by forming all p^k possible linear combinations of the k row vectors of $\mathbf{G}$; this, in turn, can be done by premultiplying $\mathbf{G}$ by all p^k possible k-dimensional row vectors over $\mathbb{Z}_p$.

An alternative way of describing an (n, k) linear code is by means of an $(n - k) \times n$ matrix whose $n - k$ row vectors constitute a basis for the null space of $\mathbf{G}$ (see Sec. 10-10). This matrix, denoted by $\mathbf{H}$, is referred to as a *parity-check matrix* of the code. Thus, an n-dimensional vector $\mathbf{v}$ is in the code if and only if

$$\mathbf{v}\mathbf{H}^{TR} = \mathbf{0}$$

(Note that, while the row space of $\mathbf{H}$ is the null space of $\mathbf{G}$, it is also true that the row space of $\mathbf{G}$ is the null space of $\mathbf{H}$.)

Example 11-3

The (5, 3) linear binary code of Example 11-2 can be described by a generator matrix whose rows are the row vectors $\mathbf{u}_2$, $\mathbf{u}_3$ and $\mathbf{u}_5$:

$$\mathbf{G} = \begin{bmatrix} 0 & 0 & 1 & 0 & 1 \\ 0 & 1 & 0 & 1 & 0 \\ 1 & 0 & 0 & 1 & 1 \end{bmatrix} \tag{11-1}$$

The code itself can be generated by forming all eight linear combinations of the row vectors of **G**. This, in turn, can be done by premultiplying **G** by an 8×3 matrix whose rows constitute all possible 3-tuples of elements from $\mathbb{Z}_2$:

$$\begin{bmatrix} 0 & 0 & 0 \\ 0 & 0 & 1 \\ 0 & 1 & 0 \\ 0 & 1 & 1 \\ 1 & 0 & 0 \\ 1 & 0 & 1 \\ 1 & 1 & 0 \\ 1 & 1 & 1 \end{bmatrix} \begin{bmatrix} 0 & 0 & 1 & 0 & 1 \\ 0 & 1 & 0 & 1 & 0 \\ 1 & 0 & 0 & 1 & 1 \end{bmatrix} = \begin{bmatrix} 0 & 0 & 0 & 0 & 0 \\ 1 & 0 & 0 & 1 & 1 \\ 0 & 1 & 0 & 1 & 0 \\ 1 & 1 & 0 & 0 & 1 \\ 0 & 0 & 1 & 0 & 1 \\ 1 & 0 & 1 & 1 & 0 \\ 0 & 1 & 1 & 1 & 1 \\ 1 & 1 & 1 & 0 & 0 \end{bmatrix}$$

Alternatively, the same code can be described by its parity-check matrix. To construct this matrix, we first determine the null space of **G**—that is, all vectors $(a_1, a_2, a_3, a_4, a_5)$ such that

$$\begin{bmatrix} 0 & 0 & 1 & 0 & 1 \\ 0 & 1 & 0 & 1 & 0 \\ 1 & 0 & 0 & 1 & 1 \end{bmatrix} \begin{bmatrix} a_1 \\ a_2 \\ a_3 \\ a_4 \\ a_5 \end{bmatrix} = \begin{bmatrix} 0 \\ 0 \\ 0 \\ 0 \\ 0 \end{bmatrix}$$

Since, in $\mathbb{Z}_2$, $\Sigma\, a_i = 0$ if and only if an even number of a_i in the sum are 1, it follows that the number of 1s (or the *parity*) in those positions of $(a_1, a_2, a_3, a_4, a_5)$ that correspond to 1s in any row of **G** must be even. Specifically, the number of 1s in positions 3 and 5 must be even; the number of 1s in positions 2 and 4 must be even; the number of 1s in positions 1, 4, and 5 must be even. By trial and error we can find that the only 5-dimensional vectors that satisfy these conditions are (0, 0, 0, 0, 0), (0, 1, 1, 1, 1), (1, 0, 1, 0, 1), and (1, 1, 0, 1, 0). These four vectors can be readily verified to form a subspace of $\mathbb{Z}_2^5$, with the second and third vectors constituting a basis. Thus, a parity-check matrix of the given code is

$$\mathbf{H} = \begin{bmatrix} 0 & 1 & 1 & 1 & 1 \\ 1 & 0 & 1 & 0 & 1 \end{bmatrix}$$

□

Let $\mathbf{G}_1$ and $\mathbf{G}_2$ be generator matrices for the linear codes K_1 and K_2, respectively. If $\mathbf{G}_1$ and $\mathbf{G}_2$ are identical except for column permutation, then the codes K_1 and K_2 are said to be *equivalent*. Now, the generator matrix $\mathbf{G} = [g_{ij}]_{k \times n}$ of any (n, k) linear code can always be transformed into its row-reduced echelon canonical form $\hat{\mathbf{G}} = [\hat{g}_{ij}]_{k \times n}$, with no effect on the generated code (see Sec. 10-10). Since $\hat{\mathbf{G}}$ consists of a $k \times k$ identity matrix with additional $n - k$ columns interspersed, we can state:

THEOREM 11-1

Every (n, k) linear code is equivalent to one whose generator matrix is of the form

$$\mathbf{G} = (\mathbf{I}_k \,|\, \mathbf{Q}) = \left[\begin{array}{cccc|cccc} 1 & 0 & \dots & 0 & q_{11} & q_{12} & \dots & q_{1(n-k)} \\ 0 & 1 & \dots & 0 & q_{21} & q_{22} & \dots & q_{2(n-k)} \\ \cdot & \cdot & & \cdot & \cdot & \cdot & & \cdot \\ \cdot & \cdot & & \cdot & \cdot & \cdot & & \cdot \\ \cdot & \cdot & & \cdot & \cdot & \cdot & & \cdot \\ 0 & 0 & \dots & 1 & q_{k1} & q_{k2} & \dots & q_{k(n-k)} \end{array}\right]$$

(This matrix is called a *standard generator matrix*.)

In many applications the message set consists of all possible p^k k-tuples from $\mathbb{Z}_p^k$. If each such k-tuple (or k-dimensional row vector) $\mathbf{m} = (a_1, a_2, \dots, a_k)$ is encoded into the n-dimensional code vector $\mathbf{mG}$, where $\mathbf{G}$ is a $k \times n$ standard generator matrix, the result is a vector of the form

$$\begin{aligned} \mathbf{mG} &= (a_1, a_2, \dots, a_k)(\mathbf{I}_k \,|\, \mathbf{Q}) \\ &= (a_1, a_2, \dots, a_k, b_1, b_2, \dots, b_{n-k}) \end{aligned} \tag{11-2}$$

where $$b_j = \sum_{i=1}^{k} a_i q_{ij} \qquad (j = 1, 2, \dots, n - k)$$

(with operations as per rules of $\mathbb{Z}_p$). When this is done with all possible k-tuples $\mathbf{m}$, the result is the (n, k) linear code generated by $\mathbf{G}$. Thus, with the encoding scheme described by (11-2), each code vector in the code generated by $\mathbf{G}$ has the attractive property that the message it represents can be obtained directly from its leftmost k symbols. When $\mathbf{G}$ is not standard, but still in a row-reduced echelon canonical form, it is still true that the message can be recovered directly from k symbols of its code vector—except that these k symbols are not necessarily the leftmost ones. In either case, it is common to say that each code vector consists of k *information symbols* (constituting the originally transmitted information—that is, the message), and $n - k$ *check symbols* (symbols added to each message in order to invest the code—as we shall see later—with some error-correcting and -detecting capabilities). A code in which each code vector has this structure is called a *systematic code*.

Another advantage of a code generated by a standard generator matrix is the simplicity with which its parity-check matrix can be constructed.

THEOREM 11-2

Let $\mathbf{G} = (\mathbf{I}_k \mid \mathbf{Q})$ be a generator matrix of an (n, k) linear code (where $\mathbf{Q} = [q_{ij}]_{k\times(n-k)}$). Then a parity-check matrix of this code is given by

$$\mathbf{H} = (-\mathbf{Q}^{TR} \mid \mathbf{I}_{n-k})$$

$$= \left[\begin{array}{cccc|cccc} -q_{11} & -q_{21} & \cdots & -q_{k1} & 1 & 0 & \cdots & 0 \\ -q_{12} & -q_{22} & \cdots & -q_{k2} & 0 & 1 & \cdots & 0 \\ \cdot & \cdot & & \cdot & \cdot & \cdot & & \cdot \\ \cdot & \cdot & & \cdot & \cdot & \cdot & & \cdot \\ \cdot & \cdot & & \cdot & \cdot & \cdot & & \cdot \\ -q_{1(n-k)} & -q_{2(n-k)} & \cdots & -q_{k(n-k)} & 0 & 0 & \cdots & 1 \end{array}\right]$$

Proof Straightforward matrix multiplication verifies that $\mathbf{GH}^{TR} = \mathbf{0}$. Since the null space of $\mathbf{G}$ is $(n - k)$-dimensional, and since the $n - k$ row vectors of $\mathbf{H}$ are linearly independent (Why?), the row vectors of $\mathbf{H}$ constitute a basis of the null space of $\mathbf{G}$. □

Example 11-4

Consider the (5, 3) linear binary code of Examples 11-2 and 11-3. Transforming its generator matrix (11-1) into the row-reduced echelon canonical form, we obtain

$$\mathbf{G} = \begin{bmatrix} 1 & 0 & 0 & 1 & 1 \\ 0 & 1 & 0 & 1 & 0 \\ 0 & 0 & 1 & 0 & 1 \end{bmatrix}$$

This generator matrix is already in the standard $(\mathbf{I}_k \mid \mathbf{Q})$ form, and no column reordering is necessary. To form a systematic code, every message $a_1a_2a_3$ is encoded into $(a_1, a_2, a_3)\mathbf{G}$. For example, the message 101 is encoded into the vector

$$(1, 0, 1)\begin{bmatrix} 1 & 0 & 0 & 1 & 1 \\ 0 & 1 & 0 & 1 & 0 \\ 0 & 0 & 1 & 0 & 1 \end{bmatrix} = (1, 0, 1, 1, 0)$$

where the first three elements immediately yield the message. Using Theorem

11-2, the parity-check matrix of the code can be constructed by inspection of **G** (noting that, in $\mathbb{Z}_2$, $-1 = 1$):

$$\mathbf{H} = \begin{bmatrix} 1 & 1 & 0 & 1 & 0 \\ 1 & 0 & 1 & 0 & 1 \end{bmatrix}$$

(It can be readily verified that $\mathbf{GH}^{TR} = \mathbf{0}$.) □

PROBLEMS

1. The following is an (n, k) linear binary code for a message set of cardinality 8:

$$\mathbf{u}_1 = 000000,\ \mathbf{u}_2 = 000111,\ \mathbf{u}_3 = 011001,\ \mathbf{u}_4 = 011110,$$
$$\mathbf{u}_5 = 101011,\ \mathbf{u}_6 = 101100,\ \mathbf{u}_7 = 110010,\ \mathbf{u}_8 = 110101$$

(a) What are the values of n and k?
(b) Construct a generator matrix **G** for this code.
(c) Construct a parity-check matrix **H** for this code.

2. Let K be a code that is equivalent to the code of Problem 1 and whose generator matrix is a standard one.
(a) Construct a generator matrix for K.
(b) Construct a parity-check matrix for K.
(c) List the code vectors of K.

3. Let the code K of Problem 2 be a systematic code for the message set $M = \mathbb{Z}_2^3$.
(a) What are the positions for the information symbols and the check symbols in each code vector?
(b) Find the code vector into which each message is encoded.

4. Consider the matrix

$$\mathbf{G} = \begin{bmatrix} 0 & 1 & 1 & 0 & 0 & 1 & 1 \\ 1 & 0 & 1 & 0 & 1 & 0 & 1 \\ 0 & 0 & 0 & 1 & 1 & 1 & 1 \end{bmatrix}$$

(a) Find the linear binary code generated by **G**.
(b) If the code generated by G is used as a systematic code for $\mathbb{Z}_2^3$, what are the positions for the information and check symbols in each code vector?

5. The following is a parity-check matrix for a linear binary code:

$$\mathbf{H} = \begin{bmatrix} 1 & 0 & 1 & 1 & 0 \\ 0 & 0 & 0 & 1 & 1 \\ 1 & 1 & 0 & 0 & 0 \end{bmatrix}$$

Construct a generator matrix for the code and list all the code vectors.

6. Let K be an (n, k) linear binary code.
   (a) Show that all code vectors in K which contain an even number of 1s form a subspace of $\mathbf{Z}_2^n = \langle \mathbb{Z}_2^n; + \rangle$.
   (b) Show that in K either all code vectors, or exactly half of all code vectors, have an even number of 1s.

11-3. ERROR CORRECTION WITH LINEAR CODES

The *weight* of a vector $\mathbf{v}$ over $\mathbf{Z}_p$, denoted by $w(\mathbf{v})$, is the number of nonzero elements in $\mathbf{v}$. The least nonzero weight of all code vectors of a linear code is called the *minimum weight* of the code. Denoting the distance between two vectors $\mathbf{v}_1$ and $\mathbf{v}_2$ by $d(\mathbf{v}_1, \mathbf{v}_2)$ (see Sec. 11-1), we have

$$d(\mathbf{v}_1, \mathbf{v}_2) = w(\mathbf{v}_1 - \mathbf{v}_2) \tag{11-3}$$

THEOREM 11-3
The minimum distance of a linear code equals its minimum weight.

Proof If $\mathbf{v}_1$ and $\mathbf{v}_2$ are code vectors, so is $\mathbf{v}_1 - \mathbf{v}_2$ (since a linear code is a vector space). Hence, using (11-3), the distance between any two distinct code vectors is the weight of a third (necessarily nonzero) code vector—which implies the theorem. □

For example, in the code of Example 11-2, the minimum distance and minimum weight equal 2.

THEOREM 11-4
Let K be a linear code with the parity-check matrix $\mathbf{H}$. Then K has minimum weight (hence, minimum distance) of at least w if and only if every combination of $w - 1$ or fewer column vectors of $\mathbf{H}$ is linearly independent.

Proof Let K be an (n, k) linear code with the parity-check matrix $\mathbf{H} = (\mathbf{h}_1, \mathbf{h}_2, \ldots, \mathbf{h}_n)$, where the $\mathbf{h}_i$ denote column vectors of $\mathbf{H}$. A vector $\mathbf{v} = (a_1, a_2, \ldots, a_n)$ is in K if and only if $\mathbf{v}\mathbf{H}^{TR} = \mathbf{0}$; hence, if and only if

$$(a_1, a_2, \ldots, a_n)\begin{bmatrix} \mathbf{h}_1^{TR} \\ \mathbf{h}_2^{TR} \\ \cdot \\ \cdot \\ \cdot \\ \mathbf{h}_n^{TR} \end{bmatrix} = a_1\mathbf{h}_1^{TR} + a_2\mathbf{h}_2^{TR} + \ldots + a_n\mathbf{h}_n^{TR} = \mathbf{0} \tag{11-4}$$

Now, suppose $w - 1$ or fewer column vectors $\mathbf{h}_i$ are linearly dependent. Then it is possible to write (11-4) with $w - 1$ or fewer nonzero coefficients a_i; hence,

K has a code vector with weight less than w. Thus, if the minimum weight of K is w, every combination of $w - 1$ or fewer column vectors of $\mathbf{H}$ must be linearly independent.

Conversely, suppose the minimum weight of K is less than w. Then (11-4) can be written with $w - 1$ or fewer nonzero coefficients a_i; hence, there is a combination of $w - 1$ or fewer column vectors $\mathbf{h}_i$ that are linearly dependent. Thus, if every combination of $w - 1$ or fewer column vectors of $\mathbf{H}$ are linearly independent, the minimum weight of K is at least w. □

Theorem 11-4 implies that in order to construct a linear code with minimum distance $w = l + 1$ (for the detection of l or fewer errors) or $w = 2l + 1$ (for the correction of l or fewer errors), it is necessary and sufficient to guarantee that every combination of $w - 1$ or fewer column vectors in the parity-check matrix of the code be linearly independent. For example, to construct an (n, k) linear code over $\mathbb{Z}_p$ (for a message set of cardinality p^k) with minimum distance 3 (for single-error correction), it is necessary and sufficient to have an $(n - k) \times n$ parity-check matrix where all pairs of column vectors are linearly independent. When $p = 2$, this reduces to the requirement that all column vectors be non-null and distinct (such as in $\mathbf{H}$ of Example 11-4).

In a linear code, if a code vector $\mathbf{u}$ is received as the vector $\mathbf{v}$, then the vector $\mathbf{v} - \mathbf{u}$ is referred to as the *error pattern* of $\mathbf{u}$. For example, if the code vector 00101 of Example 11-2 is received as 01100, then its error pattern is 01001.

THEOREM 11-5

With a linear code K, an error can be detected if and only if the error pattern is not a code vector in K.

Proof Let the code vector be $\mathbf{u}$ and the error pattern be $\mathbf{e} \neq \mathbf{0}$, so that the received vector is $\mathbf{v} = \mathbf{u} + \mathbf{e}$. If $\mathbf{e} \in K$, then $\mathbf{u} + \mathbf{e} \in K$ and the error pattern $\mathbf{e}$ results in a legitimate code vector; hence, its presence in $\mathbf{v}$ cannot be detected. Conversely, if $\mathbf{e} \notin K$, then $\mathbf{v} - \mathbf{u} \notin K$ and, hence, $\mathbf{v} \notin K$; thus, the presence of $\mathbf{e}$ can be detected by the fact that $\mathbf{v}$ is not a code vector. □

Let $K = \{\mathbf{u}_0, \mathbf{u}_1, \ldots, \mathbf{u}_{p^k-1}\}$ be an (n, k) linear code over $\mathbb{Z}_p$—that is, a k-dimensional subspace of the vector space $\mathcal{Z}_p^n = \langle \mathbb{Z}_p^n; + \rangle$. By definition of vector spaces and subspaces (see Sec. 10-9), $\langle K; + \rangle$ is an additive Abelian subgroup of the Abelian group $\mathcal{Z}_p^n$. Consequently, we can define a coset partition relative to K in $\mathcal{Z}_p^n$ (see Sec. 9-6). By Algorithm 9-1, this partition (which consists of $p^n/p^k = p^{n-k}$ cosets) is given by

$$\{K + \mathbf{v}_0, K + \mathbf{v}_1, K + \mathbf{v}_2, \ldots, K + \mathbf{v}_{p^{n-k}-1}\}$$

where $\mathbf{v}_0 = \mathbf{0}$ and $\mathbf{v}_i$ $(i = 1, 2, \ldots, p^{n-k} - 1)$ is any n-dimensional vector such that

$$\mathbf{v}_i \notin (K + \mathbf{v}_0) \cup (K + \mathbf{v}_1) \cup \ldots \cup (K + \mathbf{v}_{i-1})$$

Based on this partition, we can construct a decoding table for K where the rows represent the p^{n-k} cosets and where the column headed by the code vector $\mathbf{u}_i$ contains the vectors $\mathbf{v}_0 + \mathbf{u}_i, \mathbf{v}_1 + \mathbf{u}_i, \ldots, \mathbf{v}_{p^{n-k}-1} + \mathbf{u}_i$. Specifically, if $\mathbf{u}_0 = \mathbf{v}_0 = \mathbf{0}$, the proposed decoding table has the form shown in Table 11-2. According to this table, which is referred to as a *standard decoding table*, every received vector $\mathbf{v}_i + \mathbf{u}_j$ is decoded into the message whose code vector is $\mathbf{u}_j$ (or, for short, "into the code vector $\mathbf{u}_j$"). The vectors $\mathbf{0}, \mathbf{v}_1, \mathbf{v}_2, \ldots,$ $\mathbf{v}_{p^{n-k}-1}$, which appear in the leftmost column of the standard decoding table, are called *coset leaders*. These coset leaders are by no means unique; in fact, by Theorem 9-18, if $\mathbf{v}'_i \in K + \mathbf{v}_i$, then $K + \mathbf{v}'_i = K + \mathbf{v}_i$; hence, *any* vector of a coset can serve as a coset leader.

Table 11-2 A STANDARD DECODING TABLE

$\mathbf{0}$	$\mathbf{u}_1$	$\mathbf{u}_2$	$\cdots$	$\mathbf{u}_{p^k-1}$
$\mathbf{v}_1$	$\mathbf{v}_1 + \mathbf{u}_1$	$\mathbf{v}_1 + \mathbf{u}_2$	$\cdots$	$\mathbf{v}_1 + \mathbf{u}_{p^k-1}$
$\mathbf{v}_2$	$\mathbf{v}_2 + \mathbf{u}_1$	$\mathbf{v}_2 + \mathbf{u}_2$	$\cdots$	$\mathbf{v}_2 + \mathbf{u}_{p^k-1}$
$\vdots$	$\vdots$	$\vdots$		$\vdots$
$\mathbf{v}_{p^{n-k}-1}$	$\mathbf{v}_{p^{n-k}-1} + \mathbf{u}_1$	$\mathbf{v}_{p^{n-k}-1} + \mathbf{u}_2$	$\cdots$	$\mathbf{v}_{p^{n-k}-1} + \mathbf{u}_{p^k-1}$

THEOREM 11-6

Let $\mathbf{u}_j$ be a code vector received as the vector $\mathbf{v}$ and decoded via a standard decoding table as the code vector $\mathbf{u}_k$. Then $\mathbf{u}_k = \mathbf{u}_j$ if and only if the error pattern of $\mathbf{u}_j$ is a coset leader.

Proof If $\mathbf{u}_k = \mathbf{u}_j$, then $\mathbf{v} = \mathbf{v}_i + \mathbf{u}_j$ for some coset leader $\mathbf{v}_i$. Hence, $\mathbf{v} - \mathbf{u}_j = \mathbf{v}_i$, which implies that the error pattern of $\mathbf{u}_j$ equals some coset leader. Conversely, if $\mathbf{v} - \mathbf{u}_j = \mathbf{v}_i$, where $\mathbf{v}_i$ is some coset leader, then $\mathbf{v} = \mathbf{v}_i + \mathbf{u}_j$. Hence, $\mathbf{v}$ is decoded as $\mathbf{u}_j$, which implies that $\mathbf{u}_k = \mathbf{u}_j$. □

From Theorem 11-6, then, the following can be concluded: With a linear code associated with a standard decoding table, it is possible to correct all error patterns which appear as coset leaders, but no other error patterns.

Example 11-5

Table 11-3 shows a standard decoding table for the (5, 2) linear binary code of Example 11-2. The error patterns to which this code is immune are 00001, 00010, and 10000 (each of which results in a single error). For example, the code vector 01010, received as 01000 (that is, subject to the error pattern

Table 11-3 STANDARD DECODING TABLE FOR EXAMPLE 11-5

00000	00101	01010	01111	10011	10110	11001	11100
00001	00100	01011	01110	10010	10111	11000	11101
00010	00111	01000	01101	10001	10100	11011	11110
10000	10101	11010	11111	00011	00110	01001	01100

00010), will be correctly decoded as 01010. On the other hand, if 01010 is received as 01110 (that is, subject to the error pattern 00100), it will be decoded incorrectly as 01111. □

Let K be an (n, k) linear *binary* code, where each of the 2^k code vectors can be transmitted with equal probability. It can be shown that if K is used with a binary symmetric channel and associated with a standard decoding table, then the average probability of correct decoding is the greatest if the coset leaders are chosen to have the least weight in their respective cosets. The code of Example 11-5 is an illustration of such a choice.

Let K be an (n, k) linear code over $\mathbb{Z}_p$, with the parity-check matrix $\mathbf{H} = [h_{ij}]_{(n-k)\times n}$. The *syndrome* of any received vector $\mathbf{v}$ is defined as the $(n-k)$-dimensional vector $\mathbf{v}\mathbf{H}^{TR}$. Clearly, the syndrome of any code vector is $\mathbf{0}$.

THEOREM 11-7

Let $\mathbf{v}$ and $\mathbf{v}'$ be two vectors in a standard decoding table of a linear code. Then $\mathbf{v}$ and $\mathbf{v}'$ are in the same coset if and only if their syndromes are equal.

Proof By Theorem 9-16, $\mathbf{v}$ and $\mathbf{v}'$ are in the same coset if and only if $\mathbf{v} - \mathbf{v}'$ is a code vector; hence, if and only if $(\mathbf{v} - \mathbf{v}')\mathbf{H}^{TR} = \mathbf{0}$ and, hence, if and only if $\mathbf{v}\mathbf{H}^{TR} = \mathbf{v}'\mathbf{H}^{TR}$. □

Theorem 11-7 implies that every row in a standard decoding table can be uniquely associated with a specific syndrome. We can thus devise a greatly simplified decoding procedure which, instead of employing a p^n-entry decoding table, employs a $2p^{n-k}$-entry *syndrome table*, where only coset leaders and their syndromes are listed:

Coset leader	Syndrome
$\mathbf{0}$	$\mathbf{0}$
$\mathbf{v}_1$	$\mathbf{v}_1\mathbf{H}^{TR}$
$\mathbf{v}_2$	$\mathbf{v}_2\mathbf{H}^{TR}$
$\vdots$	$\vdots$
$\mathbf{v}_{p^{n-k}-1}$	$\mathbf{v}_{p^{n-k}-1}\mathbf{H}^{TR}$

This procedure consists of these steps:

(i) Given a received vector $\mathbf{v}$, compute its syndrome $\mathbf{vH}^{TR}$.

(ii) Look up the coset leader $\mathbf{v}_i$ listed against $\mathbf{vH}^{TR}$ in the syndrome table. (This is, presumably, the error pattern of the transmitted code vector.)

(iii) The transmitted code vector is $\mathbf{u} = \mathbf{v} - \mathbf{v}_i$.

When n is large, the advantage of decoding via a syndrome table, as compared to decoding via a decoding table, is appreciable. (For example, when $p = 2$, $k = 6$, and $n = 10$, the syndrome table consists of 32 entries, as compared with 1024 in the decoding table.)

Example 11-6

The following is a syndrome table for the linear binary code of Example 11-5 (whose parity-check matrix is given in Example 11-3).

Coset leader	Syndrome
00000	(0, 0)
00001	(1, 1)
00010	(1, 0)
10000	(0, 1)

For example, the code vector 01010, received as 01011, is decoded as follows. The syndrome of 01011 is computed to be (1, 1); the coset leader for syndrome (1, 1) is 00001; hence, the transmitted code vector is 01011 with the rightmost symbol altered; that is 01010 (correct decoding, since the error pattern is a coset leader). □

A linear binary code whose set of coset leaders (chosen to have the least weight in their cosets) is precisely the set of all error patterns of weight w or less (where w is some fixed nonnegative integer), is called a *perfect code.* A linear code whose set of coset leaders (chosen in the same manner) consists of all error patterns of weight w or less, some of weight $w + 1$, and none of weight greater than $w + 1$, is called a *quasi-perfect code.* The code of Example 11-5 is quasi-perfect (with $w = 0$).

PROBLEMS

1. Let $\mathbf{M}$ be an $m \times n$ matrix over $\mathrm{Z}_p = \langle \mathbb{Z}_p; +, \cdot \rangle$, where no two column vectors are linearly dependent.

(a) What is the largest value of n (in terms of p and m)?

(b) If $m = n - k$, what is the largest value of k (in terms of p and n)?

2. Using shortest possible code vectors, construct a single-error correcting linear code:
(a) Over $\mathbb{Z}_2$, for a message set of cardinality 16.
(b) Over $\mathbb{Z}_3$, for a message set of cardinality 27.

3. The generator matrix of a linear binary code is given by

$$\mathbf{G} = \begin{bmatrix} 0 & 1 & 1 & 1 & 0 \\ 0 & 0 & 1 & 0 & 1 \\ 1 & 1 & 0 & 0 & 0 \end{bmatrix}$$

List all the error patterns which can be detected (but not necessarily corrected) with this code.

4. How many error patterns can be detected (but not necessarily corrected) with an (n, k) linear code over $\mathbb{Z}_p$? How many can be corrected (using a standard decoding table)?

5. Construct a standard decoding table (with least-weight coset leaders) for the linear binary code generated by

$$\mathbf{G} = \begin{bmatrix} 1 & 1 & 1 & 0 & 1 & 0 \\ 0 & 1 & 1 & 1 & 0 & 1 \\ 1 & 1 & 0 & 0 & 0 & 1 \end{bmatrix}$$

Is the code perfect or quasi-perfect? Using the decoding table, decode the received vector 111001.

6. Construct a syndrome table for the code of Problem 5. Decode the received vector 111001 by means of this table.

7. Let $\mathbf{M}$ be a $p^k \times n$ matrix over $\mathbb{Z}_p = \langle \mathbb{Z}_p; +, \cdot \rangle$, whose rows are all the code vectors of an (n, k) linear code over $\mathbb{Z}_p$. Assuming that no column in $\mathbf{M}$ consists entirely of 0s:
(a) Show that if U_j is the set of all code vectors with 0 in column j of $\mathbf{M}$, then $\langle U_j; + \rangle$ is a subspace of the code.
(b) Show that, in any column j of $\mathbf{M}$, each element of $\mathbb{Z}_p$ appears exactly p^{k-1} times.
(c) Show that the sum of the weights of all code vectors is $n(p - 1)p^{k-1}$.

11-4. HAMMING CODES

The *Hamming codes* (named after the American mathematician Richard W. Hamming) are perfect, linear, binary codes whose encoding and decoding procedures are quite simple. When used with a binary symmetric channel, an (n, k) Hamming code is *optimal*, in the sense that it exhibits a probability of erroneous decoding not greater than that of any other (n, k) linear binary code.

A Hamming code K_m (for any positive integer m) is a $(2^m - 1, 2^m - m - 1)$ linear binary code (of cardinality 2^{2^m-m-1}) whose $m \times (2^m - 1)$ parity-check matrix $\mathbf{H}$ consists of all non-null m-dimensional column vectors over $\mathbb{Z}_2$. Since the columns of $\mathbf{H}$ are distinct, all pairs of columns in $\mathbf{H}$ are linearly independent; hence, K_m has minimum distance 3 (see Theorem 11-4). Thus, a Hamming code is a single-error correcting code. By Theorem 11-6, a decoding table can be constructed for K_m where the set of coset leaders contains all $2^m - 1$ distinct error patterns of weight 1. Since there are $2^{2^m-1}/2^{2^m-m-1} = 2^m$ distinct coset leaders, including the null vector of weight 0, the set of coset leaders constitutes precisely the set of all error patterns of weight 1 or less. Thus, K_m is a perfect code.

Let us consider a message set of cardinality 2^{2^m-m-1}, where each message is a $(2^m - m - 1)$-tuple from $\mathbb{Z}_2^{2^m-m-1}$, of the form $(a_1, a_2, \ldots, a_{2^m-m-1})$. Each such message can be encoded into a code vector of a Hamming code K_m in this manner: Let the parity-check matrix of K_m be $\mathbf{H} = [h_{ij}]_{m\times(2^m-1)} = (\mathbf{h}_1, \mathbf{h}_2, \ldots, \mathbf{h}_{2^m-1})$, where the $\mathbf{h}_i$ are the column vectors of $\mathbf{H}$. Referring to $\mathbf{h}_i$ as *column* $\#i$, assume that columns $\#i_1, \#i_2, \ldots, \#i_m$ are those that have 1 in positions $1, 2, \ldots, m$ (from the top), respectively, and 0 elsewhere, and that columns $\#j_1, \#j_2, \ldots, \#j_{2^m-m-1}$ are all the remaining columns. The code vector assigned to the message $(a_1, a_2, \ldots, a_{2^m-m-1})$ is then a $(2^m - 1)$-dimensional vector $\mathbf{u}$ consisting of the symbols $c_1, c_2, \ldots, c_m$ in positions $i_1, i_2, \ldots, i_m$, respectively, and of the symbols $a_1, a_2, \ldots, a_{2^m-m-1}$ in positions $j_1, j_2, \ldots, j_{2^m-m-1}$, respectively. The symbols $a_1, a_2, \ldots, a_{2^m-m-1}$ are the information symbols that constitute the transmitted message, while symbols $c_1, c_2, \ldots, c_m$ are check symbols which are chosen so as to satisfy the vector equation

$$\mathbf{uH}^{TR} = \mathbf{u}\begin{bmatrix} \mathbf{h}_1^{TR} \\ \mathbf{h}_2^{TR} \\ \cdot \\ \cdot \\ \cdot \\ \mathbf{h}_{2^m-1}^{TR} \end{bmatrix} = \underbrace{(0, 0, \ldots, 0)}_{m}$$

This equation implies the following set of m simultaneous equations:

$$c_i + \sum_{\nu=1}^{2^m-m-1} a_\nu h_{ij_\nu} = 0 \quad (i = 1, 2, \ldots, m)$$

from which we immediately have

$$c_i = \sum_{v=1}^{2^m-m-1} a_v h_{ijv} \quad (i = 1, 2, \ldots, m)$$

Example 11-7

The Hamming code K_3 has $2^{2^3-3-1} = 2^4 = 16$ code vectors over $\mathbb{Z}_2$, each of dimension $2^3 - 1 = 7$. A parity-check matrix $\mathbf{H}$ for this code consists of all non-null 3-dimensional column vectors over $\mathbb{Z}_2$:

$$\mathbf{H} = \begin{bmatrix} 0 & 0 & 0 & 1 & 1 & 1 & 1 \\ 0 & 1 & 1 & 0 & 0 & 1 & 1 \\ 1 & 0 & 1 & 0 & 1 & 0 & 1 \end{bmatrix}$$

The message 1011, for example, is encoded into the code vector

$$\mathbf{u} = (c_1, c_2, 1, c_3, 0, 1, 1)$$

where the check symbols c_1, c_2, and c_3 must satisfy the vector equation

$$\mathbf{u}\mathbf{H}^{TR} = (c_3, c_2, 1, c_1, 0, 1, 1) \begin{bmatrix} 0 & 0 & 1 \\ 0 & 1 & 0 \\ 0 & 1 & 1 \\ 1 & 0 & 0 \\ 1 & 0 & 1 \\ 1 & 1 & 0 \\ 1 & 1 & 1 \end{bmatrix} = (0, 0, 0)$$

and, hence, the three equations

$$c_1 + 0 = 0$$
$$c_2 + 1 = 0$$
$$c_3 + 0 = 0$$

Thus, $c_1 = 0$, $c_2 = 1$, $c_3 = 0$, and

$$\mathbf{u} = (0, 1, 1, 0, 0, 1, 1)$$ □

Suppose the code vector $\mathbf{u}$ of a Hamming code K_m is transmitted and is subjected to a single error. The received vector is thus $\mathbf{v} = \mathbf{u} + \mathbf{e}$, where $\mathbf{e}$ is an error pattern with 1 in the error position—say the ith—and 0s in all other positions. If $\mathbf{H} = (\mathbf{h}_1, \mathbf{h}_2, \ldots, \mathbf{h}_{2^m-1})$ as previously defined, then the syndrome of $\mathbf{v}$ is given by

$$\mathbf{v}\mathbf{H}^{TR} = (\mathbf{u} + \mathbf{e})\mathbf{H}^{TR} = \mathbf{u}\mathbf{H}^{TR} + \mathbf{e}\mathbf{H}^{TR} = \mathbf{e}\mathbf{H}^{TR}$$

$$= (0, 0, \ldots, 0, 1, 0, \ldots, 0)\begin{bmatrix} \mathbf{h}_1^{TR} \\ \mathbf{h}_2^{TR} \\ \cdot \\ \cdot \\ \cdot \\ \mathbf{h}_{2^m-1}^{TR} \end{bmatrix} = \mathbf{h}_i^{TR} \tag{11-5}$$

Thus, if the syndrome is found to equal the ith column vector of $\mathbf{H}$, the error position can be immediately deduced to be the ith, and the ith position corrected accordingly. The identification of the error position becomes even simpler when the columns of $\mathbf{H}$ are so ordered (from left to right) so as to constitute the binary representation of the integers $1, 2, \ldots, 2^m - 1$. When this is done, $\mathbf{h}_i^{TR}$ in (11-5) is simply the binary representation of the position i in which the error occurred. Notice that when $\mathbf{v} = \mathbf{u}$ (no error), the syndrome is $\mathbf{0}$ and coincides with no column vector of $\mathbf{H}$ and, hence, with no error position whatsoever.

Example 11-8

The parity-check matrix of the Hamming code K_3 of Example 11-7 is constructed in the manner proposed above; that is, the columns are ordered so as to constitute the binary representations of the integers $1, 2, \ldots, 7$.

Suppose the code vector 0110011 is received as 0110001 (with a single error in the 6th position). The syndrome in this case is

$$(0, 1, 1, 0, 0, 0, 1)\begin{bmatrix} 0 & 0 & 1 \\ 0 & 1 & 0 \\ 0 & 1 & 1 \\ 1 & 0 & 0 \\ 1 & 0 & 1 \\ 1 & 1 & 0 \\ 1 & 1 & 1 \end{bmatrix} = (1, 1, 0)$$

which is the binary representation of the integer 6. The 6th position of the received vector is then altered to 0 to yield the correct code vector 0110011. □

Given a Hamming code K_m, a new code $\tilde{K}_m$ can be produced from it in this manner: To each of the $(2^m - 1)$-dimensional code vectors add a new check symbol (at the right end) so as to make the total number of 1s even. The cardinality of $\tilde{K}_m$ is still the same as that of K_m, but now the weight of each 2^m-dimensional code vector in $\tilde{K}_m$ is one greater than the weight of the corresponding code vector in K_m. The minimum weight—and, hence, the minimum distance—of $\tilde{K}_m$ is, therefore, 4. Thus, with $\tilde{K}_m$, we can detect all double errors as well as correct all single ones.

The parity-check matrix $\tilde{\mathbf{H}}$ of $\tilde{K}_m$ can be constructed directly from the parity-check matrix $\mathbf{H}$ of K_m as

$$\tilde{\mathbf{H}} = \left[\begin{array}{cccc|c} & & & & 0 \\ & & & & 0 \\ & & \mathbf{H} & & \vdots \\ & & & & 0 \\ \hline 1 & 1 & \dots & 1 & 1 \end{array}\right]$$

If $\tilde{\mathbf{v}} = (\mathbf{v} \mid c)$ (where c is the new check symbol) is a received code vector, and N denotes the number, modulo 2, of 1s in $\tilde{\mathbf{v}}$, we have

$$\tilde{\mathbf{v}}\tilde{\mathbf{H}}^{TR} = (\mathbf{v} \mid c)\left[\begin{array}{cccc|c} & & & & 1 \\ & & & & 1 \\ & & \mathbf{H}^{TR} & & \vdots \\ \hline 0 & 0 & \dots & 0 & 1 \end{array}\right] = (\mathbf{v}\mathbf{H}^{TR} \mid N)$$

When $\tilde{\mathbf{v}}$ contains no errors, $\mathbf{v}\mathbf{H}^{TR} = \mathbf{0}$ and $N = 0$ (since the number of 1s in each code vector is even) and, hence, $\tilde{\mathbf{v}}\tilde{\mathbf{H}}^{TR} = \mathbf{0}$. If $\tilde{\mathbf{v}}$ contains a single error, $N = 1$ (since the number of 1s in $\tilde{\mathbf{v}}$ must now be odd). If the single error occurs among the leftmost $2^m - 1$ symbols of $\tilde{\mathbf{v}}$ (that is, in $\mathbf{v}$), then the leftmost m symbols of $\tilde{\mathbf{v}}\tilde{\mathbf{H}}^{TR}$ (that is, the vector $\mathbf{v}\mathbf{H}^{TR}$) constitute the binary representation of the error position. If the single error occurs in the rightmost symbol (that is, in c), then the leftmost m symbols of $\tilde{\mathbf{v}}\tilde{\mathbf{H}}^{TR}$ are 0, but we still have $N = 1$. If $\tilde{\mathbf{v}}$ contains two errors, $\tilde{\mathbf{v}}\tilde{\mathbf{H}}^{TR} \neq \mathbf{0}$ and $N = 0$ (since the number of 1s is again even); these errors, however, are uncorrectable. Thus, single-error correction and double-error detection can be performed directly by inspection of the syndrome of $\tilde{\mathbf{v}}$.

It can be shown that the modified Hamming code $\tilde{K}_m$ is quasi-perfect.

Example 11-9

The Hamming code $\tilde{K}_3$, like K_3 (see Example 11-7) has 16 code vectors over $\mathbb{Z}_2$, each of dimension $2^3 = 8$. A parity-check matrix $\tilde{\mathbf{H}}$ for this code is given by

$$\tilde{\mathbf{H}} = \left[\begin{array}{ccccccc|c} 0 & 0 & 0 & 1 & 1 & 1 & 1 & 0 \\ 0 & 1 & 1 & 0 & 0 & 1 & 1 & 0 \\ 1 & 0 & 1 & 0 & 1 & 0 & 1 & 0 \\ \hline 1 & 1 & 1 & 1 & 1 & 1 & 1 & 1 \end{array}\right]$$

The message 1011, for example, which in K_3 is encoded into (0, 1, 1, 0, 0, 1, 1), is now encoded into

$$\tilde{\mathbf{u}} = (0, 1, 1, 0, 0, 1, 1, 0)$$

If $\tilde{\mathbf{u}}$ is received as $\tilde{\mathbf{v}} = (0, 1, 1, 0, 0, 0, 1, 0)$ (with a single error in the 6th position), we have

$$\tilde{\mathbf{v}}\tilde{\mathbf{H}}^{TR} = (1, 1, 0, 1)$$

which indicates a single error (by virtue of the rightmost 1) in position 6 [corresponding to (1, 1, 0)].

If $\tilde{\mathbf{u}}$ is received as $\tilde{\mathbf{v}} = (0, 1, 1, 0, 0, 1, 1, 1)$ (with a single error in the rightmost position), we have

$$\tilde{\mathbf{v}}\tilde{\mathbf{H}}^{TR} = (0, 0, 0, 1)$$

which indicates a single error in the rightmost check symbol.

If $\tilde{\mathbf{u}}$ is received as (0, 0, 1, 0, 0, 0, 1, 0) (with errors in positions 2 and 6), we have

$$\tilde{\mathbf{v}}\tilde{\mathbf{H}}^{TR} = (1, 0, 0, 0)$$

which indicates a double error (with positions unknown). □

A single-error correcting linear binary code, where the dimension n of the code vectors is not necessarily of the form $2^m - 1$, can be produced as follows: Construct the $m \times (2^m - 1)$ parity-check matrix $\mathbf{H}$ of the Hamming code K_m, where m is such that $2^m - 1$ is the least integer that equals or exceeds n. Form the matrix $\hat{\mathbf{H}}$ from $\mathbf{H}$ by deleting from the latter $2^m - 1 - n$ columns. $\hat{\mathbf{H}}$ is then a parity-check matrix of an $(n, n - m)$ linear binary code (of cardinality 2^{n-m}) whose minimum distance is 3 (since the columns of $\hat{\mathbf{H}}$ are still non-null and distinct). It can be shown that when the $2^m - 1 - n$ columns of $\hat{\mathbf{H}}$ are judiciously deleted, the resulting code can be made quasi-perfect and optimal. By a similar deletion operation, a single-error correcting, double-error detecting linear binary code can be produced from the modified Hamming code $\tilde{K}_m$, so as to result in code vectors of any specified dimension n.

PROBLEMS

1. List all code vectors of the Hamming code K_3 and the modified Hamming code $\tilde{K}_3$.

2. Construct parity-check matrices for the Hamming code K_4 and the modified Hamming code $\tilde{K}_4$. Select five messages to illustrate encoding and error correction and detection with these two codes.

3. The $m \times n$ matrix $\hat{\mathbf{H}}$ is the "truncated" parity-check matrix of the Hamming code $\hat{K}_m$ (as described at the end of Sec. 11-4). What are the values of m and n if $\hat{\mathbf{H}}$ is to be the parity-check matrix of a single-error correcting linear binary code of cardinality 1000ν ($\nu = 1, 2, \ldots, 10$)?

4. Construct a parity-check matrix for a single-error correcting linear binary code for a message set M, where $\#M$ equals:

   (a) 2000 (b) 1000 (c) 500 (d) 100
   (e) 20 (f) 10 (g) 8

5. Define a *generalized Hamming code* $K_m^{(p)}$ as a single-error correcting linear code over $\mathbb{Z}_p$, whose parity-check matrix consists of the greatest number of m-dimensional column vectors over $\mathbb{Z}_p$ such that no two column vectors are linearly dependent.
   (a) Find the values of n and k for which $K_m^{(p)}$ is an (n, k) linear code.
   (b) Construct the parity-check matrix of $K_3^{(3)}$.
   (c) What are the positions of the information symbols and the check symbols in each code vector of $K_3^{(3)}$?
   (d) What is the code vector of the message $111 \ldots 1$ ($n - k$ times) in $K_3^{(3)}$?
   (e) Formulate an algorithm for correcting and detecting errors in $K_3^{(3)}$.

REFERENCES

BERLEKAMP, E., *Algebraic Coding Theory*. New York: McGraw-Hill, 1968.

GILL, A., *Linear Sequential Circuits*. New York: McGraw-Hill, 1967.

GOLOMB, S. W., *Shift Register Sequences*. San Francisco: Holden-Day, 1967.

HARRISON, M. A., *Lectures on Linear Sequential Machines*. New York: Academic Press, 1970.

LIN, S., *An Introduction to Error-Correcting Codes*. Englewood Cliffs, NJ: Prentice-Hall, 1970.

PETERSON, W. W., and E. J. WELDON, *Error-Correcting Codes*. Cambridge, MA: MIT Press, 1972.

12 GRAPHS

In this chapter we expand our discussion of graphs, which were first introduced in Chapter 2 as diagrammatic representations of relations over sets. After studying problems related to connectivity and traversability properties of graphs (and, in particular, of *Euler graphs*), we introduce special families of graphs that are of particular interest in applications: *Trees* (with treatment of *binary trees* and *economy subgraphs*), *bipartite graphs* (with emphasis on *matching* problems), and *planar graphs* (with discussion of *duality* and the *four-color conjecture*) are studied in detail. *Directed graphs* are then taken up and their properties scrutinized in relation to those of nondirected graphs. The last section in this chapter illustrates the application of graph theory to the solution of puzzles and to the determination of winning strategies in two-person games. In this connection, the *principle of optimality* is introduced and its usefulness demonstrated.

12-1. PRELIMINARY DEFINITIONS

We define a *graph* G as a pair (V, E), where:

(a) $V = \{v_1, v_2, \ldots, v_n\}$ is a finite nonempty set of *vertices* (also known as *nodes, junctions,* or *points*). V is called the *vertex set* of G.

(b) E is a set of (unordered) pairs of distinct vertices (that is, a set of sets of the form $\{v_i, v_j\}$ where $v_i \neq v_j$). These pairs are called *edges* (also known as *branches, arcs,* or *lines*), and E is called the *edge set* of G.

Diagrammatically, G is described by n small circles representing the vertices $v_1, v_2, \ldots, v_n$, with a line drawn between the circles representing v_i

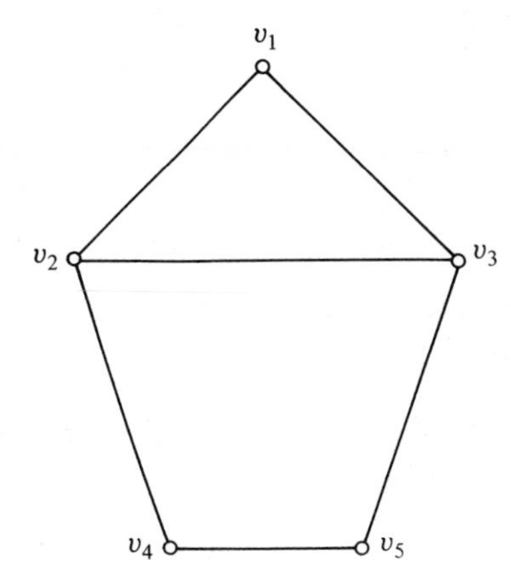

Fig. 12-1 A graph

and v_j if and only if $\{v_i, v_j\}$ is an edge of G. For example, Fig. 12-1 shows a graph whose vertex set is

$$V = \{v_1, v_2, v_3, v_4, v_5\}$$

and whose edge set is

$$E = \{\{v_1, v_2\}, \{v_1, v_3\}, \{v_2, v_3\}, \{v_2, v_4\}, \{v_3, v_5\}, \{v_4, v_5\}\}$$

A convenient way of specifying the graph $G = (V, E)$ is by means of an *adjacency matrix* $\mathbf{A} = [a_{ij}]_{n \times n}$, where

$$a_{ij} = \begin{cases} 1 & \text{if } \{v_i, v_j\} \in E \\ 0 & \text{otherwise} \end{cases}$$

Clearly, $\mathbf{A}$ has 0s along its principal diagonal, and is symmetrical about this diagonal. The adjacency matrix of the graph of Fig. 12-1, for example, is given by

$$\mathbf{A} = \begin{bmatrix} 0 & 1 & 1 & 0 & 0 \\ 1 & 0 & 1 & 1 & 0 \\ 1 & 1 & 0 & 0 & 1 \\ 0 & 1 & 0 & 0 & 1 \\ 0 & 0 & 1 & 1 & 0 \end{bmatrix}$$

A graph with n vertices and m edges is referred to as an (n, m) *graph*. The (1, 0) graph is called the *trivial graph*.

If $e = \{v_i, v_j\}$ is an edge of G, then the vertices v_i and v_j are said to be *adjacent* (to each other) and e and v_i (or e and v_j) are said to be *incident* (with each other). Two distinct edges that are incident with the same vertex are said to be *adjacent* (to each other). In Fig. 12-1, for example, v_2 and v_3 are adjacent vertices, $\{v_2, v_4\}$ is incident with v_4, and $\{v_3, v_5\}$ and $\{v_4, v_5\}$ are adjacent edges.

An (n, m) graph in which every one of the n vertices ($n \geq 2$) is adjacent to each of the other $n - 1$ vertices, is called a *complete graph*. Thus, in a complete graph, $m = \binom{n}{2} = n(n - 1)/2$. Figure 12-2 shows a complete 5-vertex graph. The *complement* of a graph G consists of the vertices of G and all edges needed to make G complete. For example, Fig. 12-3 shows the complement of the graph of Fig. 12-1.

The *degree* of a vertex v_i in a graph G, denoted by $\deg(v_i)$, is the number of edges in G which are incident with v_i. For example, the degrees of vertices v_1, v_2, v_3, v_4, and v_5 of the graph of Fig. 12-1 are 2, 3, 3, 2, and 2, respectively. Since every edge is incident with exactly two vertices, each edge of G contributes exactly 2 to the sum of the degrees of all vertices of G. We thus have:

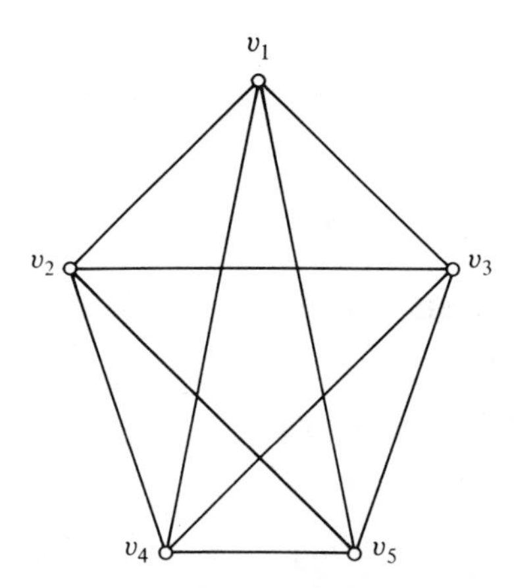

Fig. 12-2 A complete graph

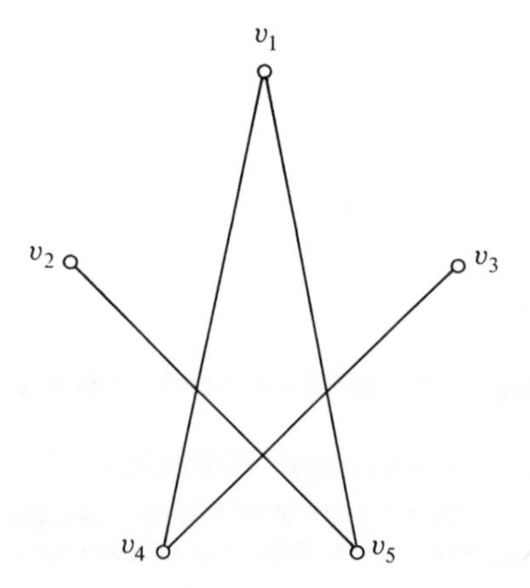

Fig. 12-3 Complement of graph of Fig. 12-1

THEOREM 12-1

Let G be an (n, m) graph with the vertex set $\{v_1, v_2, \ldots, v_n\}$. Then

$$\sum_{i=i}^{n} \deg(v_i) = 2m$$

A graph in which all vertices have the same degree d is said to be *regular* (of degree d). Figure 12-4 shows a regular graph of degree 3.

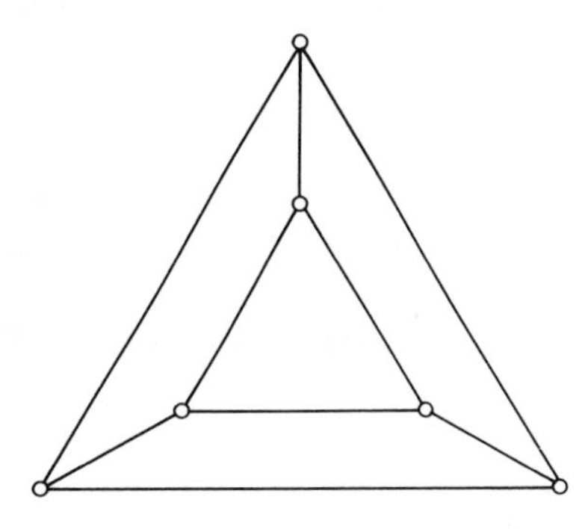

Fig. 12-4 Regular graph of degree 3

Let G and $\bar{G}$ be two graphs with the vertex set V and $\bar{V}$, respectively. $\bar{G}$ is said to be *isomorphic* to G if there exists a bijection $h\colon V \longrightarrow \bar{V}$ such that $\{v_i, v_j\}$ is an edge in G if and only if $\{h(v_i), h(v_j)\}$ is an edge in $\bar{G}$. Clearly, if $\bar{G}$ is isomorphic to G, then G is isomorphic to $\bar{G}$, and we can simply say that G and $\bar{G}$ are isomorphic (to each other). For example, Fig. 12-5 shows two isomorphic graphs where the isomorphism h is given by $h(v_i) = \bar{v}_i$ $(i = 1, 2, \ldots, 6)$. Since isomorphic graphs are identical except, possibly, for their vertex labels, any graph-theoretic statement that is true for a graph G is also true for any graph isomorphic to G.

Consider the graphs $G = (V, E)$ and $\tilde{G} = (\tilde{V}, \tilde{E})$. If $\tilde{V} \subset V$ and $\tilde{E} \subset E$, then $\tilde{G}$ is called a *subgraph* of G (a *proper subgraph* of G if $\tilde{E} \neq E$). If $\tilde{V} = V$ and $\tilde{E} \subset E$, $\tilde{G}$ is called a *spanning subgraph* of G. For example, the graph of Fig. 12-3 is a spanning subgraph of the graph of Fig. 12-2.

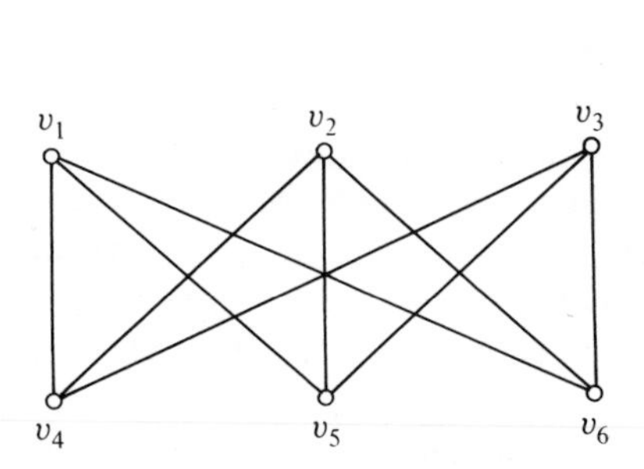

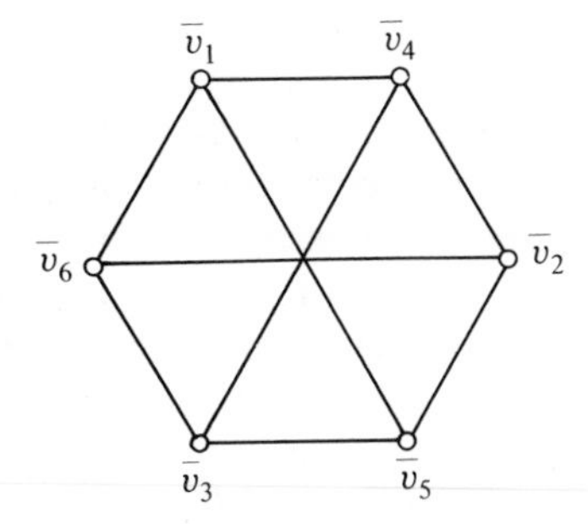

Fig. 12-5 Isomorphic graphs

A sequence of l edges $\{v_{i_0}, v_{i_1}\}, \{v_{i_1}, v_{i_2}\}, \ldots, \{v_{i_{l-1}}, v_{i_l}\}$ is called a *path* of *length l* which *connects* v_{i_0} to v_{i_l}. It is denoted either by

$$\{v_{i_0}, v_{i_1}\}\{v_{i_1}, v_{i_2}\} \ldots \{v_{i_{l-1}}, v_{i_l}\}$$

or, more concisely, by $v_{i_0}v_{i_1} \ldots v_{i_l}$. The path $v_{i_0}v_{i_1} \ldots v_{i_l}$ is said to be an *open path* if $v_{i_0} \neq v_{i_l}$, and a *loop* if $v_{i_0} = v_{i_l}$. An open path $v_{i_0}v_{i_1} \ldots v_{i_l}$ is said to be a *proper path* if $v_{i_0}, v_{i_1}, \ldots, v_{i_l}$ are distinct. A loop $v_{i_0}v_{i_1} \ldots v_{i_{l-1}}v_0$ is called a *cycle* if $v_{i_0}, v_{i_1}, \ldots, v_{i_{l-1}}$ are distinct. In Fig. 1, $v_1v_3v_5v_4v_2v_3v_1v_2$ is an open path of length 7; $v_1v_2v_3v_5v_4v_2v_1$ is a loop of length 6; $v_1v_3v_2v_4$ is a proper path of length 3; and $v_1v_2v_3v_1$ is a cycle of length 3.

Two vertices v_i and v_j in a graph are said to be *connected* if there exists a path that connects v_i to v_j. A graph G is said to be *connected* if every two vertices in G are connected. Any connected subgraph of G which is not a proper subgraph of any other connected subgraph of G, is called a *component* (or *connected component*) of G. In simple words, the components of a graph are its "separate parts." For example, Fig. 12-6 shows the six components (subgraphs $G_1, G_2, \ldots, G_6$) of a 14-vertex graph.

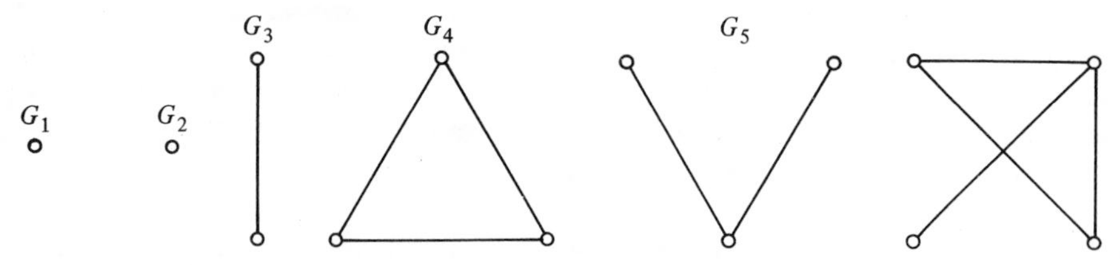

Fig. 12-6 Components of a graph

A *separating edge* of a graph G is an edge $\{v_i, v_j\}$ such that the set of vertices connected to v_i and the set of vertices connected to v_j are disjoint. Thus, the removal of a separating edge e from G transforms the component of G which contains e into two separate components. Any edge in G which is not a separating edge must appear in some cycle of G. (Why?)

Several modified versions of our original definition of a graph exist and are of use in various applications. When more than one edge is permitted between two adjacent vertices (that is, when cycles of length 2 are permitted), the result is called a *multigraph*. When cycles of length 1 are permitted around any vertex, the result is called a *pseudograph*. A graph, multigraph, or pseudograph in which each edge points (by means of an arrow) from one vertex to another is called a *directed graph*. The relation graphs introduced in Sec. 2-2 and the transition diagrams introduced in Sec. 7-9 are examples of directed graphs. Flow charts used in describing computer programs are also directed graphs (with the various instructions playing the role of vertices).

The ordering diagrams introduced in Sec. 2-7 and the rooted trees introduced in Sec. 3-10 are examples of ordinary graphs.

Unless otherwise specified, the term "graph" is restricted in the remainder of this chapter to that defined in the beginning of this section. However, the reader should note that many of the definitions and theorems developed specifically for graphs are applicable, with trivial or no modifications, to the other variants listed above.

PROBLEMS

1. Is the graph shown in Fig. 12-A isomorphic to the graphs of Fig. 12-5?

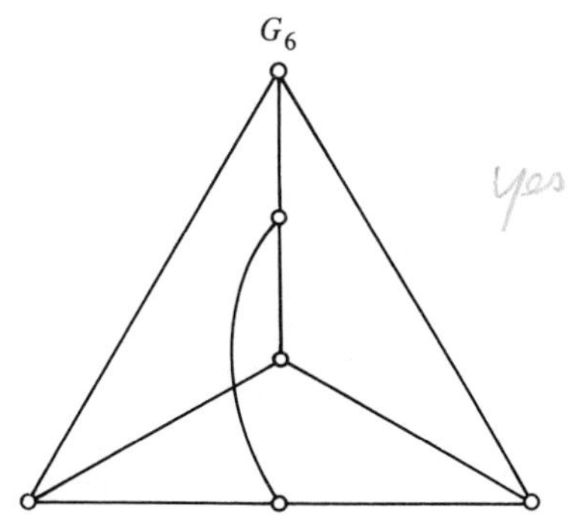

Fig. 12-A

2. Figure 12-B shows two 8-vertex graphs. Are these graphs isomorphic? Prove your answer.

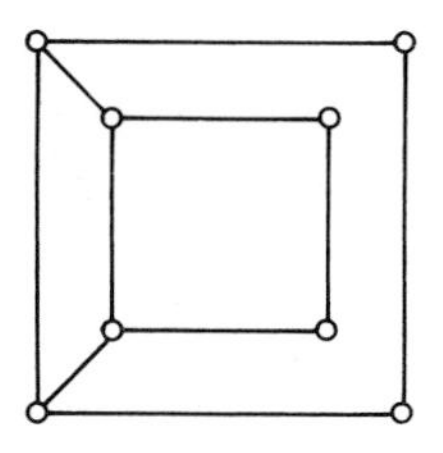

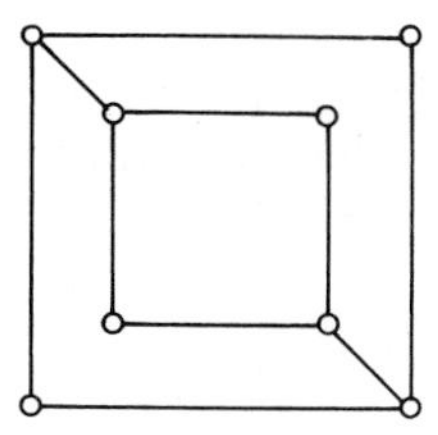

Fig. 12-B

3. Prove that, in any graph, the number of vertices of odd degree is even. (Compare with Problem 5, Sec. 3-9!)

4. Let G be a complete graph with 4 vertices.
   (a) Construct all the subgraphs of G.
   (b) Construct all the spanning subgraphs of G.
   (c) What is the number of subgraphs of G such that no two subgraphs are isomorphic?

5. In the graph of Fig. 12-C, enumerate all proper paths and all cycles. Does this graph contain a separating edge?

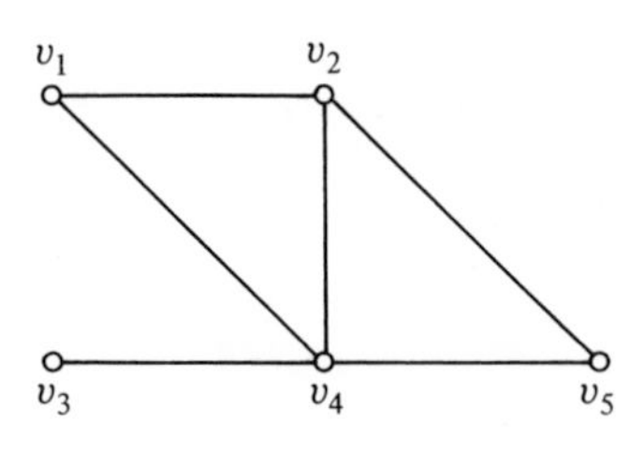

Fig. 12-C

6. The graph G is specified by this adjacency matrix:

$$\mathbf{A} = \begin{bmatrix} 0 & 0 & 1 & 1 & 0 & 0 \\ 0 & 0 & 0 & 0 & 1 & 1 \\ 1 & 0 & 0 & 0 & 0 & 0 \\ 1 & 0 & 0 & 0 & 0 & 0 \\ 0 & 1 & 0 & 0 & 0 & 1 \\ 0 & 1 & 0 & 0 & 1 & 0 \end{bmatrix}$$

Is G a connected graph?

7. Show that any edge in a graph G which is not a separating edge must appear in some cycle of G.

8. Show that if every vertex of a graph G is of degree 2, then every component of G consists of a cycle.

9. A graph G whose complement is isomorphic to G is called *self-complementary*. Show that the graph of Fig. 12-D is self-complementary. Can you find another self-complementary 5-vertex graph?

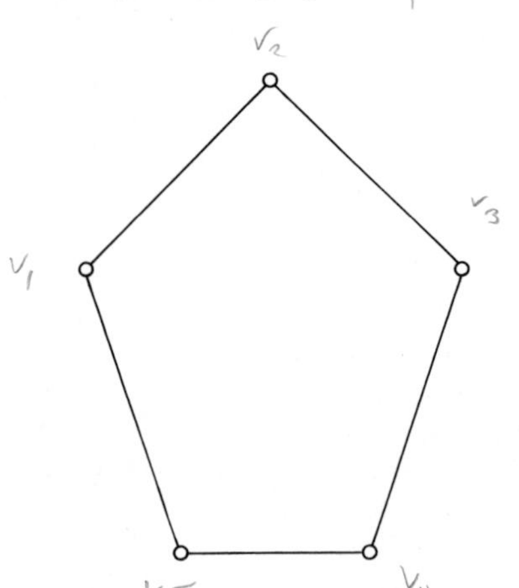

Fig. 12-D

10. Let G be a graph where the longest proper path has length L. Show that any two proper paths in G whose length is L must have at least one vertex in common.

12-2. CONNECTIVITY AND TRAVERSABILITY

When vertices v_i and v_j in a graph are connected by one or more paths, the length of any shortest such path is called the *geodesic* of v_i and v_j.

THEOREM 12-2

Let G be a graph with the vertex set $V = \{v_1, v_2, \ldots, v_n\}$. Then, for any connected vertices $v_i, v_j \in V$ ($v_i \neq v_j$), the geodesic is a proper path whose length is at most $n - 1$.

Proof Consider any path

$$\alpha = v_{i_0}v_{i_1} \ldots v_{i_\mu} \ldots v_{i_\nu} \ldots v_{i_{l-1}}v_{i_l}$$

connecting v_{i_0} to v_{i_l} ($v_{i_0} \neq v_{i_l}$). If α has any repeated vertices, say, $v_{i_\mu} = v_{i_\nu}$, then the subpath $v_{i_{\mu+1}} \ldots v_{i_{\nu-1}}$ can be deleted from α to result in a shorter path

$$\beta = v_{i_0}v_{i_1} \ldots v_{i_\mu}v_{i_{\nu+1}} \ldots v_{i_{l-1}}v_{i_l}$$

which connects v_{i_0} to v_{i_l}. If β has any repeated vertices, the process can be repeated to result in yet a shorter path. Finally, a proper path is obtained; it connects v_{i_0} to v_{i_l} and is certainly shorter than any nonproper one. Thus, no proper path can be a geodesic. Now, in any proper path $v_{i_0}v_{i_1} \ldots v_{i_l}$ of length l, the vertices $v_{i_0}, v_{i_1}, \ldots, v_{i_l}$ must be distinct, which implies that $l + 1 \leq n$ and, hence, $l \leq n - 1$. □

The length of a geodesic of v_i and v_j is called the *distance* between v_i and v_j and is denoted by $d(v_i, v_j)$. Thus, the distance between any two distinct vertices in an n-vertex graph cannot exceed $n - 1$.

When it is too cumbersome to specify a graph diagrammatically, its adjacency matrix provides a convenient means for answering various questions regarding connectivity between vertices. In what follows, multiplication of matrices is carried out in the usual manner, with all additions and multiplications performed as per rules of ordinary integer arithmetic.

THEOREM 12-3

Let G be a graph with the vertex set $\{v_1, v_2, \ldots, v_n\}$ and adjacency matrix $\mathbf{A}$. Then the (i, j) entry of $\mathbf{A}^l$ ($l = 1, 2, \ldots$) is the number of paths of length l connecting v_i to v_j.

Proof (by induction on l) (*Basis*). When $l = 1$, $\mathbf{A}^l = \mathbf{A}$ and the assertion follows from the definition of $\mathbf{A}$. (*Induction step*). Let $a_{ij}^{(k)}$ denote the (i, j) entry of $\mathbf{A}^k$, and hypothesize that the assertion is true for $l = k$. Since $\mathbf{A}^{k+1} = \mathbf{A}^k\mathbf{A}$, we have

$$a_{ij}^{(k+1)} = \sum_{h=1}^{n} a_{ih}^{(k)} a_{hj}$$

But $a_{ih}^{(k)} a_{hj}$ is the number of paths of length $k + 1$ which connect v_i to v_j with v_h as the next-to-last vertex in the path. Hence, $a_{ij}^{(k+1)}$ is the total number of paths of length $k + 1$ which connect v_i to v_j with *any* vertex being the next-to-last. The theorem, therefore, is true for $k + 1$. □

From Theorem 12-3 we can immediately conclude that $d(v_i, v_j)$ is the least integer l such that the (i, j) entry of $\mathbf{A}^l$ is nonzero. Also, from Theorem 12-2 it can be concluded that, if the (i, j) entry of $\mathbf{A}^l$ is 0 for $l = 1, 2, \ldots, n - 1$ $(i \neq j)$, then v_i and v_j are not connected by *any* path (and, hence, must belong to distinct components of G).

Example 12-1

The following is the adjacency matrix of a graph G with the vertex set $\{v_1, v_2, v_3, v_4, v_5\}$:

$$\mathbf{A} = \begin{bmatrix} 0 & 1 & 0 & 0 & 0 \\ 1 & 0 & 1 & 0 & 0 \\ 0 & 1 & 0 & 0 & 0 \\ 0 & 0 & 0 & 0 & 1 \\ 0 & 0 & 0 & 1 & 0 \end{bmatrix}$$

In this case we have:

$$\mathbf{A}^2 = \begin{bmatrix} 1 & 0 & 1 & 0 & 0 \\ 0 & 2 & 0 & 0 & 0 \\ 1 & 0 & 1 & 0 & 0 \\ 0 & 0 & 0 & 1 & 0 \\ 0 & 0 & 0 & 0 & 1 \end{bmatrix} \quad \mathbf{A}^3 = \begin{bmatrix} 0 & 2 & 0 & 0 & 0 \\ 2 & 0 & 2 & 0 & 0 \\ 0 & 2 & 0 & 0 & 0 \\ 0 & 0 & 0 & 0 & 1 \\ 0 & 0 & 0 & 1 & 0 \end{bmatrix} \quad \mathbf{A}^4 = \begin{bmatrix} 2 & 0 & 2 & 0 & 0 \\ 0 & 4 & 0 & 0 & 0 \\ 2 & 0 & 2 & 0 & 0 \\ 0 & 0 & 0 & 1 & 0 \\ 0 & 0 & 0 & 0 & 1 \end{bmatrix}$$

From these matrices we can conclude, for example, that v_1 and v_2 are connected by two paths of length 3; that the distance between v_1 and v_3 is 2; that G has no loops of length 3; and that there is no path of length 4 or less connecting v_3 and v_4, and, hence, that v_3 and v_4 belong to two different components of G. □

A common problem arising in applications of graph theory is: Given a graph G, is it possible to find a loop that traverses each edge of G exactly once? Such a loop is called an *Euler loop* (after the Swiss mathematician

Leonhard Euler, [pronounced oi'ler] 1707–1783), and a graph that exhibits it is called an *Euler graph*. Clearly, every Euler graph must be connected. An example of an Euler graph is the graph of Fig. 12-2, where an Euler loop is $v_1v_2v_4v_5v_3v_2v_5v_1v_4v_3v_1$. The graph of Fig. 12-1 is not an Euler graph—a fact confirmed by:

THEOREM 12-4

A connected graph G is an Euler graph if and only if every vertex of G has an even degree.

Proof Suppose G is an Euler graph, and let α be an Euler loop in G. Each time α crosses a vertex in G it adds 2 to its degree. Since, in α, every edge of G appears exactly once, the degree of every vertex must be a multiple of 2.

Conversely, assume that every vertex of G has an even degree. If G is trivial, the theorem is trivially true. Otherwise—since every vertex is of degree at least 2—G must contain at least one cycle, say, σ_1. The removal of the edges of σ_1 from G results in a spanning subgraph G_1 of G, in which every vertex still has an even degree. If G_1 has at least one edge left, then G_1 must contain at least one cycle, say, σ_2, whose removal results in the spanning subgraph G_2 in which every vertex has an even degree. After a finite number of repetitions of this process, all the edges of G are removed. At this point the set of removed edges can be partitioned into the cycles σ_1, σ_2, etc. Now, let σ_i be any of these cycles. If this is the only cycle in the partition, it is clearly an Euler loop. Otherwise, there must be another cycle, say, σ_j, which shares a vertex v with σ_i. If σ_i and σ_j are the only cycles in the partition, then the Euler loop is the loop that starts at v, circles around σ_i, circles around σ_j, and terminates in v. Otherwise, there must be a third cycle, say, σ_k, which shares a vertex v' with one of the other two—say, with σ_j. If σ_i, σ_j, and σ_k are the only cycles in the partition, then the Euler loop is the loop which starts at v, circles around σ_i, proceeds along σ_j to v', circles around σ_k, and returns to v via σ_j. This process can be continued to construct an Euler loop from an arbitrary number of cycles in the partition. Thus, G is an Euler graph. □

Example 12-2

The graph of Fig. 12-2 is an Euler graph, since every one of its vertices has an even degree (namely, 4). Its 10 edges can be partitioned (as described in the proof of Theorem 12-4) into the cycles

$$\sigma_1 = v_1v_2v_3v_1, \qquad \sigma_2 = v_3v_4v_5v_3, \qquad \sigma_3 = v_1v_4v_2v_5v_1$$

These cycles, with v_3 and v_5 as the vertices common to σ_1 and σ_2 and to σ_2 and σ_3, respectively, are shown in Fig. 12-7. From this figure, an Euler loop for the graph is seen to be $v_3v_1v_2v_3v_4v_5v_2v_4v_1v_5v_3$. □

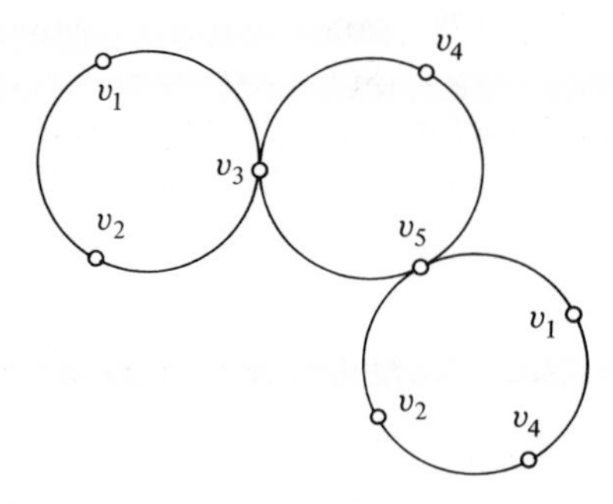

Fig. 12-7 For Example 12-2

Theorem 12-4 was used by Euler (who proved it) to solve the famous *Königsberg bridge problem.* The city of Königsberg is located on the banks and on two islands of the Pregel river. The various parts of the city are connected by seven bridges, as shown in Fig. 12-8(a). The problem faced by Euler was: Is it possible to start at any of the four land regions of the city, cross each bridge exactly once, and return to the starting point? The key to the solution of this problem is the representation of the city by a multigraph, where the regions are represented by vertices and the bridges by edges—as shown in Fig. 12-8(b). The problem now reduces to that of deciding whether or not the multigraph is an Euler graph (where multiple edges between adjacent vertices are permitted). Recognizing that Theorem 12-4 is valid for multigraphs as well as for graphs (the proof being the same), the answer is immediately seen to be negative (since none of the vertices in the multigraph is of even order).

An *Euler path* in a graph G is an *open* path that traverses each edge of G exactly once.

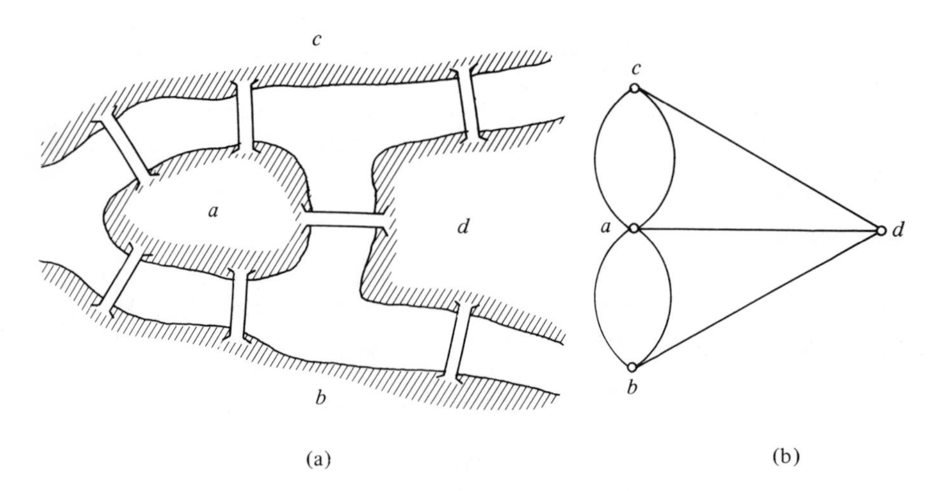

Fig. 12-8 The Königsberg bridge problem

THEOREM 12-5

A connected graph G has an Euler path, connecting the vertices v_i and v_j, if and only if v_i and v_j are the only vertices in G which have an odd degree.

Proof Let G' be the graph produced by adding the edge $\{v_i, v_j\}$ to G. Then G has an Euler path if and only if G' has an Euler loop; hence, if and only if all the vertices of G' have even orders and, hence, if and only if all the vertices of G except v_i and v_j have even orders. □

Another common problem arising in the theory and applications of graphs is: Given a graph G, is it possible to find a cycle that traverses each *vertex* of G exactly once? Such a cycle is called a *Hamilton cycle* (after the Irish mathematician Sir William Hamilton, 1805–1865), and a graph which exhibits it is called a *Hamilton graph.* Clearly, every Hamilton graph must be connected. The graphs shown in Figs. 12-1, 12-2, 12-4, and 12-5 are all Hamilton graphs (where the Hamilton cycles are trivially apparent). Unfortunately, no elegant characterization of Hamilton graphs has been discovered, although several conditions which are either necessary or sufficient (but not both) for a graph to be a Hamilton graph are known. In most practical cases, trial-and-error solutions have to be resorted to.

A problem akin to that of determining Hamilton cycles is the *traveling salesman problem.* Suppose a traveling salesman is assigned to visit certain towns before returning to his home office. How should he plan his trip so as to accomplish it as "inexpensively" as possible (in terms of distance, say)? This situation can be viewed graphically by representing the towns and home office by vertices, and the interconnecting roads by edges, labeled with the lengths of the corresponding roads. The problem then reduces to that of locating a loop in the graph which touches every vertex and which exhibits the least possible total length. Again, no elegant general solution has been discovered for this problem.

PROBLEMS

1. Graphs G_1 and G_2 are characterized by these adjacency matrices $\mathbf{A}_1$ and $\mathbf{A}_2$, respectively:

$$\mathbf{A}_1 = \begin{bmatrix} 0 & 0 & 1 & 1 & 1 & 0 \\ 0 & 0 & 0 & 0 & 1 & 1 \\ 1 & 0 & 0 & 1 & 0 & 0 \\ 1 & 0 & 1 & 0 & 1 & 0 \\ 1 & 1 & 0 & 1 & 0 & 1 \\ 0 & 1 & 0 & 0 & 1 & 0 \end{bmatrix} \qquad \mathbf{A}_2 = \begin{bmatrix} 0 & 0 & 0 & 1 & 1 & 0 & 0 \\ 0 & 0 & 0 & 0 & 0 & 1 & 1 \\ 0 & 0 & 0 & 1 & 0 & 0 & 0 \\ 1 & 0 & 1 & 0 & 1 & 0 & 0 \\ 1 & 0 & 0 & 1 & 0 & 0 & 0 \\ 0 & 1 & 0 & 0 & 0 & 0 & 0 \\ 0 & 1 & 0 & 0 & 0 & 0 & 0 \end{bmatrix}$$

(a) Evaluate $\mathbf{A}_1^l$ $(l = 1, 2, \ldots, 6)$ and $\mathbf{A}_2^l$ $(l = 1, 2, \ldots, 7)$.
(b) List the distances between every two vertices within G_1 and G_2.
(c) List all cycles in G_1 and G_2.

2. Let G be a graph with the adjacency matrix $\mathbf{A} = [a_{ij}]_{n \times n}$ $(n > 1)$, where $a_{12} = a_{21} = 1$. Let $\bar{\mathbf{A}} = [\bar{a}_{ij}]_{n \times n}$, where

$$\bar{a}_{ij} = \begin{cases} 0 & \text{if } \bar{a}_{ij} = \bar{a}_{12} \text{ or } \bar{a}_{ij} = \bar{a}_{21} \\ a_{ij} & \text{otherwise} \end{cases}$$

Show that $\{v_1, v_2\}$ is a separating edge in G if and only if the $(1, 2)$ entries in $\mathbf{A}^2, \mathbf{A}^3, \ldots, \mathbf{A}^{n-1}$ are 0.

3. Characterize graphs with adjacency matrices $\mathbf{A}$ such that the principal diagonal of $\mathbf{A}^l$ is 0 for all even l.

4. Let $\mathbf{A}$ be the adjacency matrix of the n-vertex graph G. Show that if G has a Hamilton cycle, then the principal diagonal of $\mathbf{A}^n$ has no 0s. Show that the converse of this statement is false.

5. Let G be a graph with the vertex set $\{v_1, v_2, \ldots v_n\}$. Define the *edge matrix* $\mathbf{E} = [e_{ij}]_{n \times n}$ as

$$e_{ij} = \begin{cases} \{\eta_{ij}\} & \text{if } v_i \text{ and } v_j \text{ are adjacent} \\ \varnothing & \text{otherwise} \end{cases}$$

If $\mathbf{X} = [x_{ij}]_{n \times n}$ and $\mathbf{Y} = [y_{ij}]_{n \times n}$ are matrices with sets as entries, define the product $\mathbf{Z} = \mathbf{XY} = [z_{ij}]_{n \times n}$ as

$$z_{ij} = \bigcup_{h=1}^{n} x_{ih} \cdot y_{hj}$$

where $\cdot$ denotes set concatenation (see Sec. 8-1).

Starting with these definitions, show that the (i, j) entry of $\mathbf{E}^l$ represents the set of all paths of length l which connect v_i to v_j.

6. The following facts are known about persons a, b, c, d, e, f, and g:

Person a speaks English.
Person b speaks English and Sapnish.
Person c speaks English, Italian, and Russian.
Person d speaks Japanese and Spanish.
Person e speaks German and Italian.
Person f speaks French, Japanese, and Russian.
Person g speaks French and German.

Show that every pair of persons among these seven can communicate (with the aid, if necessary, of a chain of interpreters taken from the remaining five).

7. Find an Euler loop in the graph shown in Fig. 12-E.

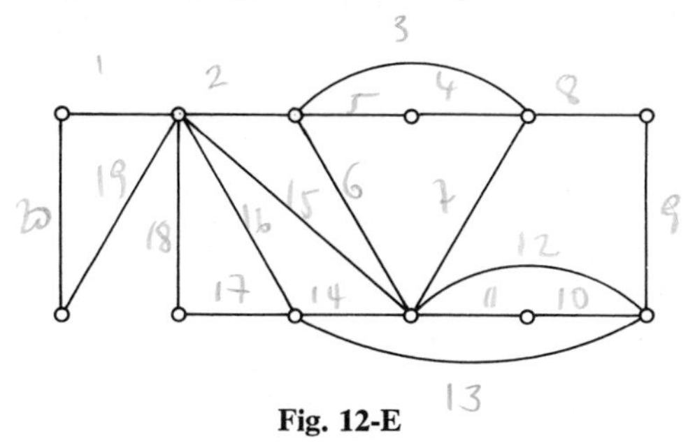

Fig. 12-E

8. Find an Euler path in the graph shown in Fig. 12-F.

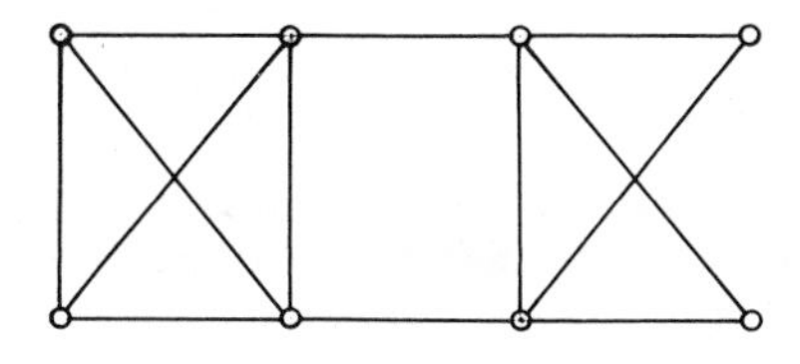

Fig. 12-F

9. Figure 12G shows four graphs. Determine which are Euler graphs and which are Hamilton graphs. Construct Euler loops and Hamilton cycles in the appropriate cases.

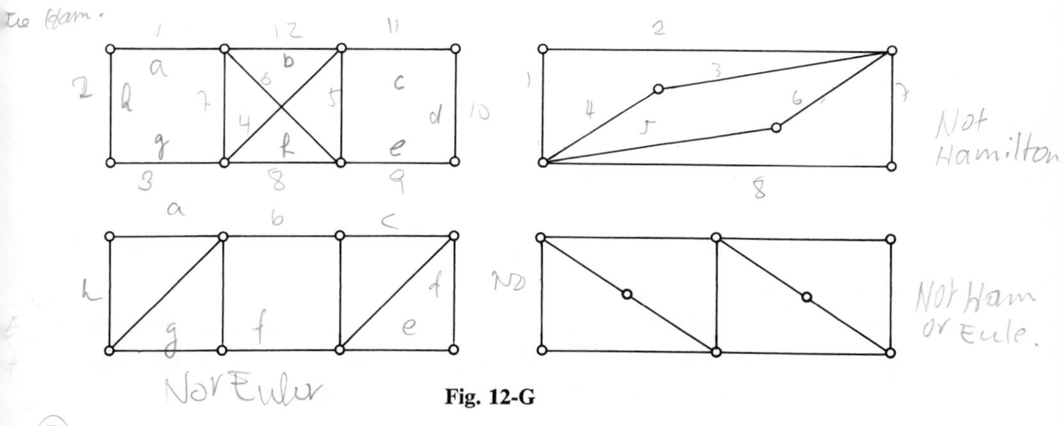

Fig. 12-G

10. A traveling salesman lives in town a and is supposed to visit towns b, c, and d before returning to a. Find the shortest route that accomplishes this trip, if the distances between the four towns are as indicated in Fig. 12-H.

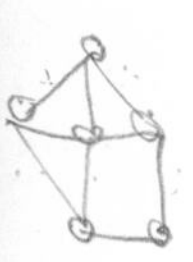

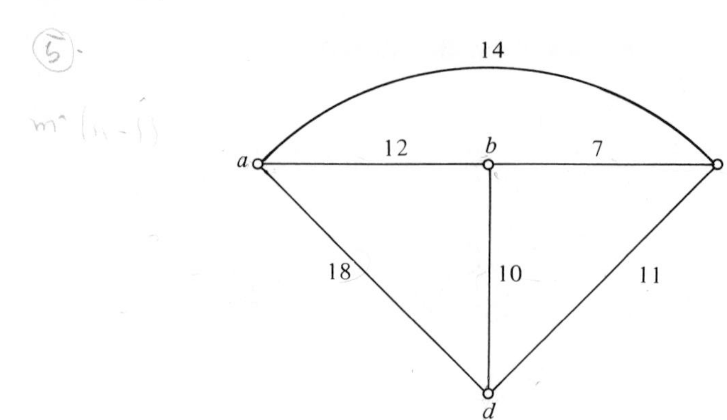

Fig. 12-H

12-3. TREES

A connected graph that contains no cycles is called a *tree*. A graph that contains no cycles—that is, a graph in which every component is a tree—is called a *forest*. Figure 12-9 shows a forest with two trees.

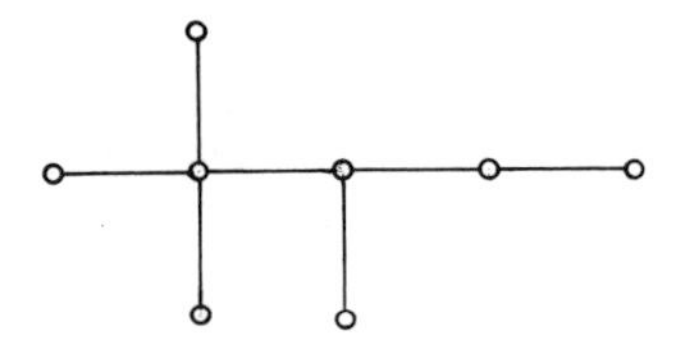

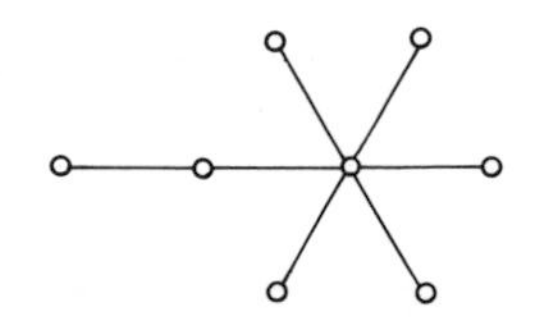

Fig. 12-9 A 2-tree forest

THEOREM 12-6

Let T be a tree where v_i and v_j are arbitrary distinct vertices. Then v_i and v_j are connected by exactly one proper path. Also, in the graph formed by adding the edge $\{v_i, v_j\}$ to T, there is exactly one cycle.

Proof Since T is connected, there must be a path—and, hence, a proper path—connecting v_i and v_j. Let α be such a proper path. If β is another proper path connecting v_i and v_j, then $\alpha\beta$ contains a cycle, which contradicts the definition of a tree. Thus, α is the only proper path connecting v_i and v_j.

If the edge $\{v_i, v_j\}$ is added to T, then $\alpha\{v_i, v_j\}$ forms a cycle in the new graph. The edge $\{v_i, v_j\}$ cannot form a cycle with any other proper path since, as just shown, no proper path exists between v_i and v_j other than α.

□

THEOREM 12-7

In an (n, m) tree, $m = n - 1$.

Proof (by induction on n) (*Basis*). For $n = 1$ and $n = 2$, the theorem is trivially true. (*Induction step*). Hypothesize that the theorem is true for all (i, m) trees where $i < n$. If T is an (n, m) tree, the removal of any edge from T transforms it into a 2-component graph, since T cannot contain any cycles. These two components must be trees—say, the (n_1, m_1) tree T_1 and the (n_2, m_2) tree T_2. By induction hypothesis, $m_1 = n_1 - 1$ and $m_2 = n_2 - 1$. Since $n = n_1 + n_2$ and $m = m_1 + m_2 + 1$, we have $m = n - 1$. □

If G is an (n, m) forest consisting of t trees, Theorem 12-7 can be readily generalized to yield the relationship:

$$m = n - t$$

A *spanning tree* T_G of a connected graph G is any tree that includes all the vertices of G. The following shows how such a tree can be constructed.

ALGORITHM 12-1

Given a connected graph G, to construct its spanning tree T_G:

(i) Let G be G_1. Set i to 1.

(ii) If G_i has no cycles, then $T_G = G_i$. Otherwise,

(iii) Locate any cycle σ_i in G_i and remove any edge e_i from σ_i. Call the remaining graph G_{i+1}. Since e_i is a cycle edge in G_i, G_{i+1} includes all the vertices of G_i; moreover, if G_i is connected, so is G_{i+1}.

(iv) Increment i by 1 and return to step (ii). □

In each step of the proposed procedure, an additional cycle of G is "broken." Since the number of cycles is, of course, finite, for some i we ultimately obtain a graph G_i which includes all vertices of G and contains no cycles; thus, G_i, by definition, is T_G. In general, T_G is not unique, since, at each iteration of the algorithm, neither σ_i nor e_i are necessarily unique.

If G is an (n, m) connected graph, then, by Theorem 12-7, T_G is an $(n, n - 1)$ graph. Hence, the number of edges that must be removed before T_G is obtained must be $m - (n - 1)$. This number is called the *cycle rank* (or the *cyclotomic number*) of G. Thus, the cycle rank of G is the least number of edges that must be removed from G in order to break all its cycles. Each one of these removed edges is referred to as a *chord* of G. (If one views G as an electric circuit where current circulates, then the set of chords of G represents the smallest set of branches that must be severed in order to stop the flow of current in every branch of the circuit.)

Example 12-3

Figure 12-10 illustrates Algorithm 12-1 with a (6, 9) graph G. T_G, in this case, has $6 - 1 = 5$ edges, and the cycle rank of G (which equals the number of chords) is $9 - (6 - 1) = 4$. □

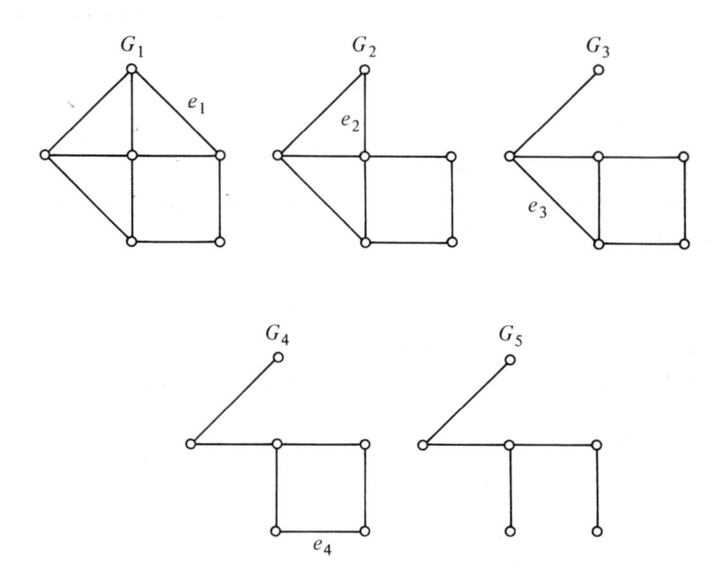

Fig. 12-10 Illustrating Algorithm 12-1

A *rooted tree* T with the vertex set V is a tree that can be defined recursively as follows: T has a specially designated vertex $v_1 \in V$ called the *root* of T; the subgraph of T, consisting of the vertices $V - \{v_1\}$, is partitionable into the subgraphs $T_1, T_2, \ldots, T_r$, each of which is itself a rooted tree. Each one of these r-rooted trees is called a *subtree of* v_1 (and their roots are sometimes referred to as the "sons" of v_1). Figure 12-11 shows a rooted tree where the root v_1 has three subtrees (consisting of the sets of vertices $\{v_2, v_5, v_6\}$, $\{v_3\}$, and $\{v_4, v_7, v_8, v_9, v_{10}, v_{11}\}$) which are themselves rooted trees (with the roots v_2, v_3, and v_4, respectively).

A *binary tree* is a rooted tree where each vertex v has at most two subtrees; if both subtrees are present, one is called the *left subtree* of v and the other the *right subtree* of v; if only one subtree is present, it can be designated either as the left subtree or the right subtree of v. (In the diagram, the left and right subtrees of v are drawn, respectively, to the left of and below or to the right of and below the vertex v). Figure 12-12 shows a 10-vertex binary tree.

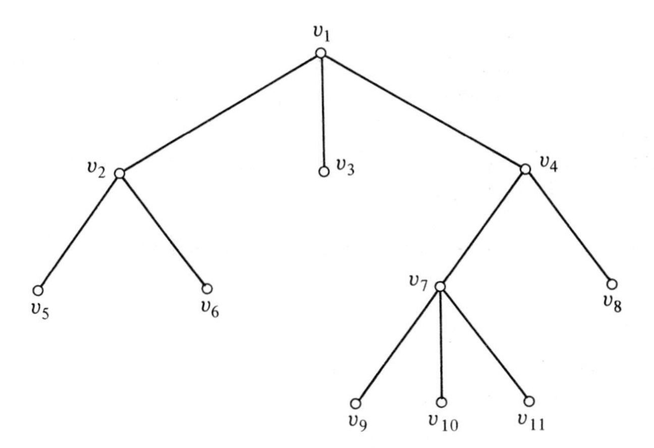

Fig. 12-11 A rooted tree

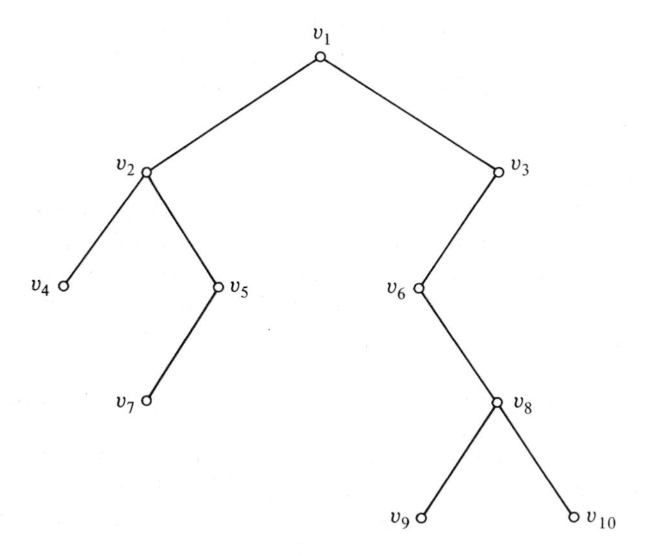

Fig. 12-12 A binary tree

In this tree, the left and right subtrees of v_1 have the roots v_2 and v_3, respectively; v_5 has only a left subtree.

A problem which often arises in computer applications is that of systematically traversing a binary tree so that every vertex is visited at least once. The following are the most common algorithms that accomplish this task (all of which are described recursively):

Pre-order traversal.

(i) Visit the root. (ii) Perform pre-order traversal on the root's left subtree. (iii) Perform pre-order traversal on the root's right subtree.

Post-order traversal.

(i) Perform post-order traversal on the root's left subtree. (ii) Visit the root. (iii) Perform post-order traversal on the root's right subtree.

End-order traversal.

(i) Perform end-order traversal on the root's left subtree. (ii) Perform end-order traversal on the root's right subtree. (iii) Visit the root.

For example, the paths traversed in the tree of Fig. 12-12 by these three methods are given by:

Pre-order: $v_1v_2v_4v_5v_7v_3v_6v_8v_9v_{10}$
Post-order: $v_4v_2v_7v_5v_1v_6v_9v_8v_{10}v_3$
End-order: $v_4v_7v_5v_2v_9v_{10}v_8v_6v_3v_1$

Another task of interest is that of representing an arbitrary rooted tree T by a binary tree T', so as to preserve the ordering of the subtrees of each vertex. One way of accomplishing it is: Suppose the r subtrees of vertex v_i in T have the roots $v_{i1}, v_{i2}, \ldots, v_{ir}$ (ordered left to right); then, in T', v_{i1} is the left son of v_i, v_{i2} is the right son of v_{i1}, v_{i3} is the right son of $v_{i2}, \ldots, v_{ir}$ is the right son of $v_{i(r-1)}$.

EXAMPLE 12-4

Figure 12-13 shows the binary tree representation T' of the rooted tree T of Fig. 12-11. It is seen that a path in T' which constitutes the "left boundary" of a subtree (for example, $v_1v_2v_5$ or $v_4v_7v_9$), constitutes a similar boundary in T. Also, any path in T' which constitutes the "right boundary" of a subtree (for instance, $v_2v_3v_4$ or $v_9v_{10}v_{11}$) constitutes the set of sons of the same vertex in T (for example, v_2, v_3, and v_4 are the sons of v_1). Thus, it is a simple task to reconstruct the original tree T from its binary representation T'. □

We now consider this problem: G is a complete n-vertex graph ($n \geq 2$), where each edge is associated with a fixed positive real number representing *cost*; find a connected spanning subgraph $\bar{G}$ of G which, of all such subgraphs, exhibits the least possible total cost. This subgraph is referred to as an *economy subgraph* of G. The following shows how it can be constructed.

ALGORITHM 12-2

Let G be a complete n-vertex graph ($n \geq 2$). To construct an economy subgraph $\bar{G}$ of G:

(i) Set i to 1. Let G_1 be any least-cost edge in G, say, e_1.

(ii) If G_i is a spanning tree of G, then $G_i = \bar{G}$. Otherwise:

(iii) Add to G_i any least-cost edge, say, e_i, which, when added to G_i, forms a cycle-free graph. Call the resulting graph G_{i+1}.

(iv) Increment i by 1 and return to step (ii). □

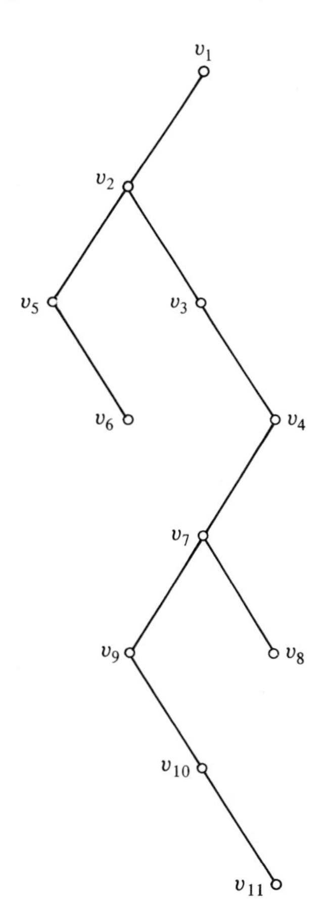

Fig. 12-13 Binary tree representation for tree of Fig. 12-11

Since Algorithm 12-2 terminates when G_i becomes a spanning tree of G, it must terminate when $i = n - 1$. The economy subgraph produced by the algorithm is, therefore, the graph G_{n-1}. To prove that G_{n-1} is indeed an economy subgraph $\bar{G}$ of G, we first note that any $\bar{G}$ must be a tree (since, otherwise, $\bar{G}$ would contain a cycle in which one edge can always be deleted to result in a less costly connected spanning subgraph). Now, assume that $\bar{G}$ is distinct from G_{n-1} and that, in the sequence of edges $e_1, e_2, \ldots, e_{n-1}$ of G_{n-1}, the edge $e_j = \{v, v'\}$ is the first to be absent from $\bar{G}$. Let α be the path in $\bar{G}$ which connects v to v'. Since G_{n-1} has no cycles, at least one edge in α—say, $\bar{e}_j$—is not in G_{n-1}. (See Fig. 12-14.) Removing $\bar{e}_j$ from $\bar{G}$ and adding e_j to $\bar{G}$ results in a graph $\bar{G}^{(1)}$ that is still a spanning tree of G. Since $\bar{G}$ is an economy subgraph, we have $c(e_j) \geq c(\bar{e}_j)$, where $c(e)$ denotes the cost associated with the edge e. But e_j, by construction, is any least-cost edge which, when added to $e_1, e_2, \ldots, e_{j-1}$, forms no cycles. Since $\bar{e}_j$, too, forms no cycles with $e_1, e_2, \ldots, e_{j-1}$, we must have $c(e_j) \leq c(\bar{e}_j)$; hence, $c(e_j) = c(\bar{e}_j)$.

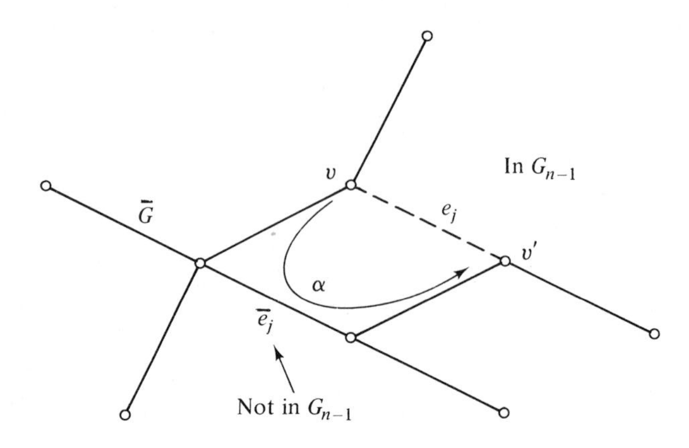

Fig. 12-14 For Algorithm 12-2

Thus, the total costs of $\bar{G}$ and $\bar{G}^{(1)}$ are identical and, hence, $\bar{G}^{(1)}$ is another economy subgraph of G. Now, G_{n-1} and the economy subgraph $\bar{G}^{(1)}$ have the edges $e_1, e_2, \ldots, e_j$ in common, and the preceding argument can be repeated with respect to G_{n-1} and $\bar{G}^{(1)}$. Finally, an economy subgraph $\bar{G}^{(k)}$ of G is obtained in which all $n - 1$ edges are the same as those of G_{n-1}. Hence, G_{n-1} is an economy subgraph of G.

Example 12-5

We wish to construct a railroad network that will interconnect, as inexpensively as possible, the five cities v_1, v_2, v_3, v_4, and v_5. The costs (in mil-

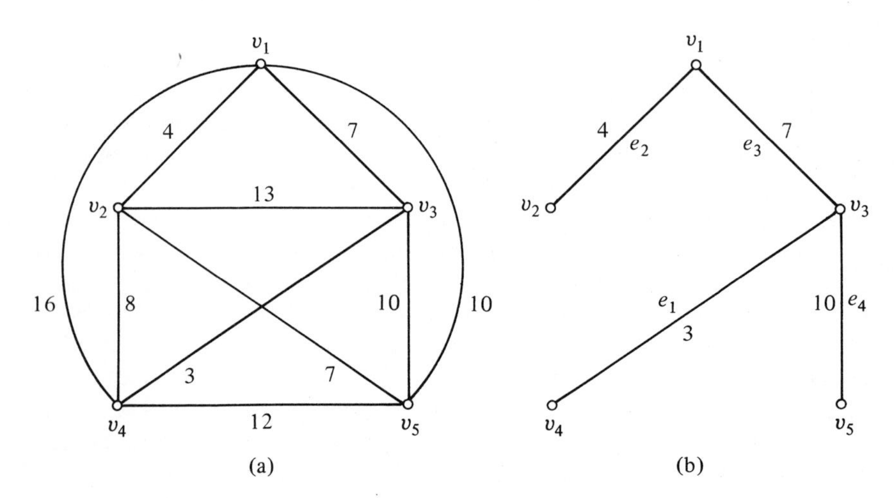

Fig. 12-15 Graph and its economy subgraph, for Example 12-5

lions of dollars) of interconnecting the various pairs of cities are given by:

$$c(v_1, v_2) = 4, \qquad c(v_1, v_3) = 7, \qquad c(v_1, v_4) = 16,$$
$$c(v_1, v_5) = 10, \qquad c(v_2, v_3) = 13, \qquad c(v_2, v_4) = 8,$$
$$c(v_2, v_5) = 7, \qquad c(v_3, v_4) = 3, \qquad c(v_3, v_5) = 10,$$
$$c(v_4, v_5) = 12$$

The desired network can be regarded as an economy subgraph $\bar{G}$ of a graph G consisting of the five vertices $v_1, \ldots, v_5$, where the cost of any edge $\{v_i, v_j\}$ is $c(v_i, v_j)$ as listed above. The graph G is shown in Fig. 12-15(a). Figure 12-15(b) shows the economy subgraph $\bar{G}$, where the labels e_1, e_2, etc. indicate the order in which the edges were selected in accordance with Algorithm 12-2. □

PROBLEMS

1. Construct all possible 5-vertex trees that are not isomorphic to each other.
2. Show that graph G is a forest if and only if every edge in G is a separating edge.
3. Let G be a connected graph where the edge e is incident with the vertex v. Show that, if v has degree 1, then e is contained in every spanning tree of G.
4. Show that any edge of a connected graph G is an edge of some spanning tree of G.
5. Prove or disprove this statement: Any edge of a connected graph G is a chord for some spanning tree of G.
6. Show that a connected graph with m edges has at most $m + 1$ vertices.
7. What is the cycle rank of a complete n-vertex graph?
8. Find spanning trees for the graph of Fig. 12-I. What are their cycle ranks?

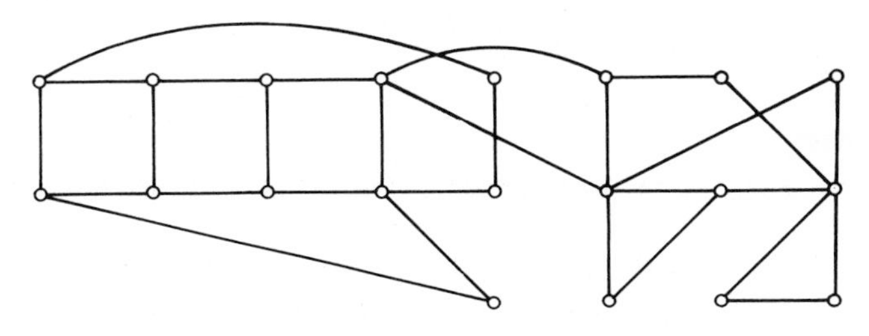

Fig. 12-I

9. Traverse the binary tree of Fig. 12-13 in pre-order, post-order, and end-order.

10. Regarding the binary trees of Figs. 12-12 and 12-13 as general rooted trees, represent them by binary trees, as described in the preceding section.

11. The vertices $a, b, \ldots, h$ in Fig. 12-J designate eight cities, as they actually appear on a map. Construct a connected tree of *minimal total length* which connects all eight cities.

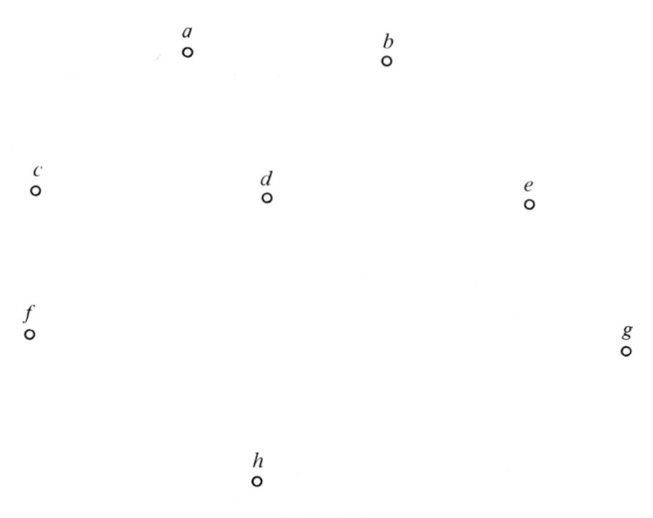

Fig. 12-J

12-4. BIPARTITE GRAPHS

A *bipartite graph* G is a graph whose vertex set V can be partitioned into two subsets, V_1 and V_2, in such a way that every edge of G is of the form $\{v_i, v_j\}$ where $v_1 \in V_1$ and $v_j \in V_2$. That is, every edge of G connects some vertex in V_1 to some vertex in V_2. The sets V_1 and V_2 are called the *complementary vertex subsets* of G. Figure 12-16 shows a 7-vertex bipartite graph, where the complementary vertex subsets are $V_1 = \{v_1, v_2, v_3, v_4\}$ and $V_2 = \{v_5, v_6, v_7\}$.

THEOREM 12-8

A graph G is a bipartite graph if and only if all its loops have even length.

Proof If G is a bipartite graph, its vertex set V can be partitioned into the subsets V_1 and V_2 such that if $\{v_i, v_j\}$ is an edge, then $v_i \in V_1$ and $v_j \in V_2$. Let $v_{i_0}v_{i_1} \ldots v_{i_{l-1}}v_{i_0}$ be any loop of length l in G. With no loss in generality, we can assume that $v_{i_0} \in V_1$. Hence, $v_{i_0}, v_{i_2}, v_{i_4}, \ldots, \in V_1$ and $v_{i_1}, v_{i_3}, v_{i_5}, \ldots \in V_2$. Thus, $l - 1$ must be odd and l must be even.

Conversely, assume that the length of each loop in G is even and that G

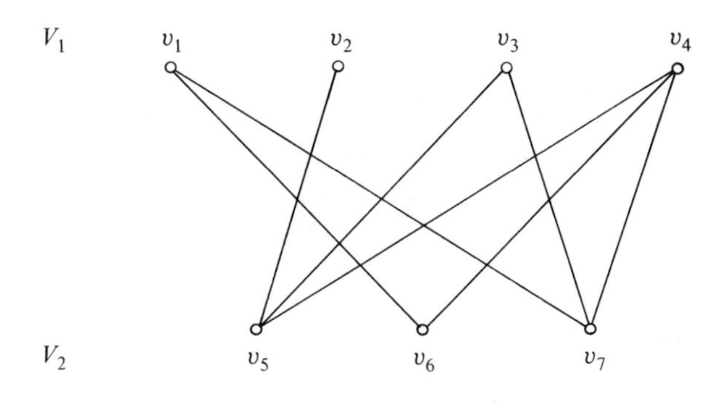

Fig. 12-16. A bipartite graph

is connected. Define the subsets of V,

$$V_1 = \{v_i \mid \text{distance between } v_i \text{ and some fixed vertex } v \text{ is even}\}$$
$$V_2 = V - V_1$$

Hypothesize that an edge $\{v_i, v_j\}$ exists such that $v_i, v_j \in V_1$. Then the loop consisting of the geodesic of v and v_i (of even length), the edge $\{v_i, v_j\}$, and the geodesic of v_j and v (of even length) must have an odd length—a contradiction. Hence, no edge in G has the form $\{v_i, v_j\}$ where $v_i, v_j \in V_1$. Next, hypothesize that an edge $\{v_i, v_j\}$ exists such that $v_i, v_j \in V_2$. Then the loop consisting of the geodesic of v and v_i (of odd length), the edge $\{v_i, v_j\}$, and the geodesic of v_j and v (of odd length) must have an odd length—again a contradiction. In conclusion, every edge in G must be of the form $\{v_i, v_j\}$ where $v_i \in V_1$ and $v_j \in V_2$. Thus, G is a bipartite graph with the complementary vertex subsets V_1 and V_2.

If the length of every loop in G is even but G is *not* connected, the preceding argument can be repeated for each component of G, to result in the same conclusion. □

Let G be a bipartite graph with the complementary vertex subsets V_1 and V_2, where $V_1 = \{v_1, v_2, \ldots, v_q\}$. A *matching* V_1 *to* V_2 is a subgraph of G which consists of q edges $\{v_1, v'_1\}, \{v_2, v'_2\}, \ldots, \{v_q, v'_q\}$, where $v'_1, v'_2, \ldots, v'_q$ are any q distinct element in V_2. Figure 12-17 shows a bipartite graph with the complementary vertex subsets V_1 and V_2, and a possible matching V_1 to V_2 (indicated in heavy lines). Clearly, not all bipartite graphs have a matching: a necessary condition is that $\#V_2$ be at least as large as $\#V_1$. This condition, however, is not sufficient, as demonstrated by the bipartite graph of Fig. 12-18, which has no matching.

The problem of finding a matching in a bipartite graph is sometimes known as the *marriage problem*: Given a group of young men and a group of

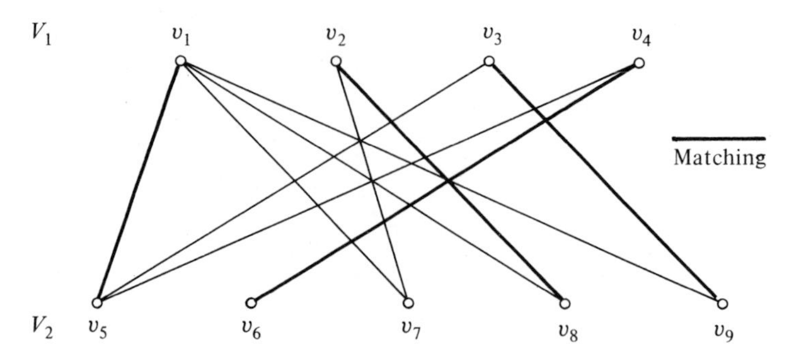

Fig. 12-17 A bipartite graph and its matching

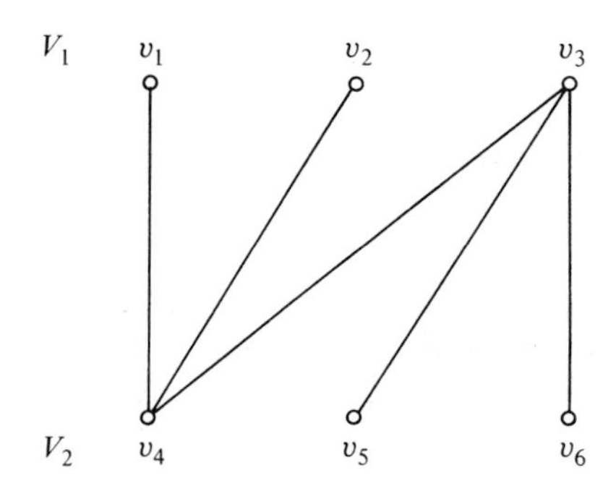

Fig. 12-18 A bipartite graph with no matching

young women, some of whom know each other already, is it possible to pair the two groups in such a way that the man and woman in each pair are already acquainted? The following theorem states the necessary and sufficient condition under which a solution to this problem exists.

THEOREM 12-9

Let G be a bipartite graph with the complementary vertex subsets V_1 and V_2. Then G has a matching V_1 to V_2 if and only if every k vertices in V_1 ($k = 1, 2, \ldots, \#V_1$) are connected to at least k vertices in V_2. (This condition is referred to as the *diversity condition*.)

Proof Let $\#V_1 = q$. If G has a matching V_1 to V_2, then the q vertices of V_1 must be connected to q distinct vertices in V_2, and the diversity condition must, obviously, be satisfied.

Conversely, assume that G satisfies the diversity condition. We shall prove (by induction on q) that a matching V_1 to V_2 can be constructed. (*Basis*). When $q = 1$ or $q = 2$, and the diversity condition is satisfied, the existence of a matching V_1 to V_2 is trivially apparent. (*Induction step*). Assuming that G satisfies the diversity condition, hypothesize that, for $i = 1, 2, \ldots, q - 1$, a matching U_1 (where $\#U_1 = i$) to U_2 exists whenever the diversity condition is satisfied for any bipartite graph whose complementary vertex subsets are U_1 and U_2. If every k vertices in V_1 ($k = 1, 2, \ldots, q$) are connected to *more* than k vertices in V_2, a matching V_1 to V_2 can be constructed as follows: Assign any edge $\{v_i, v_j\}$ (where $v_i \in V_1$ and $v_j \in V_2$) to the matching; clearly, in the bipartite graph whose complementary vertex subsets are $V_1 - \{v_i\}$ and $V_2 - \{v_j\}$ the diversity condition is still satisfied and, by induction hypothesis, a matching can be constructed; this matching, together with $\{v_i, v_j\}$, is the sought matching V_1 to V_2. With the hypothesis still in effect, let us assume now that, for some $k = k_0 \leq q - 1$, there exists a set $U_1 \subset V_1$ of k_0 vertices that are connected to *exactly* k_0 vertices in $U_2 \subset V_2$. In this case, a matching V_1 to V_2 can be constructed as follows: First, by induction hypothesis, a matching U_1 to U_2 can be constructed. Now, suppose the bipartite graph whose complementary vertex subsets are $U'_1 = V_1 - U_1$ (where $\#U'_1 = q - k_0$) and $U'_2 = V_2 - U_2$ does not satisfy the diversity condition. Then, for some $k \leq q - k_0$, there exists a set $U''_1 \subset U'_1$ of k vertices that are connected to $k' < k$ vertices in V_2. Then the set $U_1 \subset U''_1$ is a subset of $k_0 + k$ vertices in V_1 which are connected to $k_0 + k' < k_0 + k$ vertices in V_2, which contradicts the diversity condition for G. Thus, the bipartite graph whose complementary vertex subsets are U'_1 and U'_2 must satisfy the diversity condition and, hence, a matching U'_1 to U'_2 can be constructed. Since $U_1 \cup U'_1 = V_1$ and $U_2 \cup U'_2 = V_2$, this completes the construction of a matching V_1 to V_2. □

Note that, inherent in the preceding proof, is a recursive procedure for constructing a matching for any bipartite graph that satisfies the diversity condition.

The following theorem provides a sufficient (but not necessary) condition for the existence of a matching for a bipartite graph. This condition can be very easily tested in any given graph, and should be considered before the more complex diversity condition is examined.

THEOREM 12-10

Let G be a bipartite graph with the complementary vertex subsets V_1 and V_2. Then G has a matching V_1 to V_2 if, for some $t > 0$:

(a) With each vertex in V_1 there are *at least* t incident edges.
(b) With each vertex in V_2 there are *at most* t incident edges.

Proof If (a) holds, then the total number of edges incident with any set of k vertices in V_1 ($k = 1, 2, \ldots, \#V_1$) is at least kt. By (b), these edges must be incident with at least k vertices in V_2. Thus, every k vertices in V_1 ($k = 1, 2, \ldots, \#V_1$) are connected to at least k vertices in V_2. By Theorem 12-9, then, G has a matching V_1 to V_2. □

For example, the bipartite graph of Fig. 12-19 satisfies conditions (a) and (b) of Theorem 12-10 with $t = 3$ and, hence, has a matching.

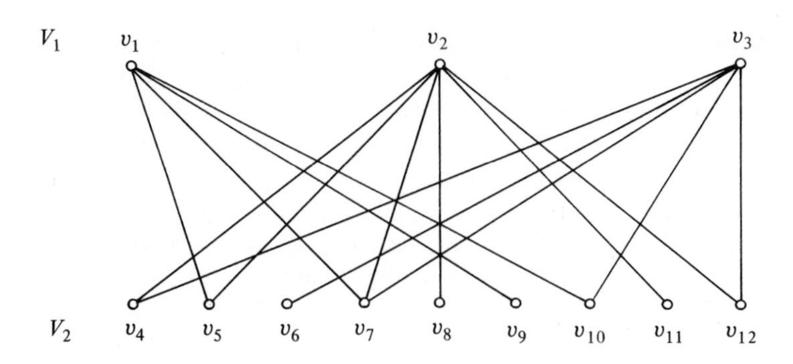

Fig. 12-19 Illustrating Theorem 12-10, with $t = 3$

Theorem 12-10 is closely related to the so-called *committee chairmanship problem*: Given a set of q committees $C = \{c_1, c_2, \ldots, c_q\}$, with P_i $(i = 1, 2, \ldots, q)$ being the membership of committee c_i, select q chairmen for the q committees such that the chairman of c_i is from P_i $(i = 1, 2, \ldots, q)$ and such that no person is a chairman of more than one committee. This situation can be represented by a bipartite graph with the complementary vertex subsets C and $P = P_1 \cup P_2 \cup \ldots \cup P_q$, where an edge $\{c_i, p\}$ is included if and only if $p \in P_i$. In this representation, the problem reduces to that of finding whether or not the graph has a matching C to P and, hence, whether or not it satisfies the diversity condition. In view of Theorem 12-10, the committee chairmanship problem is solvable if, for some $t > 0$, each committee has at least t members and every person is a member of at most t committees.

PROBLEMS

1. Is the graph of Fig. 12-K a bipartite graph? If so, find its complementary vertex subsets.

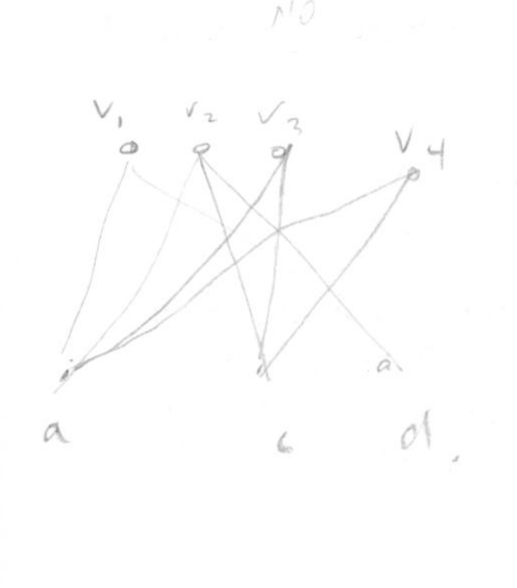

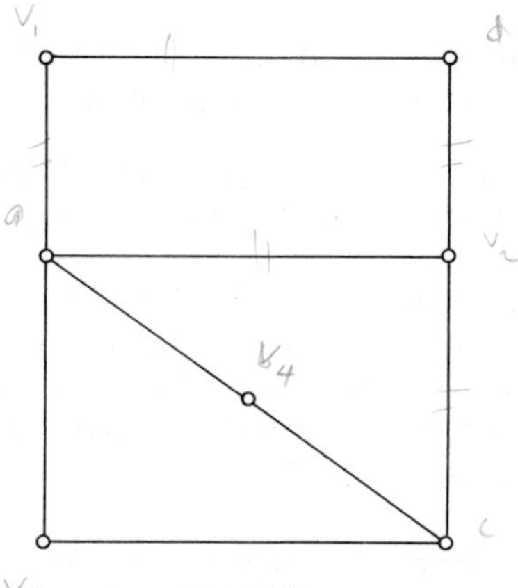

Fig. 12-K

2. The following facts are known about six inmates (named $p_1, p_2, \ldots, p_6$) in a Foreign Legion stockade:
p_1 speaks Chinese, French, and Japanese.
p_2 speaks German, Japanese, and Russian.
p_3 speaks English and French.
p_4 speaks Chinese and Spanish.
p_5 speaks English and German.
p_6 speaks Russian and Spanish.
Could these six inmates be locked in *two* separate cells such that no two inmates in the same cell be able to understand each other?

3. For the bipartite graph of Fig. 12-L:
(a) Verify that the "*t* condition" (Theorem 12-10) is satisfied.
(b) Verify that the diversity condition (Theorem 12-9) is satisfied.
(c) Construct a matching V_1 to V_2.

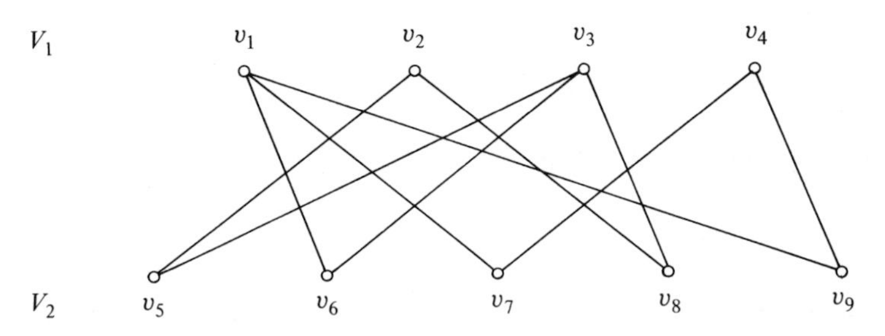

Fig. 12-L

4. Four instructors—Mr. Jones, Ms. McCarthy, Mr. Smith, and Mr. Taylor—are available for teaching four courses: Mathematics 1A, Physics 2B, Electrical Engineering 3C, and Computer Science 4D. Mr. Jones is trained in physics and electrical engineering, Ms. McCarthy in mathematics and computer science, Mr. Smith in mathematics, physics, and electrical engineering, and Mr. Taylor in electrical engineering. How can the four instructors be assigned to the four courses, so that no instructor will be obliged to teach a subject in which he is not trained?

12-5. PLANAR GRAPHS

When drawing a graph on a sheet of paper, we often find it convenient (or even necessary) to permit edges to intersect at points other than vertices. Such points are referred to as *crossovers*, and the intersecting edges will be said to *cross over* (each other). For example, the graph of Fig. 12-20(a) exhibits three crossovers: $\{v_1, v_4\}$ crosses over $\{v_2, v_3\}$, and $\{v_1, v_5\}$ crosses over both $\{v_2, v_3\}$ and $\{v_3, v_4\}$. A graph G is said to be *planar* if it can be drawn on a plane without any crossovers; otherwise, G is said to be *nonplanar*. The graph of Fig. 12-20(a) is planar, since it can be redrawn—as done in Fig. 12-20(b)—without any crossovers.

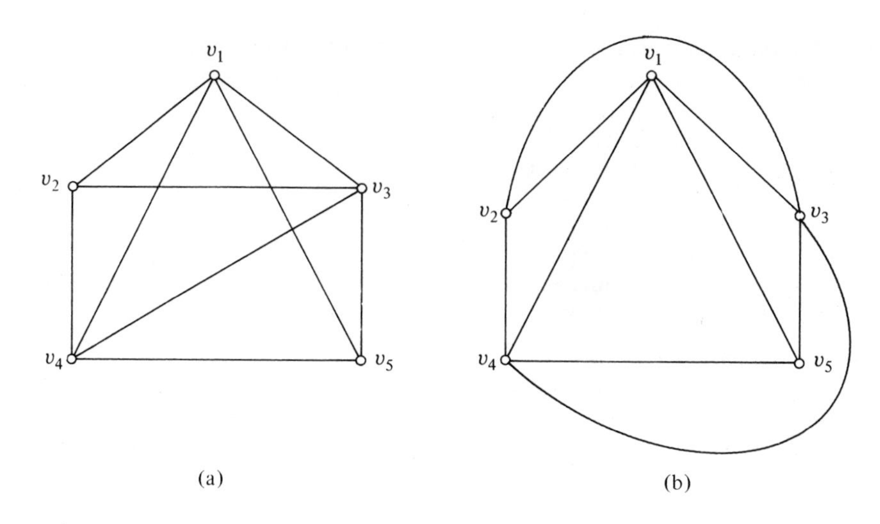

Fig. 12-20 A planar graph

Let G be a graph drawn on a plane, and let

$$\sigma = v_1 \ldots v_2 \ldots v_3 \ldots v_4 \ldots v_1$$

be any cycle in G. Also, let $\alpha = v_1 \ldots v_3$ and $\alpha' = v_2 \ldots v_4$ be any proper paths in G which do not cross over (see Fig. 12-21 that indicates various possible configurations). It is clear that G is a nonplanar graph (in which α and

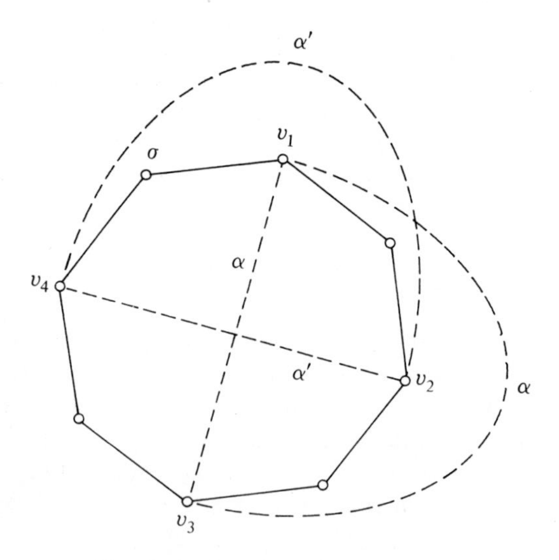

Fig. 12-21 A cycle and crossovers

α' cross over) if and only if α and α' are either *both inside* or *both outside* σ. This simple fact is often helpful in verifying the nonplanarity of given graphs by inspection.

Example 12-6

An electric circuit consists of two sets of three junctions, with wires connecting each junction in one set with every junction in the other set (see Fig. 12-22). Is it possible to lay out the circuit so that the wires do not cross over? (The avoidance of crossovers is essential if the circuit is to be manufactured in the form of a "printed circuit," where the wires are actually exposed metallic conduits embedded on a plasic board; a crossover in this case would cause a short-circuit.) Clearly, this question is equivalent to that of determining whether the graph of Fig. 12-22 is planar.

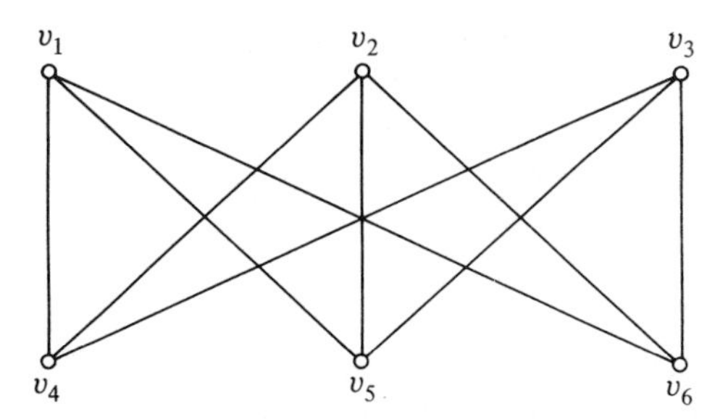

Fig. 12-22 For Example 12-6

Now, note the cycle

$$\sigma = v_1 v_6 v_3 v_5 v_2 v_4 v_1$$

and the edges $\{v_1, v_5\}$, $\{v_6, v_2\}$, and $\{v_3, v_4\}$. Evidently, each one of these edges is either entirely inside or entirely outside σ. Thus, at least two of the three edges are on the same side of σ. The graph, therefore, is nonplanar (and the circuit, therefore, cannot be laid out without crossovers). □

Let G be a graph with the vertex set V and the edge set E. An *elementary contraction* of G is formed by deleting an edge $\{v_i, v_j\}$ from E, replacing every occurrence of v_i and v_j in E with a new symbol w, deleting v_i and v_j from V, and finally adding w to V. Diagrammatically, an elementary contraction of G is effected by "merging" two adjacent vertices v_i and v_j in G (after eliminating the edge connecting them), and calling the composite vertex w. The graph G is said to be *contractible* to a graph $\bar{G}$ if $\bar{G}$ can be formed from G by a sequence of elementary contractions. For example, the graph G_1 of Fig. 12-23 is contractible to the graph G_2 of the same figure [by merging the vertices v_i and v_i' ($i = 1, 2, 3, 4, 5$) and labeling the composite vertex w_i].

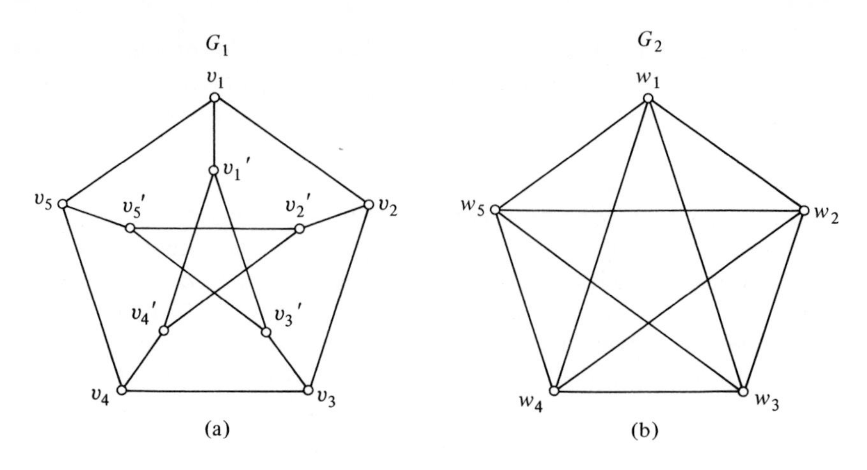

Fig. 12-23 Contracting G_1 to G_2

Using the contraction operation defined above, the Polish mathematician Kuratowski formulated (in 1930) a strong condition for the planarity of graphs. It is stated here without proof:

THEOREM 12-11

A graph is planar if and only if none of its subgraphs is contractible to the graph of Fig. 12-24(a) or to the graph of Fig. 12-24(b).

For example, the graph G_1 of Fig. 12-23 is contractible to the graph of Fig. 12-24(b) and, hence, is nonplanar.

Note that the graph of Fig. 12-24(a) is nonplanar since it is isomorphic to the graph of Fig. 12-22 which has already been shown to be nonplanar. That

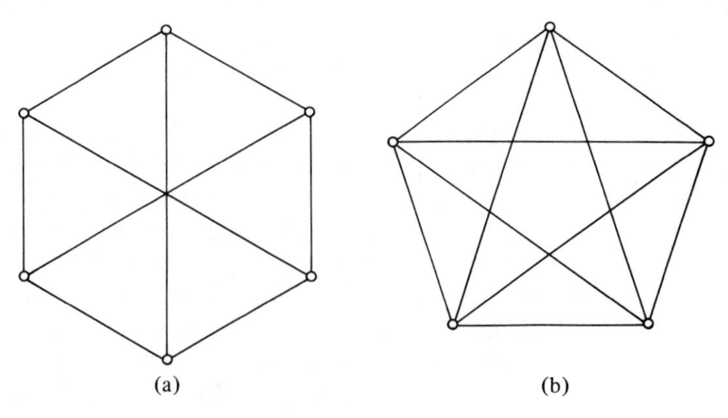

Fig. 12-24 For Theorem 12-11

the graph of Fig. 12-24(b) is also nonplanar can be demonstrated in a manner analogous to that employed in Example 12-6.

A graph that constitutes a single cycle is called a *polygon*. A *polygonal graph* is a planar graph defined recursively as follows:

(*Basis*). A polygon is a polygonal graph.

(*Induction step*). Let $G = (V, E)$ be a polygonal graph. Also, let

$$\alpha = v_i v_{i_1} v_{i_2} \ldots v_{i_{l-1}} v_j$$

be any proper path (of length $l \geq 1$) which does not cross over G and where $v_i, v_j \in V$ but $v_{i_\nu} \notin V$ ($\nu = 1, 2, \ldots, l-1$). Then the graph composed of G and α—that is, the graph $(\bar{V}, \bar{E})$, where

$$\bar{V} = V \cup \{v_{i_1}, v_{i_2}, \ldots, v_{i_{l-1}}\}$$
$$\bar{E} = E \cup \{\{v_i, v_{i_1}\}, \{v_{i_1}, v_{i_2}\}, \ldots, \{v_{i_{l-1}}, v_j\}\}$$

is also a polygonal graph.

Thus, a polygonal graph is a planar graph (or multigraph, since cycles of length 2 are permitted) which partitions the plane into regions, each of which is bounded by a polygon. Figure 12-25 is an example of a polygonal graph. (Another example is the map of the continental U.S.A., with the state boundaries shown.)

Each of the regions defined by a polygonal graph G (which, in Fig. 12-25, are denoted by F_1, F_2, F_3, etc.) is called a *face* of G. The peripheral polygon that contains all the faces of G ($v_1v_2v_3v_4v_5v_6v_7v_1$ in Fig. 12-25) is called the *maximal cycle* of G. It is convenient to regard the infinite region outside the maximal cycle (F_9 in Fig. 12-25) as another face—usually referred to as the

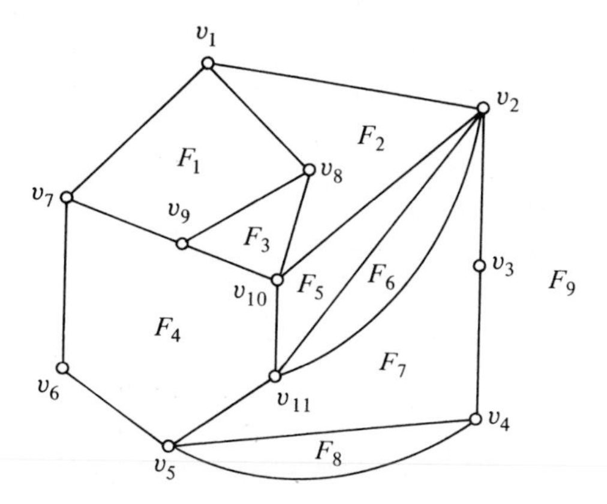

Fig. 12-25 A polygonal graph

infinite face of G. (In fact, if G is drawn on a sphere, there is really no distinction between the infinite face and any other face of G). Two faces are said to be *adjacent* if they share a common edge (such as F_1 and F_2 in Fig. 12-25).

THEOREM 12-12

Let G be an (n, m) polygonal graph with k faces (including the infinite face). Then

$$n - m + k = 2$$

Proof (by induction on k) (*Basis*). The least number of faces (including the infinite face) is $k = 2$. In this case, G is a polygon and, hence, $n = m$. Thus, $n - m + k = k = 2$. (*Induction step*). Hypothesize that the theorem holds for graphs with $k - 1$ faces. By the recursive definition of a polygonal graph, G can be constructed by first constructing an $(\bar{n}, \bar{m})$ graph $\bar{G}$ with $\bar{k} = k - 1$ faces, and then adding a proper path of some length $l \geq 1$ which shares exactly two vertices with $\bar{G}$. Thus,

$$\begin{aligned} n - m + k &= (\bar{n} + l - 1) - (\bar{m} + l) + (\bar{k} + 1) \\ &= \bar{n} - \bar{m} + \bar{k} \end{aligned}$$

But, by induction hypothesis, $\bar{n} - \bar{m} + \bar{k} = 2$. Hence, $n - m + k = 2$. □

Given a polygonal graph G with the faces $F_1, F_2, \ldots, F_k$ (including the infinite face), the *dual* $\tilde{G}$ of G is a graph obtained from G as follows: For any face F_i in G, assign a vertex f_i to $\tilde{G}$; then, for every edge that is common to F_i and F_j in G, assign an edge $\{f_i, f_j\}$ to $\tilde{G}$. In practice, this can be most conveniently accomplished by drawing each vertex f_i inside the face F_i, and letting every edge common to F_i and F_j be crossed over by an edge connecting f_i and f_j. As an illustration of this method, Fig. 12-26 shows a polygonal graph (solid lines) and its dual (dotted lines). From this construction method it is evident that the dual of every polygonal graph must also be a polygonal graph. Moreover, if $\tilde{G}$ is the dual of G, then G is the dual of $\tilde{G}$.

A graph G whose dual $\tilde{G}$ is isomorphic to G is said to be *self-dual*. Figure 12-27 shows an example of a self-dual graph.

A famous conjecture associated with polygonal graphs is the *four-color conjecture*, which states: *Every polygonal graph can be colored with four distinct colors so that no two adjacent faces (including the infinite face) have the same color.* This conjecture has been fascinating mathematicians ever since it was first introduced more than a century ago—but it has never been proved. In the course of hunting for the illusive proof, however, many important results have been discovered in graph theory and related areas.

That *at least* four colors are needed to color arbitrary polygonal graphs

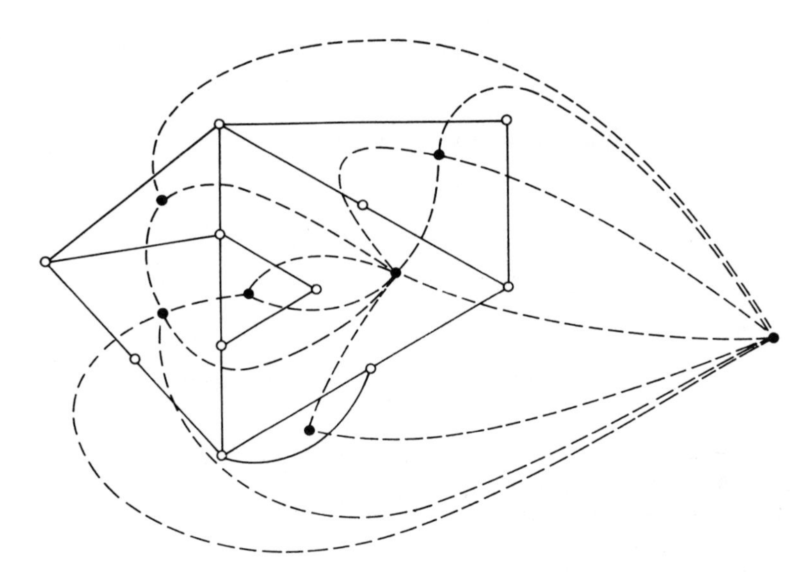

Fig. 12-26 A polygonal graph and its dual

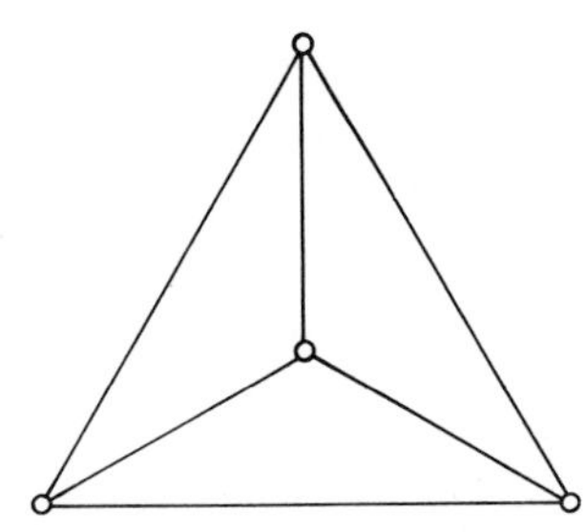

Fig. 12-27 A self-dual graph

(with no adjacent faces identically colored) is obvious from the graph of Fig. 12-27. It has also been shown that coloring in the required manner is always possible with *five* colors. From various partial results that have been derived throughout the years, there is strong circumstantial evidence that the conjecture is correct, or that, if any counterexample exists, it must be extremely complex (with no fewer than 40 faces!).

PROBLEMS

1. Show that the graph of Fig. 12-24(b) is nonplanar.

2. Draw all 6-vertex nonplanar graphs such that no two graphs are isomorphic to each other.

3. Use Theorem 12-11 to show that the graph of Fig. 12-M is nonplanar.

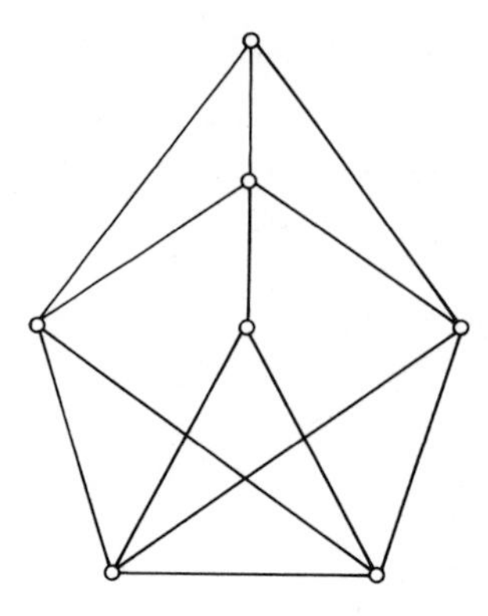

Fig. 12-M

4. Verify the formula $n - m + k = 2$ of Theorem 12-12 for:
   (a) The polygonal graph of Fig. 12-25.
   (b) A graph with $(r + 1)^2$ vertices which describes a grid of r^2 squares (such as a chessboard).

5. Construct the dual of the graph shown in Fig. 12-N.

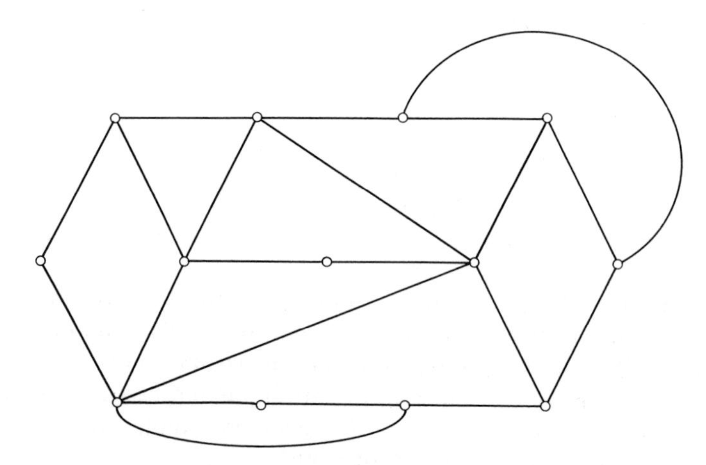

Fig. 12-N

6. Show that if an (n, m) graph is self-dual, then $m = 2(n - 1)$.

7. Using four colors, color the graph of Fig. 12-25 such that no two adjacent faces (including the infinite face) have the same color.

8. Using three colors, color the graph of Fig. 12-O in the manner prescribed in Problem 7.

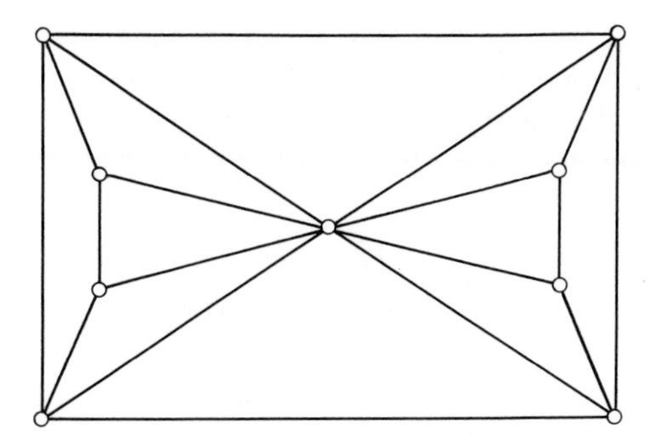

Fig. 12-O

12-6. DIRECTED GRAPHS

As defined in Sec. 12-1, a *directed graph* G is a graph, multigraph, or pseudograph in which every edge is associated with an arrow that points from one vertex to another (or, possibly, to itself). Thus, each edge that points from vertex v_i to vertex v_j in G is conveniently specified as an *ordered* pair of vertices (v_i, v_j) (instead of the unordered pair $\{v_i, v_j\}$). Figure 12-28 shows a directed graph $G = (V, E)$, where

$$V = \{v_1, v_2, v_3, v_4, v_5\}$$
$$E = \{(v_1, v_2), (v_2, v_1), (v_2, v_3), (v_2, v_4), (v_3, v_1), (v_5, v_2), (v_5, v_3), (v_5, v_5)\}$$

To avoid confusion, ordinary graphs henceforth are called *nondirected graphs*.

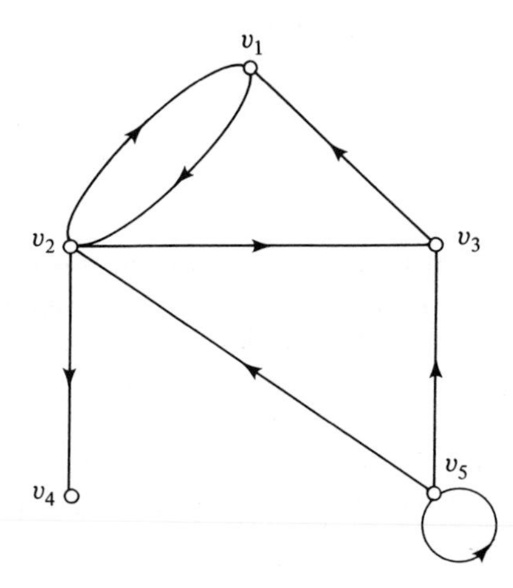

Fig. 12-28 A directed graph

From Chapter 2 we know that a directed graph G with the vertex set $V = \{v_1, v_2, \ldots, v_n\}$ and the edge set E is the graphical representation of a relation E on V. In that chapter we saw that such a graph can be represented nondiagrammatically by the matrix $\mathbf{A} = [a_{ij}]_{n \times n}$, where

$$a_{ij} = \begin{cases} 1 & \text{if } (v_i, v_j) \in E \\ 0 & \text{otherwise} \end{cases}$$

This matrix is referred to here as the *adjacency matrix* of G. (It is analogous to the adjacency matrix introduced in Sec. 12-1 for nondirected graphs, except that, in the present case, $\mathbf{A}$ is no longer necessarily symmetrical about the principal diagonal and may have nonzero entries along this diagonal.) For example, the adjacency matrix of the directed graph of Fig. 12-28 is given by

$$\mathbf{A} = \begin{bmatrix} 0 & 1 & 0 & 0 & 0 \\ 1 & 0 & 1 & 1 & 0 \\ 1 & 0 & 0 & 0 & 0 \\ 0 & 0 & 0 & 0 & 0 \\ 0 & 1 & 1 & 0 & 1 \end{bmatrix}$$

Let G and $\bar{G}$ be two directed graphs with the vertex sets V and $\bar{V}$, respectively. $\bar{G}$ is said to be *isomorphic* to G if there exists a bijection $h\colon V \longrightarrow \bar{V}$ such that (v_i, v_j) is an edge of G if and only if $(h(v_i), h(v_j))$ is an edge of $\bar{G}$. As in the case of nondirected graphs, isomorphic directed graphs are identical except, possibly, for vertex labeling.

The terms *subgraph, proper subgraph,* and *spanning subgraph* have the same connotation for directed graphs as for nondirected graphs.

A sequence of edges $(v_{i_0}, v_{i_1}), (v_{i_1}, v_{i_2}), \ldots, (v_{i_{l-1}}, v_{i_l})$ in a directed graph is called a *directed path* of length l which connects v_{i_0} to v_{i_l}. It is denoted by

$$(v_{i_0}, v_{i_1})(v_{i_1}, v_{i_2}) \ldots (v_{i_{l-1}}, v_{i_l})$$

or, more concisely, by $v_{i_0}v_{i_1} \ldots v_{i_l}$. If in the definitions of the terms *open path, loop, proper path, cycle, connected vertices, connected graph, component, geodesic,* and *distance* (see Secs. 12-1 and 12-2) we replace the term *path* with *directed path,* we obtain the definitions of these terms appropriate to directed graphs. Thus, for example, vertices v_i and v_j in a directed graph are said to be *connected* if there exists a *directed* path that connects v_i to v_j, and so forth.

Theorems 12-2 and 12-3, by trivial modifications in the proofs, can be extended to the case of directed graphs:

THEOREM 12-13

Let G be a directed graph with the vertex set $V = \{v_1, v_2, \ldots, v_n\}$. Then, for any connected vertices $v_i, v_j \in V$ $(v_i \neq v_j)$, the geodesic is a proper path whose length is at most $n - 1$.

THEOREM 12-14

Let G be a directed graph with the vertex set $\{v_1, v_2, \ldots, v_n\}$ and adjacency matrix $\mathbf{A}$. Then the (i, j) entry of $\mathbf{A}^l$ $(l = 1, 2, \ldots)$ is the number of directed paths of length l connecting v_i to v_j.

Given a *nondirected* graph (or multigraph, or pseudograph) G, it is interesting to know whether G can be transformed into a *connected directed* graph $\hat{G}$. That is, is it possible—by attaching an arrow to each edge of G—to obtain a directed graph $\hat{G}$ where every vertex is connected to every other vertex by a directed path? First, it is clear that $\hat{G}$ cannot be connected unless G is connected. Also, G cannot contain any separating edge (see Sec. 12-1); for if G contains a separating edge $e = \{v_i, v_j\}$, then, with the arrow attached to e, either v_i is connected to v_j, or v_j is connected to v_i—but not both. Thus, for $\hat{G}$ to be connected, G must be connected and must not contain any separating edge. That these conditions are also sufficient for the construction of $\hat{G}$, follows from this algorithm:

ALGORITHM 12-3

G is a connected nondirected graph (or multigraph, or pseudograph) with no separating edges and with the vertex set $V = \{v_1, v_2, \ldots, v_n\}$. To transform G into a connected directed graph $\hat{G}$:

(i) Let $V_1 = \{v_1\}$. Set i to 1.

(ii) Let G_i denote the graph G together with all the arrows attached in preceding operations (none when $i = 1$). In G_i, every pair of vertices belonging to V_i are connected by some directed path (trivially true when $i = 1$). If $V_i = V$, attach arbitrary arrows to all those edges in G_i to which no arrows have been previously attached. The resulting directed graph is the desired graph $\hat{G}$. If $V_i \neq V$:

(iii) (See Fig. 12-29.) Let $\{v_{i_0}, v_{i_1}\}$ be any edge in G_i such that $v_{i_0} \in V_i$ and $v_{i_1} \in V - V_i$ (there must be at least one such edge, since G is connected). Find a cycle or a proper path $\alpha_i = v_{i_0}v_{i_1} \ldots v_{i_{l-1}}v_{i_l}$, where $v_{i_l} \in V_i$ and $\{v_{i_0}, v_{i_1}, \ldots, v_{i_{l-1}}\} = \tilde{V}_i \subset V - V_i$ (such a path must exist, since $\{v_{i_0}, v_{i_1}\}$ is not a separating edge). Attach arrows to the edges $\{v_{i_0}, v_{i_1}\}$, $\{v_{i_1}, v_{i_2}\}, \ldots, \{v_{i_{l-1}}, v_{i_l}\}$ so as to make α_i a directed path $v_{i_0}v_{i_1} \ldots v_{i_l}$. Let G_{i+1} denote the graph G_i together with the l arrows attached to α_i, and denote $V_{i+1} = V_i \cup \tilde{V}_i$. Now, in G_i, every pair of vertices belonging to V_i is con-

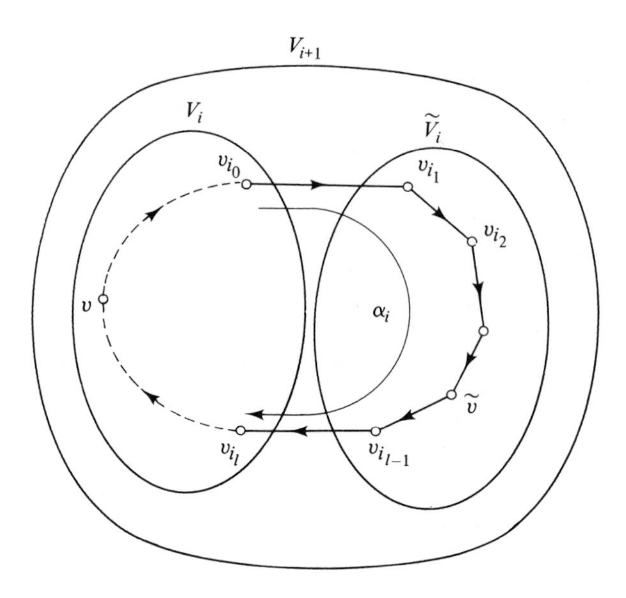

Fig. 12-29 For Algorithm 12-3

nected by some directed path. Hence, for every $v \in V_i$ and $\tilde{v} \in \tilde{V}_i$ in G_{i+1}, there are directed paths $v \ldots v_{i_0} \ldots \tilde{v}$ and $\tilde{v} \ldots v_{i_l} \ldots v$. Thus, every pair of vertices belonging to V_{i+1} is connected by some directed path.

(iv) Increment i by 1 and return to step (ii). □

Example 12-7

Figure 12-30(a) represents the street map of the downtown section of a certain town. Can every street on this map be made a one-way street in such a way that a driver can drive legally from any point on the map to any other point? Since the map constitutes a connected nondirected graph with no separating edges, the answer is in the affirmative. The determination of the street directions can be made via Algorithm 12-3:

$$V_1 = \{v_1\} \qquad \alpha_1 = v_1v_2v_5v_4v_1$$

$$V_2 = \{v_1, v_2, v_4, v_5\} \qquad \alpha_2 = v_2v_3v_6v_9v_8v_{10}v_7v_4$$

$$V_3 = \{v_1, v_2, \ldots, v_{10}\}$$

The map, complete with the one-way indications, is shown in Fig. 12-30(b). □

A *complete directed graph* is a complete nondirected graph (see Sec. 12-1) with an arrow attached to each edge. Such a graph is sometimes called a *tournament* because it describes the progress of a so-called "round-robin tournament," in which there are n players who compete in pairs—each player

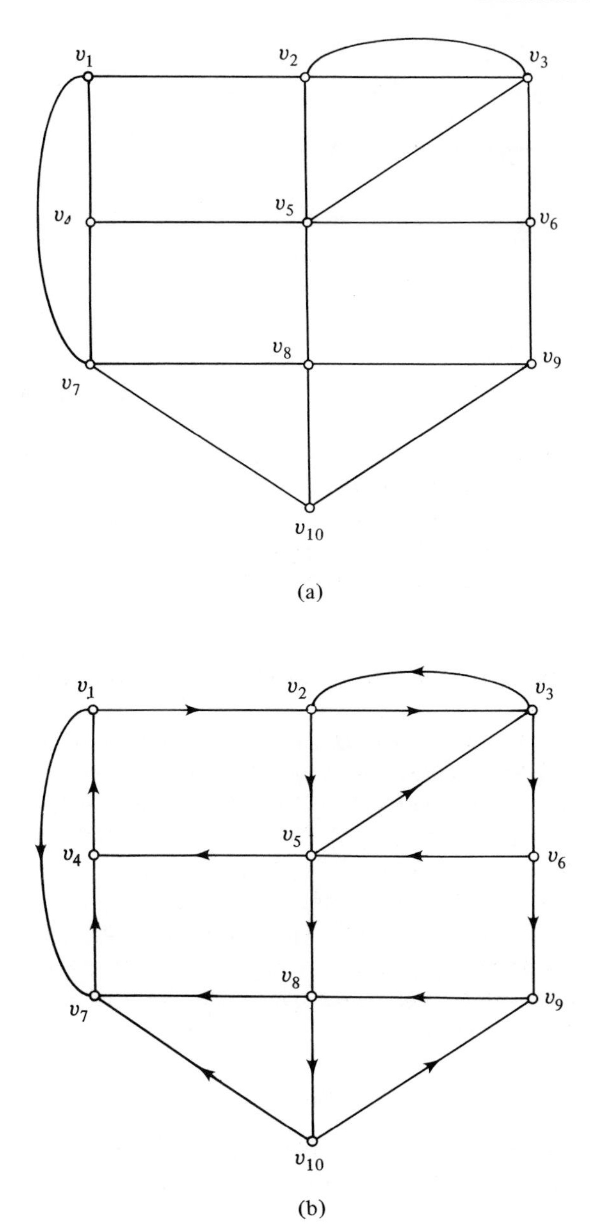

Fig. 12-30 For Example 12-7

competing in turn with each of the other players. In the graph, the n players are represented by n vertices, and a match between players v_i and v_j, where v_i is the winner (no draws are allowed), is represented by an edge (v_i, v_j). Figure 12-31 is an example of a 5-vertex tournament. In this tournament, v_1

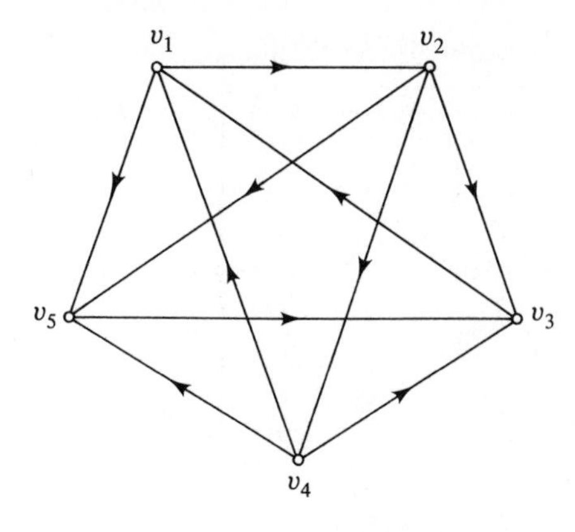

Fig. 12-31 A tournament

won against v_2 and v_5, but lost against v_3 and v_4; v_2 won against v_3, v_4, and v_5, but lost against v_1; and so forth.

In what follows, a *proper spanning path* in a directed graph G refers to a directed path that encounters every vertex of G exactly once. For example, the path $v_4v_5v_3v_1v_2$ in the graph of Fig. 12-31 is a proper spanning path.

THEOREM 12-15

Every tournament $G = (V, E)$ has a proper spanning path.

Proof (by induction on $\#V$) (*Basis*). The theorem is trivially true when $\#V = 1$ and $\#V = 2$. (*Induction step*). Hypothesize that the theorem is true for all k-vertex tournaments. Let G be a $(k + 1)$-vertex tournament with the vertex set $V = \{v_1, v_2, \ldots, v_{k+1}\}$. Let G_k denote the subgraph of G which has the vertex set $V - \{v_{k+1}\}$ and which is a k-vertex tournament. By induction hypothesis, G_k has a proper spanning path, say, $v_{i_1}v_{i_2} \ldots v_{i_k}$ $(v_{i_1}, v_{i_2}, \ldots, v_{i_k} \in V - \{v_{k+1}\})$. Now, G contains either the edge (v_{k+1}, v_{i_1}) or the edge (v_{i_1}, v_{k+1}). In the former case, $v_{k+1}v_{i_1}v_{i_2} \ldots v_{i_k}$ is a proper spanning path in G. In the latter case (see Fig. 12-32), let ν be the least integer such that (v_{k+1}, v_{i_ν}) is an edge in G. Then

$$v_{i_1}v_{i_2} \ldots v_{i_{\nu-1}}v_{k+1}v_{i_\nu}v_{i_{\nu+1}} \ldots v_{i_k}$$

is a proper spanning path in G. If no such ν exists, then the proper spanning path is $v_{i_1}v_{i_2} \ldots v_{i_k}v_{k+1}$. Under all circumstances, then, G has a proper spanning path. □

Example 12-8

Figure 12-33 illustrates how a proper spanning path for the tournament G of Fig. 12-31 can be determined, following the algorithm suggested with the proof of Theorem 12-15. (In Fig. 12-33, the heavy lines indicate a proper

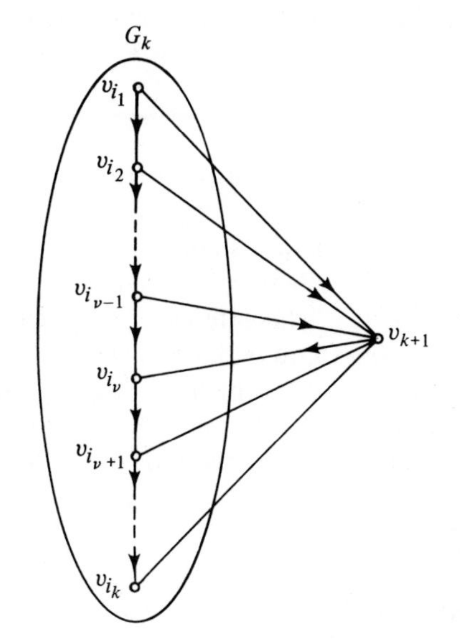

Fig. 12-32 For Theorem 12-15

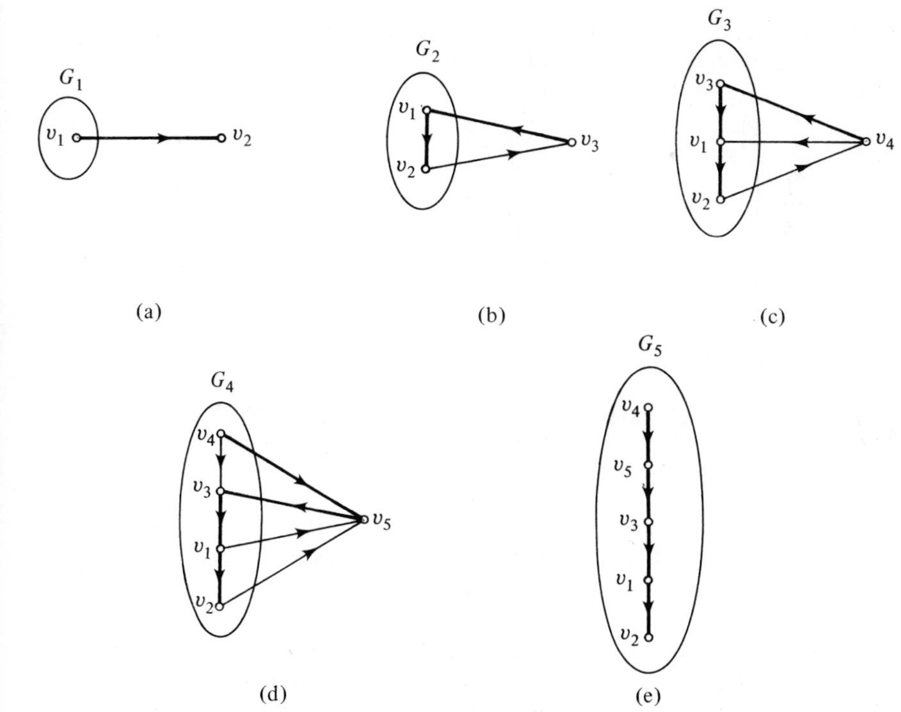

Fig. 12-33 For Example 12-8

spanning path for G_k—that is, for the subgraph of G which consists of the vertices $v_1, v_2, \ldots, v_k$ and all their interconnecting edges.) □

Theorem 12-15 implies that, after every round-robin tournament, it is possible to list the n players in some order, say, $v_{i_1}, v_{i_2}, \ldots, v_{i_n}$, where, for $\nu = 1, 2, \ldots, n-1$, v_{i_ν} emerged as a victor in his contest against $v_{i_{\nu+1}}$. We may infer from this ordering that v_{i_1} is the "best" player (since v_{i_1} can beat v_{i_2}, who can beat $v_{i_3}, \ldots$, who can beat v_{i_n}). This inference, however, is generally false, since the ordering has nothing to do with the *number* of victories credited to each player. For example, in the tournament of Fig. 12-31, the ordering is v_4, v_5, v_3, v_1, v_2 (see Example 12-8); however, v_4 cannot be considered the most successful player (noting that v_2, with his three victories, is just as successful).

PROBLEMS

1. In the street map shown in Fig. 12-P, make all the streets one-way so as to leave every junction connected to every other junction.

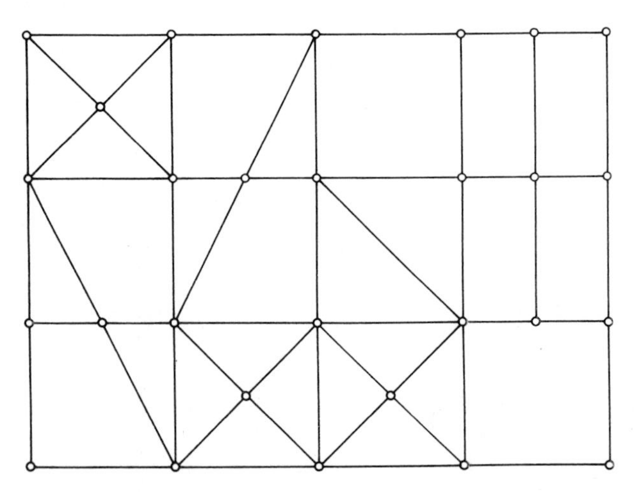

Fig. 12-P

2. Given a street map that constitutes an arbitrary nondirected connected graph G (with or without separating edges), propose an algorithm for converting as many streets as possible into one-way streets, so as to leave every junction connected to every other junction.

3. Find a proper spanning path in the tournament shown in Fig. 12-Q.

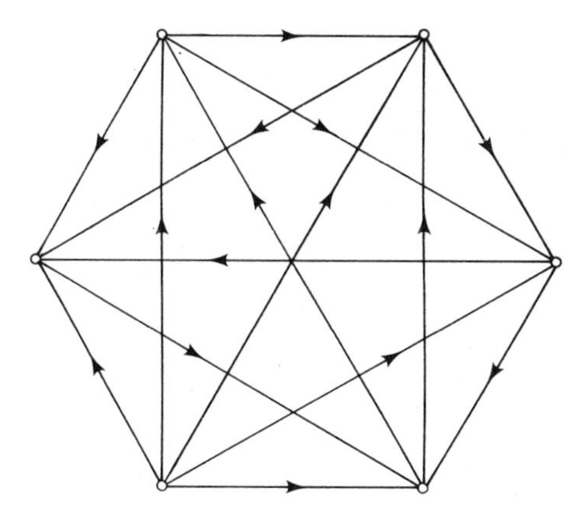

Fig. 12-Q

4. The following facts are known about five students named Bill, John, Kathy, Pat, and Sam. Bill is strong in one subject in which John and Sam are weak; John is strong in one subject in which Kathy, Pat, and Sam are weak; Kathy is strong in one subject in which Bill is weak; Pat is strong in one subject in which Bill, Kathy, and Sam are weak; Sam is strong in one subject in which Kathy is weak. Can you arrange the five students in a circle in such a way that each student can tutor the one to the left in a subject in which the latter is weaker?

12-7. PUZZLES AND GAMES

In many cases, a puzzle or a solitary game (that is, a game without adversary) can be described by means of an *initial configuration*, one or more *winning configurations*, and a set of *permissible moves*. The player's goal is to proceed from the initial configuration to any winning configuration via a finite sequence of *intermediate configurations* created by successively executing permissible moves. The number of distinct configurations and permissible moves is assumed to be finite.

It is often convenient to depict such a puzzle or game by a directed graph G, where the vertices represent configurations (the *initial vertex* representing the initial configuration, and the *winning vertices* representing the winning configurations) and where an edge (v_i, v_j) is included if and only if configuration v_j can succeed configuration v_i by executing some permissible move. In graphical terms, solving the puzzle (or winning the game) is equivalent to finding a directed path in G which starts in the initial vertex and terminates in any winning vertex. Such a path will be referred to as a *solution path*. Clearly, if a solution path at all exists, there must exist one which is proper—that is, which contains no cycles. In particular, any *shortest* solution path (which is usually the one of greatest interest to the player) must be proper.

Example 12-9

A man (denoted by m) wishes to transfer his three pets—a dog (denoted by d), a cat (denoted by c), and a rabbit (denoted by r)—from the left bank of a river to the right bank. He wishes to do so by swimming back and forth between the banks, carrying with him one pet at a time. However, because of the less-than-harmonious relationship among the pets, he cannot leave the dog alone with the cat, nor the cat alone with the rabbit. How can he accomplish the transfer under these constraints?

This puzzle is depicted by the graph of Fig. 12-34, where the vertices represent the various "populations" permitted on the two banks after each crossing [in the form (L, R), where L is the "population" on the left bank and R is the "population" on the right bank]. The initial (leftmost) vertex represents the configuration where the man and his three pets (the set $\{m, d, c, r\}$) appear on the left bank, with the right bank "unpopulated" (the set $\varnothing$). This configuration can lead only to the one where the dog and rabbit (the set $\{d, r\}$) remain on the left bank and the man and cat (the set $\{m, c\}$) appear on the right bank. This configuration, in turn, can lead to the one where the man, dog, and rabbit (the set $\{m, d, r\}$) are back on the left bank and the cat (the set $\{c\}$) remains on the right bank. It can also lead to the preceding configuration (everybody is on the left bank), but this alternative can be ignored, since it simply recreates a situation already considered and contributes nothing to the solution.

The remainder of the graph can be completed in the same manner, until the winning vertex, representing the configuration where everybody appears on the right bank, is reached. A solution to the puzzle is provided by any solution path in the graph—that is, by any sequence of moves which corresponds to a directed path connecting the initial vertex to the winning vertex. Using the upper path, for example, we see that the transfer can be accomplished by the following sequence of moves: Carry the cat to the right bank; return to the left bank; carry the rabbit to the right bank; return with the cat to the left bank, carry the dog to the right bank; return to the left bank; carry the cat to the right bank. □

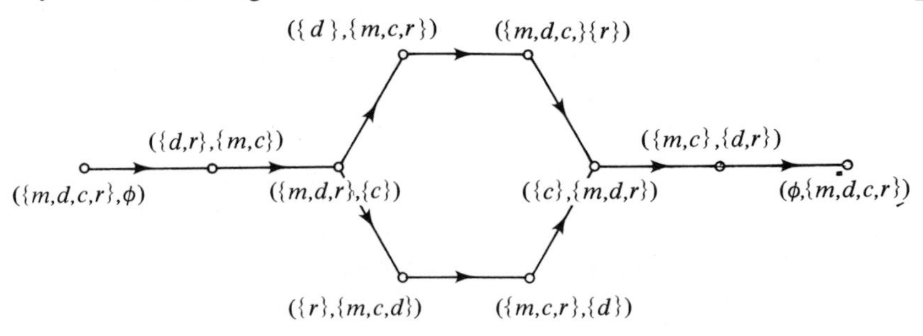

Fig. 12-34 For Example 12-9

Example 12-10

We are given three jugs, a, b, and c, with capacities 8, 5, and 3 quarts, respectively. Jug a is filled with wine, and our objective is to divide its contents into two equal parts by pouring the wine from one jug to another (without resorting to any measuring devices other than the three jugs). Because of this constraint, each "move" consists of either completely filling or completely emptying one of the three jugs.

This puzzle is depicted by the graph of Fig. 12-35, where the vertices represent the various amounts of wine which jugs b and c can contain after each pouring [in the form (B, C), where B stands for the contents of b, and C for the contents of c]. The initial vertex [labeled (0, 0)] represents the initial configuration, where both b and c are empty. Subsequently, either b or c can be filled, which results in the configurations (5, 0) and (0, 3), respectively. Configuration (5, 0) can lead to (5, 3) (if c is filled from a) or to (2, 3) (if c is filled from b). The remainder of the graph can be completed in the same manner, again—as in Example 12-9—ignoring all moves which recreate a configuration already attained.

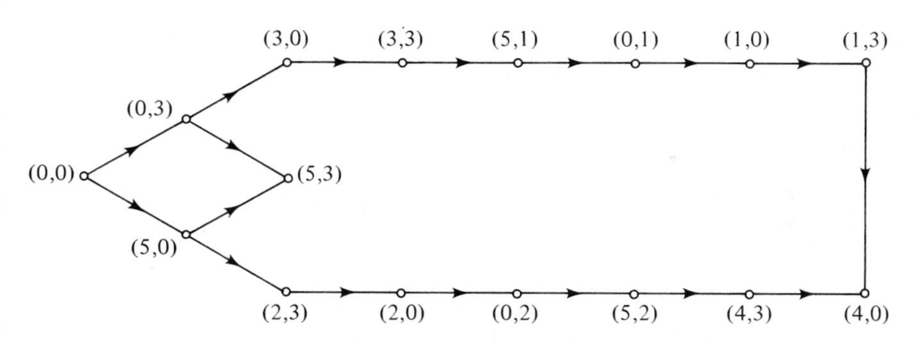

Fig. 12-35 For Example 12-10

The winning configuration, corresponding to jugs a and b filled with 4 quarts each, is represented by the lower right vertex, labeled (4, 0). The directed path connecting the initial vertex to the winning vertex provides the solution to the puzzle: Fill b from a; fill c from b; empty c into a; empty b into c; fill b from a; fill c from b; empty c into a. □

In many puzzles and games (as well as decision processes encountered in engineering, economics, and other disciplines), each move—and, hence, each edge in the representative graph G—is associated with some *cost*, which we assume to be a nonnegative real number. A common objective is to find a solution path in G whose total cost is the least possible. Such a path is called an *optimal path* of G and represents the most economical way of achieving a solution to the puzzle (or a victory in the game). Clearly, if a solution at all

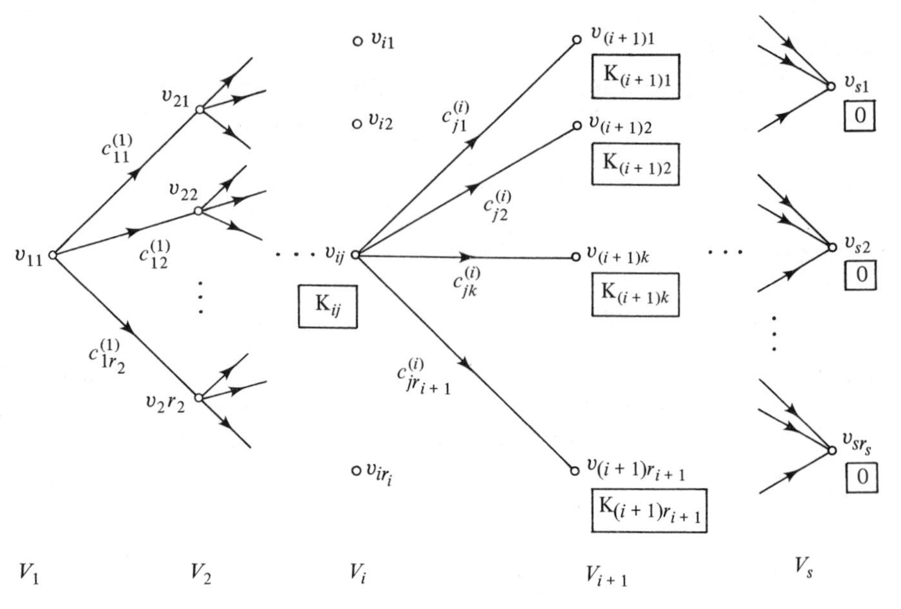

Fig. 12-36 Structure of a decision graph

exists, an optimal path in G is a *proper* directed path connecting the initial vertex to some winning vertex. (This assertion is not true when costs per move are permitted to be *negative* as well as positive.) Thus, insofar as a search for optimal paths is concerned, any edge in G which causes a path to close into a loop can be deleted. (This practice was followed in Example 12-9 and 12-10 where all moves can be assumed to have identical nonnegative costs.) The result is a special directed graph, called a *decision graph.*

Specifically, a decision graph has the following structure (see Fig. 12-36, ignoring the K_{ij} entries enclosed in boxes): The vertex set is partitionable into $s \geq 2$ subsets called *stages,* and denoted by $V_1, V_2, \ldots, V_s$ (where, for $i = 1, 2, \ldots, s$, $V_i = \{v_{i1}, v_{i2}, \ldots, v_{ir_i}\}$); V_1 consists of a single initial vertex v_{11}, and V_s consists of the winning vertices $v_{s1}, v_{s2}, \ldots, v_{sr_s}$; all edges that point away from vertices in V_i ($i = 1, 2, \ldots, s - 1$) must point towards vertices in V_{i+1}; finally, every edge $(v_{ij}, v_{(i+1)k})$ is associated with a real number $c_{jk}^{(i)} \geq 0$ which is the cost of the move from the configuration v_{ij} to the configuration $v_{(i+1)k}$ (the cost of an absent edge can be taken as infinity). An optimal path in decision graph G, then, is any directed path of length $s - 1$, say,

$$\alpha = v_{11}v_{2j_2}v_{3j_3} \ldots v_{(s-1)j_{s-1}}v_{sj_s}$$

whose total cost, denoted by

$$K_\alpha = c_{1j_2}^{(1)}c_{j_2j_3}^{(2)}c_{j_3j_4}^{(3)} \ldots c_{j_{s-1}j_s}^{(s-1)}$$

is not greater than the cost of any other path of length $s - 1$ in G. Figure

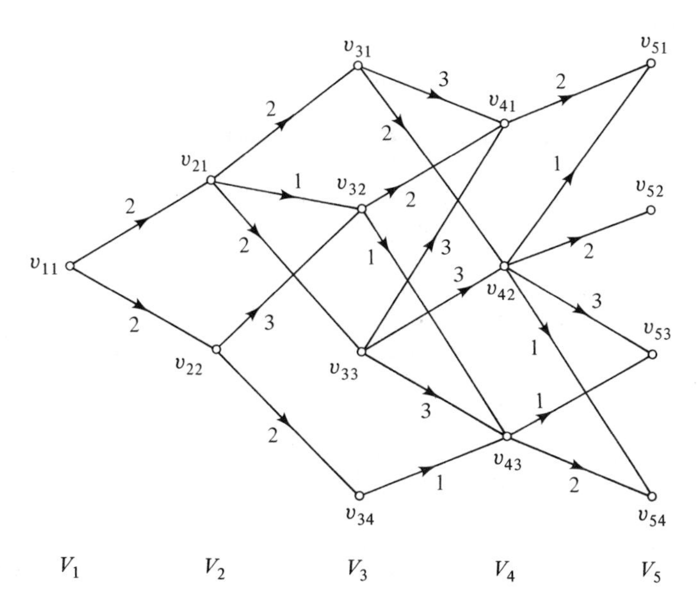

Fig. 12-37 A decision graph

12-37 shows an example of a 5-stage decision graph, exhibiting the optimal path $v_{11}v_{21}v_{32}v_{43}v_{53}$ with the total cost 5.

In any but the simplest decision graphs, determination of optimal paths by exhaustive enumeration of all directed paths that connect v_{11} to the v_{sj_s} is highly impractical. A more efficient procedure is based on the so-called *principle of optimality*. In stating a simplified version of this principle, an *optimal v_{ij}-path* in a decision graph G refers to a directed path α connecting v_{ij} to any winning vertex v_{sj_s} such that the total cost of α is not greater than that of any other such path in G. (That is, an optimal v_{ij}-path is an optimal path with v_{ij} serving as an initial vertex.)

THEOREM 12-16 (*The principle of optimality*).
Let

$$\alpha = v_{ij}v_{(i+1)j_{i+1}}v_{(i+2)j_{i+2}} \cdots v_{(s-1)j_{s-1}}v_{sj_s}$$

be an optimal v_{ij}-path in a decision graph G. Then

$$\alpha' = v_{(i+1)j_{i+1}}v_{(i+2)j_{i+2}} \cdots v_{(s-1)j_{s-1}}v_{sj_s}$$

is an optimal $v_{(i+1)j_{i+1}}$-path in G.

Proof Hypothesize that α' is not an optimal $v_{(i+1)j_{i+1}}$-path in G. Then there exists another directed path in G,

$$\hat{\alpha}' = v_{(i+1)j_{i+1}}v_{(i+2)k_{i+2}} \cdots v_{(s-1)k_{s-1}}v_{sk_s}$$

such that $K_{\hat{\alpha}'} < K_{\alpha'}$. Now, the total cost of the path

$$\hat{\alpha} = v_{ij}v_{(i+1)j_{i+1}}v_{(i+2)k_{i+2}} \cdots v_{(s-1)k_{s-1}}v_{sk_s}$$

is
$$K_{\hat{\alpha}} = c^{(i)}_{jj_{i+1}} + K_{\hat{\alpha}'} < c^{(i)}_{jj_{i+1}} + K_{\alpha'} = K_{\alpha}$$

Since the total cost of α exceeds that of $\hat{\alpha}$, α cannot be an optimal v_{ij}-path—a contradiction. Hence, α' must be an optimal $v_{(i+1)j_{i+1}}$-path. □

Given a decision graph G, let us assume that, for $k = 1, 2, \ldots, r_{i+1}$, we have already constructed an optimal $v_{(i+1)k}$-path $\alpha_{(i+1)k}$, with the total cost $K_{(i+1)k}$ (these are the entries appearing in the boxes in Fig. 12-36). Then, by Theorem 12-16, an optimal v_{ij}-path (for some fixed j) must consist of an edge $(v_{ij}, v_{(i+1)k})$ followed by an optimal $v_{(i+1)k}$-path $\alpha_{(i+1)k}$, where k is some integer between 1 and r_{i+1}; the particular k which makes the path optimal is that which results in the least total cost $c^{(i)}_{jk} + K_{(i+1)k}$. This observation facilitates the recursive construction of a subgraph of G, denoted by G_{opt}, which, for every vertex v_{ij}, exhibits an optimal v_{ij}-path (and no other path). In particular, it exhibits an optimal v_{11}-path; hence, an optimal path of G.

ALGORITHM 12-4

Given a decision graph G, with stages $V_1, V_2, \ldots, V_s$ (see Fig. 12-36), to construct G_{opt}:

(i) Let $K_{s\mu} = 0$ ($\mu = 1, 2, \ldots, r_s$). Set i to $s - 1$.

(ii) For $j = 1, 2, \ldots, r_i$, let k be the particular value of ν which minimizes the quantity $c^{(i)}_{j\nu} + K_{(i+1)\nu}$ (if more than one such ν exists, any of these will do); add the edge $(v_{ij}, v_{(i+1)k})$ to all edges previously constructed, and let

$$K_{ij} = c^{(i)}_{jk} + K_{(i+1)k}$$

The directed path connecting v_{ij} to any $v_{s\mu}$ is an optimal v_{ij}-path.

(iii) If $i = 1$, the graph consisting of all edges previously constructed is G_{opt}. The directed path in G_{opt} which connects v_{11} to any $v_{s\mu}$ is the optimal path of G; the total cost of this path is K_{11}. If $i \neq 1$:

(iv) Subtract 1 from i and return to step (ii). □

Example 12-11

Figure 12-38 illustrates Algorithm 12-4 for the decision graph of Fig. 12-37. The last step of the construction (which proceeds right to left, from the winning vertices towards the initial vertex) reveals that $v_{11}v_{21}v_{32}v_{43}v_{53}$ is an optimal path of G; $K_{11} = 5$ is the total cost of this path. □

We shall conclude this chapter with a discussion of *two-person games*. In such games, there are two players—*a* and *b*—who take turns making moves (starting at some initial configuration); the player who first reaches a winning

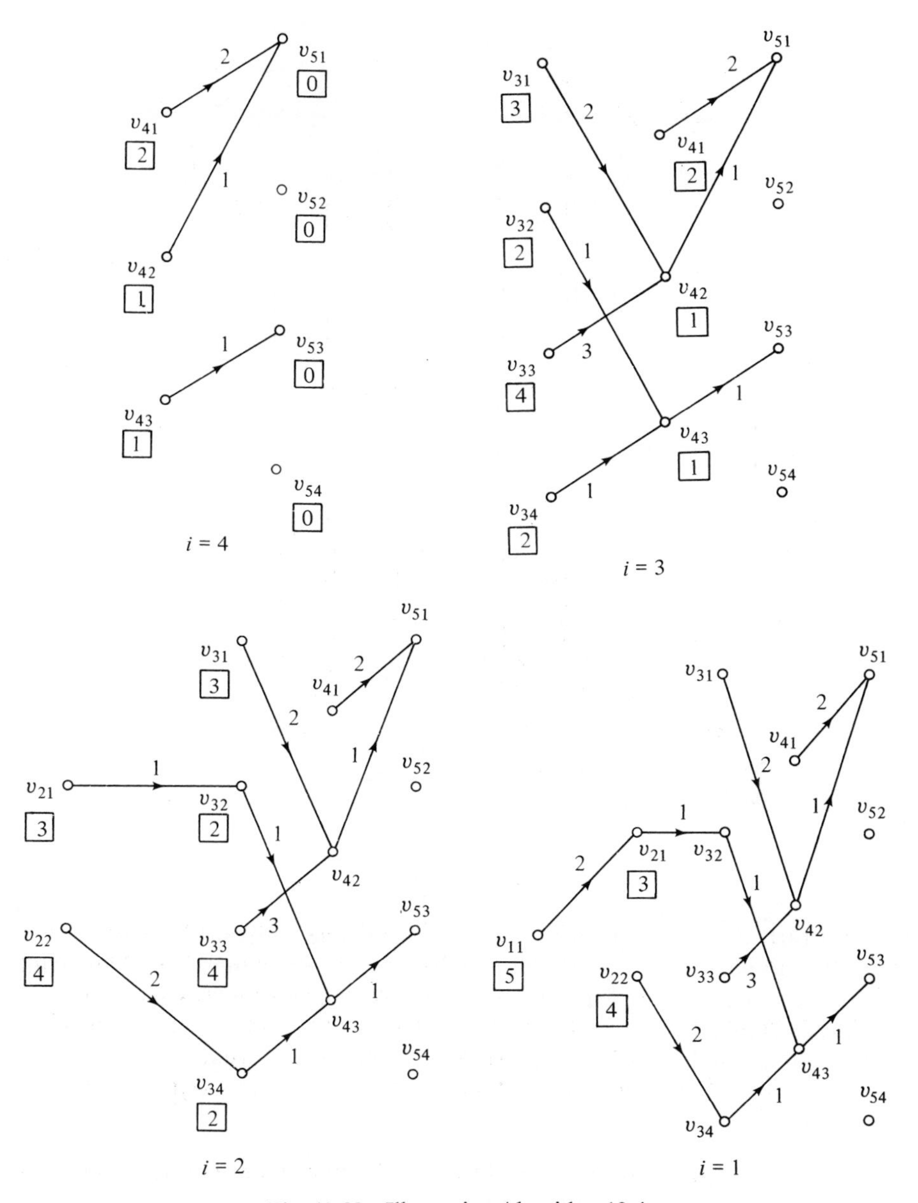

Fig. 12-38 Illustrating Algorithm 12-4

configuration is the winner in the game. Some configurations are usually recognized as *draw configurations* which, when entered, terminate the game with no winner declared. Familiar examples of two-person games are chess, checkers, and tic-tac-toe. A convenient way of depicting such games is by means of a directed bipartite graph with the complementary vertex sets A and B (see Sec. 12-4), where A represents all possible configurations when it

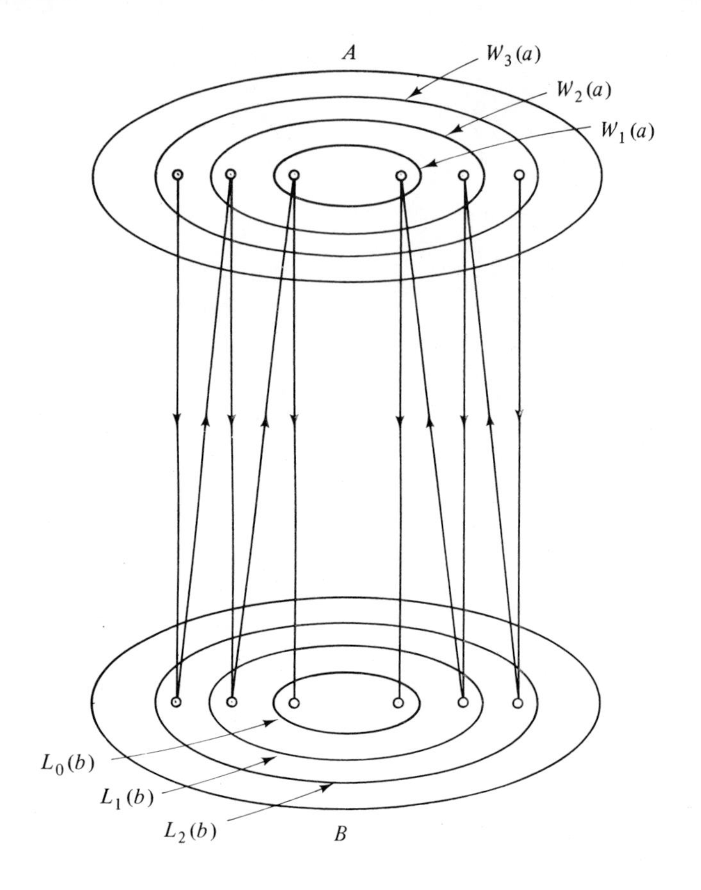

Fig. 12-39 Depiction of two-person game

is a's turn to move, and B represents all possible configurations when it is b's turn to move (see Fig. 12-39). A game consists of a directed path (starting at A if a moves first, and at B if b moves first) which meanders back and forth between A and B, and terminates either in a winning vertex (located in B if a is the winner and in A if b is the winner) or in a draw vertex.

Let us now define the sets of vertices $L_0(b)$, $W_1(a)$, $L_1(b)$, $W_2(a)$, . . . as follows (see Fig. 12-39):

$L_0(b)$ = set of all vertices in B which represent winning configurations for a

$W_1(a)$ = set of all vertices in A from which $L_0(b)$ can be reached in one move

$L_1(b)$ = union of $L_0(b)$ and the set of all vertices in B from which only $W_1(a)$ can be reached in one move

$W_2(a)$ = set of all vertices in A from which $L_1(b)$ can be reached in one move

$L_2(b)$ = union of $L_0(b)$, $L_1(b)$ and the set of all vertices in B from which only $W_2(a)$ can be reached in one move

.
.
.

$W_k(a)$ = set of all vertices in A from which $L_{k-1}(b)$ can be reached in one move

$L_k(b)$ = union of $L_0(b)$, $L_1(b)$, ..., $L_{k-1}(b)$ and the set of all vertices in B from which only $W_k(a)$ can be reached in one move

From these definitions it follows that if a starts in some vertex in $W_k(a)$, it can move into some vertex in $L_{k-1}(b)$, from which b has no choice but to move into some vertex in $W_{k-1}(a)$, from which a can move into some vertex in $L_{k-2}(b)$, from which b has no choice but to move into some vertex in $W_{k-2}(a)$, etc. Eventually, a must find itself in some vertex in $W_1(a)$, from which it can move into some vertex in $L_0(b)$ and win. Thus, $W_k(a)$ represents the set of all configurations at which a is assured victory within at most k moves, and $L_k(b)$ represents the set of all configurations at which b is assured defeat within at most k moves. It should be emphasized, however, that this assurance is predicated on a's knowledge of all the $W_i(a)$ and $L_i(b)$ sets; a's failure to move from $W_i(a)$ to $L_{i-1}(b)$ for $i = k, k - 1, \ldots, 1$, may indeed result in a's defeat.

From the definitions of the $W_i(a)$ and $L_i(b)$ sets we have:

$$W_1(a) \subset W_2(a) \subset W_3(a) \subset \ldots$$
$$L_0(b) \subset L_1(b) \subset L_2(b) \subset \ldots$$

Since A and B are finite, there must be some subset $W(a)$ of A which includes all the $W_i(a)$ and, hence, some subset $L(b)$ of B which includes all the $L_i(b)$. $W(a)$ represents the set of all configurations at which a is assured victory within a finite number of moves, and $L(b)$ represents the set of all configurations at which b is assured defeat within a finite number of moves (barring erroneous moves by a). By a similar analysis (with the roles of a and b interchanged) we can identify a subset $W(b)$ of B which represents the set of all configurations at which b is assured victory within a finite number of moves, and a subset $L(a)$ of A which represents the set of all configurations at which a is assured defeat within a finite number of moves (barring erroneous moves by b). The remaining vertices in A and B represent configurations from which the game can terminate only in a draw (see Fig. 12-40).

This result is rather surprising, inasmuch as it implies that the outcome of every two-person game is completely pre-determined: all we have to do in order to name the winner (if any) is find out in which region [that is, in which of the subsets $W(a)$, $L(a)$, $D(a)$, $W(b)$, $L(b)$, $D(b)$] the initial vertex is located. The fortunate fact that makes many games (such as chess and checkers)

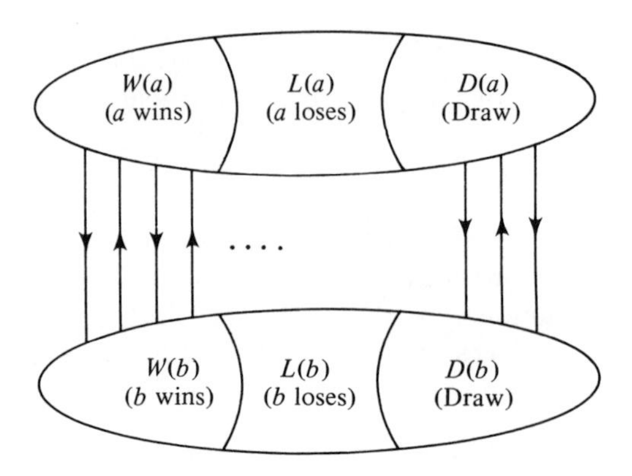

Fig. 12-40 *W*, *L*, and *D* subsets of two-person game

interesting, despite their theoretical triviality, is their enormous number of configurations and possible moves—a fact that makes any attempt to determine the *W*, *L*, and *D* subsets completely futile. Ignorant of these sets, players of such games are bound to make "erroneous" moves, with the result that the outcome is indeed unpredictable. The following is an example of a game where complete graphical representation *is* practicable and, hence, the outcome is readily pre-determined.

Example 12-12

Consider this two-person game: Players a and b take turns removing matches from a pile of 10 matches; each time the player whose turn comes up has the choice of removing either one or two matches; the player who removes the last match is the winner.

The directed bipartite graph describing this game is shown in Fig. 12-41, where the vertices labeled j_a and j_b represent the configurations where j matches remain in the pile and the player whose turn is next is a and b, respectively. From the rules of the game it follows that $L_0(b) = \{0_b\}$. The $L_i(a)$ and $W_i(b)$ sets for this game can be determined directly from the graph:

$$L_0(b) = \{0_b\} \qquad W_1(a) = \{1_a, 2_a\}$$
$$L_1(b) = \{0_b, 3_b\} \qquad W_2(a) = \{1_a, 2_a, 4_a, 5_a\}$$
$$L_2(b) = \{0_b, 3_b, 6_b\} \qquad W_3(a) = \{1_a, 2_a, 4_a, 5_a, 7_a, 8_a\}$$
$$L_3(b) = \{0_b, 3_b, 6_b, 9_b\} \qquad W_4(a) = \{1_a, 2_a, 4_a, 5_a, 7_a, 8_a, 10_a\}$$
$$L_4(b) = L_3(b) \qquad W_5(a) = W_4(a)$$

Hence,

$$W(a) = \{1_a, 2_a, 4_a, 5_a, 7_a, 8_a, 10_a\}$$
$$L(b) = \{0_b, 3_b, 6_b, 9_b\}$$

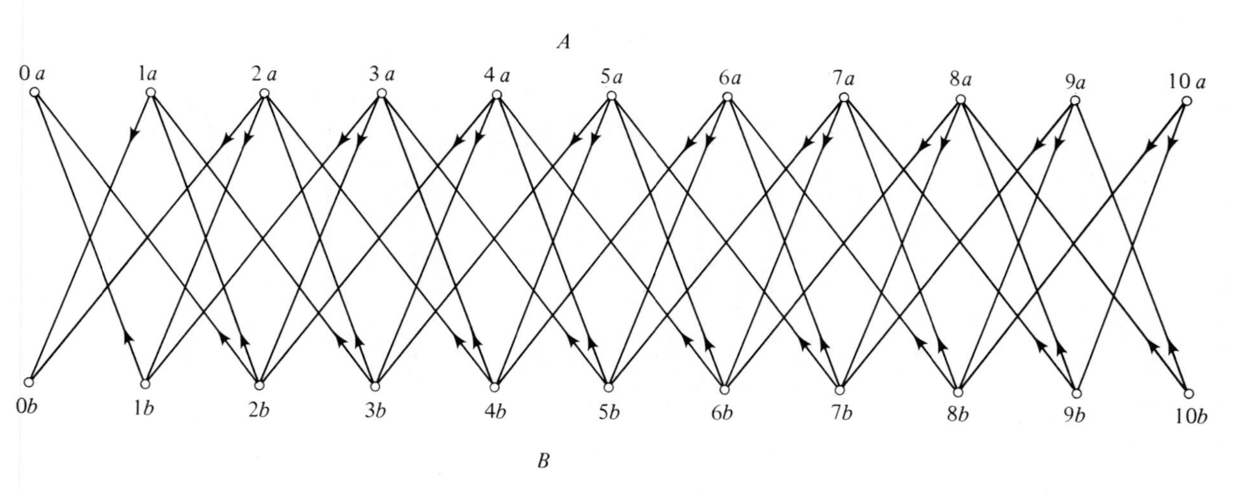

Fig. 12-41 For Example 12-12

By symmetry, we also have:

$$W(b) = \{1_b, 2_b, 4_b, 5_b, 7_b, 8_b, 10_b\}$$
$$L(a) = \{0_a, 3_a, 6_a, 9_a\}$$

[and, hence, $D(a) = D(b) = \varnothing$].

Since 10_a is the initial configuration when a makes the first move, and 10_b the initial configuration when b makes the first move, we can conclude that the player who makes the first move is always the winner (provided he makes no "erroneous" moves). □

PROBLEMS

1. Solve the puzzle of Example 12-10 with jugs a, b, and c having the capacities 12, 7, and 4 quarts, respectively.
2. The following is known as the *puzzle of the jealous husbands*: Three married couples on a journey wish to cross a river by means of a boat that cannot hold more than two persons at a time. The husbands, being rather old-fashioned, would not permit their wives to remain without them in the company of other men. How can the crossing be executed subject to these constraints?
3. Solve the puzzle of the jealous husbands (Problem 2) when the number of couples is four, and when the boat can hold three persons at a time.
4. Find an optimal path in the decision graph shown in Fig. 12-R.

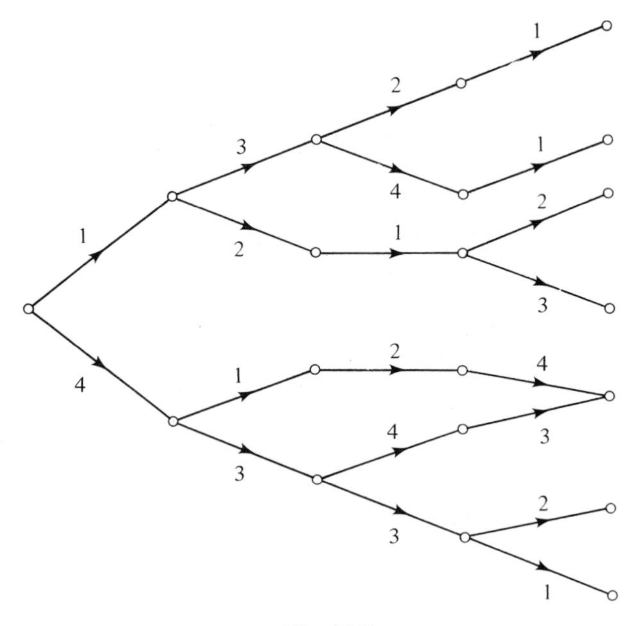

Fig. 12-R

5. Figure 12-S shows a grid of roads connecting towns a and b, with distances (in miles) indicated between each pair of adjacent junctions. What is the shortest route between the two towns?

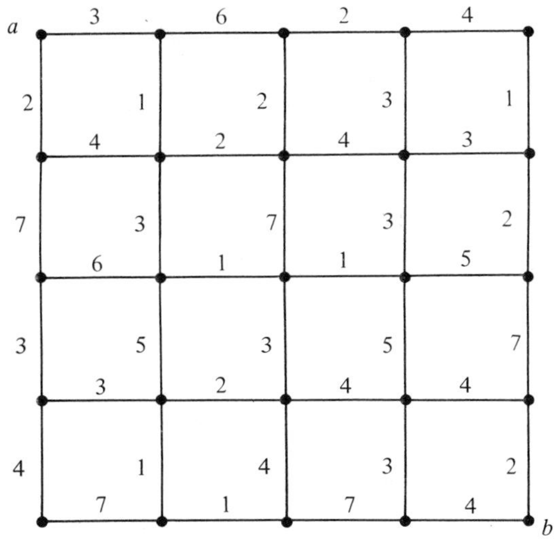

Fig. 12-S

6. Draw the directed bipartite graph describing the match game of Example 12-12, when the initial pile contains 15 matches. Determine all the W, L, and D sets for this case.

7. A simplified version of the two-person game *nim* is played as follows: There are three piles of playing sticks on the table; the players take turns selecting a pile and picking up at least one stick (possibly all sticks) from the pile. The person who picks up the last stick on the table is the winner. Construct the W, L, and D sets for a game of nim, where each pile contains 5 sticks.

REFERENCES

BERGE, C., *Theory of Graphs and Its Applications.* New York: Wiley, 1962.

BUSACKER, R. G., and T. L. SAATY, *Finite Graphs and Networks.* New York: McGraw-Hill, 1965.

GROSSMAN, I., and W. MAGNUS, *Groups and Their Graphs.* New York: Random House, 1964.

HARARY, F., *Graph Theory.* Reading, MA: Addison-Wesley, 1969.

ORE, O., *Graphs and Their Uses.* New York: Random House, 1963.

INDEX

A

Abbott, J. C. 178
Abelian group 271
Absorption laws 14, 104, 132, 149, 161, 188
Abstract algebra 94
Accepted set 241, 244, 260
Accepting state 241, 260
Ackerman's function 83
Addition 92
Additive group 271, 298
Additive identity 288
Additive inverse 297
Additive order 299
Adjacency matrix 361, 396
Adjacent edges 362
Adjacent faces 392
Adjacent vertices 361
Adjoining of root 317
Algebraic system 94
Algebra
  abstract 94
  binary 264
  Boolean 161
  cardinality of 94
  extension of 95
  isomorphism of 107
  of propositions 131
  quotient 112
  of sets 101
  switching 189
Algorithm 94, 261
  Euclidean 85, 312
  generating 233
Algorithm (cont'd.)
  recognition 232, 233
Alphabet 229
Ambiguous language 235
AND gate 184
Antisymmetry 36
Arbib, M. A. 262
Arc 360
Argument 50, 67, 134
Argumental 134
Artin, E. 337
Associative laws 13, 96, 99, 131, 148, 161, 188, 266, 271, 288, 327
Associative operation 96
Assymmetry 36
Atom 165
Automaton 233
  connected finite-state 247
  finite-state 240, 260
  linear bounded 260
  nondeterministic finite-state 243
  pushdown 260
Automorphism 109
Auxiliary head 260
Auxiliary memory 260
Axiom 94

B

Base 88
Base field 317
Basis
  of induction 71, 72
  of vector space 329

Berge, C. 415
Binary code 339
Binary operation 91
Binary relation 27
Binary scaler 219
Binary symmetric channel 341
Binary tree 376
Bipartite graph 382
Birkhoff, G. 24, 48, 90, 178, 286, 337
Block 15
Block code 339
Boolean algebra 161
Boolean arithmetic 31
Boolean expression 173
Boolean function 173
Boolean ring 291
Booth, J. L. 227, 262
Branch 360
Burnside, W. 286
Busacker, R. G. 415

C

Cancellation law 99, 275, 288
Canonical factorization 57
Cantor's paradox 5
Cardinality
  of algebra 94
  of set 2, 64
Cardinality classes 64
Carmichael, R. D. 286
Cartesian product 26
Cell 259
Chain 158
Chain rule 136
Channel 338
Characteristic 323
Characteristic function 68
Check symbol 346
Chevalley, C. 286
Chord 375
Church's thesis 261
Circular relation 42
Classes
  cardinality 64
  equivalence 39
  polynomial residue 309
  residue 41
Closure 93
Code
  binary 339
  block 339
  distance in 341
  equivalence of 395
  generalized Hamming 359
  Hamming 353
  linear 342
Code (cont'd.)
  linear binary 343
  optimal 354
  perfect 352
  quasi-perfect 352
  systematic 346
Code vector 342
Code word 339
Codomain 50
Coefficient 302
Column equivalence 332
Column rank 332
Column-reduced echelon canonical form 333
Column space 332
Column vector 331
Combinational network 180
Committee chairmanship problem 386
Common divisor 84, 311
Common multiple 88
Communication channel 338
Commutative laws 13, 96, 99, 131, 148, 161, 188, 267. 271, 288
Commutative operation 96
Commutative ring 288
Comparator gate 226
Complement
  of element 6, 92, 159
  of graph 362
Complementary transmission functions 201
Complementary vertex sets 382
Complementation 92
Complemented lattice 159
Complement laws 13, 161, 188
Complete directed graph 398
Complete graph 362
Complex number 2, 319
Component 364
Composite function 54
Composite relation 30
Composition
  of functions 54
  of relations 30
Compound proposition 123
Computable function 261
Concatenation 230
Conclusion 126
Conditional rules 136
Congruence relation 111, 115
Conjunction 123
Conjunctive normal form 133
Conjunctive simplification 136
Connected component 364
Connected finite-state automaton 247
Connected graph 364
Consequence 134
Consequent 126

Constant polynomial 302
Context-free grammar 237
Context-free language 237
Context-sensitive grammar 237
Context-sensitive language 237
Contractible graph 389
Contradiction 122
Contrapositive 130
Contrapositive inference 136
Converse
  of proposition 130
  of relation 27
Coordinate 26
Coset 281
Coset leader 350
Coset partition 284
Countable set 65
Counterexample 138
Crossover 387
Cycle 62, 367, 371
Cycle rank 375
Cyclic group 274
Cyclic monoid 267
Cyclotomic number 375

D

Dagger 207
Davis, M. 262
Decision graph 406
Decoder 339
Decoding 341
Decoding table 340
Defining condition 2
Definition 94
Degree
  of polynomial 302
  of vertex 362
Delay unit 211
De Morgan's laws 14, 75, 104, 132, 159, 161, 188
Denumerable set 65
Derivation 237
Derivative 304, 314
Descendent 29
Destination 338
Detachment 136
Deterministic recognizer 260
Diagonalization argument 67
Diagram
  Hasse 43
  ordering 43
  transition 215
  Venn 7
Dickson, L. E. 286
Dimension 329
Direct descendent 29
Directed graph 78, 364, 395
Directed path 396
Direct product 118
Disjoint sets 6
Disjunction 122
Disjunctive addition 136
Disjunctive normal form 133
Disjunctive simplification 136
Distance
  in code 341
  in graph 367
Distributive lattice 156
Distributive laws 13, 97, 99, 131, 156, 161, 188, 288, 327
Distributive operation 97
Diversity condition 384
Division theorem 83, 305
Divisor 84, 305
Domain 27, 50, 94
Draw configuration 409
Double-error correction 341
Double-error detection 341
Double induction 74
Dual 103, 132, 148
Dual graphs 392
Duality 103, 132, 148
Dual transmission functions 202

E

Economy subgraph 378
Edge 28, 360
  adjacent 362
  matrix 372
  separating 364
Edge matrix 372
Edge set 360
Element 1
  complement of 6, 92, 159
  exponent of 96
  greatest 144
  inverse of 97, 270
  least 43, 72, 144
  minimal 170
  order of 276
  primitive 325
  right inverse of 270
Elementary contraction 389
Elementary operation 332
Empty set 2
Empty string 230
Encoder 338
Endmarker 259
Endomorphism 109
End-order traversal 378
Entry 28, 331
Epimorphism 107

Equality
  of functions 50
  of sets 3
  of transmission functions 187
Equivalence
  of codes 395
  of combinational networks 180
  of grammars 237
  of propositions 128
Equivalence class 39
Equivalence kernel 57, 111
Equivalence partition 39
Equivalence relation 39
Equivalence rules 136
Error correction 340
Error detection 340
Error pattern 349
Euclidean algorithm 85, 312
Euler graph 369
Euler loop 368
Euler path 370
Exclusive OR gate 210
EXCOR gate 210
Existential proposition 139
Existential quantifier 139
Expansion theorem 194
Exponent
  of element 96
  of group 278
Extended next-state function 218, 241
Extension
  of algebra 95
  of function 51
Extension field 317
Extension ring 317

F

Face 391, 392
Factor 84, 305
Factorial 77
Factor ring 293
Fallacy 138
Falsity set 122
Fibonacci numbers 78
Field 296
  base 317
  extension 296
  Galois 323
  proper subfield 316
  skew 301
  splitting 321
  subfield 316
Final state 241
Finite algebraic system 94
Finite induction 71, 72
Finite set 2, 64
Finite-state automaton 240, 260
Finite-state control 260
Finite-state machine 214
Forest 374
Formal language 232
Four-color conjecture 392
Full adder 180
Function 50
  Ackerman's 83
  Boolean 173
  characteristic 68
  complementary transmission 201
  composite 54
  composition of 54
  computable 261
  dual transmission 202
  equality of 50, 187
  extended next-state 218, 241
  extension of 51
  idempotent 55
  identity 58
  inverse of 58, 60
  linearly separable transmission 207
  minimization of transmission 199
  next state 215, 241
  output 215
  Peano's successor 71
  recursive 251
  self-dual transmission 295
  single-valued 50
  symmetric 183
  synthesis of transmission 195
  well-defined 50
Functional completeness 205
Functional quasi-completeness 209

G

Gaal, L. 337
Galois field 323
Gate 183
  AND 184
  comparator 226
  exclusive OR 210
  EXCOR 210
  inhibitor 209
  NAND 206
  NOR 207
  NOT 184
  OR 180, 184
  quasi-universal 209
  standard 185
  threshold 207
  universal 206
Generalized Hamming code 359
Generated proposition 133

Generated set 7, 82
Generating algorithm 233
Generator 267, 274, 292
Generator matrix 343
Geodesic 367
Gill, A. 227, 325, 359
Ginsburg, S. 263
Ginzburg, A. 263
glb 143
Golomb, S. 359
Grammar 236, 237
Graph 28, 360
  bipartite 382
  complete 362
  complete directed 398
  connected 364
  contractible 389
  decision 406
  directed 78, 364, 395
  distance in 367
  dual 392
  Euler 369
  Hamilton 371
  isomorphism of 363, 396
  nondirected 395
  nonplanar 387
  planar 387
  polygonal 391
  regular 363
  self-complementary 365
  self-dual 392
  subgraph 363, 378
Gratzer, G. 178
Greatest common divisor 84, 311
Greatest element 144
Greatest lower bound 143
Grossman, I. 415
Group 270
  Abelian 271
  additive 271, 298
  cyclic 274
  exponent of 278
  multiplicative 271, 299
  order of 271
  permutation 272
  semigroup 266
  subgroup 279, 281
  symmetric 272
  symmetries of the square 274

H

Half adder 180
Hall, M. 286
Halmos, P. R. 24, 178
Halting problem for Turing machines 261
Hamilton cycle 371
Hamilton graph 371
Hamming code 353
Harary, F. 415
Harrison, M. A. 227, 263, 359
Hartmanis, J. 227
Hasse diagram 43
Hennie, F. C. 227
Herstein, I. 24, 286, 337
Hohn, F. E. 227
Homomorphic image 106
Homomorphism 106
Hopcroft, J. E. 263
Hypothesis 71, 73, 126

I

Ideal 292
Idempotent 99, 267
Idempotent function 55
Idempotent laws 14, 102, 132, 149, 161, 188
Identity 97, 264
  additive 288
  multiplicative 288
  right 264
Identity function 58
Identity laws 13, 97, 99, 131, 161, 188, 267, 271, 288, 296, 327
Identity matrix 331
Identity permutation 61
Identity relation 28
Image 50, 295
Implication 125
Implication rules 136
Incident vertices 361
Inclusion 3
Indeterminate 302
Index 285
Index set 4
Indirect proof 137
Induction 71, 72
Induction hypothesis 71, 73
Induction step 71, 73
Induction variable 72
Infinite face 392
Infinite set 2, 64
Information symbol 346
Inhibitor gate 209
Initial configuration 403
Initial state 215, 241, 260
Initial vertex 403
Injection 52
Input alphabet 214, 240
Input head 260
Input symbol 214, 240
Input tape 259

Integer 1, 2, 65
Integral domain 99, 289
Intermediate configuration 403
Intersection 6
Inverse
  additive 297
  of element 97, 270
  of function 60
  of matrix 334
  of proposition 130
Inverse image 50
Inverse laws 99, 271, 296
Invertible element 97, 270
Involution law 14, 102, 132, 159, 161, 188
Irreducible polynomial 305
Irreflexivity 36
Isomorphism
  of algebras 107
  of graphs 363, 396
Iteration 231

J

Jealous husbands puzzle 414
Join 145
Junction 360

K

Kain, R. Y. 263
Kemeny, J. G. 90, 141
Kernel 293
Kohavi, Z. 228
Königsberg bridge problem 370
Korfhage, R. R. 228, 263
Kuratowski's theorem 390

L

Lagrange's theorem 285
Language 232
  ambiguous 235
  context-free 237
  context sensitive 237
  formal 232
  natural 232
  phrase-structure 236
  recursive 232, 233
  regular 237
Lattice 145
  complemented 159
  distributive 156
  of partitions 151
  modular 151
Laws
  absorption 14, 104, 132, 149, 161, 188
  associative 13, 96, 99, 131, 148, 161, 188, 266, 271, 288, 327
  cancellation 99, 275, 288
  complement 13, 161, 188
  commutative 13, 96, 99, 131, 148, 161, 188, 267, 271, 288
  DeMorgan's 14, 75, 104, 132, 159, 151, 188
  distributive 13, 97, 99, 131, 156, 161, 188, 288, 296, 327
  idempotent 14, 102, 132, 149, 161, 188
  identity 13, 97, 99, 131, 161, 188, 267, 271, 288, 396, 327
  inverse 99, 271, 296
  involution 14, 102, 132, 159, 161, 188
  negation 131
  null 14, 103, 132, 161, 188
  propositional 131
  of sets 13, 14
Leading coefficient 302
Leading term 302
Least common multiple 88
Least element 43, 72, 144
Least upper bound 143
Lederman, W. 286
Left congruence relation 116
Left coset 281
Left coset partition 284
Left identity 264
Left inverse
  of element 270
  of function 58
Left-invertible element 270
Left subtree 376
Left zero 265
Length
  of string 229
  of path 28, 241, 364
Lin, S. 359
Line 360
Linear binary code 343
Linear bounded automaton 260
Linear code 342
Linear combination 329
Linear dependence 329
Linear independence 329
Linearly separable transmission function 207
Linear ordering 43
Ljapin, S. 286
Logic network 180
Loop 29, 364
Lub 143

M

Machine
  finite state 214
  halting problem for Turing 261
MacLane, S. 24, 48, 90, 178, 286, 337
Magnus, W. 415
Mapping 50, 52
Marriage problem 383
Matching 383
Matrix 28, 331
Maximal cycle 391
Maximal ideal 292
Maximal subgroup 279
Maximum likelihood decoding 341
Maxset 19
Maxset canonical form 20
Maxset normal form 20
Maxterm 174, 191
Maxterm normal combinational network 196
Maxterm normal form 175, 191
McCluskey, E. J. 228
McCoy, N. H. 337
Meet 145
Membership table 11
Memory 211, 260
Message 338, 339
Message set 339
Miller, R. E. 228
Minimal element 170
Minimization of transmission function 199
Minimum distance 341
Minimum polynomial 326
Minimum weight 348
Minset 16
Minset canonical form 18
Minset normal form 18
Minsky, M. L. 263
Minterm 174, 191
Minterm normal combinational network 196
Minterm normal form 175, 191
Mirkil, H. 90, 141
Modular lattice 151
Modulo 41, 107, 112
Monic polynomial 302
Monoid 266
Monomorphism 107
Multigraph 364
Multiple 84, 305
Multiplicative group 271, 299
Multiplicative identity 288
Multiplicative inverse 297
Multiplicative order 299
Mutually exclusive propositions 123

N

NAND gate 206
Natural language 232
Natural number 1
Necessity 127, 129
Negation
  arithmetic 92
  logical 122
Negation laws 131
Nelson, R. J. 263
Nering, E. D. 337
Network
  combinational 180
  logic 180
  maxterm normal combinational 196
  minterm normal combinational 196
  sequential 211
  standard combinational 185
  switching 180
Next-state function 215, 241
Nim 415
Node 360
Noise 339
Nondeterministic finite-state automaton 243
Nondeterministic recognizer 260
Nondirected graph 395
Nonplanar graph 387
Nonsingular matrix 332
Nonterminal 236
Nontransitivity 36
NOR gate 207
Normal subgroup 281
NOT gate 184
Nullity 336
Null laws 14, 103, 132, 161, 188
Null matrix 331
Null set 2
Null space 336
Null string 230
Null vector 327
Number
  complex 2, 319
  cyclotomic 375
  Fibonacci 78
  natural 1
  prime 2
  rational 2, 65, 100
  real 2, 66

O

One 144
Onto mapping 52
One-to-one mapping 52
One-way recognizer 260

Open path 364
Operation 91
  associative 96
  binary 91
  commutative 96
  distributive 97
  elementary 332
  order of 91
  regular 249
  unary 91
Operation table 92
Optimal code 354
Optimality 407
Optimal path 406
Ore, O. 415
OR gate 180, 184
Order
  of cycle 62
  of element 276
  of group 271
  of operation 91
  of set 2
Ordered tuple 25
Order(ing)
  additive 299
  linear 43
  multiplicative 299
  partial 43
  total 43
  well 43
Ordering diagram 43
Output alphabet 214
Output function 215
Output symbol 214

P

Pair 26
Paley, H. 24, 48
Paradox 3, 5
Parity checker 219
Parity check matrix 343
Parsing 234
Partial ordering 43
Partition 15, 151
  coset 284
  equivalence 39
  lattice of 151
  left coset 284
  right coset 281
  trivial 40
Path 28, 363
  directed 396
  Euler 370
  length of 28, 241, 364
  open 364
  optimal 406
Path (cont'd.)
  proper 364
  proper spanning 400
  solution 403
Peano postulates 71
Peano's successor function 71
Perfect code 352
Permissible move 403
Permutation 61, 78
Permutation group 272
Peterson, W. W. 325, 359
Phrase-structure grammar 236
Phrase-structure language 236
Pigeonhole principle 65
Planar graph 387
Point 360
Polish notation 92
Polygon 391
Polygonal graph 391
Polynomial 157, 302, 305
Polynomial ideal 308
Polynomial residue class 309
Poset 142
Post-order traversal 378
Post's correspondence problem 261
Postulate 94
Power 96, 267, 276
Power set 4
Premise 134
Pre-order traversal 377
Prime factorization 76
Prime factorization theorem 87, 314
Prime ideal 295
Prime number 2
Primitive element 325
Primitive polynomial 325
Principal diagonal 36, 331
Principal ideal ring 292
Problem
  committee chairmanship 386
  Königsberg bridge 370
  marriage 383
  Post's correspondence 261
  recursively unsolvable 261
  traveling salesman 371
Product 271, 303
Production 236
Projection 26, 51
Proof
  by contradiction 137
  indirect 137
  by induction 72
Proper ideal 292
Proper path 364
Proper refinement 15
Proper spanning path 400
Proper subfield 316
Proper subgraph 363

Proper subgroup 279
Proper subring 290
Proper subset 4
Proper subsystem 95
Proposition 121
  compound 123
  converse of 130
  existential 139
  equivalence of 128
  generated 133
  inverse of 130
  mutually exclusive 123
  universal 139
  valid 134
Propositional calculus 131
Propositional laws 131
Pseudograph 364
Pushdown automaton 260
Puzzles 79, 414

Q

Quantifier 139
Quasi-perfect code 352
Quasi-universal gate 209
Quotient 84, 306
Quotient algebra 112
Quotient ring 293
Quotient set 40

R

Radix 88
Range 27
Rank
  of matrix 332
  of relation 40
Rational number 2, 65, 100
Realization of transmission function 195
Real number 2, 66
Received encoded message 339
Received message 339
Received vector 342
Received word 340
Recognition algorithm 232
Recognition quasi-algorithm 233
Recognizer 259
  deterministic 260
  nondeterministic 260
  one- and two-way 260
Recognizer 259
Recursive definition 77
Recursive function 261
Recursive language 232
Recursively enumerable language 233
Recursively unsolvable problem 261
Recursive procedure 83
Reducible polynomial 305
Reductio ad absurdum 137
Refinement 15
Reflexive transitive closure 37
Reflexivity 36
Regular grammar 237
Regular graph 363
Regular language 237
Regular operation 249
Regular set 249
Relation 26
  binary 27
  circular 42
  composite 30
  composition of 30
  congruence 111, 115, 116
  equivalence 39
  identity 28
  rank of 40
  universal 28
Relation matrix 28
Relation on set 28
Relatively prime 87
  in pairs 87
Remainder 84, 306
Residue class 41
Restriction 51
Right coset 281
Right coset partition 284
Right identity 264
Right inverse
  of element 270
  of matrix 337
Right-invertible element 270
Right subtree 376
Right zero 265
Ring 288
  Boolean 291
  commutative 288
  extension 317
  factor 293
  of polynomials 304
  principal idea 292
  quotient 293
  subring 290
Rogers, H. 263
Root
  of polynomial 307
  of tree 78, 376
Rooted tree 78, 376
Row equivalence 332
Row rank 332
Row-reduced echelon canonical form 333
Row space 332
Row vector 331

Rules 136
Russell's paradox 3
Rutherford, D. E. 178

S

Saaty, T. L. 415
Scalar 327
Self-complementary graph 365
Self-dual graph 392
Self-dual set 105
Self-dual transmission function 295
Semigroup 266
Separating edge 364
Sequential network 211
Serial adder 212
Set 1
  accepted 241, 244, 260
  cardinality of 2, 64
  complementary vertex 382
  countable 65
  denumerable 65
  disjoint 6
  edge 360
  empty 2
  equality of 3
  falsity 122
  finite 2, 64
  generated 7, 82
  index 4
  infinite 2, 64
  maxset 19, 20
  message 339
  minset 16, 18
  null 2
  order of 2
  power 4
  quotient 40
  regular 249
  relation on 28
  self-dual 105
  solution 122
  state 215, 240
  subset 3, 4
  superset 4
  truth 122
  uncountable 65
  universal 6
  vertex 360
  void 2
Set laws 13, 14
Sheffer stroke 206
Single-error correction 341
Single-valued function 50
Singular matrix 332
Skew field 301
Snell, J. L. 90, 141
Solution path 403
Solution set 122
Source 338
Space
  basis of vector 329
  column 332
  null 336
  row 332
  subspace 328
  vector 327, 329
Spanning of vector space 329
Spanning subgraph 363
Spanning tree 375
Splitting field 321
Square matrix 36, 331
Standard combinational network 185
Standard decoding table 350
Standard gate 185
Standard generator matrix 345
Start symbol 236
State 214, 240
  accepting 241, 260
  final 241
  initial 215, 241, 260
State set 215, 240
Stearns, R. E. 227
Stochastic matrix 93
Stoll, R. R. 141
String 229, 230
Subfield 316
Subgraph 363
Subgroup 279
Subring 290
Subset 3
Subspace 328
Substitution property 110, 115
Subsystem 95
Subtree 376
Sufficiency 127, 129
Sum 271, 302
Superset 4
Suppes, P. 141
Surjection 52
Switching algebra 189
Switching network 180
Symbol 229
  check 346
  information 346
  input 214, 240
  output 214
  start 236
Symmetric function 183
Symmetric group 272
Symmetry 36
Syndrome 351
Syndrome table 351
Synthesis of transmission function 195

System 94, 95
Systematic code 346

T

Tables
  decoding 340
  membership 11
  operation 92
  standard decoding 350
  transition 215
  truth 124, 180
Tautology 122
Terminal 236
Theorem 94
  division 83, 305
  expansion 194
  Kuratowski's 390
  LaGrange's 285
  prime factorization 87, 314
Thompson, G. L. 90
Threshold 207
Threshold gate 207
Topological sorting 45
Total ordering 43
Tournament 398
Tower of Hanoi puzzle 79
Transformation 50
Transition diagram 215
Transition table 215
Transitive closure 33
Transitivity 36
Transmission function 185
Transmitted encoded message 338
Transpose 331
Transposition 64
Traveling salesman problem 371
Tree 374
  binary 376
  rooted 78, 376
  spanning 375
  subtree 376
  syndrome 351
Trivial epimorphic image 295
Trivial ideal 292
Trivial partition 40
Trivial subgroup 279
Truth set 122
Truth table 124, 180
Truth value 123
Tuple 25
Turing machine 261
Two-person game 409
Two-sided inverse 58
Two-way recognizer 260

U

Ullman, J. D. 263
Unary operation 91
Uncountable set 65
Union 6, 92
Universal gate 206
Universal proposition 139
Universal quantifier 139
Universal relation 28
Universal set 6
Universal Turing machine 261
Universe 6
Unordered tuple 25

V

Valid proposition 134
Value 50, 307
Van der Waerden, B. L. 24, 48, 90, 286, 337
Vector 327
  code 342
  column 331
  null 327
  received 342
  row 331
Vector space 327
Venn diagram 7
Vertex 28, 360
Vertex set 360
Void set 2

W

Well-defined function 50
Well ordering 43
Weichsel, A. 24
Weight 348
Whitesitt, J. E. 228
Wielandt, H. 286
Winning configuration 403
Winning vertex 403
Wood, P. E. 228

Z

Zero 144, 265
Zero polynomial 302

INDEX TO THEOREMS, EXAMPLES, AND ALGORITHMS

Theorem Number	Page	Theorem Number	Page
1–1	4	4–3	97
1–2	5	4–4	97
1–3	16	4–5	101
1–4	18	4–6	102
1–5	20	4–7	102
1–6	23	4–8	103
1–7	23	4–9	104
2–1	32	4–10	104
2–2	33	4–11	108
2–3	37	4–12	111
2–4	39	4–13	112
2–5	41	4–14	113
3–1	59	4–15	116
3–2	59	4–16	116
3–3	60	4–17	119
3–4	64	6–1	143
3–5	66	6–2	144
3–6	71	6–3	146
3–7	72	6–4	148
3–8	73	6–5	148
3–9	77	6–6	149
3–10	83	6–7	149
3–11	86	6–8	149
3–12	87	6–9	153
3–13	87	6–10	156
3–14	88	6–11	159
4–1	93	6–12	159
4–2	96	6–13	159

Theorem Number	Page	Theorem Number	Page
6–14	160	10–11	297
6–15	162	10–12	297
6–16	163	10–13	297
6–17	165	10–14	297
6–18	166	10–15	298
6–19	166	10–16	299
6–20	166	10–17	299
6–21	167	10–18	299
6–22	168	10–19	300
6–23	171	10–20	300
6–24	174	10–21	304
7–1	193	10–22	305
7–2	199	10–23	307
7–3	202	10–24	307
8–1	230	10–25	309
8–2	238	10–26	310
8–3	241	10–27	311
8–4	244	10–28	313
8–5	246	10–29	314
8–6	249	10–30	315
8–7	252	10–31	316
8–8	254	10–32	317
8–9	257	10–33	318
9–1	264	10–34	319
9–2	265	10–35	321
9–3	267	10–36	322
9–4	268	10–37	323
9–5	270	10–38	324
9–6	275	10–39	324
9–7	275	10–40	325
9–8	275	10–41	327
9–9	276	10–42	328
9–10	277	11–1	345
9–11	277	11–2	346
9–12	277	11–3	348
9–13	278	11–4	348
9–14	279	11–5	349
9–15	280	11–6	350
9–16	282	11–7	351
9–17	283	12–1	363
9–18	283	12–2	367
9–19	283	12–3	367
9–20	285	12–4	369
9–21	285	12–5	371
9–22	285	12–6	374
10–1	289	12–7	375
10–2	290	12–8	382
10–3	290	12–9	384
10–4	293	12–10	385
10–5	294	12–11	390
10–6	294	12–12	392
10–7	294	12–13	397
10–8	294	12–14	397
10–9	295	12–15	400
10–10	295	12–16	407

Example Number	Page	Example Number	Page
1–1	9	5–8	136
1–2	12	5–9	136
1–3	18	5–10	137
1–4	20	5–11	138
1–5	21	5–12	140
1–6	22	5–13	141
2–1	26	6–1	143
2–2	27	6–2	145
2–3	30	6–3	146
2–4	32	6–4	153
2–5	36	6–5	163
2–6	40	6–6	164
2–7	41	6–7	168
2–8	43	6–8	171
2–9	44	6–9	173
2–10	44	6–10	175
2–11	45	6–11	176
3–1	50	7–1	180
3–2	50	7–2	182
3–3	54	7–3	185
3–4	55	7–4	189
3–5	61	7–5	193
3–6	63	7–6	194
3–7	65	7–7	201
3–8	65	7–8	212
3–9	65	7–9	213
3–10	69	7–10	216
3–11	74	7–11	218
3–12	75	7–12	223
3–13	75	8–1	233
3–14	76	8–2	237
3–15	77	8–3	238
3–16	78	8–4	242
3–17	78	8–5	247
3–18	78	8–6	251
3–19	79	9–1	268
3–20	82	9–2	271
3–21	85	9–3	272
3–22	86	9–4	280
3–23	89	9–5	281
4–1	95	9–6	283
4–2	100	9–7	284
4–3	100	10–1	292
4–4	107	10–2	294
4–5	108	10–3	298
4–6	111	10–4	303
4–7	114	10–5	306
4–8	118	10–6	309
5–1	123	10–7	310
5–2	127	10–8	310
5–3	127	10–9	312
5–4	129	10–10	318
5–5	129	10–11	319
5–6	133	10–12	320
5–7	135	10–13	324

Example Number	Page
10–14	326
10–15	328
10–16	328
10–17	329
10–18	330
10–19	332
10–20	333
10–21	335
11–1	340
11–2	343
11–3	344
11–4	346
11–5	350
11–6	352
11–7	355
11–8	356
11–9	357
12–1	368
12–2	369
12–3	376
12–4	378
12–5	380
12–6	389
12–7	398
12–8	400
12–9	404
12–10	405
12–11	408
12–12	412

Algorithm Number	Page
1–1	18
1–2	20
1–3	22
1–4	22
2–1	45
5–1	133
6–1	176
6–2	176
7–1	191
7–2	192
7–3	192
7–4	198
9–1	284
10–1	320
12–1	375
12–2	378
12–3	397
12–4	408